Springer Series on

Atoms+Plasmas

26

Editor: I. I. Sobel'man

Springer
Berlin
Heidelberg
New York
Barcelona
Hong Kong
London
Milan
Paris
Singapore
Tokyo

Springer Series on

Atoms+Plasmas

Editors: G. Ecker P. Lambropoulos I. I. Sobel'man H. Walther

H. F. Beyer V. P. Shevelko (Eds.)

Atomic Physics with Heavy Ions

With 131 Figures

Springer

Dr. Heinrich F. Beyer

Gesellschaft für Schwerionenforschung, Planckstr. 1,
D-64291 Darmstadt, Germany

Dr. Viatcheslav P. Shevelko

Lebedev Physical Institute, Russian Academy of Sciences,
Leninskii Prospekt 53, 117924 Moscow, Russia

Library of Congress Cataloging-in-Publication Data

Atomic physics with heavy ions / H.F. Beyer and V.P. Shevelko, eds. p.cm. – (Springer series on atoms + plasmas ; 26) Includes bibliographical references and index. ISBN 3-540-64875-5 (hardcover: alk. paper)
1. Heavy ions. 2. Heavy ion collisions. I. Beyer, H.F. (Heinrich F.), 1950- . II. Shevelko, V.P. (Viacheslav Petrovich) III. Series. 99-11675
QC702.7.H42A86 1999 CIP
539.7'234--dc21

ISSN 0177-6495
ISBN 3-540-64875-5 Springer-Verlag Berlin Heidelberg New York

Cover concept: eStudio Calamar Steinen
Cover production: *design & production* GmbH, Heidelberg
Typesetting: Camera ready from authors.
SPIN 10659940 57/3142 - 5 4 3 2 1 0 - Printed on acid-free paper

Preface

This book is devoted to one of the most active domains of atomic physics – atomic physics of heavy positive ions. During the last 30 years, this terrain has attracted enormous attention from both experimentalists and theoreticians. On the one hand, this interest is stimulated by rapid progress in the development of laboratory ion sources, storage rings, ion traps and methods for ion cooling. In many laboratories, a considerable number of complex and accurate experiments have been initiated, challenging new frontiers. Highly charged ions are used for investigations related to fundamental research and to more applied fields such as controlled nuclear fusion driven by heavy ions and its diagnostics, ion–surface interaction, physics of hollow atoms, x-ray lasers, x-ray spectroscopy, spectrometry of ions in storage rings and ion traps, biology, and medical therapy.

On the other hand, the new technologies have stimulated elaborate theoretical investigations, especially in developing QED theory, relativistic many-body techniques, plasma-kinetic modeling based on the Coulomb interactions of highly charged ions with photons and various atomic particles – electrons, atoms, molecules and ions.

The idea of assembling this book matured while the editors were writing another book, *X-Ray Radiation of Highly Charged Ions* by H.F. Beyer, H.-J. Kluge and V.P. Shevelko (Springer, Berlin, Heidelberg 1997) covering a broad range of x-ray and other radiative phenomena central to atomic physics with heavy ions. At that time we felt that there was a lack of comprehensive textbooks covering the various subfields of atomic physics involving heavy ions. The chapters of the present book, written by experts in a variety of subfields, describe the state-of-the-art methods and highlights from the individual subdisciplines, related to the primary subject. This book was conceived as an up-to-date tool for graduate students and researchers working in the field of highly charged ions or related disciplines.

The book is arranged as 18 chapters devoted to the physics of highly charged ions, quantum electrodynamics, physics of electron–ion, ion–atom and ion–ion collisions, and x-ray spectroscopy.

The material of the book is divided into three parts. Part A, entitled *Experimental Methods*, deals with heavy-ion storage rings and the Electron-Beam Ion Trap (EBIT). These devices provide the basis for high-precision experiments with heavy ions in different charge states including radioactive ions and antiparticles. The principles of operation of ion storage rings including electron and laser cooling are also discussed with special emphasis paid to the Coulomb interaction within ion beams, interaction of projectiles with residual gas and electron–ion recombination mechanisms.

In Part B, *Fundamental Properties*, relativistic, radiative and QED effects are considered in relation to the binding energies, wavelengths, Lamb shifts and transition probabilities in ions of high nuclear charge. The use of modern accelerators combined with electron and laser cooling can be applied to the study of fundamental properties of heavy ions with high precision. Applicability of the relativistic many-body techniques and high orders of QED theory is discussed as a tool for calculating Lamb shifts, hyperfine structure, transition probabilities and other fundamental properties of heavy ions.

Part C, *Collisional and Radiative Processes*, is devoted to radiative and collisional elementary processes involving heavy ions – excitation, ionization, and electron capture. These processes are mainly responsible for charge-changing reactions occurring in high-temperature astrophysical and laboratory plasmas which define beam lifetimes of the stored ions, their heating, energy losses, the charge-state balance, and finally, characteristic electromagnetic radiation.

We hope the reader will find this book as instructive and exciting as we have found it while assembling the material. We appreciate the constructive and fruitful cooperation of all contributors, to whom we express our gratitude. We are also grateful to many other colleagues who helped us to systematize the material. Our special thanks are addressed to Dr. C. Ascheron of Springer-Verlag for his continuous interest in this book.

Darmstadt *H.F. Beyer*
Moscow *V.P. Shevelko*
May 1998

Table of Contents

Part C. Collisional and Radiative Processes

List of Contributors

B. Ban-d'Etat
Centre Interdisciplinaire de Recherches
avec les Ions Lourds, UMR 11
rue Claude Bloch, BP 5133
F-14 070 Caen
France
ban@ganaben.in2p3.fr

H.F. Beyer
Gesellschaft für Schwerionenforschung
GSI
Planckstraße 1
D-64 291 Darmstadt
Germany
h.beyer@gsi.de

D. Böhne
Gesellschaft für Schwerionenforschung
GSI
Planckstraße 1
D-64 291 Darmstadt
Germany

F. Bosch
Gesellschaft für Schwerionenforschung
GSI
Planckstraße 1
D-64 291 Darmstadt
Germany
f.bosch@gsi.de

J.P. Briand
Laboratoire de Physique Atomique
et Nucléaire
Université Pierre et Marie Curie
4 place Jussieu
F-75 252 Paris, Cedex 05
France
briand@lpnax1.in2p3.fr

J.R. Crespo López-Urrutia
Fakultät für Physik
Albert Ludwigs-Universität
Hermann Herder-Straße 3
D-79 104 Freiburg
Germany
crespojr@uni-freiburg.de

L.J. Curtis
Physics Department
University of Toledo
Toledo, OH 43606
USA
ljc@physics.utoledo.edu

J. Eichler
Bereich Theoretische Physik
Hahn-Meitner-Institut
D-14 109 Berlin
Germany
eichler@hmi.de

B. Franzke
Gesellschaft für Schwerionenforschung
GSI
Planckstraße 1
D-64 291 Darmstadt
Germany
b.franzke@gsi.de

P. Indelicato
Laboratoire Kastler-Brossel
Université Pierre et Marie Curie
4 place Jussieu
F-75 252 Paris, Cedex 05
France
paul@spectro.jussieu.fr

R.K. Janev
International Atomic Energy Agency
Wagramerstraße 5
PO Box 100
A-1400 Wien
Austria
janev@ripcsr01.iaea.or.at

Th. Kühl
Gesellschaft für Schwerionenforschung
GSI
Planckstraße 1
D-64 291 Darmstadt
Germany
t.kuehl@gsi.de

D. Liesen
Gesellschaft für Schwerionenforschung
GSI
Planckstraße 1
D-64 291 Darmstadt
Germany
d.liesen@gsi.de

I. Martinson
Fysika Institutionen
Lunds Universitet
Sölvegatan 14
S-223 62 Lund
Sweden
indrek.martinson@fysik.lu.se

F. Melchert
Strahlenzentrum der
Justus Liebig-Universität
Leihgesterner Weg 217
D-35 392 Gießen
Germany
frank.melchert@strz.uni-giessen.de

A. Müller
Strahlenzentrum der
Justus Liebig-Universität
Leihgesterner Weg 217
D-35 392 Gießen
Germany
alfred.mueller@strz.uni-giessen.de

V.G. Palchikov
VNIIFTRI
Mendeleevo
Moscow Region
141 570 Moscow
Russia
bpt97@ftri.extech.msk.su

G. Plunien
Institut für Theoretische Physik
Technische Universität Dresden
Mommsenstraße 13
D-01 062 Dresden
Germany
plunien@physik.tu-dresden.de

L.P. Presnyakov
P.N. Lebedev Physical Institute
Leninsky pr. 53
117 924 Moscow
Russia
presn@sci.lpi.ac.ru

E. Salzborn
Institut für Kernphysik
Justus Liebig-Universität
Leihgesterner Weg 217
D-35 392 Gießen
Germany
erhard.salzborn@strz.uni-giessen.de

T. Schenkel
Department of Physics and Space Technology, Lawerence Livermore National Laboratory
Livermore, CA 94551
USA
schenkel2@llnl.gov

D. Schneider
Department of Physics and Space Technology, Lawerence Livermore National Laboratory
Livermore, CA 94551
USA
schneider2@llnl.gov

V.M. Shabaev
Department of Physics
St-Petersburg State University
Ulianovskaya 1, Petrodvorets
St. Petersburg 198 904
Russia
shabaev@pobox.spbu.ru

V.P. Shevelko
P.N. Lebedev Physical Institute
Leninsky pr. 53
117 924 Moscow
Russia
shev@sci.lpi.ac.ru

B.M. Smirnov
Institute for High Temperatures
Izhorskaya 13/19
127 412 Moscow
Russia
smirnov@termo.msk.su

G. Soff
Institut für Theoretische Physik
Technische Universität Dresden
Mommsenstraße 13
D-01 062 Dresden
Germany
soff@physik.tu-dresden.de

J. Steiger
Department of Physics and Space Technology, Lawerence Livermore National Laboratory
Livermore, CA 94551
USA
steiger3@llnl.gov

Th. Stöhlker
Gesellschaft für Schwerionenforschung
GSI
Planckstraße 1
D-64 291 Darmstadt
Germany
t.stoehlker@gsi.de

H. Tawara
National Institute for Fusion Science
NIFS
Toki 509-52
Japan
tawara@dptawara.nifs.ac.jp

A. Wolf
Max Planck-Institut
für Kernphysik
Postfach 103 980
69 029 Heidelberg
Germany
andreas.wolf@mpi-hd.mpg.de

Part A

Experimental Methods

I Heavy-Ion Storage Rings

Andreas Wolf

Summary. In heavy-ion atomic physics, fast beams, with energies ranging from keV up to the relativistic regime, are widely used. A new generation of atomic physics experiments with fast-ion beams has become possible at heavy-ion storage rings, machines similar to synchrotron accelerators, but mostly used for ring-internal experiments at fixed, moderate ion energies. At these storage rings, a central role is played by phase-space cooling techniques, such as electron cooling and laser cooling, ensuring low divergence, low momentum spread, and a small size of the circulating beam. The possibility of ion-beam accumulation by phase-space cooling also brings a large gain in beam intensity and makes feasible new types of experiments in areas such as ion–electron and ion–laser interactions. The chapter discusses the principles of ion storage rings with emphasis on cooling techniques and on the performance of these machines in atomic physics experiments. Existing heavy-ion storage rings are presented and the typical instrumentation for some classes of atomic physics experiments at these devices is discussed.

1. Introduction

Accelerated ion beams are very commonly used in heavy-ion physics. In particular, by acceleration and subsequent stripping of bound electrons in the interaction of fast ions with solid or gaseous targets, virtually any element can now be obtained in any desired charge state at sufficiently powerful heavy-ion beam facilities. Numerous experiments have been performed with fast-moving heavy ions colliding with fixed targets or in similar "single pass" arrangements, which are generally characterized by a short time interval between the production of the system under study and its destruction in the experimental setup or a beam dump. The short transit time of such experiments often presents a limitation for studying the internal properties of the accelerated ions or their behavior in certain environments. This clearly holds for processes of various types occurring at a small rate, such as internal relaxation over time intervals of the order of milliseconds or longer, but also for reactions with small cross sections or on targets with a limited density (for example photons, free electrons, or thin atomic targets).

The short time scale inherent in single-pass fast-beam experiments can be avoided by storing accelerated ion beams. In contrast to trapping techniques for heavy ions at rest, such as those discussed in the following chapter, storage of fast beams preserves the advantages offered by energetic ions: easy detection, low perturbation by interactions with rest-gas molecules in the experimental setup, low Doppler spread and possibly large Doppler shift in the interaction with collinear photon beams, to name a few examples. In addition, storage devices for ion *beams* in general provide extended regions without any confining and reflecting fields where the ions perform only a free, well-predictable drift motion. These regions are well suited for introducing controlled external interactions using photon, electron or atom targets. The recirculation of the ions during long storage limes (typically seconds up to hours) makes it possible to maintain a high steady-state density of ions for experimental purposes without a continuous supply of ions from a source, and the development of a given ensemble of ions can be followed over extended periods of time.

Almost all existing storage devices for fast ion beams are synchrotron storage rings, where charged particles move on a stationary closed orbit defined by a number of purely magnetic deflection and focusing elements. This technique, in connection with repeated acceleration in a radiofrequency cavity, is used extensively in synchrotron accelerators, which can now produce heavy particle beams with energies in the TeV range. Storage rings for heavy particles of a fixed energy are also widely used in high-energy physics, mainly for providing high center-of-mass interaction energies in colliding beam machines; much of the basic knowledge on the operation of heavy-particle storage rings has been gained in particular at the Intersecting Storage Rings (ISR) at CERN (Geneva) [1] and later at the antiproton accumulators [2] at CERN and the Fermi National Accelerator Laboratory (FNAL) near Chicago. The development of beam cooling techniques for heavy particles – stochastic and electron cooling – was initially motivated primarily by the goal of accumulating intense antiproton beams. Even before the large-scale projects applying beam cooling of antiprotons were realized in the Western hemisphere, much pioneering work on these techniques was performed at the Institute of Nuclear Study (now the Budker Institute) in Novosibirsk [3]. In the years 1972–1982 three test facilities for beam cooling techniques were operated: the NAP-M ring [4] in Novosibirsk, the ICE ring [5] at CERN, Geneva, and the Fermilab cooler ring [6].

Cooler storage rings for hadrons and heavy ions over a wide range of low and intermediate energies (with an upper limit of ~ 1 GeV times the mass number A) were conceived soon after phase-space cooling methods for heavy particle beams were shown to work. Built as a cooler, stretcher and decelerator for antiprotons at CERN and operating since 1982, the Low-Energy Antiproton Ring (LEAR) [7] can be regarded as a pioneering project for such low-energy storage rings. A list of the existing cooler rings, which came into operation between 1987 and 1993, is given in Table 1. Several of the cooler rings listed are used for protons and light ions only and the main interest of experiments there is directed to nuclear physics; work performed at these rings has recently been reviewed by *Meyer* [8]. The storage rings in our list that are used for heavy-ion atomic physics are the TSR (Test Storage Ring) in Heidelberg, the ASTRID ring in Aarhus, the ESR (Experimental Storage Ring) at GSI, Darmstadt, and CRYRING (a ring operated in connection with a CRYEBIS electron beam ion source) in Stockholm. Some atomic and

Table 1. Cooler rings for light and heavy ions with their magnetic deflection power $B\rho$, circumference C, start of operation, type (LI: light ions, HI: heavy ions, see text) and a reference for further information

Ring name	Laboratory	$B\rho$ [Tm]	C [m]	Year	Type	Ref.
LEAR	CERN, Geneva (Switzerland)	6.7	78.0	1982	LI, HI	[7]
Cooler	IUCF, Bloomington, Indiana (USA)	3.6	86.8	1987	LI	[9]
TSR	MPI for Nucl. Phys. Heidelberg (Germany)	1.7	55.4	1988	HI	[10]
CELSIUS	Svedberg Lab., Uppsala Univ. (Sweden)	7.0	82.0	1988	LI	[11]
TARN II	INS, Tokyo (Japan)	5.8	77.8	1989	LI	[12]
ASTRID	ISA, Univ. of Aarhus (Denmark)	2.0	40.0	1990	HI	[13]
ESR	GSI, Darmstadt (Germany)	10.0	108.4	1990	HI	[14]
CRYRING	MSL, Stockholm Univ. (Sweden)	1.4	51.6	1992	HI	[15]
COSY	Jülich Research Center (Germany)	12.0	184.0	1993	LI	[16]

molecular physics work has also been performed at the LEAR ring (CERN, Geneva), the TARN II ring at the University of Tokyo, and at the cooler ring of the Indiana University Cyclotron Facility (IUCF) in Bloomington. The devices differ largely with respect to their ion beam injectors. For the light-ion cooler rings mostly cyclotrons are used as injectors. Of the heavy-ion machines, two rings (ESR and TSR) are connected to larger heavy-ion accelerator facilities. Thus, the ESR can be fed with highly charged ions up to bare uranium from the SIS synchrotron at GSI and even with rare and radioactive isotopes produced in the fragment separator [17] following the SIS. The injector of the TSR is a 12-MV tandem accelerator followed by a rf postaccelerator. At CRYRING, ions from the CRYEBIS source are pre-accelerated in an RFQ and then further accelerated in the machine itself, operating as a synchrotron. ASTRID operates in a similar way, but with a small electrostatic pre-accelerator and small ion sources predominantly for singly charged ions; this machine presently is the only one pursuing an experimental program on stored negative ions [18].

2. Layout and Performance of Heavy-Ion Storage Rings

2.1 Layout and Basic Parameters

In a heavy-ion storage ring the stored particles move in an ultrahigh-vacuum system (10^{-12}–10^{-10} mbar pressure) on a closed orbit defined by reflecting and focusing fields. In addition to reflecting and focusing elements field-free regions are provided for further beam manipulation and for experimental equipment. As an example of a typical layout the cooler ring TSR is shown in Fig. 1. The beam guiding elements and the desired drift sections determine the circumference C of the closed orbit.

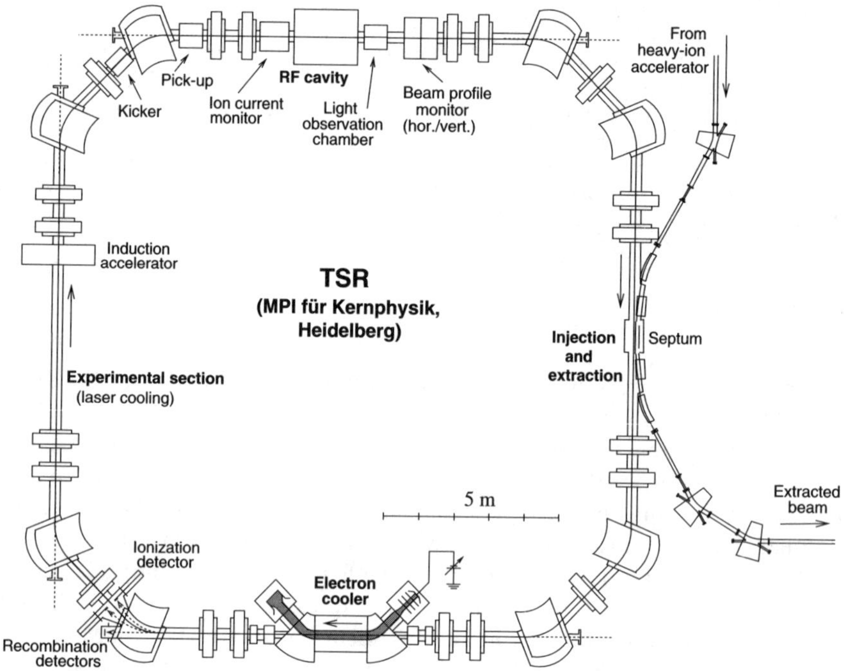

Fig. 1. Layout and principal elements of the heavy-ion storage ring TSR. Unlabeled components are dipole and quadrupole magnets. Dotted lines indicate laser beam paths

For magnetic deflection, the bending radius ρ in the reflecting dipoles, an essential factor for the machine circumference, depends on the ion momentum p, the ion charge qe, and the magnetic field strength B (limited to a maximum of 1–1.5 T for the conventional magnets used at all existing facilities) according to

$$B\rho = p/qe = 3.11\,\mathrm{Tm} \cdot (\mathrm{A}/q)\,\beta\gamma\,, \tag{1}$$

where β determines the beam velocity $v = \beta c$, $\gamma = (1 - \beta^2)^{-1/2}$, and $B\rho$ is referred to as the ion beam (magnetic) rigidity. The maximum beam rigidity of a storage ring, its *bending power*, in general indicates the machine size (see Table 1). The ion velocity or, equivalently, the specific energy E/M at a given bending power depends strongly on the charge-to-mass ratio q/A and decreases as the ion mass $M = \mathrm{A}m_\mathrm{u}$ increases, where m_u denotes the atomic-mass unit. Heavy-ion storage rings are operated in a large range of q/A, varying from values near 0.5 for bare nuclei (except protons) and heavy few-electron ions, down to 10^{-2} for heavy singly charged atomic ions and to even lower values for complex molecular ions. In the non-relativistic approximation (sufficient at most machines as soon as $q/\mathrm{A} \lesssim 0.2$) the relation

$$E/M = \tfrac{1}{2}(qe/M)^2\,(B\rho)^2 = 48.2\,\mathrm{MeV/u}\cdot(q/\mathrm{A})^2\,(B\rho/\mathrm{Tm})^2 \qquad (2)$$

gives the beam energy. For very heavy ions, higher specific energies with a given machine size can be reached by electric instead of magnetic deflection. For example, in a cylindrical deflector with field strength $\mathcal{E}$ the ion energy on the nominal orbit of a bending radius ρ is given by $E = \frac{1}{2}qe\mathcal{E}\rho$. In comparing magnetic and electric machines, we can define the electric bending power $\mathcal{E}\rho$; a bending power of $\mathcal{E}\rho = 1\,\mathrm{MV}$ corresponds to a convenient bending radius of $1\,\mathrm{m}$ in a realistic field of $10\,\mathrm{kV/cm}$. The non-relativistic specific energy is then

$$E/M = \tfrac{1}{2}(qe/M)\,\mathcal{E}\rho = 0.5\,\mathrm{MeV/u}\cdot(q/\mathrm{A})\,\mathcal{E}\rho/\mathrm{MV}\,. \qquad (3)$$

Comparison of (2) and (3) shows that for $q/\mathrm{A} < 0.01$ (singly charged ions with $A > 100$), a "convenient" electric deflection ($\mathcal{E}\rho = 1\,\mathrm{MV}$) allows ions of higher specific energy to be stored than with a magnetic deflection at $B\rho = 1\,\mathrm{Tm}$. As a new addition to the storage ring family, an electrostatic ring for high-mass ions, ELISA [19], is presently under construction at the University of Aarhus. Moreover, storage of keV ion beams between electrostatic mirrors in a linear trap arrangement has recently been demonstrated [20].

As can be seen in Fig. 1 the essential beam transport elements of a storage ring (dipole and quadrupole magnets and possibly higher-order multipole lenses) consume an appreciable fraction of the ring circumference. Free space between these components is occupied by the beam injection system, the radiofrequency cavity, diagnostic installations, and the electron cooling device, discussed below. The remaining space is then taken by experimental installations, such as a gas target facility, a laser interaction region, other targets, and detectors. Also the cooling and diagnostic devices often serve in experiments; for example, the electron cooler is extensively used for ion–electron collision experiments.

2.2 Beams and Lifetimes

Ions already used in heavy-ion storage rings cover a wide range of masses and charges; indicative examples are in given in Table 2. Also the ion energies and velocities vary widely; extreme examples are velocities of $\sim 0.75\,c$ for 500-MeV/u U^{92+} ions in the ESR and $\sim 2 \times 10^{-5}\,c$ for 200-keV $C_{60}{}^+$ ions in ASTRID; slightly relativistic beams at $\beta \sim 0.1$ are typical. Correspondingly the revolution period also varies over a wide range, from 0.5 μs (U^{92+}, ESR) to 150 μs ($C_{60}{}^+$, ASTRID); the common typical values are of the order of a few microseconds. Usually ion storage rings are operated with continuous coasting beams with electrical ion currents as high as 2–3 mA for U^{92+} in the ESR and, for example, up to 18 mA for C^{6+} ions in the TSR. These currents are based on stacking of injected ion beams (see Sect. 3), which to be useful requires beam lifetimes of $\gtrsim 1$ min. Without stacking typical currents in the μA range are reached. Using a radiofrequency cavity in the ring at a constant frequency equal to (some integral multiple of) the revolution frequency, it is also possible to produce a bunched circulating beam with a stationary energy.

Table 2. Examples for ion beams used in heavy-ion storage rings with beam energy and reported lifetimes

Ion	Energy	Ring	Lifetime	Loss mechanism	Ref.
C^{6+}	6.1 MeV/u	TSR	2 h	RR	[21]
O^{7+}	8.9 MeV/u	TSR	420 s	stripping	[22]
Cu^{26+}	8.9 MeV/u	TSR	122 s	capture	[21]
U^{92+}	300 MeV/u	ESR	1 h	RR/capture	[23]
D^-	2 MeV/u	ASTRID	2 s	detachment	[24]
H_3^+	11 MeV/u	CRYRING	2 s	fragmentation	[25]
$C_{60}{}^+$	58 keV	ASTRID	11 s	fragmentation	[26]

A tutorial introduction to beam dynamics in ion storage rings has been given by Pollock [27]. The cross section of the evacuated beam pipe, the apertures of the magnetic elements and their fiducial volumes define the acceptance of the ring, denoting the region of phase space in which ions can be stored on stable orbits. Stable orbits within the acceptance are usually possible only for ions with a definite charge-to-mass ratio. (In some rings the large acceptance permits simultaneous storage of ions with different adjacent charges or masses.) In most cases, therefore, any "reactive" collision of stored ions with residual gas molecules (electron loss or capture, dissociation in the case of molecules) leads to beam loss, which imposes strict vacuum requirements for the operation of heavy-ion storage rings.

Typical storage times reached in practice range from a few seconds to many hours. The decay rates for beams of moderate energy ($E/M = 6\,\mathrm{MeV/u}$, $\beta \sim 0.1$), typical for operation with highly charged ions at smaller rings, are shown for different loss processes as a function of the ion charge state in Fig. 2, assuming a typical pressure and rest-gas composition. Generally, charge changing collisions yield the highest loss rates. For bare ions,

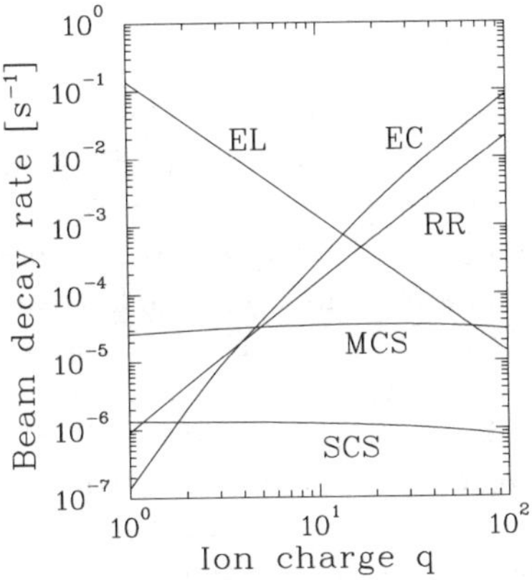

Fig. 2. Beam loss rates for heavy ions ($q/\mathrm{A} \sim 0.5$) at an energy of 6 MeV/u; residual gas total pressure of 1×10^{-10} mbar with a composition of 95% H_2, 4% of molecules containing C, N and O, and 1% of Ar. MCS: Multiple and SCS: Single Coulomb scattering on rest gas nuclei (angular acceptance ± 5 mrad); EC: Electron capture (bare nuclei) and EL: Electron loss (hydrogen-like ions) by rest gas collisions; RR: Radiative recombination in the electron cooler (bare nuclei, electron density 3×10^7 cm^{-3}, $k_{\mathrm{B}}T = 0.1$ eV). Formulae given in Ref. [28] were used

capture in the rest gas dominates for high charge, while recombination with free electrons in the electron cooler determines the beam lifetime at lower charge. Note that the beam lifetimes in practice sensitively depend on the rest gas composition and therefore Fig. 2 can only be indicative. There is also a strong dependence on the beam velocity, in particular regarding electron capture in the rest gas; thus, for the fast beams used at the ESR ($\beta \sim 0.3$–0.7) recombination in the electron cooler generally becomes the lifetime limitation. Since this latter process takes place in merged beams (see Sect. 4.1) it depends weakly on the average beam velocity, and the RR curve in Fig. 2, scaled down according to the lower electron density in the high-energy beam of the ESR, roughly gives the beam loss rate for fast highly charged ions in this machine (lifetime ~ 1 h at $n_{\mathrm{e}} = 3 \times 10^5$ cm^{-3} for U^{92+}).

For ions with remaining bound electrons stripping in rest gas collisions must be considered; it becomes the dominant loss process in lower charge states owing to the lower electron binding energies. Beam loss rates due to stripping are plotted in Fig. 2 in a very simple approximation for hydrogen-like ions only. The plot illustrates the general trend that for a given isoelectronic sequence the beam decay rate starts out large at low q (where stripping is most important), decreases as q grows, and then increases again when electron capture dominates. On the other hand, at a given q increasing the ion mass yields higher stripping rates as more electrons with lower binding energies are present; moreover, the ion beam velocity v becomes lower for a given bending power of the machine, and the stripping rates roughly increase $\propto v^{-1}$ up to a maximum reached at about the orbital velocities of the bound electrons ($v \sim c/137$ for $q \sim 1$). In the operation of storage rings with complex heavy ions at low beam velocity, the beam decay rates, due to stripping, are quite generally in the range of 0.1–1 s^{-1}; this also holds for loss by dissociation of molecular ions. Hence, convenient beam lifetimes are reached also for heavy and slow, singly charged ions.

3. Beam Cooling Techniques

For heavy particles circulating in a storage ring, natural dissipative processes are practically absent in contrast to electrons where emission of synchrotron radiation plays a significant role. A single ion therefore performs an undamped oscillatory motion around the nominal ring orbit in the transverse direction where focusing forces are present, and a free motion in the longitudinal direction. Scattering of the ions on residual gas nuclei and, often more importantly, intrabeam scattering between the stored ions lead to a *heating* of the ion motion in the co-moving frame, while for *cooling* this motion suitable dissipative forces have to be introduced deliberately, applying beam cooling methods [29].

Similar to the situation in an ion trap the long-term "thermodynamical" behavior of the stored ions becomes important in a storage ring. For the statistical motion of particles seen in a reference frame where the average velocity of ions in a beam vanishes (the co-moving reference frame or "beam rest frame") temperatures T are introduced so that $\frac{1}{2}k_{\mathrm{B}}T$ (with Boltzmann's constant k_{B}) represents the average kinetic energy for a given degree of freedom. In a storage ring, moving along with an ensemble of ions in a slice of the stored beam, one would see the transverse temperature and the particle density vary periodically as imposed by the focusing forces; quantities remaining quasi-constant and changing only due to cooling and heating are the beam emittances, related to the conserved action of the ions' phase-space motion. The beam emittances roughly represent the product of beam size and divergence in each transverse direction; together with the relative momentum spread $\Delta p/p$ they define the quality of a *coasting* (unbunched) circulating beam.

Beam cooling techniques, which allow *Liouville's theorem* to be circumvented and the transverse size as well as the divergence and momentum spread of a circulating beam to be reduced at the same time, strongly enhance the performance of ion storage rings. The brilliance of the stored beams becomes largely independent of the ion source conditions. Moreover, cooling regains phase space which can be filled by more ions while the circulating beam remains in the ring; provided there is a sufficiently long beam lifetime, high circulating currents can be accumulated (stacked) by repeated cycles of ion injection and cooling. The currents given in Sect. 2.2 were partly obtained by stacking.

At increasing density and decreasing temperature of the beam strong heating processes due to the collisions between the circulating ions set in and determine the beam quality in equilibrium with the cooling. The final limits of the phase-space density, ultimately determined by the Coulomb repulsion between the ions, present a challenging field of studies at ion storage rings, both experimentally and theoretically, raising issues such as crystalline ion beams [30].

3.1 Electron Cooling

Electron cooling [3, 31, 32] is based on the friction experienced by the ions when they pass through an electron beam which has the same average velocity as the ion beam in a collinear overlap region of the two beams. The friction is caused by small-angle Coulomb scattering of electrons on a moving ion and always tends to reduce the ion velocity in the co-moving reference frame of the electron beam, where one can consider the ions as being embedded in a stationary electron gas.

Electron beams with velocities matched to stored ion beams up to roughly 0.6 GeV/u are presently produced by electrostatic acceleration at voltages up to 300 keV. The electrons are accelerated from a cathode and transported at a constant velocity into the interaction region, where the electron beam and the circulating ion beam merge over a length of typically a few meters. After demerging the electrons are dumped so that the cooling electron beam is continuously renewed. The resulting arrangement (see Fig. 1) is called an electron cooler and also serves as a collinear electron target in a large class of storage ring experiments on electron–ion collisions.

3.1.1 Electron beams for cooling and for collision experiments. In all electron coolers presently used at ion storage rings, the electron beam is guided over its whole path by a magnetic field parallel to the beam axis to counteract the radial space charge force (typical field strength 0.02–0.1 T). The electron densities are in the range of 10^6–10^9 cm^{-3}; the density profile of the electron beam is nearly homogeneous over the beam cross section (typical radius a few cm). Because of the space charge potential a slight ($\sim$ 1%) variation of the average electron velocity occurs across the electron beam, with a minimum of the nearly parabolic velocity profile on the beam axis. The electron beam diameter is chosen about equal to the ion beam diameter before cooling (and much larger than its diameter after cooling).

Both for electron cooling and for collision experiments in the merged electron–ion beam setup, the distribution of the electron velocity $\boldsymbol{v}_\mathrm{e}$ is of particular importance. Starting with a given thermal energy at the source (generally a thermocathode, $k_\mathrm{B}T_\mathrm{e} \approx 0.1$ eV) the acceleration of the electrons leads to a strong reduction of their longitudinal temperature $T_{\mathrm{e}\parallel}$; invoking Liouville's theorem, this temperature reduction can be viewed as caused by the longitudinal expansion of the electron gas on acceleration. Longitudinal temperatures $k_\mathrm{B}T_{\mathrm{e}\parallel}$ of the order of 0.1 meV are encountered frequently. A further reduction of the longitudinal temperature is prevented by heating processes [33]. As long as the strength of the magnetic guiding field is kept constant along the electron beam path, the transverse electron temperature is fixed by the electron source to $k_\mathrm{B}T_{\mathrm{e}\perp} \approx 0.1$ eV. (The electron beam optics can generally be designed to prevent the excitation of additional transverse electron motion.) Since the collisional energy exchange between the transverse and the longitudinal degrees of freedom in the electron beam is small [33], the thermal velocity distribution of the electrons in general is anisotropic

($T_{e_\parallel} \ll T_{e_\perp}$). An appropriate expression for this distribution is the "flattened Maxwellian", defined for the electron velocity $\boldsymbol{v}'_\mathrm{e}$ in the co-moving reference frame of the electron beam ($\langle v'_{e_\parallel} \rangle$=0):

$$f(\boldsymbol{v}'_e) = \frac{m^{3/2}}{(2\pi k_\mathrm{B})^{3/2} T_{e_\perp} T_{e_\parallel}^{1/2}} \exp\left(-\frac{m {v'_{e_\perp}}^2}{2k_\mathrm{B} T_{e_\perp}} - \frac{m {v'_{e_\parallel}}^2}{2k_\mathrm{B} T_{e_\parallel}} \right) , \tag{4}$$

where m is the electron mass. Again in accordance with Liouville's theorem, also the transverse temperature can be lowered performing transverse expansion of the electron beam; this is achieved by applying at the cathode a higher magnetic field strength (B_c) than in the interaction region (B_0) [34, 35]. In the magnetic expansion taking place in the transition region (adiabatically with respect to the cyclotron motion of the electrons [35]), the beam radius increases by a factor of $(B_\mathrm{c}/B_0)^{1/2}$ and the temperature decreases by a factor of B_c/B_0. Field ratios $B_\mathrm{c}/B_0 \approx 100$ were used so far and transverse temperatures near 1 meV have been obtained. Recent studies, aiming at the use of photocathodes for producing intense cold electron beams, indicate that such devices offer transverse energies much below 0.1 eV already at the electron source [36].

3.1.2 Electron cooling performance. Electron cooling acts both longitudinally and transversely and yields an ion beam of small momentum spread and small transverse emittances. In the co-moving reference frame of the electron beam the cooling process is described by the friction force vector $\boldsymbol{F}(\boldsymbol{v}'_\mathrm{i})$, where $\boldsymbol{v}'_\mathrm{i}$ is the ion velocity in this frame. The friction force originates from the momentum transfer in Coulomb collisions of an ion with a large number of electrons, mostly distant collisions where the electron is scattered only by a small angle. For high ion velocities, $v'_\mathrm{i} \gg \langle v'^2_\mathrm{e} \rangle^{1/2}$, the resulting force is equivalent to the familiar electronic energy loss ($\mathrm{d}E/\mathrm{d}x$) of charged particles in matter, which implies that $F \propto v'^{-2}_\mathrm{i}$. The friction thus increases for decreasing ion rest-frame velocity v'_i, reaching a maximum when this velocity becomes $\sim \langle v'^2_\mathrm{e} \rangle^{1/2}$; for $v'_\mathrm{i} \ll \langle v'^2_\mathrm{e} \rangle^{1/2}$, the cooling force $\boldsymbol{F}(\boldsymbol{v}'_\mathrm{i})$ describes a linear (viscous) friction which cools the ions towards a stable point at $v'_\mathrm{i} = 0$. Restricting $\boldsymbol{v}'_\mathrm{i}$ to any fixed direction x and varying the velocity coordinate $v_{\mathrm{i}x}$, one obtains the variation of the cooling force component $F_x(v'_{\mathrm{i}x})$ as shown in

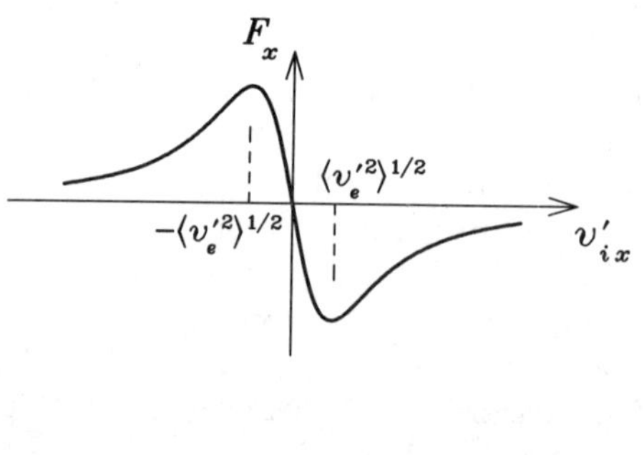

Fig. 3. Schematic shape of the electron cooling force as a function of the ion velocity in the electron-beam rest-frame after averaging over the thermal electron velocities

Fig. 3. The thermal electron velocity spread $\langle v_{\rm e}'^2\rangle^{1/2}$ determines the position and size of the maximum in the cooling force as well as the size of the friction coefficient. The friction coefficient represents the slope of the friction force at the stable point ($v_{\rm i}' = 0$) and yields the cooling rate on an ion of mass M according to $\lambda_x = -(1/M)\partial F_x/\partial v_{{\rm i}x}$. The model of independent multiple collisions with electrons in the screened Coulomb field of the ion yields for the cooling decrement the value

$$\begin{aligned}\lambda &= \frac{\eta_L n_{\rm e} q^2 e^4 L_{\rm C}}{4\pi\varepsilon_0^2 m M} \frac{k_f}{\langle v_{\rm e}'^2\rangle^{3/2}} \\ &= 1.64\times10^{-17}\,{\rm cm^3 s^{-1}}\cdot\frac{q^2}{A}\,\eta_L n_{\rm e} L_{\rm C}\,\frac{k_f c^3}{\langle v_{\rm e}'^2\rangle^{3/2}}\,,\end{aligned} \tag{5}$$

where k_f is a factor of order unity describing the result of averaging over the electron velocities for the specific form of the distribution function $f(\boldsymbol{v}_{\rm e}')$, η_L (typically ~ 0.03) represents the ratio of the electron cooler overlap length to the ring circumference, and $L_{\rm C}$ is the Coulomb logarithm. It represents the logarithmic ratio of the maximum to the minimum impact parameter taken into account for the multiple scattering description, the maximum impact parameter being determined by the screening length of the ion in the surrounding electron gas. At the typical thermal velocity of $\langle v_{\rm e}'^2\rangle^{1/2} \sim 10^7$ cm/s ($k_{\rm B}T_{\rm e} = 0.1\,{\rm eV}$) and electron densities of $n_{\rm e} = 3\times10^7\ {\rm cm^{-3}}$, its value is $L_{\rm C} \sim 9 - \ln q$.

As long as the thermal electron velocity is calculated in a straightforward manner, counting both the transverse and longitudinal velocity components, the predictions for the cooling rates from (5) and the corresponding values of the cooling force lie considerably below most experimental results. Experimentally [31, 32] the maximum both for the longitudinal and the transverse cooling force is reached, quite independently of the transverse electron temperature, at the thermal velocity corresponding to the *longitudinal* temperature, 0.5–1 $\times\,10^6$ cm/s. Thus, Fig. 3 and also the cooling rates predicted from (5) are found to be close to the experimental results when $\langle v_{\rm e}'^2\rangle^{1/2}$ is replaced by $\langle v_{{\rm e}_\parallel}'^2\rangle^{1/2}$.

Much higher cooling forces than expected were already observed in the early experimental studies of electron cooling (see the review in Ref. [31]) and attributed to the presence of the longitudinal magnetic field in the interaction region, which suppresses the effect of the transverse electron velocity spread in distant collisions with collision times long compared to the inverse of the cyclotron frequency $\omega_{\rm c}$ of the electrons. For these collisions the effective electron velocity spread is determined by the considerably smaller longitudinal temperature. The contribution to the cooling force from these distant collisions, referred to as the *magnetic cooling force*, results in a high maximum value of the cooling force (inversely proportional to $T_{{\rm e}_\parallel}$) at rest-frame ion velocities $v_{\rm i}' \sim \langle v_{{\rm e}_\parallel}'^2\rangle^{1/2}$. Close collisions, which proceed rapidly compared to the electron cyclotron period and therefore are only slightly influenced by the magnetic field, yield contributions to the cooling force referred to as *non-magnetic*. They are much smaller than the magnetic cooling force at $v_i' \ll \langle v_{e\perp}'^2 \rangle^{1/2}$, but they dominate for high rest-frame ion velocities, $v_i' \gtrsim \langle v_{e\perp}'^2 \rangle^{1/2}$.

At heavy-ion storage rings for moderate energies up to ~ 20 MeV/u the typical velocity spread of the injected ion beam is close to $\langle v_{e\perp}'^2 \rangle^{1/2}$; hence, the magnetic cooling force yields in fact the dominant contribution to the cooling performance and largely helps in obtaining the short cooling times and small ion beam emittances found in these machines. At higher energies, the magnetic cooling force is mainly felt when the ion beam has already been cooled down, and electron cooling times tend to become longer.

The ion-beam temperature reached by electron cooling is ultimately limited by the electron temperature, corresponding to a thermal equilibrium between the two beams. At injection, ion beams have a temperature of the order of 100 eV and hence much higher than the electron temperature; as mentioned, the ion and electron velocity spreads are then about similar. The electron temperature relevant for the equilibrium also reflects the presence of the magnetic field and hence is effectively given by $T_{e\parallel}$, the temperature of the 'magnetized' electrons which can move freely only in the longitudinal direction. Hence, the ultimate equilibrium temperature of electron cooling amounts to $\sim 10^{-4}$ eV (~ 1 K). In practice, heating processes in the ion beam, in particular intrabeam scattering, limit the ion beam temperature. The typical relative momentum spreads of electron cooled ion beams lie in the range 10^{-5}–10^{-4} and the angular divergence is ~ 0.1–0.3 mrad (beam diameter 0.5–1.5 mm); the corresponding ion temperatures range down to ~ 10 K. At the ESR, much lower longitudinal equilibrium temperatures were observed at low ion currents [37].

3.2 Stochastic Cooling

At heavy ion storage rings, stochastic cooling [2, 38] complements electron cooling which in this area is the most efficient and versatile cooling technique. Electron cooling generally becomes inefficient when the ion-beam velocities in the co-moving frame are large (hot ion beam); typically, at ion velocities in the beam rest-frame of $v_i' \sim 10^8$ cm/s the electron cooling times are already in the range of 100 s. Also, for ion beams of 1 GeV/u and more the technical effort required for producing velocity-matched, sufficiently dense electron beams increases strongly. For high-energy hot ion beams, stochastic cooling therefore provides a useful alternative.

The method of stochastic cooling is based on corrections to the transverse and longitudinal ion momentum using external fields controlled by signals picked up from the ion beam itself. The signals measured at these pickups in fact represent the statistical fluctuations of the average beam position (for the transverse directions) and of the current density (for the longitudinal direction), and these fluctuations, which would disappear if the ion beam could be regarded as a continuous fluid, contain information about the coordinates of individual particles in the beam. If, for a subensemble of the ion beam just passing a pickup, a transverse excursion from the beam axis is detected, the related oscillation around the nominal ring orbit can be stopped in the transverse field of a kicker, set to the necessary strength when the same ion ensemble arrives there. Similarly, the Schottky noise signal of the ion beam, representing the longitudinal density fluctuations, can be fed to an acceleration electrode after suitable filtering and in this way used to correct longitudinal velocity deviations of subensembles of the ion beam. A steady cooling effect is obtained by this procedure since the ion subensembles detected by the pickup stations mix in the course of time; in fact, as long as the ion beam does not become too cold, the ions change their neighborhood along the beam because of their different longitudinal velocities.

Clearly, a high bandwidth and correct tuning of the signal transmission system are critical for the performance of this cooling method. For dense and cold ion beams the statistical fluctuations become smaller and are eventually buried in the thermal noise of the signal transmission system, so that the efficiency of stochastic cooling decreases as the cooling proceeds. In this respect the properties of the method are reciprocal to electron cooling which works best for a cold ion beam and does not critically depend on the ion beam intensity.

A stochastic pre-cooling system [39] has been implemented at the storage ring ESR (GSI, Darmstadt) in order to cool down hot, high-energy ions injected into the ring at relatively low intensities from the GSI fragment separator, where exotic and radioactive nuclei can be produced in nuclear reactions. In the area of relativistic proton and light-ion beams, stochastic cooling has been used widely and in particular has served for accumulating intense antiproton beams. Several cooler rings for light ions are also equipped with stochastic cooling.

3.3 Laser Cooling

Laser cooling has been used extensively on slow atomic beams and trapped atoms and has yielded very low equilibrium temperatures in the mK and even μK regime. Its application to ion beams opens up the perspective of very low beam temperatures and consequently extremely high phase-space densities, which should finally lead to ordered (crystalline) ion beams [30]. Among singly charged ions, some species ($^9Be^+$, $^{6,7}Li^+$, $^{28}Mg^+$ and a few more) possess resonance lines suitable for continuous interaction with narrow-band light. In a laser fed in parallel or antiparallel to the ion beam (dashed lines in Fig. 1) the ions can then undergo continuously repeated cycles of absorption and subsequent spontaneous emission of photons. On the average, a momentum of $\hbar k$, where k is the photon wave number, is transferred to the ion in each cycle, since the absorption is well directed while the directions of spontaneous emission are randomly distributed. The momentum transferred per unit time represents the laser cooling force.

The frequency of the laser photons in the ion center-of-mass frame depends on the ion velocity according to the Doppler formula

$$\omega(\boldsymbol{v}_\mathrm{i}) = \omega_0(1 \mp v_{i_\parallel}/c)/[1-(v_\mathrm{i}/c)^2] \sim \omega_0(1 \mp v_{i_\parallel}/c) \;, \tag{6}$$

where ω_0 is the laser frequency in the laboratory, the $-$ $(+)$ sign holds for parallel (antiparallel) beams, respectively, and the last expression is valid in the non-relativistic approximation. An optical resonance at ω_r with a width of $\delta\omega$ via (6) translates into a laser cooling force (Fig. 4) centered at $v_\mathrm{i} \sim c\,|\omega_0 - \omega_\mathrm{r}|/\omega_0$ with a width of $\delta v \sim c\,\delta\omega/\omega_0$. For typical laser transitions ($\delta\omega/\omega_\mathrm{r} \sim 3\times10^{-8}$) the velocity width of the laser cooling force is only $\sim 10^3$ cm/s, about a factor of 10^3 smaller than that of the electron cooling force (magnetized cooling); on the other hand, at realistic laser intensities of order 100 mW over a few-mm beam, the maximum sizes of the laser and electron cooling forces become comparable. The damping rates achieved by laser cooling, related to the slope of $F_\parallel(v_{i_\parallel})$, therefore largely exceed those of electron cooling.

Laser cooling was first demonstrated with $^{6,7}\mathrm{Li}^+$ atoms using two counteracting lasers [40, 41]. More recently, a continuous, velocity-independent force F_cont was applied in addition to the laser force in order to obtain a stable longitudinal velocity v_s (Fig. 4), using either an inductive accelerator in cooling experiments on $^9\mathrm{Be}^+$ [42], or rf acceleration in experiments on $^{28}\mathrm{Mg}^+$ [43] and $^9\mathrm{Be}^+$ [44] ions. The latter method offers the advantage of maintaining laser cooling over the entire storage time of a beam. By observing the Doppler profile of the stored ion beam with a probe laser scanned over the ionic resonance line, the longitudinal beam temperature was measured; $\sim 10^7$ stored ions can typically be cooled down longitudinally to ~ 1 K. The laser cooling force basically does not act on the transverse ion velocities; however, in recent experiments also the transverse ion temperatures could be reduced by intrabeam collisions within a laser-cooled beam on a time scale of order 1 s [45].

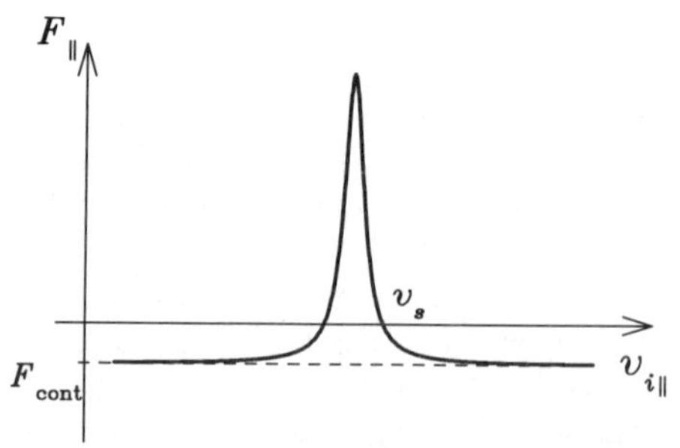

Fig. 4. Schematic shape of the laser cooling force as a function of the ion velocity. Ions are cooled towards the stable velocity v_s where $F_\parallel = 0$ and $\mathrm{d}F_\parallel/\mathrm{d}v_{i_\parallel} < 0$

4. Experimental Techniques at Heavy-Ion Storage Rings

4.1 Electron–Ion Collisions

The electron cooling device of heavy-ion storage rings offers very favorable conditions for investigating electron–ion collisions, in particular inelastic collisions which lead to a change of the ion charge. By adding suitable detectors one readily obtains a ring-internal merged-beam arrangement with a long collinear overlap region, high-current and high-quality ion and electron beams, and a high efficiency for detecting recombination and ionization products, which are separated from the circulating ion beam at the first deflection element following the interaction zone. By tuning the electron energy while keeping the ion energy fixed, the average interaction energy between electrons and ions can be varied over a wide range, starting at very low energies of the order of the electron temperature and going up to several keV.

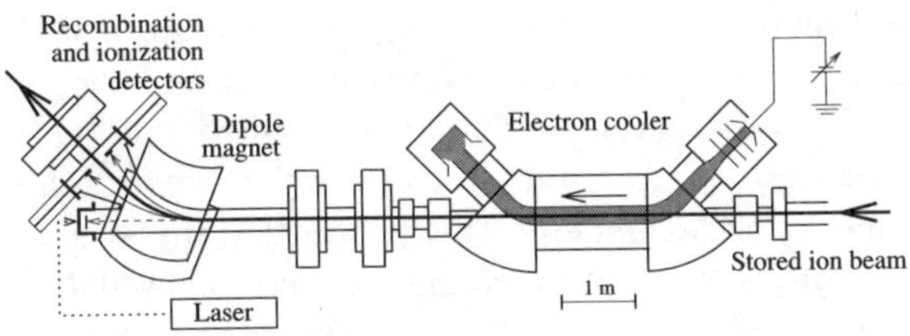

Fig. 5. Experimental equipment around the TSR electron cooler

The arrangement of the electron cooler and the related detectors for recombination and ionization products at the TSR facility [46, 47] is shown in Fig. 5. It also includes detectors in the straight extension of the beam axis for neutral products from recombination of singly charged ions. For ions of very high charge, for example few-electron very heavy ions as studied in particular at the ESR facility, the separation of up- or downcharged ions from the circulating beam becomes small so that the detectors are moved downstream along the ring to selected points with a high momentum dispersion function, providing the best charge-state separation. The low-energy rings use microchannel-plate, Channeltron, Si surface-barrier or, for high count rates, scintillation detectors [48], while at the high-energy ring ESR mainly wire chambers [49] are used. The geometry of the beam overlap region is well known as the magnetically guided electron beam usually has a homogeneous density profile and is much larger than the ion beam; thus, in connection with electron and ion beam current measurements, absolute cross sections can be determined as a function of the collision energy with high energy resolution. More detailed information can be obtained on the electron energy distribution and the space charge potential of the electron beam from Ref. [46], and on the influence of the merging regions on the effective interaction length from Ref. [50].

The principal recombination processes are non-resonant radiative recombination at low energies (near the electron beam temperature) and dielectronic recombination via autoionizing resonances at higher energies. The observed low-energy recombination rates, scaling roughly with q^2, typically amount to 10^3–$10^5\,s^{-1}$ for the achieved beam intensities. On strong dielectronic resonances similar recombination rates are observed. The low rest-gas pressure and the relatively high beam energy help to keep the background rate, caused by electron capture by circulating ions in the rest gas, small, usually lower than the radiative recombination rates for sufficiently fast ion beams. However, the capture background does limit recombination measurements on heavy ions with $q/A \lesssim 0.1$ at the smaller storage rings (magnetic rigidity $\lesssim 1.5$ Tm), where they can be stored only at low velocities.

Spontaneous recombination of electrons and protons was already observed in the first tests of the electron cooling method [4]; the shape of the emerging fast atomic hydrogen beam was used as a quasi-non-destructive diagnostic for the angular divergence of the stored proton beam. Also theoretical predictions for radiative recombination during electron cooling were developed [51]. After heavy-ion rings went into operation, high-resolution studies of dielectronic recombination started with measurements on hydrogen-like ions (O^{7+} [22]), providing the first cross-section data on dielectronic recombination for the most elementary system which can undergo this process. Many studies on other species with more complex electronic configurations followed (see Chap. XII). Important recent developments of the experimental technique include improvements of the energy accuracy [52] and the extension to the heaviest few-electron systems (U^{89+} [53]).

Apart from dielectronic recombination also non-resonant low-energy recombination was studied at storage rings for many highly charged ions. Surprisingly, the low-energy recombination rates were systematically found to exceed theoretical predictions for radiative recombination, a process generally considered to be amenable to an elementary quantum mechanical description. A more detailed discussion of this phenomenon can be found elsewhere [54] (see also Chap. XII).

Spontaneous low-energy recombination is the result of radiative transitions starting at continuum states in an energy band corresponding to the electron temperature and leading into a large number of bound levels, ranging from very high Rydberg levels (quantum number $n \lesssim 3.7q/(kT/\mathrm{eV})^{1/2}$) to the ground state of the product ion. More specific information about the recombination process is obtained by observing the emitted photons; however, bearing in mind that the photons are emitted isotropically and can be observed only with a small solid angle near the beam axis, detection of recombination radiation so far was feasible at ion storage rings only for x-rays emitted by recombination of the heaviest ions, for example fully stripped uranium (x-ray energy range up to 100 keV). Here, spectroscopy of the emitted radiation is of great interest as it yields accurate results for the binding energies of the final states, including low-lying states of hydrogen-like uranium where high-precision measurements of QED shifts become possible [55] (see also Chap. VII). For low-energy photons emitted on recombination at storage rings it appears hardly feasible to measure their spectra directly; however, equivalent information can be obtained with very high resolution using laser-stimulated recombination [56–59]. As discussed in more detail below, free–bound transitions at the frequency of the external, highly monochromatic laser field are detected in these measurements through an increase of the count rate at the recombined-ion detector. For a final bound state of well-defined energy, the stimulated recombination rate as a function of the laser wavelength then yields the distribution of free electrons present in the interaction region just before radiative recombination occurs.

So far at only one storage ring (the TSR) was the electron cooler also used to measure cross sections for electron-impact ionization [47]. In contrast to the recombination channel, the background from rest-gas collisions presents a much larger problem in ionization measurements and rest-gas stripping events mostly dominate the observed countrate. While this is true for all merged-beam measurements on electron-impact ionization, the storage-ring technique nevertheless yields large advantages because of the high stored-ion current (allowing good statistics to be accumulated within reasonable measuring times) and the high energy resolution; in particular, contributions to the ionization cross section from excitation–autoionization, including fine details due to double-autoionizing dielectronic resonances, could be studied with high accuracy.

Starting in about 1992, electron interactions were studied at storage rings also with molecular ions [60, 61]. Here in particular the long beam lifetimes allow internal relaxation of the stored species to take place, so that measurements on molecular ions in the vibrational ground state can be performed; this represents a large advantage over earlier measurements in this area since sources for molecular ions produce species with mostly high and hardly predictable internal excitation.

4.2 Atomic Lifetime Measurements

The storage of ions produced in excited states opens up the possibility of following their decay over extended periods of time so that natural lifetimes of long-lived metastable species can be measured. While with single-pass beam-foil experiments, ionic level lifetimes of up to $\sim 1\,\mu$s have been accessible for a number of years, the use of storage rings extends the accessible range to longer lifetimes up to the order of the beam storage times (from seconds up to hours). For natural decay times in the range of ~ 0.1 ms to ~ 1 s, in particular, the influence of injection transients (in the short time limit) and of the finite beam storage time (in the long time limit) can both be expected to be small and hence high-accuracy measurements of natural lifetimes become possible.

Atomic lifetime measurements at heavy-ion storage rings started with negative ions (at ASTRID) for which in a number of cases even the lowest energy level is metastable against Auger-type autodetachment (e.g., He^-, Ne^-, Be^-, Mg^-) or the electron affinity is so low that radiative detachment by black-body radiation is relatively fast (Ca^-). Such systems were studied in great detail, as reviewed by *T. Andersen* [62]; the experimental procedure basically consists in measuring the decay rate of the beam current. In some cases the magnetic reflecting and focusing fields of the storage ring were found to influence the observed lifetimes through fine-structure mixing, so that the natural lifetimes were derived from series of measurements at different beam energies (and hence different reflecting fields), followed by extrapolation to zero field.

Metastable excited states of interest in positive (singly and multiply charged) ions usually decay without change of the charge state, so that methods must be found to distinguish excited from ground state ions. The first storage-ring measurements on such systems were performed at the TSR on metastable helium-like ions $C^{4+}(2\,^3S)$ [63] with a natural lifetime near 20 ms. Here, dielectronic recombination at a certain relative energy, where it can only occur for the metastable ions, was used as a signal for the excited ion inventory in the stored beam. Later the decay of laser-induced fluorescence was used at the ESR for a lifetime measurement of the excited level involved in laser spectroscopy of the ground-state hyperfine structure in very heavy hydrogen-like ions (including Bi^{82+} [64]), as described in Chap. VIII. More recently, a laser probing technique, also based on fluorescence observation, was introduced at CRYRING and used on Xe^+ ions [65]. Finally, in measurements at the TSR, direct detection of vacuum-ultraviolet radiation emitted in the slow spontaneous decay of stored metastable ions was shown to yield high-quality decay curves [66]. With the excitation taking place in a target before injection, this method of storage-ring lifetime measurements most closely resembles the familiar beam-foil arrangement and appears to be widely applicable for all systems with suitable lifetimes and emission wavelengths for which low-noise optical detectors are available. Results of storage ring lifetime measurements are included in Chap. XIV and are further discussed there.

4.3 Internal Gas Targets for Ion–Atom Collision Experiments and X-Ray Spectroscopy

Internal gas targets operated within storage ring provide a well accessible, clean, low-pressure environment for studying ion–atom collisions and related processes, among which x-ray emission from highly charged few-electron systems has so far been studied with particular success. In a target of this type a narrow ($\sim$ few mm diam.) atomic beam with an internal pressure of the order of 10^{-4}–10^{-3} mbar, representing a target thickness of up to $\sim 10^{13}\,\mathrm{cm}^{-2}$, is crossed with the circulating ion beam. This yields a well-confined, small interaction region which in principle can be surrounded on all sides by detectors placed at small distances, if suitable windows are provided. In some cases such as x-ray spectroscopy or electron-capture studies the main purpose of the target is to supply electrons to be picked up by the ions; hence it is interesting to compare gas targets to electron coolers (cf. Sect. 3.1). These are unique in providing free electrons but offer lower target thicknesses of the order of $10^{10}\,\mathrm{cm}^{-2}$ and much more restricted possibilities of access, limited essentially to lines of sight near the beam axis. Thus, where high density and good access are important, gas targets may be preferable.

In the operation of ring-internal gas targets, cooling of the circulating ion beam plays an important role. In particular, electron cooling provides a very well-defined constant energy, important especially for Doppler corrections in x-ray spectroscopy, and also an ion beam of well-defined transverse momentum (small divergence) so that small amounts of momentum transfer for example in charge-changing reactions can be analyzed. In this context one has to bear in mind that a thickness of $10^{13}\,\mathrm{cm}^{-2}$ at an internal target exceeds by at least 10^3 the column density of residual gas atoms encountered by the circulating ion beam on one turn around the ring; multiple scattering and energy loss of the ions therefore increase correspondingly. A rough estimate for the energy loss in a H_2 target of $10^{13}\,\mathrm{cm}^{-2}$ is $(q^2/\beta^2)\cdot 10^{-4}\,\mathrm{eV}$ per passage, yielding for uncooled 300 MeV/u U^{92+} ions ($\beta = 0.65$) and 10^6 revolutions per second an energy loss rate of $\sim 2\,\mathrm{MeV/s}$, or a relative energy change of $\sim 10^{-3}$ after 100 s of storage. In contrast, electron cooling can keep the ion energy fixed and assures an energy spread of the order of 10^{-5}. Note that for a lower ion beam velocity the energy loss in the gas target can come close to the maximum that can be compensated by the friction force in the electron cooler, with a realistic value of $\sim q^2 \cdot 0.1\,\mathrm{eV/m}$ (electron density $10^7\,\mathrm{cm}^{-3}$).

Considering the loss processes discussed in Sect. 2.2 one can see that with an internal gas target of $> 3\times 10^{10}\,\mathrm{cm}^{-2}$ (the equivalent background-gas thickness at 10^{-10} mbar and 100 m circumference) the beam lifetime will be limited by charge changing processes in the target; for experiments in atomic (as opposed to nuclear) physics these are often the processes of interest so that the stored ion beam is used up efficiently. Clearly, optimal reaction rates are obtained if the target-dependent beam lifetime in the ring and the repetition rate of the injector are matched to each other.

Of the heavy-ion storage ring facilities, only the ESR has used an internal gas target [67] for some time. Results obtained at this facility were recently discussed in a review by *Mokler* and *Stöhlker* [68]. At CRYRING a gas target is under construction [69]. At the ESR the average ion numbers stored in practical operation amount to $\sim 2\times 10^9$ for lower charge states (e.g., Ar^{18+}) and to $\sim 10^8$ for the highest ones (e.g., U^{92+}) [68]; with a typical revolution frequency of $10^6\,\mathrm{s}^{-1}$ this yields luminosities of $\sim 2\times 10^{28}$ and $\sim 10^{29}\,\mathrm{cm}^{-2}\,\mathrm{s}^{-1}$, respectively. Internal gas targets are more commonly used in light-ion storage rings for nuclear physics [8] for which considerably higher luminosities have been reported.

At the gas target used in the ESR (see Chaps. VII and XI), detectors for x rays and for recoil ions are placed around the interaction region; in addition detectors for projectile ions undergoing charge exchange are provided at the downstream ring dipoles. Experiments on ion–atom collisions and x-ray spectroscopy using heavy few-electron ions at the ESR internal gas target are discussed in detail in other chapters of this book. Atomic processes in the gas target were also employed for the identification of isotopes circulating in the ESR; in particular, the conversion of stored bare $^{163}Dy^{66+}$ ions to one-electron ions with the same q and A, $^{163}Ho^{66+}$, by bound-state β decay was detected [70] by stripping the Ho ions in the gas target, so that the Ho^{67+} products could be counted efficiently at the next downstream ring-dipole. New experiments with internal gas targets planned at ESR and CRYRING involve in particular recoil-ion spectroscopy, and a study of the Thomas peak in electron capture is planned at CRYRING [69].

A new approach to atomic targets for ion storage rings is made by the development of a laser trap arrangement, providing a dense cloud of alkali atoms suspended in a magnetooptic trap [71]. Column densities in the range of $10^{11}\,\mathrm{cm}^{-2}$ at a target temperature of only $\sim 100\,\mu\mathrm{K}$ are envisaged. Presently, a ring-internal target consisting of a magnetooptic trap for Cs atoms is under construction at the TSR. The much lower temperatures as compared to the more conventional gas targets discussed above (with temperatures down to $\sim 10\,\mathrm{K}$) are of interest for advanced stages of recoil-ion spectroscopy; the main motivation of the present development at the TSR, however, is to use the trapped target cloud, which can be compressed to a size below 0.1 mm and easily moved across the ion beam, for precision diagnostics of the stored ion beam profile [72].

4.4 Laser Experiments

Heavy-ion storage rings provide long straight sections where the circulating ions can interact with collinear laser beams. In collinear geometry and with continuous laser sources the interaction times can amount to a large fraction of the total storage time (typically 0.1–0.2) so that in particular laser cooling becomes possible. In the discussion of this cooling method in Sect. 3.3 it was already recalled that the laser frequency in the center-of-mass frame of an ion is determined by its velocity through the Doppler formula (6). Hence, for absolute spectroscopic measurements the average velocity of the stored ions must be known absolutely; reciprocally, velocity calibration can be performed by detecting laser excitation of resonances for which the frequencies are already accurately known. In addition, the ion velocity spread leads to Doppler broadening. It is useful to bear in mind that the relative Doppler shift for a given relative change in velocity increases with the beam energy, scaling as $\propto (\gamma - 1) \mp \beta\gamma$. For fast ion beams the transformation of the light intensity (I in the co-moving frame and I_0 in the laboratory) according to $I = I_0(1 \mp \beta)/(1 \pm \beta)$ (upper/lower signs for parallel/antiparallel beams) also becomes relevant. Crossed or tilted laser beams have not so far been used at heavy-ion storage rings.

With present laser sources the accessible transition energies are limited to the order of 30 eV for the fastest beams at the ESR (maximal Doppler shifts $\omega/\omega_0 \lesssim 4$ for $\beta \lesssim 0.9$) and to $\lesssim 8.5$ eV for the fastest beams at the smaller rings ($\beta \lesssim 0.15$). Experiments using "conventional" excitation via allowed electric dipole transitions starting at the ground state or at longlived metastable states are generally limited to low charge states, for which the energies of such transitions are not too high. So far, most of these experiments were performed with singly charged ions, typically stored at rather low velocities ($\beta \lesssim 0.03$) because of the low specific charge of these species, and concerned in particular laser cooling (see Sect. 3.3), laser spectroscopy, and laser probing. In all the measurements a signature of the laser interaction was obtained from the fluorescence of the excited ions, emitted typically after short lifetimes (< 100 ns) and observed sideways over a short (a few cm long) section of the interaction zone. As an example of laser spectroscopy let us mention the precise verification of the relativistic Doppler shift on an electron and laser cooled $^7Li^+$ beam, providing an Ives–Stillwell-type test of the special theory of relativity. The measurements [73] performed at the TSR have already exceeded the precision of earlier laser-spectroscopic relativity tests of this type, and further improvements seem possible (see also Chap. VIII). Laser probing of level populations in Xe^+ [65] and other heavy, low-charged ions was performed at CRYRING.

For growing ion charge, the energy difference from the stable ground state or from metastable levels to configurations that can be reached by electric dipole transitions rapidly increases. The potential of boosting the photon energies available from present laser sources by large Doppler shifts, in particular at the ESR, in fact seems to leave some room for future measurements using "conventional" laser excitation even on multicharged ions. Laser experiments at heavy-ion storage rings were already extended to ions in high charge states by using novel types of transitions at frequencies in the near-visible spectral range. Considering excitation from the ground state, higher-order multipole transitions between terms within the ground configuration can be considered. In particular, magnetic dipole transitions between the hyperfine sublevels of the 1s ground configuration in hydrogen-like systems reach the visible and near-visible spectral range for high nuclear charge. Using beams of $^{208}Bi^{82+}$ and $^{207}Pb^{81+}$ ions at velocities of $\sim 0.6\,c$ in the ESR, the hyperfine splitting for these elementary systems could be measured by laser spectroscopy [64]. In order to obtain sufficient excitation rates and detection sensitivities, these experiments used high-power short ($\sim$ 20-ns) laser pulses and time-resolved fluorescence observation, which also yielded the decay rates (in the 100 μs range) of the excited hyperfine levels (cf. Sect. 4.2).

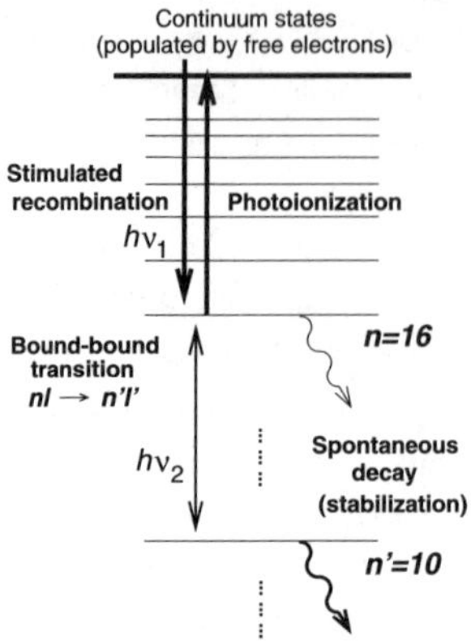

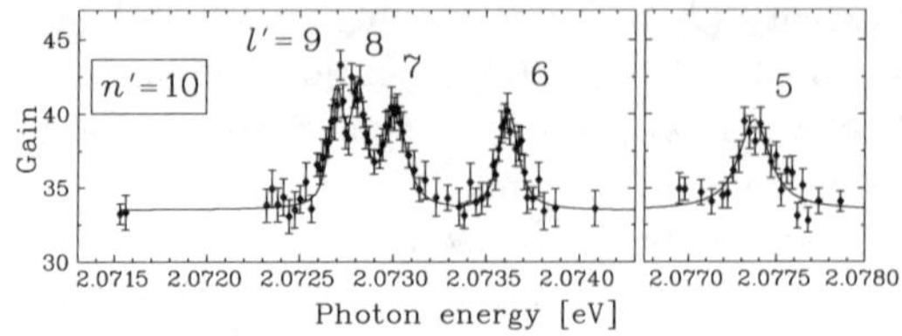

Fig. 6. Laser-stimulated recombination $O^{5+} + e \rightarrow O^{4+}$ (photon energy $h\nu_1$) followed by induced bound–bound transitions ($h\nu_2$) between $O^{4+}(1s^2 2s\, nl)$ Rydberg levels [74]. Left: Level scheme; Top: Recombination rate observed as ν_2 is tuned over resonances $nl \rightarrow n'l'$, $l = l' + 1$

Another method which can be used at heavy-ion storage rings to reach excited states of highly charged ions by laser transitions starts from the continuum, performing laser-stimulated electron–ion recombination [56] in a merged-beam electron target, such as the electron cooler. In the equilibrium state during electron cooling, when the ion average velocity is locked to that of the electrons by the friction force, the free electrons moving in the field of an ion occupy energy levels just above the ionization limit in a band representing the thermal electron-energy spread (see Fig. 6). A laser field of appropriate frequency can resonantly couple a contiunuum level within the occupied band to a bound state, so that stimulated electron capture into this level can take place. If the captured electron remains bound, this process gives rise to a laser-induced count at the recombination detector. To reach a sufficient transition rate and to discriminate the laser-induced signal against the background of spontaneous recombination, high-power laser pulses (typical length ~ 20 ns) are normally used in these experiments; a laser-stimulated recombination signal time-correlated to the laser pulse can then be recorded at the detector. The capture of an electron by stimulated photon emission is basically reversible through photoionization (similar to dielectronic capture and autoionization) as long as the laser field is present; thus, laser-stimulated recombination (like dielectronic recombination) is in fact only completed if, before reionization, the system stabilizes by a spontaneous radiative decay to a lower state or moves out of the laser field.

Two types of experiments on laser-stimulated recombination have so far been performed at ion storage rings. In the first type, by using a single laser ($h\nu_1$ in Fig. 6) and tuning its frequency, the distribution of free electrons over continuum energies near the ionization threshold was investigated [57–59]. The ratio of the laser-stimulated recombination rate (during the laser pulse) to the spontaneous background reached a maximum of up to 16 for C^{6+} [59]; the typical photon energy in these experiments was near 2.5 eV. In the second type, a strong fixed-frequency laser (photon energy near 1 eV) created a population in a bound level nl and transitions further down were induced by a tuneable second laser ($h\nu_2$) [74, 75]. The bound–bound transition opens up an additional channel for radiative stabilization, favoring recombination against competing processes such as field ionization of Rydberg levels [75] or re-ionization in the laser field [74]. In this way resonance signals from transitions between highly excited levels of stored highly charged ions were studied spectroscopically (Fig. 6) at a precision much beyond that obtained by beam–foil spectroscopy, the standard method used for such measurements with single-pass accelerated ion beams.

5. Conclusion

Over the last ten years, heavy-ion storage rings have proved to be powerful instruments for atomic-physics studies, yielding new experimental results in several previously hardly accessible domains of heavy-ion atomic physics, maybe the most prominent ones being high-resolution studies of electron–ion recombination, x-ray spectroscopy of highly charged few-electron systems and laser-studies of such ions, as well as lifetime measurements of long-lived metastable ionic states. Many ongoing activities will yield further improvements and extensions of the experimental techniques, accompanied by further development of phase-space cooling techniques and the study of the properties of cold ion beams.

Acknowledgement. This review arose from the fruitful interaction with many colleagues from the fields of atomic physics and accelerator physics at the Max-Planck-Institut für Kernphysik and at several institutions collaborating in joint storage-ring experiments. Part of it was written during a stay in the Molecular Physics Group (Dept. of Particle Physics) at the Weizmann Institute of Science, Rehovot, Israel, whose hospitality is greatly appreciated. The work was partly supported by the German Federal Minister for Education, Science, Research and Technology (BMBF) and by the Human Capital and Mobility Programme of the European community.

References

1. K. Johnsen: Nucl. Instrum. Methods. **108**, 205 (1973)
2. S. van der Meer: Rev. Mod. Phys. **57**, 689 (1985)
3. G. I. Budker, A. N. Skrinsky: Sov. Phys. Usp. **21**, 277 (1978)
4. G. I. Budker, N. S. Dikansky, V. I. Kudelainen, I. N. Meshkov, V. V. Parkhomchuk, A. N. Skrinsky, B. N. Sukhina: Part. Accel. **7**, 197 (1976)
5. M. Bell, J. Chaney, H. Herr, F. Krienen, P. Møller-Petersen: Nucl. Instrum. Methods **190**, 237 (1981)
6. T. Ellison, W. Kells, V. Kerner, F. Mills, R. Peters, T. Rathbun, D. Young, P. McIntyre: IEEE Trans. Nucl. Sci. **30**, 2636 (1983)
7. S. Baird, M. Chanel, D. Möhl, G. Tranquille: Part. Accel. **26**, 223 (1990)
8. H. O. Meyer: Annu. Rev. Nucl. Part. Sci. **47**, 235 (1997)

9. R. E. Pollock: IEEE Trans. Nucl. Sci. **30**, 2056 (1983)
10. E. Jaeschke, G. Bisoffi, M. Blum, A. Friedrich, C. Geyer, M. Grieser, B. Holzer, H. W. Heyng, D. Habs, M. Jung, D. Krämer, A. Noda, W. Ott, R. E. Pollock, R. Repnow, F. Schmitt, M. Steck: Part. Accel. **32**, 97 (1990)
11. S. Kullander, C. Ekström, A. Johansson, D. Reistad, L. Westerberg: Phys. Scr. **T32**, 221 (1996)
12. T. Tanabe, T. Watanabe: Nucl. Instrum. Methods **A278**, 241 (1989); A. Noda, Y. Hattori, A. Itano, M. Kanazawa, T. Katayama, M. Kodaira, M. Mutoro, S. Sasaki, N. Takahashi, M. Takanaka, T. Tanabe, N. Tokuda, M. Yoshzawa, Y. Hirao: IEEE Trans. Nucl. Sci. **32**, 2684 (1985)
13. S. P. Møller: in *Conference Record of the 1991 IEEE Particle Accelerator Conference*, San Francisco, 1991, L. Lizana, J. Chew, eds. IEEE Catalog No. 91CH3030-7 (IEEE, New York, 1991), p. 1741
14. B. Franzke, K. Beckert, F. Bosch, H. Eickhoff, B. Franczak, A. Gruber, O. Klepper, F. Nolden, P. Raabe, H. Reich, P. Spädtke, M. Steck, J. Struckmeier: in *Proceedings of the 1993 Particle Accelerator Conference*, Washington DC, 1993, ed. by S. T. Corneliussen, IEEE Catalog No. 93CH3279-7 (IEEE, Piscataway, 1993), p. 1645
15. K. Abrahamsson, G. Andler, L. Bagge, E. Beebe, P. Carlé, H. Danared, S. Egnell, K. Ehrstén, M. Engström, C. J. Herrlander, J. Hilke, J. Jeansson, A. Källberg, S. Leontein, L. Liljeby, A. Nilsson, A. Paál, K.-G. Rensfelt, U. Rosengård, A. Simonsson, A. Soltan, J. Starker, M. af Ugglas: Nucl. Instrum. Methods. B **79**, 269 (1993)
16. R. Mayer et al.: in *Proceedings of the Fourth European Particle Accelerator Conference*, London, 1994, ed. by V. Suller and Ch. Petit-Jean-Genaz (World Scientific Publishers, Singapore, 1994), p. 165
17. H. Geissel, P. Armbruster, K. H. Behr, A. Brünle, K. Burkhard, M. Chen, H. Folger, B. Franczak, H. Keller, O. Klepper, B. Langenbeck, F. Nickel, E. Pfend, M. Pfützner, R. Roeckl, K. Rykaczewski, I. Schall, D. Schardt, C. Scheidenberger, K.-H. Schmidt, A. Schöter, T. Schwab, K. Sümmerer, M. Weber, G. Münzenberg, T. Brohm, H.-G. Clerc, M. Fauerbach, J.-J. Gaimard, A. Grewe, E. Hanelt, B. Knödler, M. Steinder, B. Vossm, J. Weckenmann, C. Ziegler, A. Magel, H. Wollnik, J. P. Dufour, Y. Fujita, D. J. Vieira, B. Sherrill: Nucl. Instrum. Methods. B **70**, 286 (1992)
18. L. H. Andersen, T. Andersen, P. Hvelplund: Adv. At. Mol. Phys. **38** 155 (1997)
19. S. P. Møller, Nucl. Instrum. Methods. A **394**, 281 (1997)
20. D. Zajfman, O. Heber, L. Veijby-Christensen, I. Ben-Itzhake, M. Rappaport, R. Fishman, M. Dahan: Phys. Rev. A **55**, R1577 (1997); M. Dahan, R. Fishman, O. Heber, M. Rappaport, N. Altstein, D. Zajfman, W.J. van der Zande: Rev. Sci. Instrum. **69**, 76 (1998)
21. M. Grieser, D. Habs, R. v. Hahn, C. M. Kleffner, R. Repnow, M. Stampfer, E. Jaeschke, M. Steck: in same proceedings as Ref. [13], p. 2817
22. G. Kilgus, J. Berger, P. Blatt, M. Grieser, D. Habs, B. Hochadel, E. Jaeschke, D. Krämer, R. Neumann, G. Neureither, W. Ott, D. Schwalm, M. Steck, R. Stokstad, E. Szmola, A. Wolf, R. Schuch, A. Müller, M. Wagner: Phys. Rev. Lett. **64**, 737 (1990)
23. F. Bosch: in *The Physics of Electronic and Atomic Collisions*, T. Andersen, B. Fastrup, F. Folkmann, H. Knudsen, N. Andersen, eds. (AIP Press, New York, 1993), p. 432
24. L. H. Andersen, D. Mathur, H. T. Schmidt, L. Vejby-Christensen: Phys. Rev. Lett. **74**, 892 (1995)

25. M. Larsson, H. Danared, J. R. Mowat, P. Sigray, G. Sundström, L. Broström, A. Filewich, A. Källberg, S. Mannervik, K. G. Rensfeldt, S. Datz: Phys. Rev. Lett. **70**, 430 (1993)
26. P. Hvelplund, in same proceedings as Ref. [23], p. 709
27. R. E. Pollock: Annu. Rev. Nucl. Part. Sci. **41**, 357 (1991)
28. D. Habs, W. Baumann, J. Berger, P. Blatt, A. Faulstich, P. Krause, G. Kilgus, R. Neumann, W. Petrich, R. Stockstad, D. Schwalm, E. Szmola, K. Welti, A. Wolf, S. Zwickler, E. Jaeschke, D. Krämer, G. Bisoffi, M. Blum, A. Friedrich, C. Geyer, M. Grieser, H.-W. Heyng, B. Holzer, R. Ihde, M. Jung, K. Matl, W. Ott, B. Povh, R. Repnow, M. Steck, E. Steffens, D. Dutta, T. Kühl, D. Marx, S. Schröder, M. Gerhard, R. Grieser, G. Huber, R. Klein, M. Krieg, N. Schmidt, R. Schuch, J.F. Babb, L. Spruch, W. Arnold, A. Noda: Nucl. Instrum. Methods. B **43**, 390 (1989)
29. F. T. Cole, F. E. Mills: Annu. Rev. Nucl. Part. Sci. **31**, 295 (1981)
30. D. Habs, R. Grimm: Annu. Rev. Nucl. Part. Sci. **45**, 391 (1996)
31. H. Poth: Phys. Rep. **196**, 135 (1990)
32. V. V. Parkhomchuk, A. N. Skrinsky: Rep. Prog. Phys. **54** 919 (1991)
33. V.I. Kudelainen, V.A. Lebedev, I.N. Meshkov, V.V. Parkomchuk, B.N. Sukhina: Sov. Phys. JETP **56**, 1191 (1982)
34. H. Danared, G. Andler, L. Bagge, C.J. Herrlander, J. Hilke, J. Jeansson, A. Källberg, A, Nilsson, A. Paál, K.-G. Rensfelt, U. Rosengård, J. Starker, M. af Ugglas: Phys. Rev. Lett. **24**, 3775 (1994)
35. S. Pastuszka, U. Schramm, M. Grieser, C. Broude, R. Grimm, D. Habs, J. Kenntner, H.-J. Miesner, T. Schüßler, D. Schwalm, A. Wolf: Nucl. Instrum. Methods. **A369**, 11 (1996)
36. S. Pastuszka, A. S. Terekhov, D. Kratzmann, D. Schwalm, A. Wolf: Appl. Phys. Lett. **71**, 2967 (1997)
37. M. Steck, K. Beckert, H. Eickhoff, B. Franzke, F. Nolden, H. Reich, B. Schlitt, T. Winkler: Phys. Rev. Lett. **77**, 3803 (1996)
38. D. Möhl, G. Petrucci, L. Thorndahl, S. van der Meer: Phys. Rep. **58**, 73 (1980)
39. F. Nolden, G. Berthes, D. Böhne, B. Franzke, A. Schwinn: in *Proceedings of the European Particle Accelerator Conference*, Rome 1988, S. Tazzari, ed. (World Scientific Publishers, Singapore, 1989), p. 579
40. S. Schröder, R. Klein, N. Boos, M. Gerhard, R. Grieser, G. Huber, A. Karafillidis, M. Krieg, N. Schmidt, T. Kühl, R. Neumann, V. Balykin, M. Griser, D. Habs, E. Jaeschke, D. Krämer, M. Kristensen, M. Music, W. Petrich, D. Schwalm, P. Sigray, M. Steck, B. Wanner, A. Wolf: Phys. Rev. Lett. **64**, 2901 (1990)
41. J. S. Hangst, M. Kristensen, J. S. Nielsen, O. Poulsen, J. P. Schiffer, P. Shi: Phys. Rev. Lett. **67**, 1238 (1991)
42. W. Petrich, M. Grieser, R. Grimm, A. Gruber, D. Habs, H.-J. Miesner, D. Schwalm, B. Wanner, H. Wernøe, A. Wolf, R. Grieser, G. Huber, R. Klein, T. Kühl, R. Neumann, S. Schröder: Phys. Rev. A **48**, 2127 (1993)
43. J. S. Hangst, J. S. Nielsen, O. Poulsen, P. Shi, J. P. Schiffer: Phys. Rev. Lett. **74**, 4432 (1995)
44. H.-J. Miesner, M. Grieser, R. Grimm, I. Lauer, V. Luger, P. Merz, A. Peters, U. Schramm, D. Schwalm, M. Stößel: Nucl. Instrum. Methods. **A383**, 634 (1996)
45. H.-J. Miesner, R. Grimm, M. Grieser, D. Habs, D. Schwalm, B. Wanner, A. Wolf: Phys. Rev. Lett. **77**, 626 (1996)
46. G. Kilgus, D. Habs, D. Schwalm, A. Wolf, N. R. Badnell, A. Müller: Phys. Rev. A **46**, 5730 (1992)
47. J. Linkemann, A. Müller, J. Kenntner, D. Habs, D. Schwalm, A. Wolf, N. R. Badnell, M. S. Pindzola: Phys. Rev. Lett. **74**, 4173 (1995)

48. G. Miersch, D. Habs, J. Kenntner, D. Schwalm, A. Wolf: Nucl. Instrum. Methods **A369**, 277 (1996)
49. O. Klepper, F. Bosch, H. W. Daues, H. Eickhoff, B. Franczak, B. Franzke, H. Geissel, O. Gustafsson, M. Jung, W. Koenig, C. Kozhuharov, A. Magel, G. Münzenberg, H. Stelzer, J. Szerypo, M. Wagner: Nucl. Instrum. Methods. B **70**, 427 (1992)
50. A. Lampert, A. Wolf, D. Habs, J. Kenntner, G. Kilgus, D. Schwalm, M. S. Pindzola, N. R. Badnell: Phys. Rev. A **53**, 1413 (1996)
51. M. Bell, J. S. Bell: Part. Accel. **12**, 49 (1982)
52. W. Zong, R. Schuch, E. Lindroth, H. Gao, D. R. DeWitt, S. Asp, H. Danared: Phys. Rev. A **56**, 386 (1997)
53. C. Brandau, F. Bosch, G. Dunn, B. Franzke, A. Hoffknecht, C. Kozhuharov, P. H. Mokler, A. Müller, F. Nolden, S. Schippers, Z. Stachura, M. Steck, T. Stöhlker, T. Winkler, A. Wolf: Hyp. Int. **114**, 45 (1998)
54. A. Müller, A. Wolf: Hyp. Int. **109**, 233 (1997)
55. H. F. Beyer, D. Liesen, K. D. Finlayson, F. Bosch, M. Jung, O. Klepper, R. Moshammer, K. Beckert, H. Eickhoff, B. Franzke, F. Nolden, P. Spädtke, M. Steck, G. Menzel, R. D. Deslattes: Phys. Lett. A **184**, 435 (1994)
56. R. Neumann, H. Poth, A. Winnacker, A. Wolf: Z. Phys. A **313**, 253 (1983)
57. U. Schramm, J. Berger, M. Grieser, D. Habs, E. Jaeschke, G. Kilgus, D. Schwalm, A. Wolf, R. Neumann and R. Schuch, Phys. Rev. Lett. **67**, 22 (1991)
58. S. Asp, R. Schuch, D. R. DeWitt, C. Biedermann, Gao Hui, W. Zong, G. Andler, E. Justiniano: Nucl. Instrum. Methods. B **117**, 31 (1996)
59. U. Schramm, A. Wolf, T. Schüßler, D. Habs, D. Schwalm, O. Uwira, J. Linkemann, A. Müller: Hyp. Int. **108**, 273 (1997)
60. *Dissociative Recombination: Theory, Experiment and Applications III,* ed. by D. Zajfman, J. B. A. Mitchell, D. Schwalm, B. R. Rowe (World Scientific, Singapore, 1996)
61. M. Larsson: Annu. Rev. Phys. Chem. **48**, 151 (1997)
62. T. Andersen: Phys. Scr. **T59**, 230 (1995)
63. H. T. Schmidt, P. Forck, M. Grieser, D. Habs, J. Kenntner, G. Miersch, R. Repnow, U. Schramm, T. Schüssler, D. Schwalm, A. Wolf: Phys. Rev. Lett. **72**, 1616 (1994)
64. I. Klaft, S. Borneis, T. Engel, B. Fricke, R. Grieser, G. Huber, T. Kühl, D. Marx, R. Neumann, S. Schröder, P. Seelig, L. Völker: Phys. Rev. Lett. **73**, 2425 (1994)
65. S. Mannervik, L. Broström, J. Lidberg, L.-O. Norlin, P. Royen: Phys. Rev. Lett. **76**, 3675 (1996)
66. J. Doerfert, E. Träbert, A. Wolf, D. Schwalm, O. Uwira: Phys. Rev. Lett. **78**, 4355 (1997)
67. A. Gruber, W. Bourgeois, B. Franzke, A. Kritzer, C. Treffert: Nucl. Instrum. Methods **A282**, 87 (1989)
68. P. H. Mokler, T. Stöhlker: Adv. At. Mol. and Opt. Phys. **37**, 297 (1997)
69. H. T. Schmidt, H. Cederquist, R. Schuch, L. Bagge, A. Kaellberg, J. Hilke, K. G. Rensfelt, V. Mergel, M. Achler, R. Doerner, L. Spielberger, O. Jagutzki, H. Schmidt-Boecking, J. Ullrich, H. Reich, M. Unverzagt, W. Schmitt, R. Moshammer: Hyp. Int. **108**, 339 (1997)
70. M. Jung, F. Bosch, K. Beckert, H. Eickhoff, H. Folger, B. Franzke, A. Gruber, P. Kienle, O. Klepper, W. Koenig, C. Kozhuharov, R. Mann, R. Moshammer, F. Nolden, U. Schaaf, G. Soff, P. Spädtke, M. Steck, Th. Stöhlker, K. Sümmerer: Phys. Rev. Lett. **69**, 2164 (1992)
71. U. Schuenemann, I. Manek, R. Grimm, D. Habs, D. Schwalm: Hyp. Int. **99**, 127 (1996)

72. V. Luger, R. Grimm, priv. commun.
73. R. Grieser, R. Klein, G. Huber, S. Dickopf, I. Klaft, P. Knobloch, P. Merz, F. Albrecht, M. Grieser, D. Habs, D. Schwalm, T. Kühl: Appl. Phys. B **59**, 127 (1994); R. Grieser, P. Merz, G. Huber, V. Sebastian, P. Seelig, M. Grieser, R. Grimm, I. Lauer, V. Luger, H. J. Miesner, D. Schwalm, A. Dax, T. Engel, M. Gerlach, T. Kühl, H. Winter: Hyp. Int. **108**, 241 (1997)
74. T. Schüßler, U. Schramm, T. Rüter, C. Broude, M. Grieser, D. Habs, D. Schwalm, A. Wolf: Phys. Rev. Lett. **75**, 802 (1995)
75. S. Borneis, F. Bosch, T. Engel, M. Jung, I. Klaft, T. Kühl, D. Marx R. Moshammer, R. Neumann, S. Schröder, P. Seelig, L. Völker: Phys. Rev. Lett. **72**, 207 (1994)

II Physics at the Electron Beam Ion Trap

D.H.G. Schneider, J. Steiger, T. Schenkel and J.R. Crespo Lòpez-Urrutia

Summary. Two Electron Beam Ion Traps (EBIT), the low energy EBIT II (40 keV electron beam energy) and the high-energy Super EBIT (200 keV electron beam energy) at Lawrence Livermore National Laboratory produce the highest charge state ions of any element (up to U^{92+}). An EBIT confines target ions within a compressed electron beam and it can be operated as a trap or a source for highly charged ions. This provides the opportunity for advances in areas such as the precision spectroscopy of heavy ions with few electrons, the investigation of unusual states of matter such as Coulomb crystals and collision studies at low energies. Furthermore, ion/matter interaction can be studied in a hitherto inaccessible regime since in highly charged ions the potential energy can reach several hundreds of keV, exceeding the kinetic energy of the ions by far. In this contribution, the facilities are briefly described, and recently completed experiments in the areas of spectroscopy, ion–solid interactions and the trapping of highly charged ions are presented.

1. Introduction

The current availability of highest charge state atomic ion species at near rest energies from any element in the periodic table and from many isotopes, has stimulated new research initiatives in various areas of the physical sciences. The "parameter" charge for low-energy high-Z ions has been expanded following the design and construction of the Electron Beam Ion Trap (EBIT) [1, 2] as a spin-off development of the Electron Beam Ion Source (EBIS) [3]. Both designs facilitate the basic concept of trapping ions within the space–charge potential of an electron beam in combination with external electrical fields and subsequent successive ionization. The EBIT was originally designed for in-situ atomic structure studies following electron impact ionization/excitation. Its modification for use as an ion source [4] verified the original EBIS concept of producing low-emittance, highest charge state ion beams. The production of high charge state ions inside the trap (up to 92+) has been demonstrated [5] and ions with charges up to 91+ have been extracted [6, 7].

The production and confinement of ions in the electron beam in EBIT enables spectroscopic studies of the x-ray emission following radiative recombination and deexcitation [8] and the investigation of ion production via electron impact excitation/ionization processes as well as the study of ions with strong Coulomb fields [9]. Electron impact excitation/ionization cross sections for few-electron, high-Z systems (He-, Ne-, Ni-like) are effected by electron correlation and relativistic effects (e.g., Breit-interaction) [10]. The ability to produce an electronic configuration along isoelectronic sequences allows the performance of atomic structure studies for ions with Coulomb fields varying over a wide range. As the nuclear charge Z increases along an isoelectronic sequence, the atomic transition energies (proportional to Z^2) shift into the x-ray region reaching 100 keV for the Lyman alpha transition in U^{91+}. The effects of relativity, quantum electrodynamics and electron–nuclear overlap grow even faster with Z, so that these effects are most visible and most important at the highest atomic number.

The study of the interaction of highly charged high-Z ions (large Coulomb fields, large potential energy) with solid surfaces is significantly expanded by EBIT's capability to produce beams of ions with hundreds of keV potential energy [11]; some examples are shown in Fig. 1. Slow highly charged ions interact with solid surfaces in an extraordinary way, where the ion charge state development above and below the surface is determined by a complex neutralization process. The neutralization causes the formation of "hollow" atoms where high n-states of the incident ion are populated while inner shells are still vacant. The interaction at the surface–vacuum interface depends on the ion structure and the surface material. The first few layers of the solid respond violently and the ions lose their large potential Coulomb energy via inner shell vacancy decay in a few femtoseconds and Ångströms. The ions can furthermore cause very localized transitions in solid surfaces while the intense, hot electron plasma at the surface creates a highly reactive environment for precursor chemistry. One may speculate that a good understanding of this dynamical interaction process could lead to a variety of applications for highly charged ion beams towards advances in surface analysis and modification.

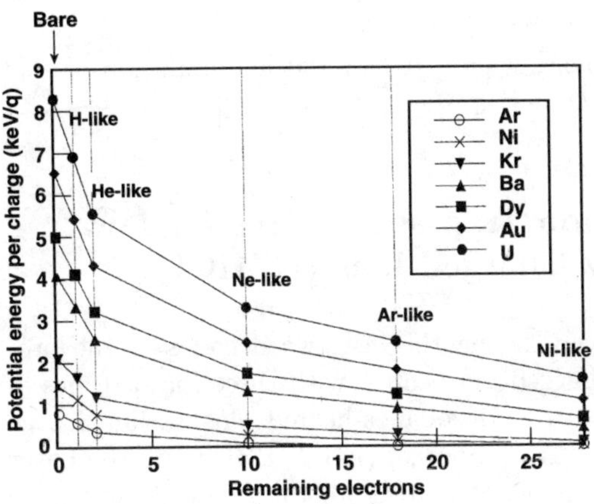

Fig. 1. Total potential energy as a function of the number of remaining electrons for different elements

The EBIT is a well-suited device that allows the study of processes relevant for plasma diagnostics under various conditions and different regimes. X-ray emission spectra are used to diagnose plasma parameters and to determine to what extent the conditions in the plasma are driven by the atomic properties of the highly charged ions they contain. Studies of slow collisions between extracted beams of highly charged, high-Z ions and gaseous (atomic, molecular) or solid targets are of interest for a variety of basic and applied physics questions (e.g., astrophysical plasmas, fusion plasmas and surface plasmas formed in ion–surface interactions). Here knowledge of charge exchange cross sections, recombination dynamics and decay processes is essential. For example, in astrophysical matter such as the sun, white dwarfs and interstellar clouds, highly charged ions *undergo* collisions with neutrals. For fusion plasmas, charge exchange and radiative deexcitation channels are important in order to account for trace elements and to diagnose plasma processes.

A new domain of collision physics and precision spectroscopy studies becomes available if the highly charged ions are extracted, retrapped, electromagnetically confined and cooled to low energies ($\leq$100 K) in a secondary trap [12]. This can be accomplished by transferring and cooling a selected ion species from EBIT into a precision ion trap. Such a system (the EBIT/RETRAP system) is able to receive and confine any highly charged ion that can be produced in an electron beam ion trap. This opens up the following by no means exhaustive research possibilities

(a) ion–atom and ion–ion collision studies at very low energies;
(b) precision lifetime measurements and laser spectroscopy on groundstate hyperfine structure transitions in high-Z ions;
(c) precision mass spectrometry;
(d) condensed-matter studies on non-neutral, strongly coupled, highly charged ion plasmas;
(e) coherent quantum control studies;
(f) applications in nanotechnology where control and manipulation of cold ions and clusters is required.

2. The Electron Beam Ion Trap as a Source for Low Energy, High-Z, Highly Charged Ions (HZHCI)

Much progress has been achieved during the last two decades in the production of low-energy, highly charged ion beams, with three major types of ion sources – the Electron Cyclotron Resonance-heated plasma Ion Source (ECRIS), the Electron Beam Ion Trap (Source) (EBIT/EBIS) and the Laser Ion Source (LIS). The first two types of sources [3, 13] augment successive electron impact ionization to achieve high charges. However, the physical parameters for the electron/atom interaction in an ECRIS and EBIT source are vastly different. The ECRIS source maintains an electron plasma with a wide distribution of electron energies resembling roughly a Maxwellian distribution. Electron energies range from 1000 eV up to 2 MeV. The plasma is confined in a magnetic field and the densities for electrons and ions are about

10^{10}cm^{-3} and 10^8cm^{-3}, respectively. In an EBIT/EBIS an almost monoenergetic electron beam provides the successive ionization of ions trapped in the space charge potential of the compressed electron beam [1, 2]. The electron beam compression is accomplished by a magnetic field, typically 3 Tesla, over a length of about 4 cm in EBIT and about 100 cm in an EBIS. The electron density is typically 10^{12} cm^{-3} while the ion density is of the order of 10^8 cm^{-3}. These differences between ECRIS and EBIT(S) determine the balance between ionization, recombination and heating. The two types of sources provide different ranges of charge states, ion beam intensities and emittances, e.g., ECRIS sources can provide ion beams of U^{60+} and fully stripped Ar with electrical currents of up to one hundred nA [14]. The ECRIS as well as the EBIT/EBIS have the capability of producing highly charged ions of practically all elements, and they can be operated over reasonably long periods of time under stable conditions. Apart from their usefulness as a first stage of an ion accelerator (e.g., for injection into cyclotrons and storage rings), these sources are especially well suited for experiments with slow highly charged ion beams, i.e., at ion kinetic energies well below 2 keV per atomic mass unit [15].

2.1 General EBIT Design and In-Situ Operation

The EBIT employs a high current density electron beam to trap, ionize and excite a population of ions. The trapped ions form an ionization–state equilibrium determined by ionization and recombination rates. The ions are nearly at rest, and, for most cases, in their ground-state configuration.

The operation scheme and the electron beam path in EBIT are shown in Fig. 2. The electron beam originates in an electron gun at ground potential with zero magnetic field at the cathode [16]. The drift tubes, floating at a high potential, accelerate the electrons on their way to the top. After passing through the trap, the electron beam terminates in a collector. Here, the magnetic field decreases rapidly, allowing the beam to expand and the electrons to be collected on the walls. Suppressor electrodes between the drift tubes and collector prevent secondary electrons emitted in the collector from entering the drift tube region.

After the successful construction and operation of two EBITs at LLNL, one of the devices was upgraded to provide a higher electron beam energy. The modified EBIT, called Super EBIT [17], achieves a higher electron beam energy by floating the electron gun at a negative voltage (e.g., 100 kV) and modifications of the drift tubes to eliminate breakdowns induced by trapped Penning electrons has extended their operating voltage range to approximately 40 kV (positive). Super-EBIT was operated successfully at electron-beam energies as high as 205 keV. Typically EBIT and Super-EBIT do not function well at electron energies below 2 keV, but can be operated at a current of 250 mA.

Atoms or low-charge ions are introduced into EBIT in different ways (Fig. 2). Atoms evaporating from the electron gun cathode naturally find their way into the beam, become ionized, and are trapped. These atoms are primarily barium, with smaller amounts of tungsten and osmium. A gas injection system is used for gaseous elements and can have an oven attached to inject Ti, Na and other metals with low melting temperatures. Ions in low charge states from metals are injected from a Metal Vapor Vacuum Arc (MEVVA) source located above the trap [2]. These ions are injected along the electron beam, and captured by raising the top drift tube voltage while the ion pulse is between the top and bottom drift tube. The MEVVA source is also used to inject ions from sintered materials comprising metal oxides in a metallic copper matrix. When only very small amounts of material are available a wire protruding into the periphery of the electron beam is used. Atoms sputtered from the coated wire tip are ionized by collisions with electrons and trapped.

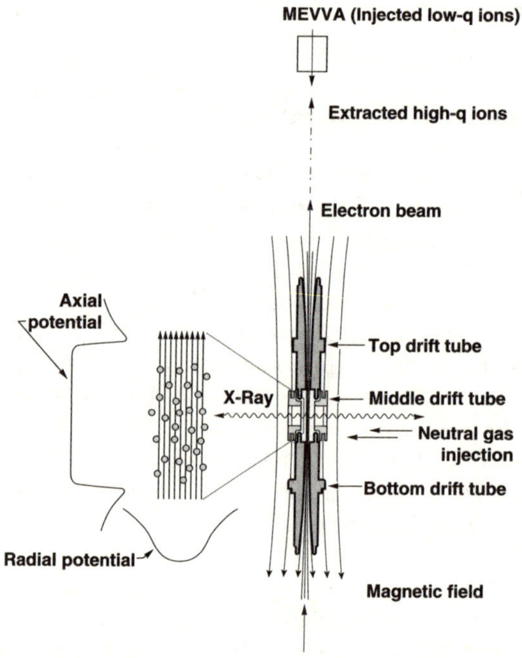

Fig. 2. Schematic for EBIT operation: The radial ion confinement in the space-charge potential of the electron beam and atom/ion injection paths are shown

The primary interactions that affect the charge balance in EBIT are electron–ion and charge exchange collisions with neutral gas atoms. Beam electrons affect the charge state distribution by ionization, radiative recombination, and dielectronic recombination. The latter is a resonant process, however, and can be neglected at most operating energies. Therefore, for electron-beam energies which do not overlap with dielectronic recombination resonances, assuming that the interactions are dominated by single-charge-changing processes, the time evolution for the population of charge state q of a given element Z is given by:

$$\begin{aligned}\frac{\mathrm{d}N_q}{\mathrm{d}t} &= \frac{\langle j_\mathrm{e}\rangle}{e}\left[\left(\sigma_{q-1}^{\mathrm{I}}\,N_{q-1}+\sigma_{q+1}^{\mathrm{RR}}\,N_{q+1}\right)-\left(\sigma_q^{\mathrm{I}}+\sigma_q^{\mathrm{RR}}\right)N_q\right]\\ &\quad +n_0\,v_0\left(\sigma_{q+1}^{\mathrm{CX}}\,N_{q+1}-\sigma_q^{\mathrm{CX}}\,N_q\right), \qquad (1)\end{aligned}$$

where N_q is the number of ions in charge state q, $\langle j_\mathrm{e}\rangle$ is the effective current density of the electron beam, and σ_q^I, σ_q^{RR}, and σ_q^{CX} are the cross sections for ionization, radiative recombination and charge-exchange recombination, respectively; n_0 is the neutral gas density and v_0 the velocity of the ions. Equilibrium is obtained by setting $\mathrm{d}N_q/\mathrm{d}t = 0$ for all q and solving the resulting equations simultaneously [6].

The observed trapping times for ions in EBIT are much longer than was originally predicted [5]. However, ion-electron collisions heat the ions. To avoid ion loss, this energy must be removed. It has been demonstrated that the ions in EBIT undergo a process of evaporative cooling, in which low-Z ions preferentially escape the trap, cooling high-charge, high-Z ions [1, 16]. In equilibrium the mean kinetic energy is the same for all ions but the confining electrical forces scale with the charge. Therefore, low-q ions have a higher probability of escaping, thereby decreasing the kinetic energy of the remaining ions. Thus the required ingredient for the long-term confinement of high-Z ions is a constant influx of low-Z ions for cooling the heavier elements. A gas injection system continuously introduces a small number of ballistic, low-Z atoms into the trap, where they are ionized by the beam and cool the higher-Z ions.

A typical EBIT experiment to study dielectronic recombination via electron–ion excitation in situ requires the following timing sequence (Fig. 3): Ions are generated in the MEVVA source and injected into the trap. The top drift tube voltage is lowered to allow the ions to enter the trap. The drift tube voltage is adjusted to minimize the velocity of the ions in the trap. As the ions are reflected off the bottom drift tube potential, the voltage of the top drift tube is quickly raised, trapping the ions. The ions are then allowed to ionize for a time until the ionization balance comes to equilibrium. The electron energy is then changed to the "probe energy", which may either be constant or a function of time. During this phase, the photons emitted by the ions are observed by a variety of detectors and the data including the electron energy are recorded with an event mode data acquisition system.

2.2 Ion Extraction

The use of the Electron Beam Ion Trap (EBIT) (which was built in 1986 for in-situ spectroscopy) as an ion source, is relatively new and started in 1990 for the study of the interaction of highly charged ions with matter [15].

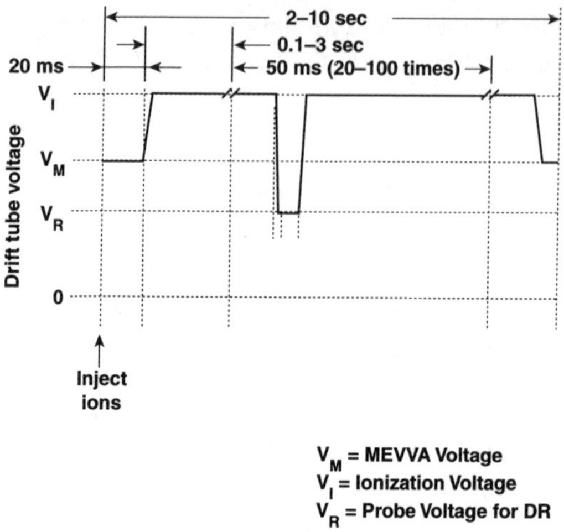

Fig. 3. Typical timing pattern for dielectronic recombination measurements

Ions can be extracted from EBIT in two modes, continuous or pulsed [4]. In both cases, the drift tube potential, which determines the ion energy, is set to the desired value for extraction. In the continuous mode, ions which have obtained enough energy through heating surmount the potential well and escape along the magnetic field lines. For a pulsed extraction, the center drift tube potential can be raised to 20–200 V above the top drift tube potential, thereby expelling the ions. The width of the extraction pulse can be varied from 20 ms for slow pulsed extraction to $10\mu s$ needed for retrapping experiments. Once outside the center drift tube the ions are accelerated into the extractor optics and analyzing magnet for momentum analysis. In the pulsed mode the evolution of the charge state distribution of all the ions in the trap can be studied by varying the time for ionization.

A charge state spectrum from extracted ions at Super EBIT is shown in Fig. 4. It demonstrates the production and extraction of Xe in its highest charge state (Xe^{54+}). Due to the high charge and a high extraction potential, the kinetic energy was in the MeV range. The estimated extraction efficiency can be as high as 80%. Table 1 shows typical numbers of ions extracted from EBIT in the continuous and slow (ms) pulsed extraction mode.

Table 1. Number of Extracted Ions from EBIT at 7 kV extraction potential

Ion	Extraction Mode	Number of Ions
Th^{70+}	Pulsed	1.0×10^5/s
Th^{56+}	Pulsed	1.3×10^5/s
Au^{63+}	Pulsed	2.0×10^5/s
Xe^{44+}	Continuous	1.2×10^7/s
Xe^{27+}	Continuous	3.6×10^6/s
N^{4+}	Continuous	5.0×10^6/s

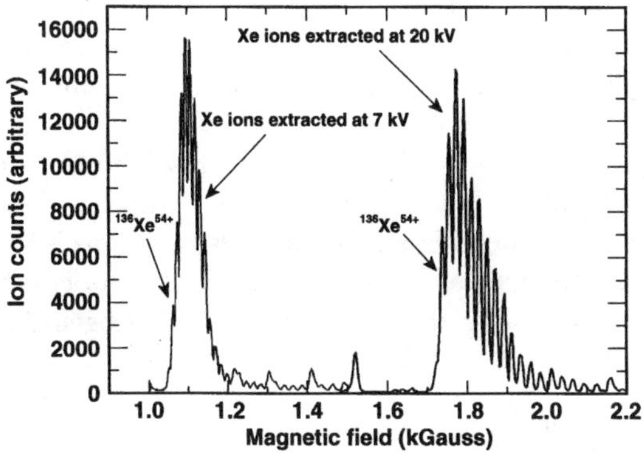

Fig. 4. Charge state spectrum of Xe ions extracted from Super EBIT. Two spectra for different extraction potentials are shown (7 kV and 20 kV). The kinetic energy for fully stripped Xe at 20kV is 1.08 MeV

3. Production and Spectroscopy of Bare and H-Like High-Z Ions

Production of very highly charged and bare uranium ions in Super EBIT requires understanding and control of electron beam heating, radiative recombination and charge exchange collisions with neutral gas atoms and molecules [5, 18].

Highly charged ions trapped in an electron beam are heated by small-angle Coulomb collisions with beam electrons. The (non-relativistic) heating rate is given by

$$\frac{\mathrm{d}E_\mathrm{i}}{\mathrm{d}t} = \pi \frac{j_\mathrm{e}}{c} \frac{{q_\mathrm{i}}^2 e^4}{E_\mathrm{e}} \frac{2\, m_\mathrm{e}}{M_\mathrm{i}} \ln \Lambda, \tag{2}$$

where E_i, q_i, and M_i are the ion energy, charge and mass, respectively; E_e and j_e are the electron beam energy and current density and $\ln \Lambda$ is the Coulomb logarithm.

For uranium ions with $q_\mathrm{i} = 90$ in a 200 keV, 200 mA electron beam, $\mathrm{d}E_\mathrm{i}/\mathrm{d}t \approx 6$ keV sec^{-1} per ion. A comparison of rates shows that in the absence of a cooling mechanism high-Z ions will "evaporate" from the beam before reaching high charge states. For the evaporative cooling of uranium, a collimated beam of neon atoms with a carefully controlled density crosses the electron beam at 90 degrees. Roughly 0.2% of the neon atoms that intercept the electron beam are ionized and captured. Direct electron beam heating of the low-charge neon ions is small. They are heated primarily by collisions with high-charge uranium ions before escaping axially (evaporating) from the trap with the removal of $q_\mathrm{Ne} \cdot V_\mathrm{A} \approx 300$ eV per neon ion (V_A is the axial trapping potential over which the ions escape).

Evaporative cooling is extremely efficient, producing a population of uranium ions that remain trapped for several minutes, or even hours under optimum conditions. Usually the trap is initially "over filled" by the injection of a large number of low-charge uranium ions from the MEVVA. The number of uranium ions remaining in EBIT is determined by the evaporative cooling power (i.e., the neutral neon density) and not by the number of uranium ions injected initially.

Once highly charged uranium ions are confined in Super EBIT's electron beam, the removal of the last few tightly bound electrons can take place. The equilibrium ionization balance for trapped highly charged uranium ions is determined by the relative rates for the ionization and recombination processes that couple adjacent ionization states. The abundance ratio for hydrogen-like and bare uranium at equilibrium is given by [5, 16]

$$\frac{N_{\mathrm{H}}}{N_{\mathrm{Bare}}} = \frac{\sigma^{\mathrm{RR}}_{\mathrm{Bare}\to\mathrm{H}}}{\sigma^{I}_{\mathrm{H}\to\mathrm{Bare}}} + \frac{e}{\langle j_{\mathrm{e}}\rangle} n_0\, v\, \frac{\sigma^{\mathrm{CX}}_{\mathrm{Bare}\to\mathrm{H}}}{\sigma^{I}_{\mathrm{H}\to\mathrm{Bare}}}, \tag{3}$$

where the second term accounts for charge exchange recombination with neutral neon atoms. Here n_0 is the neutral gas density and v is the average ion-neutral collision velocity (approximately the ion thermal velocity). The cross sections for radiative recombination (RR), ionization, and charge exchange (CX) are written in obvious notation.

The highest charge states are obtained by keeping the ratio $n_0/\langle j_{\mathrm{e}}\rangle$ small. For the production of U^{92+}, n_0 was chosen so that the charge exchange term in (3) was about 10% of the radiative recombination term. However, since $\sigma^{\mathrm{RR}}_{\mathrm{Bare}\to\mathrm{H}}/\sigma^{\mathrm{I}}_{\mathrm{H}\to\mathrm{Bare}} \approx 40$ for uranium at 198 keV, it is always true that $N_{\mathrm{Bare}} \ll N_{\mathrm{H}}$. The fractional abundance of U^{92+} could be increased by operating at a higher electron energy and higher currents. Dielectronic recombination does not contribute at 198 keV, and three-body recombination is negligible at these energies and densities.

The charge-state distribution of the trapped uranium or other ions can be derived from x-ray emission spectra such as that shown in Fig. 5 [5]. Radiative recombination of the uranium ions with the (monoenergetic) beam electrons produces a series of lines whose energy is equal to the sum of the electron energy and the binding energy of the vacant orbital into which the electron is captured. Bare and hydrogen-like ions produce resolved radiative recombination lines for K-shell capture. Radiative recombination of the other high charge states of uranium is incompletely resolved, but the ionization balance can be obtained by fitting the radiative recombination spectrum. The cross sections for radiative recombination are well known. Out of a total of 5×10^4 highly charged uranium ions, there was an average of 500 hydrogen-like and 10 bare uranium ions. For the relatively low neutral neon densities used here ($\sim 2\times 10^5 \mathrm{cm}^{-3}$ averaged over the length of the trap), the total compensation of the electron beam space charge by neon and uranium charges was less than 10%.

The ionization cross section for U^{91+} is the key to the production of U^{92+}. Cross sections (see Table 2) for various high-Z elements were determined from the equilibrium ionization balance in Super EBIT [18]. This was done

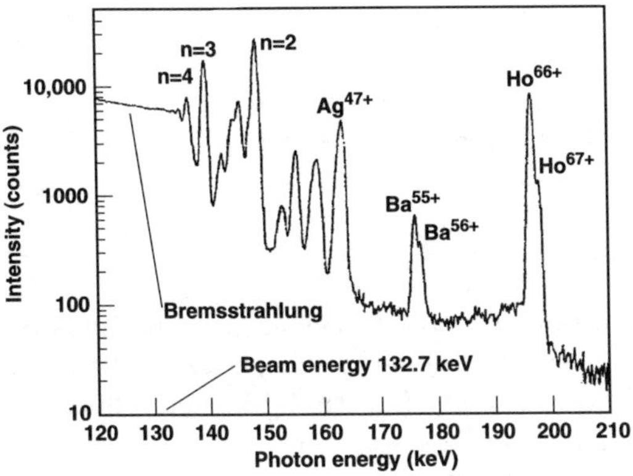

Fig. 5. X-ray spectrum from Super EBIT with the hydrogenic and bare Ho radiative recombination lines. Only emission above 102 keV photon energy is shown

by using (1) and normalizing to the known calculated radiative recombination cross section. The resulting ionization cross section agrees well with recent theoretical calculations.

A large number of x-ray spectroscopy measurements on EBIT have been performed with high precision [19, 20] for atomic structure studies on highly charged ions. EBIT provides a line source for diverse spectrometer geometries, it is a very convenient device with regard to the experimental setups and has several physical advantages [21, 22]. Due to the weak field and low density of the plasma in EBIT, various complications can be neglected (e.g., Stark effect, absorption). Essentially no Doppler shift corrections need to be applied and low ion temperatures can be achieved ($\sim$ 1 eV). Regarding the atomic structure, ions with one electron outside the closed shells and in particular the hydrogen-like ions are easier to treat theoretically and calculations of their energy levels are thought to be highly accurate even for very

Table 2. Ionization cross sections for hydrogen-like ions. Ionization potentials (I.P.) and electron energies E_e are given in keV. The quantities σ^{RR} and σ^{CX} are the total RR and effective charge exchange recombination cross sections, respectively. All cross sections are given in 10^{-24} cm^2. The relativistic distorted-wave calculations (RDWA) are from *Moores* and *Reed* [18]

Element (Z)	I.P.	E_e	σ^{RR}	σ^{CX}	Present exp.	RDWA
Mo (42)	24.6	31.5	90.33	7.2	15.9±1.6	15.1
		36.1	72.39	7.0	21.2±1.9	20.8
		64.8	26.56	3.1	30.8±2.6	31.6
		95.6	13.01	7.3	34.7±7.2	32.2
Dy (66)	63.1	95.1	67.2	6.6	4.17±0.58	4.13
		153.1	30.2	6.6	6.29±0.83	6.11
Au (79)	93.3	153.1	58.6	9.6	2.33±0.33	2.47
Bi (83)	104.1	191.6	48.5	7.7	2.37±0.19	2.40
U (92)	131.8	198.	67.5	7.5	1.55±0.27	1.44

high-Z elements. Effects due to quantum electrodynamics (QED), relativistic effects and multielectron contributions in strong Coulomb fields appear as a difference between the measured energy and the Dirac–Coulomb energy of an atomic state. For one-electron systems this difference is the Lamb shift and is commonly expressed as

$$L = \frac{\alpha}{\pi} \frac{(Z\alpha)^4}{n^3} F(Z\alpha)\, m_e c^2 \,, \tag{4}$$

where α is the fine-structure constant, n is the principal quantum number, $m_e c^2$ is the electron rest mass (in energy units), and $F(Z\alpha)$ is a slowly varying function of Z which contains all the necessary (mostly QED) corrections. At high-Z, $F(Z\alpha)$ is dominated by higher-order terms.

QED contributions to atomic energy levels increase with higher powers of the nuclear charge Z than the Dirac energies. The effects of QED are thus enhanced in the spectra of the heaviest elements, and precise measurements can provide a very sensitive test of QED theory in intense nuclear fields. Such tests complement the many existing tests of QED in weak fields, which include measurements of the lepton anomalous magnetic moment or the structure of neutral hydrogen and helium, by assessing the non-perturbative aspects of QED theory in the limit $Z\alpha \approx 1$.

Experimental studies of hydrogen-like low-Z ions provide a measurement of the lower-order self-energy corrections with extraordinary precision. For high-Z ions, the higher-order self-energy corrections and the vacuum polarization are the most important QED contributions that cannot be tested with low-Z ions. In addition, the effect of finite nuclear size has to be considered for heavy elements.

Measurements of the hyperfine splitting of the ground state of hydrogen provide a very sensitive tool to explore QED and nuclear contributions to the electron energy. The $1s$ electron of a highly charged ion probes the structure of the nucleus more deeply, due to its Coulomb field, enabling a very sensitive test of theory in strong fields. While the spontaneous $1s$ hyperfine transition for H, D, and He^+ has been observed in radio astronomy, laboratory measurements in hydrogen and low-Z hydrogen-like ions have to rely on stimulated emission using a maser setup due to the extremely long lifetime of the upper hyperfine level (1.1×10^7 years for H) [23, 24]. This technique was successful in measuring the splitting of H, D, T, and He^+. Laser pumping has recently been applied to H-like Bi^{82+} [25] and H-like Pb^{89+} [26] ions circulating at nearly 60% of the speed of light in a heavy-ion storage ring resulting in the first measurement of the hyperfine splitting in multiply charged ions. Here laser fluorescence was detected as the laser frequency was tuned in resonance with the Doppler-shifted transition.

Spectroscopy in the visible region was first performed at the NIST (National Institute of Standards and Technology) EBIT [27] and is used here for hyperfine transition spectroscopy. At Super EBIT the $F = 4$ to $F = 3$ hyperfine transition of the 1s ground level of H-like $^{165}Ho^{66+}$ and $^{187}Re^{74+}$ using passive emission spectroscopy has been measured [28]; only the Ho measurement is discussed here. The fact that the transition is excited by electron collisions and the relative ease of the technique, in principle, opens the upper half of the periodic table to scientific scrutiny.

In the experiment, the trap is simultaneously monitored by two germanium detectors. Like in the uranium case, they register the emitted x-rays which include the $K\alpha$ line from the trapped ions, the electron bremsstrahlung continuum and the radiative recombination (RR) lines produced by beam electrons captured into different open shells of the ions (Fig. 6). These spectra were used to determine the ionic species and the charge balance in the trap. With the holmium MEVVA on, at a beam energy of 132 keV and 285 mA beam current, the He-like Ho^{65} ion was the most abundant species in the trap (40 %) followed by Li-like Ho^{64+} (25%). The concentration of H-like Ho^{66+} was 6 %, that of bare Ho^{67+} around 0.5%.

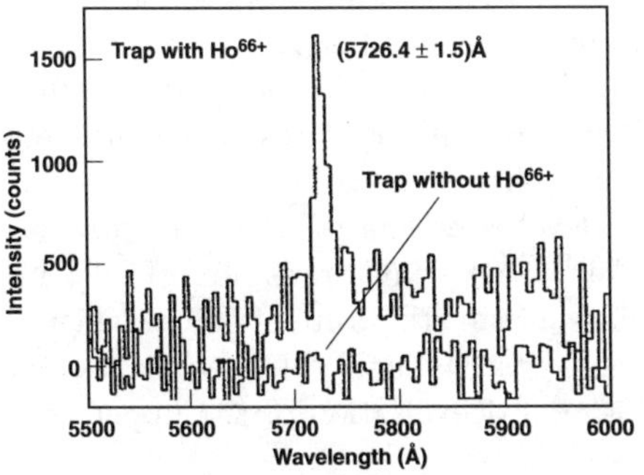

Fig. 6. The $F = 4$ to $F = 3$ hyperfine transition of hydrogenlike Ho^{66+} and its baseline

The hyperfine transition was measured using a prism spectrograph with an aperture of $f/9$, and a cryogenically cooled charge coupled device (CCD) camera detector. It allows simultaneous coverage of the whole visible spectrum, providing fast data acquisition. The trap is spatially resolved in the vertical and horizontal directions, giving information about the spatial distribution of the ions.

The electron beam and the apparatus symmetry axis are aligned vertically. The beam is stigmatically imaged by two achromats ($f1 = 655$ mm, $f2 = 250$ mm, demagnification 2.6 times) on the focal plane of the spectrograph, where the CCD detector is placed.

To obtain the spectra from the two-dimensional images, the pixel counts on the CCD detector are integrated along one dimension. The result of the addition of 18 background corrected spectra is displayed in Fig. 6. A single feature at 5726.4 ±1.5 Å with a peak height 10 times the standard deviation of the background line and a FWHM of 15 Å appeared. This feature is attributed to the hyperfine transition in Ho^{66+}.

Observation of the $F = 4$ to $F = 3$ hyperfine transition is expected for several reasons. The trap contains a suitable number of bare and H-like Ho ions. The recombination lines of these two species in the Ge detector were recorded at a rate of more than 10,000 counts/h. The bare Ho^{67+} alone accounted for roughly 2000 counts/h. Two main processes are expected to populate the upper F level of the H-like ion. The most important is collisional ionization of He-like ions, where one of the two 1s electrons is removed leaving a H-like ion behind with similar probability for population in either the $F = 4$ or $F = 3$ state. In ionization equilibrium (reached within 2 s) which was determined by time-resolved x-ray spectra, the ionization rate is balanced by the radiative recombination rate of H-like to He-like ions [29]. The ionization rate populating the $F = 4$ level can be estimated from the observed radiative recombination photons, it should produce 190 counts/h in the optical spectrograph according to the ratio of the solid angles of the Ge detector and the spectrograph. The second process is radiative recombination of beam electrons with bare Ho, also populating $F = 4$ and $F = 3$ nearly evenly. The rate estimated here from the x-ray signal is 35 counts/h. With these processes, a spontaneous emission count rate for the hyperfine transition of about 225 counts/h can be expected. This result is consistent with the observed count rate of 220 counts/h.

The radiative lifetime of the $F = 4$ level is estimated to be roughly 2–3 ms from the measured value for the Bi^{82+} hyperfine transition of 0.34 ms and assuming a Z^9 scaling law. This gives a transition rate of 500/s, which is much faster than the collisional deexcitation rate (roughly 10 s) at the present electron beam density. Therefore, collisional deexcitation of the upper level does not affect the result significantly.

Neutral atoms from the gas injector excited by the electron beam at low energies were used to generate a large number of calibration lines. Possible instrument shifts were monitored by observing the position of the NeI line at 5852.48 Å which was still weakly visible at beam energies of 132 keV. The wavelength measurement accuracy of 250 ppm is currently limited by counting statistics and calibration uncertainties. Our experimental result of 5726.4±1.5 Å in air corresponds to a wavelength of 5727.9±1.5 Å in vacuum and an energy difference between the $F = 4$ and $F = 3$ levels of 2.1645±0.0006 eV.

The energy difference Δ between two neighboring F levels in a H-like ion is [30]

$$\Delta = \frac{\alpha^4 Z^3}{n^3} \frac{m \mu_\mathrm{I}}{I} \frac{(I + j)}{j\,(j+1)} \frac{m_\mathrm{e}\, c^2}{(2l+1)} \left[A\,(1-\delta)\,(1-\varepsilon) + K_\mathrm{rad}\right] \qquad (5)$$

with the following terms: α, fine-structure constant; Z, nuclear charge; n, j, ℓ, principal, total moment, and orbital moment electron quantum numbers; μ_I, nuclear magnetic moment; $m_\mathrm{e}, m_\mathrm{p}$, electron and proton mass; I, nuclear spin; A, relativistic factor; δ, nuclear charge distribution correction; ε, nuclear magnetization distribution (Bohr–Weisskopf) correction; and K_rad, QED, radiative correction. Using this formula, but neglecting QED corrections, a vacuum wavelength of 5639 Å is predicted (air wavelength 5637.4 Å). The uncertainty of at least 40 Å is due to the uncertainties in the μ_I value quoted. This value differs significantly from the measured value, by 89 Å. Two dominant QED corrections have recently been calculated. Based on the calculations of the self-energy correction by *Persson* et al. [31], we estimate for $Z = 67$ a 19.3 meV shift. The vacuum polarization was given by *Schneider* et al. [32]. From their work one estimates this contribution to be 9.4 meV. The total net QED correction is therefore −9.9 meV, or 27 Å. Even after including the shift from QED, there is still a difference of 62 Å between the measurement and the calculated wavelength. *Shabaev's* [30] calculation uses a value for the nuclear magnetic moment of ^{165}Ho of 4173(27) μ_N tabulated by Lederer and *Shirley* [33] and based on a measurement by the atomic beam resonance method of 4094(44) μ_N [34] (without diamagnetic correction), later recalculated in [35] to 4128(27) μ_N (without diamagnetic correction). This value is cited in the literature as 4160(27) μ_N, or 4173(27) μ_N, depending on the size of the diamagnetic correction employed. The evaluation of the original data seemed nonetheless uncertain. A newer measurement was performed by *Nachtsheim* [36] resulting in 4132(5) μ_N (corrected), with a 5 times smaller uncertainty. This value is compiled in the review of Peker [37]. Using this value in (4) and adding the above mentioned QED corrections, we obtain a transition energy of 2.1675 eV and a vacuum wavelength of 5720.15 Å. This differs from our measured vacuum wavelength 5727.9± 1.5 Å by only 0.13%. We emphasize that this good agreement was reached only using the updated value for the nuclear magnetic moment and the QED corrections. In the case of Bi^{82+} *Shabaev* [30] predicts (no QED) a wavelength of 2412 Å, while the measurement at GSI was 2438.7 Å, a discrepancy of 1.1%. A more refined calculation [38, 39] for Bi^{82+} leads to a value of 2419.8 Å, with a smaller discrepancy of 0.8%, and a third one, taking nuclear many-body corrections into account, gives an even closer prediction of 2435.4 Å (0.13% discrepancy).

The question of the accurate determination of the nuclear magnetic moment of holmium remains open. A precision measurement of the hyperfine splitting can at present be used only for testing the sum of the contributions from the nuclear magnetic moment, nuclear size effects, and QED. Table 3 summarizes the present results including recently obtained results for Re [22]. Hyperfine measurements along the isoelectronic sequence and for different isotopes can be used to determine the size of the QED contributions, which follow well–known scaling laws, with high accuracy. Determinations of the nuclear magnetic moment from the ls hyperfine transition measurements thus seems within reach. Such measurements would avoid uncertainties in the evaluation of the atomic diamagnetism that may affect values of the tabulated nuclear magnetic moments.

Table 3. Measured hfs splitting of the 1s level in H-like ions

Ion	Experimental hfs splitting (eV)	References
$^{165}Ho^{66+}$	2.1645(6)	[28]
$^{185}Re^{74+}$	2.7190(18)	[28]
$^{187}Re^{74+}$	2.7450(18)	[28]
$^{209}Bi^{82+}$	5.0840(8)	[25]
$^{207}Pb^{81+}$	12.358(7)	[26]

4. Collisional Interactions of High-Z Highly Charged Ions with Solid Surfaces

The ion source development of the LLNL EBIT has made low emittance (1π mm mrad) beams of slow (0.5–20 keV/u), highly charged ions (up to U^{90+}) available for ion–solid interaction studies [4, 6, 11, 40]. Slow, highly charged ions deposit tens to hundreds of keV of potential energy on a femtosecond time scale into a surface near a target volume when they interact with solids. Power densities are $\sim 10^{14}$ W/cm^2. A fundamental challenge is the quantification and microscopic understanding of energy dissipation channels and the characterization of the response of different materials to the massive disturbance caused by the relaxing ion. In ongoing investigations, research at the LLNL EBIT facility has addressed emission of secondary electrons [41], x-rays [42], secondary ions [43], cluster ions [44], neutrals [45] and the formation of nanometer size defects on surfaces following the impact of individual ions [46]. The finding of very high secondary ion yields of several secondary ions detected per incident ion has motivated the development of highly charged ion based TOF-SIMS (time-of-flight secondary ion mass spectrometry) for sensitive surface analysis [47, 48]. The very high densities of target excitations induced by ions like Au^{69+} constitute unique physical conditions very far from thermal equilibrium. Here, one prospect is the formation of otherwise unattainable solid state phases, and the induction of local phase transitions [45], and athermal chemical reactions on surfaces.

In this section we will focus on recent studies of one fundamental aspect of the interaction, namely the energy loss of slow, highly charged ions in thin foils [49]. Energetic ions traveling in solids lose kinetic energy in collisions with target electrons and nuclei [50]. Energy loss is a function of the projectile charge state in the solid [51, 52]. Pre-charge-state equilibrium effects have been observed for intermediate energies ($\sim$ 1 MeV/u), where projectiles entering solids with charge states lower or higher than the mean equilibrium charge show increased or decreased stopping until charge state equilibrium is established [53]. The charge states of slow, highly charged ions like Xe^{44+} or Au^{69+} are far in excess of their mean equilibrium charge state in solids. The latter is neutral or 1+ for velocities much smaller then the Bohr velocity [54]. Consequently, a finite de-excitation time, τ_{eq}, of these ions inside solids should result in pre-equilibrium contributions to energy loss processes along a charge state equilibration length, x_{eq} [55]. An upper limit for τ_{eq} for highly charged ions was established in measurements of equilibrium charge state distributions of ions after passage through thin carbon foils ($\sim$ 2μg/cm^2 or $\sim$ 10 nm). For Th^{65+} ($v = 0.29\, v_{Bohr}$) τ_{eq} was less than or equal to $\sim$ 21 fs [43].

We have measured the energy loss of slow ($v = 0.3\, v_{Bohr}$), highly charged ions transmitted through thin carbon foils using time-of-flight spectrometry [1, 47]. Time-of-flight start signals were provided by secondary electrons emitted from the target by individual highly charged ions. Secondary electrons were detected by an annular micro-channelplate detector. A negative target bias of 100 V was applied to provide for strong start signals for all initial charge states. The target consisted of a $\sim$ 10 nm thick ($\sim 2 \pm 0.5$ μg/cm^2) carbon foil with a diameter 3 mm. The ion flux was typically $< 10^3$ ions/s, and the foil was exposed to a dose $< 5 \times 10^8$ ions. The energy loss of Xe^{44+} was measured at the beginning and at the end of the study and no indications of foil modification were observed. Transmitted ions were detected by a second micro-channelplate detector after a flight path of 52.5 $\pm$ 0.3 cm, and provided time-of-flight stop signals. The solid angle of the detector was 15 msr. The time resolution of the setup, including detectors and electronics, was about 1 ns. Energy loss in the foil increased ion flight times by $\sim$40–60 ns, as compared to flight times at the initial ion velocity ($\sim$ 800 ns). The fractional uncertainty in impact energies of projectiles of given mass and different charge states was $\leq 6 \times 10^{-3}$.

Energy-loss distributions, dN/dE, are shown in Fig. 7 for Au^{69+} and Au^{33+} ions. The initial energy was 454.4 $\pm$3 keV. Energy loss values reported here are average energy losses, ΔE_{ave}, determined by averaging energy loss distributions after background subtraction. Values of the most probable energy loss were found to be systematically lower than ΔE_{ave} [56]. The energy loss of Au^{69+} is significantly higher than that for Au^{33+}-projectiles. The functional dependency of the highly charged ion energy loss on projectile charge is shown in Fig. 8 (a) and (b). We used a constant velocity of 0.3 v_{Bohr} for ions of all species and charge states. Projectiles and initial kinetic energies were:

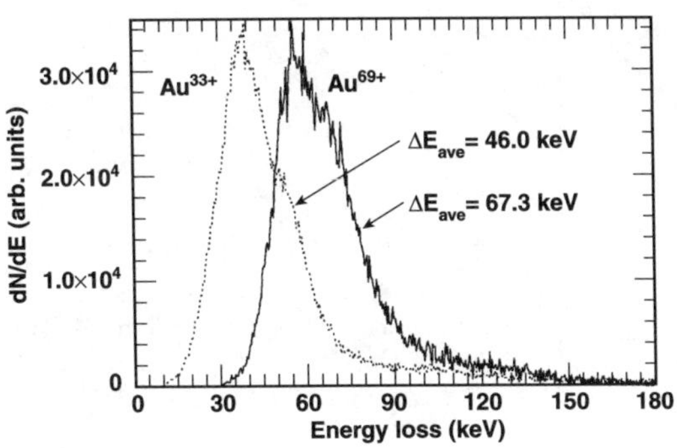

Fig. 7. Energy-loss distributions of Au^{33+} and Au^{69+} after transmission through a thin carbon foil ($\Delta x \approx$ 10.4 nm). The initial energy was 454.4 $\pm$ 0.3 keV

^{16}O, 35.5± 0.2 keV; ^{40}Ar, 92.3±0.6 keV; ^{86}Kr, 197.7±1 keV; ^{136}Xe, 312.4±2 keV and ^{197}Au, 454.4±3 keV. Average energy losses of O^{7+}, Ar^{17+} and Ar^{18+} increase slightly over low charge state values. The charge-dependent energy loss enhancements are close to the limit of significance at the given accuracy of the measurements. Using a thinner target, lower ion velocities and higher charge states, our results extend earlier studies [57], were no effect of incident ion charge on energy loss was found for ions up to Ar^{16+}, to a regime were the onset of charge effects can be observed. Predictions from density functional theory calculations of decreasing energy loss in the presence of one K-shell vacancy in oxygen ions [58] can only be tentatively compared to our results. The calculations considered only electron–hole pair excitations of target valence electrons at the Fermi level as mechanisms for energy loss. The carbon foils used in this study cannot be described as a free electron gas. Also, our experimental technique does not allow us to differentiate inelastic and elastic contributions to energy loss processes. Average impact parameters for the latter are in the order of the K-shell radius of oxygen ions (~0.14 a.u.). The observed small increase of $\Delta E_{\rm ave}$ in the presence of one K-shell vacancy in oxygen ions, from 3.5 ± 0.2 keV for O^{5+} to 4.1 ± 0.2 keV for O^{7+} can result from increased momentum transfer in both elastic and inelastic collisions. For xenon and gold ions, $\Delta E_{\rm ave}$ increases strongly with charge.

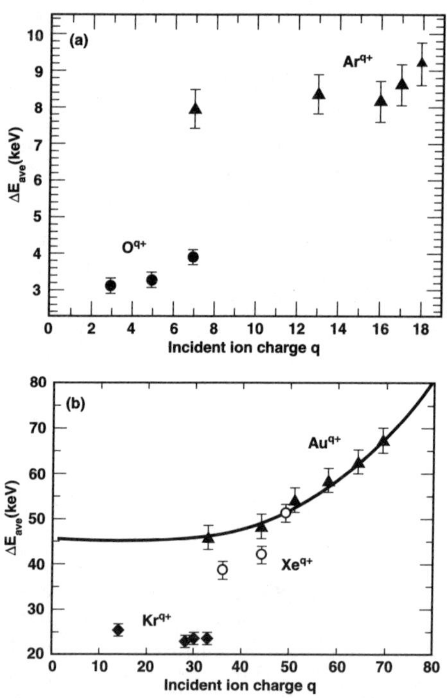

Fig. 8. Average energy loss of (a) $O^{3,5,7+}$ and $Ar^{7,13,16,17,18+}$; (b) $Kr^{14,28,30,33+}$, $Xe^{36,44,49+}$, and $Au^{33,44,51,64,69+}$ in a thin carbon foil. The dotted line is a fit to the Au^{69+} data using (6) with $n = 1$, $m = 2$ ($\alpha = 0.0165$, $\beta = 8 \times 10^{-5}$)

Energy loss in the foil is the sum of contributions to stopping processes before, $(\mathrm{d}E/\mathrm{d}x)_{\mathrm{pre}}$, and after, $(\mathrm{d}E/\mathrm{d}x)_{\mathrm{eq}}$, a dynamic charge-state equilibrium is established. Both, equilibration length, Δx_{pre}, and pre-equilibrium stopping power, $(\mathrm{d}E/\mathrm{d}x)_{\mathrm{pre}}$, are a function of the projectile charge state, q. In an attempt to estimate values for Δx_{pre}, τ_{eq} and pre-equilibrium stopping power enhancements, we assume a simple power law dependency of Δx_{pre} and $\Delta(\mathrm{d}E/\mathrm{d}x)_{\mathrm{pre}}$ on q. We note that a quadratic charge dependency of the electronic stopping power has been assumed in the framework of an effective charge theory in order to estimate charge state effects on the stopping of slow, highly charged ions in solids [52, 57]. In a rough approximation that does not include the time dependency of q in the solid, we can then write for the energy loss in the foil:

$$\begin{aligned}\Delta E(q) &= \frac{\mathrm{d}E}{\mathrm{d}x}\Big|_{\mathrm{eq}} \cdot [\Delta x - \Delta x_{\mathrm{pre}}] + \frac{\mathrm{d}E}{\mathrm{d}x}\Big|_{\mathrm{pre}} \cdot \Delta x_{\mathrm{pre}} \\ &= \frac{\mathrm{d}E}{\mathrm{d}x}\Big|_{\mathrm{eq}} \cdot [\Delta x - \alpha \cdot q^n] + \beta \cdot q^{n+m} . \end{aligned} \tag{6}$$

The average energy loss value for Au^{33+} agreed reasonably well with values calculated for gold ions in charge state equilibrium using the TRIM code [59]. Thus this value was taken to account for equilibrium energy loss contributions. With the constants α and β as free parameters, we used (6) to fit the gold data in Fig. 8(b) for $n, m \leq 2$. The shown fit was obtained for $n = 1$ and $m = 2$. Values for α and β from realistic fits allow for very tentative estimations of Δx_{pre} and $(\mathrm{d}E/\mathrm{d}x)_{\mathrm{pre}}$. Resulting increases of pre-equilibrium over equilibrium stopping powers during de-excitation range from a factor 2 along an equilibration length of ~ 4 nm ($\tau_{\mathrm{eq}} \sim 6$ fs), to a factor 5 with $\Delta x_{\mathrm{pre}} \sim 1$ nm ($\tau_{\mathrm{eq}} \sim 2$ fs). In the absence of a critical number of inner shell vacancies, e.g., for ions like Kr^{33+}, de-excitation processes at and below the surface are too fast ($\tau_{\mathrm{eq}} \leq 1$ fs, $\Delta x_{\mathrm{pre}} \leq 0.7$ nm) and pre-equilibrium stopping power increases are too weak ($\mathrm{d}E/\mathrm{d}x|_{\mathrm{pre}} < 2 \cdot \mathrm{d}E/\mathrm{d}x|_{\mathrm{eq}}$) to allow for strong contributions to energy loss processes before de-excitation is completed.

This estimate of the mean de-excitation time can be tested by varying the impact velocity, v, of ions, and thus increasing the number of collisions before a dynamic charge state equilibrium is reached. If τ_{eq} assumes values not much smaller than the time the ions spend inside the target foil, then a higher number of collisions before de-excitation is completed will result in an energy loss increase as a function of v. The dependence of ΔE_{ave} on v is shown in Fig. 9(a) for Ar^{18+} and Xe^{44+} and in (b) for Au^{69+}. For argon, even bare ions reach equilibrium so fast (≤ 1 fs) that a dependency on impact velocity cannot be observed at the present sensitivity. Consistent with the q-dependency, ΔE_{ave} increases strongly as a function of v for both Xe^{44+} and Au^{69+} projectiles. To allow for a comparison with the stopping of ions in charge state equilibrium, we use values for nuclear, S_n, and electronic stopping powers, S_{el}, as calculated by the TRIM code [59]. Resulting values, multiplied by the effective target thickness are shown in Fig. 9(b) for slow gold ions in carbon. Stopping power values calculated by TRIM cannot expected to be very accurate in this velocity regime, but are known to reproduce the velocity dependences of elastic and inelastic contributions to energy loss in conducting, monoatomic solids correctly. We can describe the interaction of slow, highly charged ions with solids qualitatively in terms of the formation and decay of hollow atoms [11, 60]. Having reached a critical distance from a metal or semi-metal surface (~3.4 nm for Au^{69+} on amorphous carbon), incoming ions begin to neutralize rapidly by resonant capture of electrons into high n Rydberg states ($n \sim 60$ for Au^{69+}) [61]. Mostly low energy (< 20 eV) electrons are emitted from the ion already above the surface by autoionization. The time available for above-surface relaxation is too short (~ 5 fs for Au^{69+} at $v = 0.3\, v_{\mathrm{Bohr}}$) to allow for complete relaxation by Auger transitions and after a few femtoseconds a neutralized but still highly excited hollow atom impinges on the surface. Now, electrons in high n states are rapidly peeled off when a screening cloud of target electrons is formed around the projec-

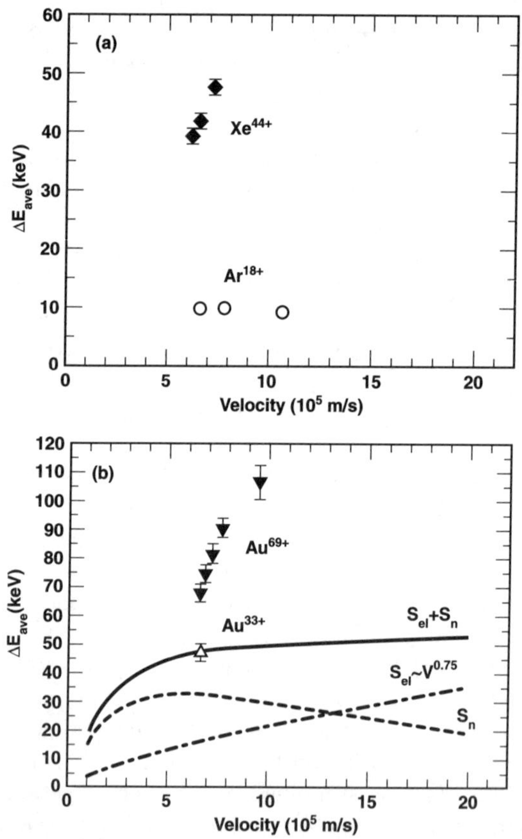

Fig. 9. Average energy loss, ΔE_{ave}, of (a) Ar^{18+} and Xe^{44+} and (b) Au^{69+} in a thin carbon foil as a function of projectile velocity. The lines are estimated energy loss values of gold ions in charge state equilibrium: S_{n} and S_{el} are the nuclear and electronic energy loss

tile. Neutralization of the projectile charge by this screening cloud requires an extreme degree of target polarization, involving over ten carbon atoms to provide the charge required to neutralize a Au^{69+} projectile. The situation is different for insulating targets were a lack of free electrons inhibits efficient screening of projectiles. It has been suggested that this reduced screening results in pre-equilibrium energy loss enhancements due to long range Coulomb interactions [52] between projectile nuclei and ionic cores in insulators like CaF_2 [62]. De-excitation of hollow atoms proceeds via Auger cascades and radiative transitions. Yields of secondary electrons [41] and x rays [42] emitted in the course of de-excitation are relatively high, but only a small fraction ($\sim$ 10%) of the available potential energy of projectiles is dissipated in emission of secondaries. In fact, most of the dissipated energy is currently not accounted for. In the asymmetric combination of a heavy projectile incident on a light target, direct filling of M-shell vacancies in gold projectiles by target electrons is strongly inhibited. Only projectile levels with principal quantum numbers $n \sim 6$–10 can be populated directly. The charge distribution in the hollow atom is characterized by electrons in the initially filled K- and L-shells and by electrons in highly excited states. Intermediate levels with $n = 3$–5 start out empty and are filled in the course of de-excitation. Screening of the projectile nuclear charge is reduced at distances from the nucleus equal to the radii of these empty levels. Averaging over all impact parameters, this can be interpreted as an increased effective charge of hollow atoms. Scaling of the electronic stopping power with the square of an effective ion charge was proposed to estimate charge state dependent electronic stopping power increases for slow, highly charged ions [52, 57]. Effective charge theory is a semi-empirical parametrization that has been applied successfully to the systematization of energy loss data for heavy ions at intermediate energies ($\sim$MeV/u). The usefulness of its conceptual adaptation to collisions in the deeply adiabatic velocity regime present in this study is consequently rather limited. At atomic transition rates in the order of 10^{15}–10^{16} s^{-1}, screening of the nuclear charge and the interaction potential between projectile and target nuclei change on the time scale of individual collisions ($\sim$0.5 fs). Momentum transfer to target electrons and nuclei in collisions with impact parameters comparable to the radii of unoccupied levels in the transient hollow atom is increased over corresponding values for projectiles in equilibrium. Increasing the impact velocity of Au^{69+} allows for the de-exciting projectile to engage in more collisions before a dynamic charge state equilibrium is established. Increasing the velocity further will extend de-excitation deeper into the target and will eventually allow projectiles to exit the foil in a highly excited state and in mean charge states in excess of equilibrium charge states. The effect is less pronounced for Xe^{44+} as the pre-equilibrium stopping increase is weaker and the mean de-excitation time is shorter than for Au^{69+}. In agreement with the charge state dependency, no indications for strong pre-equilibrium effects are observed for Ar^{18+}. A practical limit for solid target thicknesses is $\sim$ 0.5–

1 μg/cm^2 (2.5–5 nm). We can expect initially highly charged projectiles, e.g., neon-like gold ions, to exit such ultra thin targets in highly excited states when the impact velocity is $\geq 10^6$ m/s. The complexity of charge changing dynamics has, so far, limited more refined theoretical treatments of charge state effects on (electronic) stopping, such as density functional theory, to hydrogen and helium ions in free electron metals [63]. The effect of K-shell vacancies in heavier projectiles has been approached recently as described above [58]. Quantitative differentiation of contributions to energy loss processes, i.e., elastic and inelastic collisions and the energy balance in the large ($\gg q$) number of charge changing events, are the subject of ongoing research in this new regime of ion-solid interactions.

The presented data demonstrate the presence of strong pre-equilibrium energy loss enhancements for slow, very highly charged ($q > 40+$) ions in solids. The mean de-excitation time and charge state equilibration distance of slow Au^{69+}-ions in amorphous carbon are estimated to be ~5 fs and ~3 nm, respectively. The density and distribution of target excitation are determined by these characteristic time and length scales. Time and length scales for equilibration of very highly charged ions in solids are important fundamentals in the development of highly charged ions as sensitive probes in materials analysis and as tools for materials modification on nanometer scales.

5. High-Z Highly Charged Ion Confinement and Charge Exchange Facility

The combination of EBIT with a cryogenic, UHV ($P < 10^{-12}$ Torr) precision Penning trap offers the unique possibility of performing new studies in various areas of physics with trapped, high-Z, highly charged ions. This includes physics with one- or multicomponent highly charged ion plasmas at very low temperatures (< 10 K). The EBIT/RETRAP facility [12, 15, 40, 64], as depicted in Fig. 10, has been installed at LLNL where the retrapping of highly charged ions (e.g., Th^{80+}) was first developed and is now performed routinely. Different geometries for the trap electrodes can be inserted and used in the cold bore (4 K) of the device. Tuned circuits are attached to the endcap electrodes for the detection of the ion noise signal. First results include the nondestructive observation of single ions which has led to new data for electron transfer cross sections for highly charged ions at low energies. Electronic or stochastic cooling as well as ion stacking have been demonstrated and a laser laboratory adjacent to the EBIT/RETRAP facility is set up to study laser based cooling schemes and to perform laser spectroscopy (e.g., hyperfine structure spectroscopy in H-like, high-Z ions). As a test for a sympathetic highly charged ion cooling scheme, the cooling of Be^{2+} ions via laser cooled Be^{+} ions has been demonstrated. For physics experiments with trapped, highly charged ions the understanding of charge transfer processes

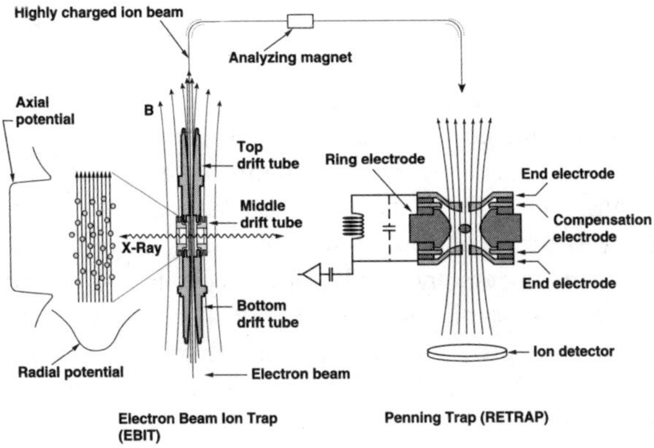

Fig. 10. Schematic of the EBIT/RETRAP system showing EBIT where the highly charged ions are produced and extracted, and RETRAP (hyperbolic Penning trap) where the ions are retrapped, including the tuned circuit for ion detection

and rates is essential since charge transfer limits the lifetime of the ions in the trap. Furthermore, high charge state and low energy electron transfer collisions are important in laboratory and astrophysical plasmas, ion sources and in certain types of x-ray lasers. General considerations important for the theory for these collisions are the scaling of the cross section with ion charge state, the velocity dependences of the cross section in different energy regions [65], and the relationship of single to multiple electron capture. Particular interest has focused on true double capture, in which both captured electrons remain on the ion after the collision [66].

At the low energies of interest here, theory has been most successful with few-electron atomic targets, like atomic hydrogen, H_2 and He. At low energies, the collision can often be treated as the temporary formation of a quasi-molecule, with molecular wave-functions and potential energies calculated as a function of the internuclear separation R of the ion and neutral target. Here a detailed quasi-molecular theoretical treatment is appropriate. An electron is transferred at certain avoided crossings of the potential energy curves. The magnitudes of the capture cross sections can vary appreciably with energy and charge. At the lowest energies, orbiting collisions are expected to occur in the attractive R^{-4} potential of the induced dipole moment of the neutral target [67]. Under these conditions, electron capture is expected to vary with velocity as v^{-1} and scale linearly with the ion charge state q. With molecular targets, any permanent dipole or quadrupole moments can be significant to cross section magnitudes.

Electron capture to ions in very high charge states q is expected to occur into closely spaced, highly excited states, under conditions where there are a large number of avoided crossings. The absorbing sphere model of *Olson* and *Salop* [65], based on Landau–Zener theory, applies under these conditions, while over-barrier models [68] apply at higher energies. An important prediction of these theories is an approximately linear dependence of the cross section for capture with charge state q. A complication with multi-electron targets is whether multiple-electron capture occurs either sequentially or in a single event. Experimental studies at low energy for q <12 indicated that the cross section for two-electron transfer depended similarly on energy and charge as the cross section of single electron transfer [69]. Multiple-electron capture into auto-ionizing states may occur, resulting in emission of electrons to the continuum. Capture of two electrons with autoionization of one may appear experimentally as single capture. A two-electron target was chosen to simplify this aspect of the study. The probability of an Auger process is independent of ion charge q, since it depends only on the correlation of two captured electrons, while radiative stabilization probability scales as q^4 for $\Delta n \neq 0$ transitions. However, both processes depend on the principal quantum number n of the states initially populated by the capture, which also depend on q. Nevertheless, the ratio of the Auger to radiative rates tends to scale as q^{-4}, indicating that radiative stabilization should dominate for $q > 10$ [70], according to an early analysis. Recent data for Xe^{q+}–He collisions ($15 < q < 42$) [66] shows that true double capture increases rapidly for $q > 28$ to a plateau for $q > 35$. This indicates that a more-detailed analysis of the double capture rate, possibly depending on the details of ion structure, is required, and that data for still higher charge states are desirable [71].

First measurements were carried out using ions with a combination of very high charge states ($35 < q < 80$) and very low energy (< 6 eV in the center-of-mass frame) at the EBIT/RETRAP facility using an open cylinder trap geometry. Here, nondestructive monitoring of the ion numbers in each charge state permitted a study of the time evolution of the confined ion charges. Charge states changed when the ions collided with H_2 molecules existing at low density in the trap region. Following capture, the reduced-charged, heavy product ions remained confined, providing information on the product ion charge states, time development and capture rates. For the measurements, one or a few ions in a single charge state were captured and confined. Their signals, dependent on charge to mass ratios, were periodically and nondestructively observed. The measurements were carried out with the following procedure. The ions were extracted from EBIT, mass-to-charge selected, and transported to RETRAP. After deceleration and catching, the endcap electrode potential provides for axial confinement, while a 4 T magnetic field confines the ions radially. The captured ions typically have a wide range of axial energy. By a temporary decrease of the trapping potential the highest energy ions were released.

The remaining ions resided in a nearly axial harmonic well, and oscillated axially with frequency

$$\nu = \frac{1}{2\pi}\sqrt{\frac{qeU_0C_2}{md^2}}, \quad d^2 = \frac{1}{2}\left(z_0^2 + \frac{1}{2}r_0^2\right), \quad C_2 = 0.5449, \tag{7}$$

where q is the ion charge, U_0 is the endcap potential, m is the ion mass, z_0 is the half length and r_0 the radius of the trap. These ions were detected by their interaction with a high impedance parallel tuned circuit consisting of the capacitance between the compensation electrodes of the trap and a large external inductance. The circuit resonance frequency was about 1.215 MHz. Voltages induced across this circuit by image currents flowing between the compensation electrodes were amplified, starting with a cryogenic preamplifier, and observed using a spectrum analyzer operated in the zero-span mode with a bandwidth of 9 kHz. When the ion axial oscillation frequency was swept through the tuned circuit resonance by a linear ramp of the trapping potential, signals due to the motions of ions appeared at ramp potentials determined by their charge state. Periodic ramps of the potential produced signals which showed the development of the charge of the particles throughout the storage interval. The data from each ramp were recorded by computer for later analysis.

A plot of the number of observations of each charge state vs. the storage duration of the particle provided information from which the charge transfer rate coefficient and mean cross section were derived [72]. Figure 11 shows signals, each taken at a different time, due to a single, stored particle. The initial signal is from a Th^{79+} ion. Between the second and third observations $t2$ and $t3$, two electrons are captured in either one double-capture collision or two single-capture collision with H_2. A single electron is captured between $t2$ and $t4$, and between $t4$ and $t5$. The particle is then Th^{75+}, and it remains so until the interval between t10 and t11, when two electrons are again captured. Note that whenever a single electron is captured, the signal moves one charge state, always lower. A second measurement technique was based on averaging the tuned circuit signal at each delay time for many measurement cycles [73]. These areas were determined either by integrating the signals between voltage limits, or by fitting the data peaks. The areas under each peak were proportional to the ion number N, q^2 and the mean axial energy E_z at each time. The initial ion axial energy (of the sample surviving the preparation) was analyzed by comparing observed signal widths to the calculated broadening due to amplitude dependence of the resonant frequency. Based on this analysis, the average axial ion energy at the start of the measurements was $4q$ eV in the laboratory frame, or approximately 2.6 eV in the center-of-mass frame. These estimates were also in accord with signal amplitudes measured for single ions. Energy also resided in the cyclotron motion of the ions. In

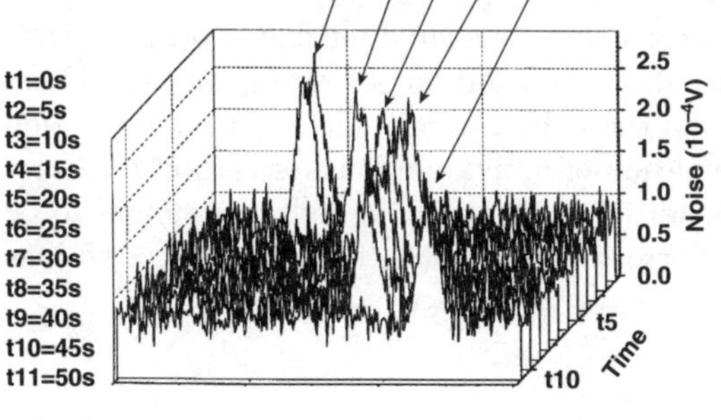

Fig. 11. Noise of the tuned circuit as a function of endcap electrode potential at different times. The peaks are due to one Th ion. It successively catches electrons from H_2 molecules, thereby reducing its initial charge from 79+ to 73+

determining the ion number from signals of product ion charge states resulting from electron capture, there is also a reduction of the signal from the decreased charge state. This dependence is explicitly proportional to q^2 for ions with a constant axial amplitude of oscillation. However, the detection sweep was adiabatic, so the axial component of the action is a constant of motion during the ramp. A $q^{3/2}$ dependence of the signals results, requiring a small charge state correction to the data.

However, there can be significant energy transfer between the axial and radial degrees of freedom of the confined ion motion, produced by ion–ion collisions or by other coupling. The initial radial energy of the ions in the cyclotron motion arises from effects associated with deceleration of the ions in the relatively weak fringe magnetic field above the trap. As the ions approach the trap, the field becomes sufficiently strong to guide the ions adiabatically along the field lines, the effective magnetic moment of the radial orbits becomes an adiabatic invariant. Estimates of the mean radial energy, based on modeling the deceleration process, indicated that the expected radial energy could be comparable to the spread of axial energies of the stored ions, depending on the details of the ion trajectories during deceleration.

Ion–ion collisions tend to equilibrate the ion energies among the degrees of freedom, with a time constant that varies as $n^{-1}q^{-4}$ at constant mean energy, where n denotes the ion number density and q is the charge state. The relatively low density of highly charged ions tended to lengthen the energy equipartition time to a significant portion of an electron transfer measurement. Transfer of energy between axial and radial degrees of freedom was found to be consistent both with the overall signal changes observed in some of the electron transfer data, and with the estimated ratio of axial and radial energies.

To compensate the data for these signal-changing effects, the sum of the heavy ion signals with different charges (following corrections for charge state) was fitted to a function of the form $F(t) = A\,(1 + B\,\exp(-t/t_1))\,\exp(-t/t_2)$. The effective time scale for energy transfer is denoted by t_1 and the much longer t_2 characterizes the time scale for ion cooling by temporary interaction with the tuned circuit. Here A is proportional to the equilibrium axial ion energy, while B denotes the fraction of axial energy differing from equilibrium. The fitted parameters were A, B, t_1 and t_2. The scaled signal for each charge state is then given by the equation $S(t) = M(t)/F(t)$, where $M(t)$ is the area under the charge state corrected, measured ion signal peaks at time t. These rescaled data were fitted to the solutions of rate equations. Cross sections obtained from analysis of these data were compared to cross section results obtained from the observation of individual ions for certain measurements taken under the same conditions. The two methods were found to be consistent within the statistical error of 10%, indicating that the corrections to the raw data were adequate.

The composition and density of the target gas are other important considerations in these measurements. The temperature of the Penning trap and its environs was held at 4.2 K by the large liquid helium reservoir. No direct density or pressure measurement was feasible. Only H_2 and He have appreciable vapor pressures at 4.2 K, but they are effectively pumped by the large area of cold surfaces in the vicinity of the trap. No He was admitted to the vacuum system, but H_2 is a common residual gas in ultra-high vacuum systems. A time-of-flight measurement was applied to the products of the charge-exchange reaction between stored Be^{2+} ions and the target gas, by a rapid dump of the trap onto the ion detector mounted below the trap. Ions of Be and hydrogen product ions were observed, but no other mass-to-charge ratios. These results provided direct evidence that the target gas was H_2, with negligible impurities. It was noted that after many days at cryogenic temperature, the neutral density at the Penning trap increased, as evidenced by shorter ion storage times associated with increased electron capture rates. This was interpreted as the effect of H_2 buildup on the cold surfaces raising the local vapor pressure.

The charge exchange cross section was determined by measuring the charge exchange rate for Ar^{11+}. The cross section for Ar^{11+} had been independently measured using a beam technique [69, 74]. With the known cross section, mean ion energy, and storage time constant, the H_2 density was determined. A detailed description of the calibration procedure can be found in [73]. Figure 12 shows a plot of representative cross section data vs. charge state together with the results of a fit to the expression $\sigma = A \cdot q^b$. Due to the low mean collision energies, the relatively high charge states, and because of the use of an H_2 target, the most appropriate theoretical model applicable to this data is the absorbing sphere model of *Olson* and *Salop* [65]. A unit probability for reaction is assumed to occur below a critical radius R_c. The velocity $v_0 = 5.2 \times 10^{-3}$ a.u. is determined by the mean ion energy, and the first ionization potential $I = 15.4$ eV for H_2 was used to calculate R_c. The predicted cross section is $\sigma_c = \pi R_c^2$, which scales approximately as q, but

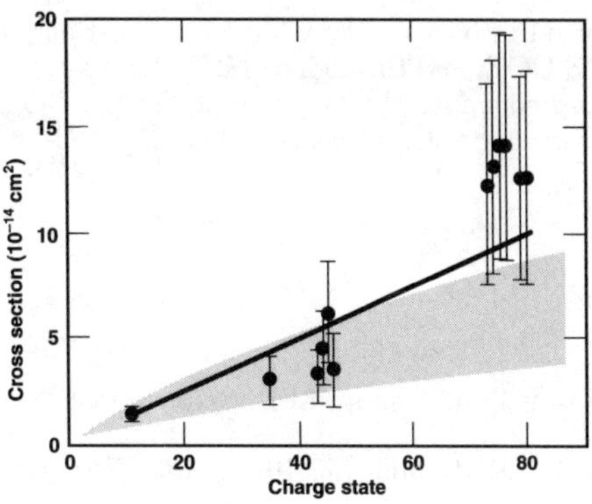

Fig. 12. Charge exchange cross sections vs. charge. The solid line is a fitted curve to the data of the form $\sigma = A \cdot q^b$. The shaded region is the charge exchange cross section range predicted by the absorbing sphere model

which yields a lower cross section at high q than a strictly linear dependence (Fig. 12). This cross section applies to the sum of single and double electron capture.

The accuracy of the absorbing sphere calculation for $q > 10$ was estimated to be only about 40% in the original work. The absorbing sphere calculation lies below the data in Fig. 12 near $q = 76$. However, when the error estimates for the data and for the theory are taken into account, no conclusion about the difference can be drawn. The data overall agree rather well with a linear dependence of the cross section on ion charge.

The relative amount of true double capture to be expected in high q collisions is still a matter of debate. The results from our measurements are listed in Table 4. The ratio of true double to total electron capture for Ar^{11+} is in good agreement with the results from [69]. For higher charges this ratio increases to 0.26 for Xe^{q+}(q =43–45). No further increase is found for higher charges (Th^{q+}, q= 73–80).

In summary, we have shown that the dominant ion loss mechanism in RETRAP is due to collisions with residual gas molecules. A careful analysis of the measured ion lifetimes by using a nondestructive detection technique leads to the determination of charge exchange cross sections for highly charged ions at low energies. The long lifetimes of even the highest charge states (~20 s for Th^{80+}) allows the performance of the above mentioned new experiments with trapped, highly charged ions.

Table 4. The ratio of true double to total electron capture. The ratios for Xe^{q+}(q = 43–45) and Th^{q+}(q = 73–80) are obtained by averaging over a narrow charge state range

Charge State q	λ_{0d}/λ_0	Charge State q	λ_{0d}/λ_0
11+	0.086(14)	(43–45)+	0.263(32)
35+	0.205(25)	(73–80)+	0.267(93)

Acknowledgement. This work was supported in part by the Office of Basic Energy Sciences, Division of Chemical Sciences US DOE, and through a NATO travel grant (SA.5-2-05(CRG9610 66)) and was performed under the auspices of the US DOE by Lawrence Livermore National Laboratory under Contract No. W-7405-ENG-48.

References

1. M.A. Levin, R.E. Marrs, J.R. Henderson, D.A. Knapp, M.B. Schneider: Phys. Scr. **T22**, 157 (1988)
2. R.E. Marrs, M.A. Levine, D.A. Knapp, J.R. Henderson: Phys. Rev. Lett. **60**, 1715 (1988)

3. E.D. Donets: in *The Physics and Technology of Ion Sources*, ed. I.G. Brown (Wiley, New York 1989), p. 245
4. D. Schneider, D. DeWitt, M.W. Clark, R. Schuch, C.L Cocke, R. Schmieder, K.J. Reed, M.H. Chen, R.E. Marrs, M.A. Levin, R. Fortner: Phys. Rev. A **42**, 3883 (1990)
5. R.E. Marrs, S.R. Elliott, D.A. Knapp: Phys. Rev. Lett. **72**, 4082 (1994)
6. D. Schneider, M.W. Clark, B.M. Penetrante, J. McDonald, D. DeWitt, J.N. Bardsley: Phys. Rev. A **44**, 3119 (1991); B.M. Penetrante, J.N. Bardsley, D. DeWitt, M. Clark, D. Schneider: Phys. Rev. A **43**, 4861 (1991)
7. J. McDonald, D. Schneider: submitted for publication (1998)
8. P. Beiersdorfer, R.E. Marrs, J.R. Henderson, D. Knapp, M.A. Levine, D.B. Platt, M.B. Schneider, D.A. Vogel, K.L. Wong: Rev. Sci. Instr. **61**, 2338 (1990)
9. P. Beiersdorfer, D. Knapp, R.E. Marrs, S.R. Elliott, M.H. Chen: Phys. Rev. Lett. **71**, 3939 (1993)
10. J.H. Scofield: Phys. Rev. A **40**, 3054 (1989); E.B. Saloman, J.H. Hubbell, J.H. Scofield: At. Data Nucl. Data Tables **38**, 1 (1988)
11. D. Schneider, M.A. Briere: Phys. Scr. **53**, 228 (1996)
12. D. Schneider, D.A. Church, G. Weinberg, J. Steiger, B. Beck, J. McDonald, D. Knapp: Rev. Sci. Instrum. **65**, 3472 (1994)
13. R. Geller: *Cyclotron Resonance Ion Sources and ECR Plasmas* (Institute of Physics, London 1996)
14. Z.Q. Xie: Rev. Sci. Instrum. **63**, 2 (1998)
15. D. Schneider: Hyp. Int. **99**, 47 (1996)
16. R.E. Marrs, P. Beiersdorfer, D. Schneider: Physics Today, American Institute of Physics (October 1994)
17. D. Knapp, R. Marrs, S. Elliot, E. Magee, R. Zasadzinsky: Nucl. Instrum. Methods. A **334**, 305 (1993)
18. R.E. Marrs, S.R. Elliot, J.H. Scofield: Phys. Rev. A **56**, 33 (1997); D.L. Moores, K.J. Reed: Phys. Rev. A **51**, R9 (1995)
19. P. Beiersdorfer: AIP Conference Proceedings 389, Woodbury, New York, R.L. Johnson, H. Schmidt-Böcking, B.F. Sonntag, eds. (1996)
20. P. Beiersdorfer, D. Knapp, R. Marrs, S.R. Elliot, M. Chen: Phys. Rev. Lett. **71**, 3939 (1993)
21. P. Beiersdorfer, V. Decaux, S.R. Elliot, K. Widmann, K. Wong: Rev. Sci. Instrum. **66**, 303 (1995)
22. D.A. Knapp, R.E. Marrs, S.R. Elliot, E.W. Magee, R. Zasadzinski: Nucl. Instrum. Methods. A **334**, 305 (1993)
23. L. Essen, R.W. Donaldson, M.J. Bangham, E.G. Hope: Nature (London) **229**, 110 (1971)
24. R.J. Gould: Astrophys. J. **423**, 522 (1994)
25. I. Klaft, S. Borneis, T Engle, B. Fricke, R. Gieser, G. Huber, T. Kühl, D. Marx, R. Neumann, S. Schröder, P. Seelig, L. Völker: Phys. Rev. Lett. **73**, 2425 (1994)
26. P. Seelig, S. Borneis, A. Dax, T. Engel, S. Faber, M. Gerlach, C. Holbrow, G. Huber, T. Kühl, D. Marx, K. Meier, P. Merz, W. Quint, F. Schmitt, M. Tomaselli, L. Völker, H. Winter, M. Würtz, K. Beckert, B. Franzke, F. Nolden, H. Reich, M. Steck, T. Winkler: submitted to Phys. Rev. Lett. (1998)
27. C.A. Morgan, F.G. Serpa, E. Takacs, E.S. Meyer, J.D. Gillaspy, J. Sugar, J.R. Roberts: Phys. Rev. Lett. 74, 1716 (1995)
28. J.R. Crespo López-Urrutia, P. Beiersdorfer, D.W. Savin, K. Widmann: Phys. Rev. Lett **77**, 826 (1996); J.R. Crespo López-Urrutia, P. Beiersdorfer, K. Widmann, B.B. Birkett, A.M. Martensson-Pendrill, M.G.H. Gustavsson: Phys. Rev. A **57**, 2 (1998)

29. D.L. Moores, K.J. Reed: Phys. Rev. B **98**, 122 (1995)
30. V.M. Shabaev: J. Phys. B **27**, 5825 (1994)
31. H. Persson, S.M. Schneider, W. Greiner, G. Soff, L. Lindgren: Phys. Rev. Lett. **76**, 1433 (1996)
32. S.M. Schneider, W. Greiner, G. Soff: Phys. Rev. A **50**, 118 (1994)
33. E. Browne, J.M. Dairiki, R.E. Doebler: in *Table of Isotopes*, C.M. Lederer and V.S. Shirley, eds. (Wiley, New York 1978), 7th edn
34. R.A. Haberstroh, T.I. Moran, S. Penselin: Z. Phys. A **252**, 421 (1997)
35. W. Dankwort, J. Ferch: Z Phys. **267**, 239 (1974)
36. G. Nachtsheim: PhD. thesis, Univesität Bonn, (Bonn 1980) (unpublished)
37. L.K. Peker: Nucl. Data Sheets **50**, 137 (1987)
38. L.N. Labzowski, W.R. Johnson, G. Soff, S.M. Schneider: Phys. Rev. A **51**, 4597 (1995)
39. M. Thomaselli, S.M. Schneider, E. Kankeleit, T. Kühl: Phys. Rev. C **51** 2989 (1995)
40. D. Schneider: Phys. Scr. **T59**, 189 (1995)
41. J. W. McDonald, D. Schneider, M.W. Clark, D. Dewitt: Phys. Rev. Lett. **68**, 2297 (1992); F. Aumayr, H. Kurz, D. Schneider, M.A. Briere, J.W. McDonald, C.E. Cunningham, HP. Winter: Phys. Rev. Lett. **71** 1943 (1994)
42. R. Schuch, D. Schneider, D.A. Knapp, D. DeWitt, J. McDonald, M.H. Chen, M.W. Clark, R.E. Marrs: Phys. Rev. Lett. **70**, 1073 (1993); J.P. Briand, G. Giardino, B. d'Etat-Ban, D. Schneider, M.A. Briere, V. Decaux, J. McDonald, S. Bardin: Phys. Rev. A **53**, 1 (1996); M.A. Briere, D. Schneider, J. McDonald, M. Reaves, C. Rühlicke, G. Weinberg, D. Knapp: Nucl. Instrum. Methods. B **90**, 231 (1994)
43. T. Schenkel, M.A. Briere, H. Schmidt-Böcking, K. Bethge, D. Schneider: Phys. Rev. Lett. **78**, 2481 (1997); T. Schenkel, A.V. Barnes, M.A. Briere, A.V. Hamza, A. Schach von Wittenau, D. Schneider: Nucl. Instrum. Methods. B **125**, 153 (1997)
44. T. Schenkel, A.V. Barnes, A.V. Hamza, D.H. Schneider: European Phys. J. D **1**, 297 (1998)
45. T. Schenkel, A.V. Hamza, A.V. Barnes, J.C. Banks, B. L. Doyle, D.H. Schneider: Phys. Rev. Lett. **80**, 4325 (1998)
46. D. Schneider, M.A. Briere, M.W. Clark, J. McDonald, J. Biersack, W. Siekhaus: Surface Science **294**, 403 (1993); C. Rühlicke, M. A. Briere, D. Schneider: Nucl. Instrum. Methods. B **99**, 528 (1995)
47. T. Schenkel, M.A. Briere, H. Schmidt-Böcking, K. Bethge, D. Schneider: Materials Science Forum 248–249, 413 (1997); G. Schiwietz, M. Briere, D. Schneider, J. McDonald, C. Cunningham: Nucl. Instrum. Methods. B **100**, 47 (1995)
48. T. Schenkel, A.V. Hamza, A.V. Barnes, D.S. Walsh, B. L. Doyle, D.H. Schneider: J. Vac. Sci. Technol. A (1998) (in press)
49. T. Schenkel, M.A. Briere, A.V. Barnes, A.V. Hamza, K. Bethge, H. Schmidt-Böcking, D.H. Schneider: Phys. Rev. Lett. **79**, 2030 (1997); T. Schenkel, A.V. Hamza, A.V. Barnes, D.H. Schneider: Phys. Rev. A **56**, R1701 (1997)
50. N. Bohr: Kgl. Danske Videnskab. Selskab. Mat.-fys. Medd. 18 (1948) No. 8; O.B. Firsov: Sov. Phys. - JETP, **36**, 1076 (1959); J. Lindhard, M. Scharff: Phys. Rev. **124**, 128 (1961)
51. J. Knipp, E. Teller: Phys. Rev. **59**, 659 (1941); P. Sigmund: Phys. Rev. A **50**, 3197 (1994)
52. J.P. Biersack: Nucl. Instrum. Methods. B **80/81**, 12 (1993)
53. K.B. Winterbon: Nucl. Instrum. Methods. **144**, 311 (1977); T. Kaneko, Y. Yamamura: Phys. Lett. A **100**, 313 (1984)
54. K. Shima, N. Kuno, M. Yamanouchi: Phys. Rev. A **40**, 3557 (1989)

55. W.N. Lennard, T.E. Jackman, D. Phillips: Phys. Lett. A **79**, 309 (1980)
56. P. Sigmund, K.B. Winterbon: Nucl. Instrum. Methods. B **12**, 1 (1985)
57. R. Herrmann, C.L. Cocke, J. Ullrich, S. Hagmann, M. Stoeckli, H. Schmidt-Böcking: Phys. Rev A **50**, 1435 (1994)
58. J.I. Juaristi, A. Arnau: Nucl. Instrum. Methods. B **115**, 173 (1996)
59. J.F. Ziegler, J.P. Biersack, U. Littmark: *The Stopping and Range of Ions in Solids* (Pergamon, New York 1985)
60. A. Arnau, F. Aumayr, P.M. Echenique, M. Grether, W. Heiland, J. Limburg, R. Morgenstern, P. Roncin, S, Schippers, R. Schuch, N. Stolterfoht, P. Varga, T.J.M. Zouros, H.P. Hippler: Surf. Sci. Rep. **229**, 1 (1997)
61. J. Burgdörfer, P. Lerner, F.M. Meyer: Phys. Rev. A **44**, 5674 (1991); J. Burgdörfer, F.W. Meyer: Phys. Rev. A **47**, R20 (1993)
62. M.A. Briere, T. Schenkel, D.H. Schneider, P. Bauer, A. Arnau: Phys. Scr. **T73**, 324 (1997)
63. A. Arnau, M. Peñalba, P.M. Echenique, F. Flores: Nucl. Instrum. Methods. B **69**, 102 (1992); Phys. Rev. Lett. **65**, 1024 (1990)
64. P.K. Ghosh: *Ion Traps* (Oxford Science Publications 1995)
65. R.E. Olson, A. Salop: Phys. Rev. A **14**, 579 (1976)
66. H. Cederquist, H. Andersson, E. Beebe, C. Biedermann, L. Broström, Å. Engström, H. Gao, R. Hutton, J.C. Levin, L. Liljeby, M. Pajek, T. Quinteros, N. Selberg, P. Sigray: Phys. Rev. A **46**, 2592 (1992)
67. G. Gioumousis, D.P. Stevenson: J. Chem. Phys. **29**, 294 (1958); P. Langevin: Ann. Chim, Phys. **5**, 245 (1905)
68. A. Niehaus: J. Phys. B **19**, 2925 (1986)
69. S. Kravis H. Saitoh, K. Okuno, K. Soejima, M. Kimura, I. Shimamura, Y. Awaya, Y. Kaneko, M. Oura, N. Shimakura: Phys. Rev. A **52**, 1206 (1995)
70. R.K. Janev, L.P. Presnyakov: Phys. Rep. **70**, 1 (1981)
71. H. Cederquist, C. Biedermann, N. Selberg, P. Hvelplund: Phys. Rev. A **51**, 2169 (1995)
72. J. Steiger, D.A. Church, G. Weinberg, B.R. Beck, J. McDonald, D. Schneider: Hyp. Int. **108**, 115 (1997)
73. B.R. Beck, J. Seiger, G. Weinberg, D.A. Church, J. McDonald, D. Schneider: Phys. Rev. Lett. **77**, 1735 (1996); G. Weinberg, B.R. Beck, J. Steiger, D.A. Church, J. McDonald, L. Gruber, D. Schneider: Phys. Rev.A (1998) (to be published)
74. J. Vancura, V.J. Marchetti, J.J. Perotti, V.O. Kostroun: Phys. Rev. A **47**, 3758 (1993)

Part B

Fundamental Properties

III Quantum Electrodynamics of Highly Charged Ions

Günter Plunien and Gerhard Soff

Summary. Quantum electrodynamics (QED) provides a relativistic description of highly charged few-electron atomic systems. Despite inherent problems associated with renormalization the predictions of bound-state QED can be tested to very high precision by measurements of the Lamb shift of electron levels in highly charged ions. Dominant corrections to the energy spectrum are due to the finite nuclear size and to QED effects: self energy and vacuum polarization of the order of α, where α is the fine-structure constant. Both radiative corrections have to be evaluated to all orders in $Z\alpha$ in the interaction with the external Coulomb potential to achieve agreement with Lamb shift data for a hydrogen-like system measured with a relative precision of about 10^{-4}. (Here Z is the nuclear charge number.) Aiming at the utmost experimental precision, it becomes essential to determine the level of accuracy at which one is leaving the framework of pure QED. The natural limitation for testing QED is set by nuclear polarization effects and by the uncertainties of nuclear parameters. In heavy systems nuclear structure becomes non-negligible at the level of relative precision of about 10^{-6}. To provide predictions for the Lamb shift taking into account this ultimate standard requires the exact evaluation of all QED radiative corrections of order α^2. This brief review is focused on QED of hydrogen-like systems. Particular emphasis is laid on the exact evaluation of the self-energy and vacuum-polarization correction as well as on the influence of nuclear effects. For hydrogen-like lead and uranium we discuss the current status of Lamb shift predictions.

1. Relativistic Description of Few-Electron Atoms

The relativistic theory of atoms is, to high precision, contained in quantum electrodynamics (QED) [1–3]. Electrons interacting via the exchange of photons are bound in the electromagnetic field generated by the atomic nucleus. To a major extent the nuclear electromagnetic field can be considered as a classical background field in which the fermion field and the free-photon field are quantized (bound-state QED). Although the fundamental interaction between electrons and photons is known, the influence of nuclear properties which must be taken into account at a certain level of accuracy, cannot be formulated from first principles. This has far-reaching consequences even in one-electron systems, for which electron–electron correlation effects are negligible. Although one might be able to evaluate pure QED corrections to atomic spectra up to any desired accuracy one has to include nuclear effects, which will inevitably set inherent limitations for tests of QED due to uncertainties of the nuclear parameters. The Lagrangian reads

$$\mathcal{L} = \mathcal{L}^0_\psi + \frac{1}{2}\left\{\mathcal{J}^\mu(x), \mathcal{A}_\mu(x)\right\} + \mathcal{L}^0_\mathcal{F} + \mathcal{L}_{\text{nuc}}\,, \tag{1}$$

where $\mathcal{L}^0_\psi$ and $\mathcal{L}^0_\mathcal{F}$ denote the standard free-field Lagrangians of the Dirac field (ψ) and the total Maxwell field ($\mathcal{F}$) respectively. The Lagranian introduced above also contains a total nuclear Lagrangian $\mathcal{L}_{\text{nuc}}$ which may be specified within particular nuclear models. The interaction term between the total Maxwell field $\mathcal{A}_\mu$ and the total electromagnetic source current $\mathcal{J}^\mu$ is written in a properly symmetrized form. The total current appears as a sum of the Dirac current j^μ_ψ and the nuclear source current j^μ_{nuc}, i.e.

$$\mathcal{J}^\mu(x) = j^\mu_{\text{nuc}} + j^\mu_\psi(x) = j^\mu_{\text{ex}}(x) + j^\mu_{\text{fluc}}(x) + j^\mu_\psi(x) \tag{2}$$

with

$$j^\mu_\psi(x) = -\frac{e}{2}\,\left[\overline{\psi}(x), \gamma^\mu \psi(x)\right]. \tag{3}$$

The latter consists of a classical, external part $\langle j^\mu_{\text{nuc}}\rangle_{\text{g.s.}} \equiv j^\mu_{\text{ex}}$ describing the nucleus in its ground state and a second quantized part j^μ_{fluc} due to internal nuclear degrees of freedom. Explicit expressions for the fluctuating current may be derived within given nuclear models. Variation with respect to the fields ψ and $\mathcal{A}^\mu$ leads to the coupled Dirac–Maxwell equations:

$$(\mathrm{i}\gamma^\mu \partial_\mu - m)\psi(x) - \frac{e}{2}\gamma^\mu \{\mathcal{A}_\mu(x), \psi(x)\} = 0\,, \tag{4}$$

$$\partial_\mu \mathcal{F}^{\mu\nu} = (\delta^\nu_\mu \,\Box - \partial^\nu \partial_\mu)\mathcal{A}^\mu = \mathcal{J}^\nu. \tag{5}$$

With the aid of the (gauge-dependent) free photon propagator $D_{\mu\nu}$, which is the Green's function of Maxwell's equation (5) one can formally solve the problem for the total radiation field $\mathcal{A}^\mu$. According to the decomposition of the total current we may write

$$\mathcal{A}^\mu(x) \quad = \quad A^\mu_{\text{free}}(x) + A^\mu_{\text{ex}}(x) + A^\mu_{\text{fluc}}(x) + A^\mu_\psi(x), \tag{6}$$

where the sum of the classical (external) part A^μ_{ex} is generated by the nucleus in its ground state and a total, second quantized part, which is given as the sum of the free radiation A^μ_{free}, the field A^μ_ψ due to the presence of electrons and the fluctuating field A^μ_{fluc} generated by the nuclear transition current j^μ_{fluc}. For later purposes we may introduce the total radiation field

$$A^\mu_{\text{rad}} = A^\mu_{\text{free}}(x) + A^\mu_{\text{fluc}}(x). \tag{7}$$

Insertion of (6) into the Dirac equation (4) leads to the form

$$\begin{aligned}(\mathrm{i}\gamma^\mu \partial_\mu - e\gamma^\mu A^{\text{ex}}_\mu(x) - m)\psi(x) \quad &= \quad \frac{e}{2}\{\gamma^\mu \psi(x), A^{\text{rad}}_\mu(x)\} \\ &+ \frac{e}{2}\int \mathrm{d}^4x'\, D_{\mu\nu}(x-x')\,\{\gamma^\mu \psi(x), j^\nu_\psi(x')\}. \end{aligned} \tag{8}$$

This equation of motion for the field operator ψ is exact provided that the time evolution of all field operators involved is governed by the total Hamiltonian describing the total many-body problem of interacting Dirac particles + radiation field + nuclear system. The part of the Hamiltonian relevant for the relativistic description of the atomic many-body problem in the presence of the external electromagnetic field of the nucleus with charge number Z including radiative corrections as well as interactions with internal nuclear degrees of freedom is given by

$$
\begin{aligned}
H_{\mathrm{QED}} &= \frac{1}{2}\int \mathrm{d}^3x\, \left[\psi^{\dagger}(x), h_{\mathrm{ex}}(x)\psi(x)\right] \\
&\quad + \frac{1}{2}\int \mathrm{d}^3x\, \mathrm{d}^4x'\, D_{\mu\nu}(x-x')\left\{j^{\mu}_{\psi}(x), j^{\nu}_{\psi}(x')\right\} \\
&\quad + \frac{1}{2}\int \mathrm{d}^3x\, \left\{j^{\mu}_{\psi}(x), A^{\mathrm{rad}}_{\mu}(x)\right\} + H_{\mathrm{counter}} \\
&= H_{\mathrm{ex}} + H_{\mathrm{e-e}} + H_{\mathrm{rad}}
\end{aligned} \tag{9}
$$

with

$$
h_{\mathrm{ex}}(x) = \left[-\mathrm{i}\boldsymbol{\alpha}\cdot(\boldsymbol{\nabla} + \mathrm{i}e\boldsymbol{A}_{\mathrm{ex}}(x)) + V_{\mathrm{ex}}(x) + \beta m\right]. \tag{10}
$$

h_{ex} denotes the one-particle Dirac Hamiltonian, $\alpha^{\mu} \equiv \gamma^{\mu}\gamma^{0}$ and $\gamma^{0} \equiv \beta$ are the Dirac matrices. The meaning of the individual terms is obvious: The first term H_{ex} defines the dominant external field problem, where the interaction with the classical electromagnetic field generated by the extended nucleus in its ground state is treated to all orders in the effective coupling $Z\alpha$. Here $\alpha = e^2$ denotes the fine-structure constant. Neglecting recoil effects due to the finite nuclear mass the external field may be considered as static. In the case of even–even nuclei and in the infinite-mass limit the external field problem treats the motion of Dirac electrons in a static external Coulomb potential V_{ex} only. The second term $H_{\mathrm{e-e}}$ accounts for the electron–electron interaction self-consistently up to order α. The third term H_{rad} describes the interaction between bound electrons and the vacuum of the total radiation field giving rise to various QED radiative corrections (A^{free}_{μ}) as well as nuclear polarization effects (A^{fluc}_{μ}) at the same level. All effective radiative effects (A^{rad}_{μ}) are evaluated perturbatively in powers of the coupling constant α. The Hamiltonian also contains H_{counter} which includes all necessary counter terms required for the renormalization of mass and charge divergencies.

Depending on the physical problem under consideration, one may decompose the Hamiltonian (9) into an unperturbed (dominant) part H_0 and a time-dependent interaction part H_{int}. For atomic problems it is convenient to work in the interaction representation assuming that the interaction is switched on and off at asymptotic times:

$$
H_0 = \lim_{|t|\to\infty} H(t) \quad \Longleftrightarrow \quad \lim_{|t|\to\infty} H_{\mathrm{int}}(t) = 0. \tag{11}
$$

While the equation of motion of an operator A is given by

$$
\frac{\mathrm{d}}{\mathrm{d}t}A = \mathrm{i}\,[A, H_0] + \frac{\partial}{\partial t}A\,, \tag{12}
$$

the time-evolution of a given (Fock) state $|\Phi\rangle$ is governed by

$$
\mathrm{i}\frac{\mathrm{d}}{\mathrm{d}t}|\Phi\rangle = H_{\mathrm{int}}(t)|\Phi\rangle. \tag{13}
$$

Equation (13) can be formally solved by means of the time-evolution operator

$$U(t,t') = \mathrm{T}\exp\left[-\mathrm{i}\int_{t'}^{t} \mathrm{d}t_1\, H_{\mathrm{int}}(t_1)\right] , \tag{14}$$

where T denotes the time-ordering. The S-matrix operator is defined as $S = U(\infty, -\infty)$. The interaction between electrons and the radiation field is treated perturbatively, i.e. $H_{\mathrm{int}} \equiv H_{\mathrm{rad}}$. A proper definition of creation and annihilation operators $\tilde{a}_n^\dagger$ and $\tilde{a}_n$ for (quasi-) particles is provided by diagonalization of the (time-independent) unperturbed part $H_0 = H_{\mathrm{ex}} + H_{\mathrm{e-e}}$. The latter should possess the representation:

$$H_0 = \frac{1}{2}\int \mathrm{d}^3x\, \left[\psi^\dagger(x), h_{\mathrm{eff}}(\boldsymbol{x})\psi(x)\right] , \tag{15}$$

where the meaning of the effective one-particle Hamiltonian h_{eff} will become clear soon. The field operator ψ may be expanded in terms of non-interacting quasi-particle annihilation and quasi-hole creation operators

$$\psi(x) = \sum_{n>\mathrm{F}} \tilde{a}_n\varphi_n(x) + \sum_{n<\mathrm{F}} \tilde{b}_n^\dagger\varphi_n(x)\, , \quad \varphi_n(x) = \varphi_n(\boldsymbol{x})\,\mathrm{e}^{-\mathrm{i}\mathcal{E}_n t}. \tag{16}$$

Here F denotes the Fermi surface fixed at the Fermi-energy E_{F} located above the highest occupied quasi-electron state. The operator $\tilde{a}_n$ annihilates a quasi-electron described by an effective spinor φ_n and the energy eigenvalue $\mathcal{E}_n > E_{\mathrm{F}}$ and similarly $\tilde{b}_n^\dagger$ creates a quasi-electron hole in the state φ_n with the energy eigenvalue $\mathcal{E}_n < E_{\mathrm{F}}$. The Dirac theory reinterprets electron holes with energies $\mathcal{E}_n < -1$ as positrons. The effective eigenstates φ_n with energy eigenvalue $\mathcal{E}_n$ are determined as self-consistent solutions of the Dirac equation (8) neglecting radiative corrections ($A_\mu^{\mathrm{rad}} = 0$). The ground state $|\mathrm{F}\rangle$ of a bound multielectron system is defined as

$$|\mathrm{F}\rangle = \prod_n^{\mathrm{F}} \tilde{a}_n^\dagger|0\rangle\, , \qquad \begin{array}{lll} \tilde{a}_n|\mathrm{F}\rangle = 0 & \text{for} & \mathcal{E}_n > E_{\mathrm{F}}\, , \\ \tilde{b}_n|\mathrm{F}\rangle = 0 & \text{for} & \mathcal{E}_n < E_{\mathrm{F}}. \end{array} \tag{17}$$

Analogously, the QED vacuum state $|\mathrm{F}_0\rangle$ of Dirac's theory is defined with respect to the Fermi-surface F_0 with a Fermi-energy E_{F_0} located above $E = -1$ and below the lowest (unoccupied) bound state. The negative energy continuum is considered as completely filled with "sea-electrons". The operator

$$\tilde{d}_n^\dagger \equiv \tilde{b}_n^\dagger = \tilde{a}_n \qquad \text{with} \quad \mathcal{E}_n < E_{\mathrm{F}_0}$$

creates a quasi-positron in the state φ_n with energy $(-\mathcal{E}_n)$.

The system of self-consistent equations for the orbitals φ_n with energy eigenvalue $\mathcal{E}_n$, e.g. of a quasi-electron, is obtained by projection with $\tilde{a}_n^\dagger|\mathrm{F}\rangle$ from the right and with $\langle\mathrm{F}|$ from the left onto the field equation (8) for ψ (see [4, 5]). In the Feynman gauge one obtains the effective eigenvalue problem:

$$
\begin{aligned}
(\mathcal{E}_n - h_{\text{ex}}(\boldsymbol{x}))\,\varphi_n(\boldsymbol{x}) \;=\; & -\alpha \int \mathrm{d}^4x' \sum_{n'} \operatorname{sign}(\mathcal{E}_{n'} - E_{\mathrm{F}}) \\
& \times \left[\frac{\alpha^\mu \varphi^\dagger_{n'}(\boldsymbol{x}')\alpha_\mu \varphi_{n'}(\boldsymbol{x}')}{|\boldsymbol{x} - \boldsymbol{x}'|} \varphi_n(\boldsymbol{x}) \right. \\
& \left. - \frac{\mathrm{e}^{\mathrm{i}|\mathcal{E}_{n'} - \mathcal{E}_n||\boldsymbol{x} - \boldsymbol{x}'|}}{|\boldsymbol{x} - \boldsymbol{x}'|} \alpha^\mu \varphi_{n'}(\boldsymbol{x}) \varphi^\dagger_{n'}(\boldsymbol{x}')\alpha_\mu \varphi_n(\boldsymbol{x}') \right] .
\end{aligned}
\tag{18}
$$

We recognize the direct term and the exchange term. Since the summation runs over the whole (one-particle) Dirac spectrum appropriate renormalizations are required to establish a well-defined iterative scheme for solving the system (19) for the effective orbitals φ_n with the energy $\mathcal{E}_n$. The self-consistent treatment of the electron-electron interaction to first order in α accounts for the resummation of an infinite number of two classes of diagrams: the (one-loop) vacuum polarization and the (one-loop) self energy. Following Ref. [5], the sum over n' may be divided into the Hartree–Fock sum (all occupied bound states) and the Uehling sum:

$$
\sum_n \operatorname{sign}(\mathcal{E}_n - E_{\mathrm{F}}) = -2 \sum_{\mathrm{F}_0 < n < \mathrm{F}}^{\mathrm{HF}} + \sum_n^{\mathrm{Ueh}} \operatorname{sign}(\mathcal{E}_n - E_{\mathrm{F}_0}). \tag{19}
$$

The latter sum remains in the case when no bound electrons are presented and describes QED effects. Neglecting the Uehling sum (no-pair approximation) the system (19) reduces to the Dirac–Hartree–Fock equations for bound-electron states. Provided the self-consistent solutions φ_n are obtained, one arrives at the diagonalized (effective) one-particle Hamiltonian (15). The effective eigenvalue problem (19) reveals that the Dirac theory is inherently a many-particle theory. Even in the case of a one-electron system the bound electron experiences the interaction with the sea-electrons and, in principle, it has to be envisaged as a "dressed" quasi particle. In view of the small coupling constant α respectively the large energy gap between typical binding energies in light atoms and the negative Dirac continuum, it may be legitimate to identify all the non-interacting (HF-like) particles with observable electrons and positrons. However, one could imagine that the situation might be different in the case of very strong fields at which the energy of innermost electron shells approaches the negative energy continuum. The self-consistent approach above does not distinguish between electrons occupying the innermost shells and the sea-electrons.

1.1 Bound-State QED

In one-electron systems the electron–electron interaction can be neglected except for radiative corrections. The QED effects are treated perturbatively by inclusion of H_{rad}. Accordingly the unperturbed Hamiltonian H_0 reduces to the external field problem (normal ordering is indicated by : :)

$$H_0 \equiv H_{\text{ex}} = \frac{1}{2}\int \mathrm{d}^3x : \left[\psi^\dagger(x), h_{\text{ex}}(\boldsymbol{x})\psi(x)\right] : . \tag{20}$$

Even–even nuclei may be described by a spherically symmetric Coulomb potential V_{ex} of an extended nucleus with charge number Z. The effects of pure QED corrections are carried by the interaction Hamiltonian

$$H_{\text{I}}(t) = -\frac{e}{2}\int \mathrm{d}^3x \left[\overline{\psi}(x), A\!\!\!/_{\mu}^{\text{free}}(\boldsymbol{x})\psi(x)\right] - \frac{\delta m}{2}\int \mathrm{d}^3x \left[\overline{\psi}(x), \psi(x)\right]. \tag{21}$$

The field operator ψ is now expanded into the set of solutions ϕ_n of the corresponding one-particle Dirac equation:

$$\psi(x) = \sum_{n>\text{F}_0} a_n\,\phi_n(x) + \sum_{n<\text{F}_0} b_n^\dagger\,\phi_n(x)\,, \quad \phi_n(x) = \phi_n(\boldsymbol{x})\,\mathrm{e}^{-\mathrm{i}E_n t}. \tag{22}$$

Here a_n and $b_n^\dagger$ denote electron annihilation and positron creation operators, respectively, defined via diagonalization of the unperturbed Hamiltonian H_{ex}. The particular interaction picture defined by (20)–(22) where the interaction with the external Coulomb potential is treated exactly to all orders in $Z\alpha$, is known as the *Furry picture.*

For deriving the general formula for the energy shift caused by the interaction (21) one utilizes the concept of adiabatic switching, i.e. $H_{\text{I}}(t) \to \lambda H_{\text{I}}^{\epsilon}(t) = \lambda \mathrm{e}^{-\epsilon|t|}\,H_{\text{I}}(t)$. The damping factor makes integrations over time (S-matrix elements or energy shifts) finite. Restoring the original interaction in the limit $\epsilon \to 0$ leads to energy conservation (δ-functions) at each vertex. The smallness parameter λ may also serve to trace back the order in the coupling constant e. This allows one to collect singularities that appear in separate terms of the energy shift expression which cancel if appropriate diagrams are evaluated simultaneously. The S-operator may be expanded in powers ν of the coupling e:

$$S_{\epsilon,\lambda} = \sum_{\nu=0}^{\infty} \frac{(-\mathrm{i}\lambda)^\nu}{\nu!}\int \mathrm{d}t_1 \cdots \int \mathrm{d}t_\nu\, \mathrm{T}\left[H_{\text{I}}^{\epsilon}(t_1)\cdots H_{\text{I}}^{\epsilon}(t_1)\right]. \tag{23}$$

According to *Gell-Mann* and *Low* [6] and *Sucher* [7] the energy shift of an N-electron state $|N\rangle = \sum_{i_1,\cdots,i_\nu} c_{i_1,\cdots,i_\nu} a_{i_1}^\dagger \cdots a_{i_\nu}^\dagger |0\rangle$ can be derived from the S-matrix element (up to an irrelevant, state-independent constant)

$$\Delta E_{\text{N}} = \lim_{\substack{\epsilon\to 0\\ \lambda\to 1}} \frac{\mathrm{i}\epsilon}{2}\,\frac{\frac{\partial}{\partial\lambda}\langle N|S_{\epsilon,\lambda}|N\rangle_{\text{c}}}{\langle N|S_{\epsilon,\lambda}|N\rangle_{\text{c}}} + \text{constant}. \tag{24}$$

The subscript c indicates that only connected diagrams have to be included because the contributions due to disconnected graphs can be eliminated using the factorization property [8] $\langle N|S_{\epsilon,\lambda}|N\rangle = \langle N|S_{\epsilon,\lambda}|N\rangle_{\text{c}}\langle 0|S_{\epsilon,\lambda}|0\rangle_{\text{c}}$. The Gell-Mann–Low–Sucher formula necessitates the evaluation of vacuum expectation values of time-ordered products involving multiple fermion and photon operators. These products can be evaluated by means of *Wick*'s theorem [9] which provides information on how to expand time-ordered products into a sum of normal-ordered terms including all possible numbers and contributions of contractions.

Let us concentrate on one-electron systems. The energy shift of a bound electron state $|A\rangle$ of the order of α is derived as a sum of completely contracted terms of the second-order S-matrix element $S^{(2)}_{\epsilon,\lambda}$ taking into account the mass-renormalization counter term provided by $S^{(1)}_{\epsilon,\lambda}$ according to

$$\Delta E^{(2)}_A = \lim_{\epsilon\to 0} \frac{\mathrm{i}\epsilon}{2} \left(2\langle A|S^{(2)}_{\epsilon,\lambda}|A\rangle_{\mathrm{c}} + \langle A|S^{(1)}_{\epsilon,\lambda}|A\rangle_{\mathrm{c}}\right). \tag{25}$$

Detailed instructions for deriving general expressions for energy shifts at a given order may be found in [3]. Contractions between pairs of fermion ψ or boson field operators A^{free}_μ lead to propagator functions. The exact electron propagator in a static external field is homogeneous in time and appears as

$$\begin{aligned}\mathrm{i}S(x,x') &= \langle 0|\mathrm{T}[\psi(x)\overline{\psi}(x')|0\rangle = \int_{C_{\mathrm{F}}} \frac{\mathrm{d}E}{2\pi}\, \mathrm{e}^{-\mathrm{i}E(t-t')} G(\boldsymbol{x},\boldsymbol{x}',E)\,\gamma^0 \\ &= \int \frac{\mathrm{d}E}{2\pi}\,\mathrm{e}^{-\mathrm{i}E(t-t')} \sum_n \frac{\phi_n(\boldsymbol{x})\overline{\phi}_n(\boldsymbol{x}')}{E - E_n + \mathrm{i}\eta\,\mathrm{sign}(E_n - E_{\mathrm{F}_0})}\,, \end{aligned}\tag{26}$$

where the temporal Fourier transform of the exact Greens' function G fulfills the Dirac equation in the external Coulomb field of the nucleus:

$$\left[E - (-\mathrm{i}\boldsymbol{\alpha}\cdot\boldsymbol{\nabla} + V_{\mathrm{ex}}(\boldsymbol{x}) + \beta m)\right] G(\boldsymbol{x},\boldsymbol{x}',E) = \delta(\boldsymbol{x}-\boldsymbol{x}'). \tag{27}$$

This Greens' function is analytic in the complex energy plane except for the bound state poles at E_n, with branch points at $|E| = 1$ and cuts along the real axis for $|E| > 1$. Assuming systems with nuclear charge numbers Z such that bound states occur only at energies $E > 0$ one might choose $E_{\mathrm{F}_0} = 0$. The energy denominator in (26) then reads $[E - E_n(1-\mathrm{i}\eta)]$. The free photon propagator appears as a time-ordered product of free photon-field operators

$$\begin{aligned}\mathrm{i}D_{\mu\nu}(x-x') &= \langle 0|\mathrm{T}\left[A_\mu(x)A_\nu(x')\right]|0\rangle = g_{\mu\nu}\,\mathrm{i}\,D(x-x') \\ &= -g_{\mu\nu}\int_{C_{\mathrm{F}}} \frac{\mathrm{d}E}{2\pi}\,\mathrm{e}^{-\mathrm{i}E(t-t')}\,\frac{\mathrm{e}^{-b|\boldsymbol{x}-\boldsymbol{x}'|}}{|\boldsymbol{x}-\boldsymbol{x}'|}\,, \quad b = -\mathrm{i}\sqrt{E^2 - \mathrm{i}\varepsilon}. \end{aligned}\tag{28}$$

The propagation function D is analytic in the complex E-plane except for branch points and cuts at $\mathrm{Re}(E) > 0$. This knowledge of the analytic properties of both the electron propagator G and the free photon propagator $D_{\mu\nu}$ will play a key role in the evaluation and renormalization of QED radiative corrections.

Coming back to QED radiative effects of the order of α, we identify the formal expression for the self-energy correction

$$\begin{aligned}\Delta E^{\mathrm{SE}}_A &= \mathrm{i}\alpha \int \mathrm{d}^3x\,\mathrm{d}^3x'\,\phi^\dagger_A(\boldsymbol{x}) \int_{C_{\mathrm{F}}} \frac{\mathrm{d}E}{2\pi}\,\alpha^\mu\,G(\boldsymbol{x},\boldsymbol{x}',E)\,\alpha^\nu \\ &\quad\times D_{\mu\nu}(\boldsymbol{x},\boldsymbol{x}',E_A - E)\,\phi_A(\boldsymbol{x}') - \delta m \int \mathrm{d}^3x\,\phi^\dagger_A(\boldsymbol{x})\beta\phi_A(\boldsymbol{x})\end{aligned}\tag{29}$$

and for the vacuum-polarization correction

$$\Delta E_A^{\mathrm{VP}} = -\mathrm{i}\alpha \int \mathrm{d}^3x\, \mathrm{d}^3x'\, \phi_A^\dagger(\boldsymbol{x})\alpha^\mu \phi_A(\boldsymbol{x}) D_{\mu\nu}(\boldsymbol{x},\boldsymbol{x}', E=0) \times \int_{C_\mathrm{F}} \frac{\mathrm{d}E}{2\pi}\, \mathrm{Tr}[\alpha^\nu G(\boldsymbol{x},\boldsymbol{x}', E)]. \tag{30}$$

In Fig. 1 the corresponding Feynman diagrams are depicted. The self energy of the electron arises due to the emission and absorption of a virtual photon by the bound electron. The vacuum-polarization correction can be viewed as an additional interaction of the bound electron with virtual electron–positron pairs induced by the Coulomb potential of the nucleus which modify the external field. It should be emphasized again: bound-state QED ($H_0 = H_{\mathrm{ex}}$) thus treats the interaction with the external field exact to all orders in $Z\alpha$, while additional effects due to the interaction with the quantized, free radiation field are treated perturbatively to any desired order in α. Note that the order of a given (bound-state) QED diagram is given by the number of free internal photon lines involved.

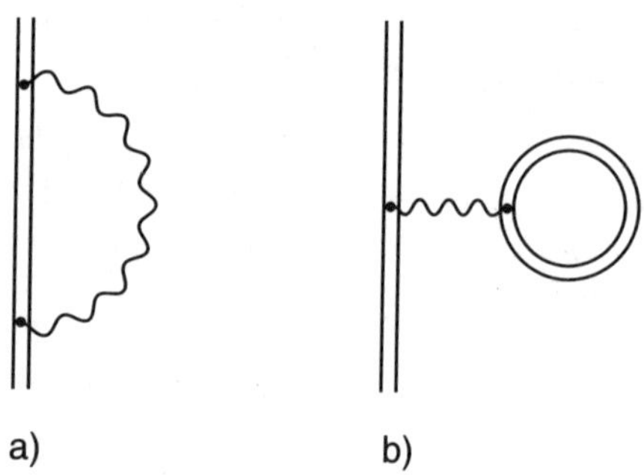

Fig. 1. Feynman diagrams for the self energy (a) and the vacuum polarization (b) of a bound electron of the order of α. The double lines indicate wave functions and propagators in the external Coulomb potential of the nucleus

2. Influence of Finite Nuclear Extension

The theory of energy levels and QED effects in high-Z atoms is based primarily on the Dirac equation for electrons in a spherically symmetric external Coulomb potential V_{ex}:

$$[-\mathrm{i}\boldsymbol{\alpha}\cdot\boldsymbol{\nabla} + V_{\mathrm{ex}}(r) + \beta - E_A]\,\phi_A(\boldsymbol{r}) = 0\,, \tag{31}$$

together with the Dirac matrices

$$\boldsymbol{\alpha} = \begin{pmatrix} 0 & \boldsymbol{\sigma} \\ \boldsymbol{\sigma} & 0 \end{pmatrix} \quad \text{and} \quad \beta = \begin{pmatrix} I & 0 \\ 0 & -I \end{pmatrix}, \tag{32}$$

where the components $\boldsymbol{\sigma}$ are the Pauli spin matrices and I is a 2×2-unit matrix. In central potentials the angular part of the wave functions ϕ_A are eigenfunctions of the square of the total angular momentum operator $\boldsymbol{J}^2$, where $\boldsymbol{J} = \boldsymbol{L} + \frac{1}{2}\boldsymbol{\sigma}$, and of the z-component J_z of the angular momentum operator. In an equivalent way to the use of these operators with quantum numbers (j, ℓ, μ) one may classify the states ϕ_A with respect to the Dirac operator $\mathcal{K} = \beta(\boldsymbol{\sigma} \cdot \boldsymbol{L} + 1)$ and J_z with eigenvalues $(-\kappa, \mu)$:

$$\mathcal{K}\phi_A(\boldsymbol{x}) \;=\; -\kappa\phi_A(\boldsymbol{x})\,, \quad J_z\phi_A(\boldsymbol{x}) = \mu\phi_A(\boldsymbol{x}). \tag{33}$$

The relation between the eigenvalues of the total angular momentum j and orbital angular momentum ℓ and κ are given by $j = |\kappa| - 1/2$ and $l = |\kappa + 1/2| - 1/2$. Accordingly, the wave functions may be written as [10]

$$\phi_{n,\kappa,\mu}(\boldsymbol{x}) = \frac{1}{r}\begin{pmatrix} G_{n,\kappa}(r)\,\chi^{\mu}_{\kappa}(\theta,\varphi) \\ \mathrm{i}\,F_{n,\kappa}(r)\,\chi^{\mu}_{-\kappa}(\theta,\varphi) \end{pmatrix}, \tag{34}$$

where $G_{n,\kappa}$ and $F_{n,\kappa}$, are functions of the radial coordinate that depend on the principal quantum number (or energy) n (E) and κ. χ^{μ}_{κ} are the spherical spinors. Insertion of the separation ansatz (34) into the Dirac equation (31) yields the radial Dirac equation

$$\frac{\mathrm{d}}{\mathrm{d}r}\begin{pmatrix} G \\ F \end{pmatrix} = \begin{pmatrix} -\kappa/r & (E + 1 + V_{\mathrm{ex}}) \\ -(E - 1 - V_{\mathrm{ex}}) & \kappa/r \end{pmatrix}\begin{pmatrix} G \\ F \end{pmatrix}. \tag{35}$$

This set of coupled differential equations has to be integrated numerically for the determination of the radial wave functions G and F and the corresponding Dirac energy eigenvalue. For pointlike nuclei the system of radial equations (35) can be solved analytically for bound states as well as for continuum states. For bound electrons states the level energies $E_{n\kappa}(\text{point})$, respectively the corresponding binding energies, are given by the Bohr–Sommerfeld formula:

$$E^{\mathrm{bind}}_{n\kappa}(\text{point}) \;=\; \left[1 + \left(\frac{Z\alpha}{n - |\kappa| + \sqrt{\kappa^2 - (Z\alpha)^2}}\right)^2\right]^{-1/2} - 1\,. \tag{36}$$

The difference $\Delta E^{\mathrm{F.S.}}_{n,\kappa}$ between the energy eigenvalue $E_{n,\kappa}$ obtained in the presence of the Coulomb potential of extended nuclei and the Bohr–Sommerfeld value $E_{n\kappa}(\text{point})$ for pointlike nuclei is by definition considered as the effect of the finite nuclear size (FS). The present experimental accuracy necessitates the employment of general potentials in any calculations of energy levels

$$V_{\mathrm{ex}}(r) = -e\left[\frac{1}{r}\int_0^r \mathrm{d}r'\,r'^2\,\rho_{\mathrm{ex}}(r') + \int_r^{\infty} \mathrm{d}r'\,r'\,\rho_{\mathrm{ex}}(r')\right] \tag{37}$$

generated by an extended nuclear charge density ρ_{ex} that is normalized to the nuclear charge Ze. An important empirical parameter that characterizes ρ_{ex} is the root mean square (rms) radius

$$r_{\mathrm{rms}} \equiv \langle r^2\rangle^{1/2} = \left[\frac{4\pi}{Ze}\int_0^{\infty} \mathrm{d}r'\,r'^2\,\rho_{\mathrm{ex}}(r')\right]^{1/2}. \tag{38}$$

A simple model for the nucleus often employed is the *uniform sphere* model. A more realistic nuclear charge distribution used frequently is the two-parameter *Fermi distribution*

$$\rho_{\mathrm{F}}(r) = \frac{\rho_0}{1 + \mathrm{e}^{(r-r_0)/a}}\,, \tag{39}$$

where r_0 denotes the half-density radius. The parameter a is related to the skin thickness $t = 4 \ln 3a$, which is defined as the radial distance over which the charge density declines from 90% to 10% of the value at the origin. Instead of r_0 one uses the rms radius which is always kept fixed when comparing different model distributions. The difference in the calculated binding energies employing a Fermi distribution and the uniform sphere model with the same rms radius might well serve as an estimate of the *nuclear-shape uncertainty* $\delta E_A^{\mathrm{shape}}$. Any uncertainty in the knowledge of the rms radius also induces a *size uncertainty* $\delta E_A^{\mathrm{size}}$ in the binding energy. Predictions for the influence of the finite nuclear size on the Lamb shift based on a Fermi distribution are usually presented together with an estimate of the shape uncertainty. At the level of accuracy of about 10^{-5} they are no longer negligible [11]. Nuclear size effects represent an important contribution to the Lamb shift which in heavy systems becomes as important as the most dominant radiative correction, i.e. the self energy of the order of α.

3. Self Energy

In Fig. 1(a) the self-energy correction to the first order in α is depicted. This section indicates the exact evaluation of the self-energy corrections for hydrogen-like systems laying particular emphasis on *Mohr*'s method [12–14] and on the partial wave renormalization approach (PWR) [15, 16]. The latter has also been fruitfully applied for evaluating radiative effects of the order of α^2 (see, e.g., [16–19]). All these calculations of the self energy are based on the expansion of the electron Green's function in terms of eigenfunctions of angular momentum. Since the unrenormalized self energy is infinite, a method of subtracting off the infinite mass renormalization term must be employed which is suitable for a numerical calculation as well. This is accomplished by constructing counter terms that can be calculated analytically and that are also expanded in terms of eigenfunctions of angular momentum so that a term-by-term subtraction can be performed. One aspect in which the various approaches differ is in the detailed form of the function employed in the term-by-term subtraction in the angular momentum expansion.

3.1 Regularization and the Singular Part

The expression (29) for the self-energy correction includes an integration over intermediate energies and diverges for arguments $x_{21} = |\boldsymbol{x}_2 - \boldsymbol{x}_1| \approx 0$.

Accordingly appropriate regularizations are required to make the integral finite and to isolate the infinite mass divergency. For instance, in coordinate space the *Pauli–Villars regularization* [20] is implemented by making use of the replacement

$$\frac{\mathrm{e}^{-bx_{21}}}{x_{21}} \rightarrow \frac{\mathrm{e}^{-bx_{21}}}{x_{21}} - \frac{\mathrm{e}^{-b'x_{21}}}{x_{21}}, \quad \begin{aligned} b' &= -\mathrm{i}\sqrt{(E_A - E)^2 - \Lambda^2 - \mathrm{i}\varepsilon} \\ b &= -\mathrm{i}\sqrt{(E_A - E)^2 - \mathrm{i}\varepsilon} \end{aligned} \tag{40}$$

in the integrand in the expression for the photon propagator in (29). At the same time, the mass renormalization term δm (free-electron self energy) is replaced by the corresponding term calculated with the regularized photon propagator, i.e. $\delta m \rightarrow \delta m(\Lambda)$. The fully regularized expression for the self energy then reads

$$\begin{aligned} E_A^{\mathrm{SE}}(\Lambda) &= -\mathrm{i}\alpha \int \mathrm{d}^3x_2\, \mathrm{d}^3x_1\, \phi_A^\dagger(\boldsymbol{x}_2)\alpha_\mu \int_{C_{\mathrm{F}}} \frac{\mathrm{d}E}{2\pi}\, G(\boldsymbol{x}_2, \boldsymbol{x}_1, E)\alpha^\mu \phi_A(\boldsymbol{x}_1) \\ &\quad \times \left[\frac{\mathrm{e}^{-bx_{21}}}{x_{21}} - \frac{\mathrm{e}^{-b'x_{21}}}{x_{21}}\right] - \delta m(\Lambda) \int \mathrm{d}^3x\, \phi_A^\dagger(\boldsymbol{x})\beta\phi_A(\boldsymbol{x}) \end{aligned} \tag{41}$$

which yields the finite physical energy shift E_A^{SE} in the limit $\Lambda \rightarrow \infty$. Singular terms arise from the high-energy region of the integration over E, which can be isolated in the first few terms in the expansion of the electron propagator $G(\boldsymbol{x}_2, \boldsymbol{x}_1, E) = \langle \boldsymbol{x}_2 | G(E) | \boldsymbol{x}_1 \rangle$ in powers of the external potential. The resolvent $G(E) = (h_0 + V_{\mathrm{ex}} - E)^{-1}$ can be expanded as

$$\begin{aligned} G(E) &= [F(E) - F(E)\, V_{\mathrm{ex}}\, F(E)] + F(E)\, V_{\mathrm{ex}}\, G(E)\, V_{\mathrm{ex}}\, F(E) \\ &= G_{\mathrm{a}}(E) + [G(E) - G_{\mathrm{a}}(E)] \end{aligned} \tag{42}$$

together with the free resolvent operator $F(E) = (h_0 - E)^{-1}$. The term in square brackets contains the leading divergent terms in the asymptotic expansion in $|E|^{-1}$. The asymptotic term $G_{\mathrm{a}}(E)$ will be derived from the high-energy region by extracting the leading terms in a form amenable to analytic calculation and evaluating the remainder $(G(E) - G_{\mathrm{a}}(E))$ numerically. The various methods of calculation can be characterized in part by the choice of $G_{\mathrm{a}}(E)$, which is made from variations of the expansion of $G(E)$ in powers of V_{ex}. The evaluation of the self energy already dates back to the 1950s, when *Brown, Langer* and *Schaefer* [21] proposed a method of numerically calculating the QED self energy, which has subsequently been corrected and implemented by *Desiderio* and *Johnson* [22]. Different variants of self-energy calculations where developed by *Mohr* et al. [12–14] and *Blundell* and *Snyderman* [23]. Mohr's method deals with the choice

$$G_{\mathrm{a}}(E) = \frac{1}{h_0 - E} - V_{\mathrm{ex}} \left.\frac{1}{(h_0 - E)^2}\right|_D \equiv F(E) - V_{\mathrm{ex}} \left.[F(E)]^2\right|_D, \tag{43}$$

where $|_D$ denotes the simplification that the off-diagonal terms, which connect the large components to the small components of the wave function, are omitted in the second term.

Before we turn the discussion to Mohr's method we sketch the idea of carrying out mass renormalization numerically within the partial wave renormalization (PWR). This approach is based on the observation that the mass counter term can be represented as the free-electron self energy. Similar to the method indicated above, an angular momentum decomposition allows for a term-by-term cancellation of the mass divergence for each partial wave. Accordingly the regulated (finite) expression for the (unrenormalized) bound self energy reads [16, 24]:

$$\Delta E_A^{\mathrm{bou}}(\Lambda) = \sum_{\ell=0}^{\infty} \int_0^{\infty} \mathrm{d}k\, k^2 \left\langle A \middle| \left[\Sigma_\ell^{\mathrm{bou}}(k, E_A; \Lambda = 0) - \Sigma_\ell^{\mathrm{bou}}(k, E_A; \Lambda) \right] \middle| A \right\rangle ,$$

$$\begin{aligned} \langle A|\Sigma_\ell^{\mathrm{bou}}(k, E_A; \Lambda)|A\rangle &= -\frac{\alpha}{\pi}\,(2\ell+1) \sum_n \frac{\langle A|\alpha^\mu\, j_\ell(kx_2)\, \boldsymbol{C}_\ell(2)|n\rangle}{\sqrt{k^2+\Lambda^2}} \\ &\times \frac{\langle n|\alpha_\mu\, j_\ell(kx_1)\, \boldsymbol{C}_\ell(1)|A\rangle}{[E_A - E_n - \mathrm{sign}(E_n)\sqrt{k^2+\Lambda^2}]} , \end{aligned} \tag{44}$$

where the Dirac notation for the bound states $\phi_A(\boldsymbol{x}) = \langle \boldsymbol{x}|A\rangle$, etc. has been introduced. Mass renormalization within the PWR approach is essentially based on expressing the regulated mass counter term $M_A(\Lambda) = \delta m(\Lambda)\langle A|\beta|A\rangle$ in terms of the regulated free-electron self-energy shift:

$$\begin{aligned} M_A(\Lambda) &= \sum_{\ell=0}^{\infty} \int_0^{\infty} \mathrm{d}k\, k^2 \sum_{\boldsymbol{p},r,\lambda,\boldsymbol{p}',r',\lambda'} \langle A|\boldsymbol{p}, r, \lambda\rangle\langle \boldsymbol{p}', r', \lambda'|A\rangle \\ &\times \langle \boldsymbol{p}, r, \lambda| \left[\Sigma_\ell^{\mathrm{free}}(k, E_p^r; \Lambda = 0) - \Sigma_\ell^{\mathrm{free}}(k, E_p^r; \Lambda)\right] |\boldsymbol{p}', r', \lambda'\rangle \end{aligned} \tag{45}$$

with insertions of free-electron states $|\boldsymbol{q}, s, \sigma\rangle$, etc. For each value of ℓ one obtains a finite term, although both sums diverge separately. The PWR approach is based on the formula

$$\begin{aligned} \Delta E_A^{\mathrm{SE}} &= \sum_{\ell=0}^{\infty} \mathrm{Re}\Big\{ \int_0^{\infty} \mathrm{d}k\, k \left\langle A \middle| \Big[\Sigma_\ell^{\mathrm{bou}}(k, E_A; \Lambda = 0) - \right. \\ &\left. \sum_{\boldsymbol{p},r,\lambda,\boldsymbol{p}',r',\lambda'} |\boldsymbol{p}, r, \lambda\rangle\langle \boldsymbol{p}, r, \lambda|\Sigma_\ell^{\mathrm{free}}(k, E_p^r; \Lambda = 0)|\boldsymbol{p}', r', \lambda'\rangle\langle \boldsymbol{p}', r', \lambda'| \Big] \middle| A \right\rangle \Big\} \end{aligned} \tag{46}$$

Taking the difference of (44) and (45) this implies a non-covariant renormalization of the self energy. In the present case all cutoff-dependent parts vanish in the limit $\Lambda \to \infty$. Further details can be found in [15, 16, 18].

3.2 Integration Contour and Evaluation

The Feynman contour C_F of integration over E in (41) extends from $-\infty$ below the negative real axis, passes through $E = 0$, and extends to $+\infty$ above the positive real axis. Figure 2 shows the contour and the singularities in the

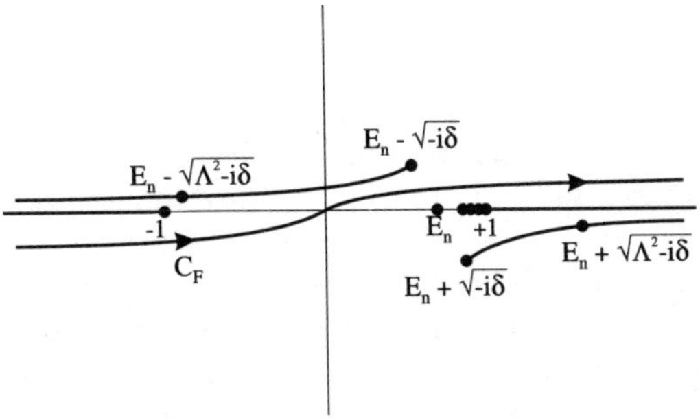

Fig. 2. The contour and the singularities of the propagator $G(\boldsymbol{x}_2, \boldsymbol{x}_1, E)$ in the complex E-plane. The dots to the left of $E = +1$ represent the poles of bound-states; E_n indicates the ground-state energy assuming $E_n > 0$

complex E plane for the calculation of the $1s_{1/2}$-ground-state self energy. We consider systems with nuclear charge numbers Z such that $E_{1s} > 0$ holds (for the known periodic system). There are branch points associated with b and b' in the photon propagator shown in Fig. 2; those associated with b are shown at $E_n \pm \sqrt{-i\delta}$. Cuts are drawn from the branch points such that the conditions $\mathrm{Re}(b) > 0$ and $\mathrm{Re}(b') > 0$ are satisfied everywhere in the cut E plane. In view of these analytic properties it is useful to deform the contour C_F to a new path in order to improve the numerical properties of the integral. In particular performing a Wick rotation the integral over E is exponentially damped along the imaginary axis. The contributions from the quarter circles from $-\infty$ to $-i\infty$ and ∞ to $i\infty$ yield no contribution to the regulated expression.

In Refs. [12–14, 25–30] numerical calculations could be improved by choosing the new contour, which extends from $-i\infty$ along the negative imaginary axis to the point E_1 just below the real axis, around the branch point in the photon propagator, and from E_2 just above the real axis, along the positive imaginary axis to $+\infty$, as shown in Fig. 3. The evaluation is then divided into a low-energy part (C_L), that arises from the integration from E_1 to E_2 and into a high-energy part (C_H) from the integration along the imaginary axis. Both parts are defined as the limit in which $\delta \to 0$ followed by the limits $E_1 \to 0$ and $E_2 \to 0$. The limit $\delta \to 0$ results in the two branch points associated with b coalescing at E_n.

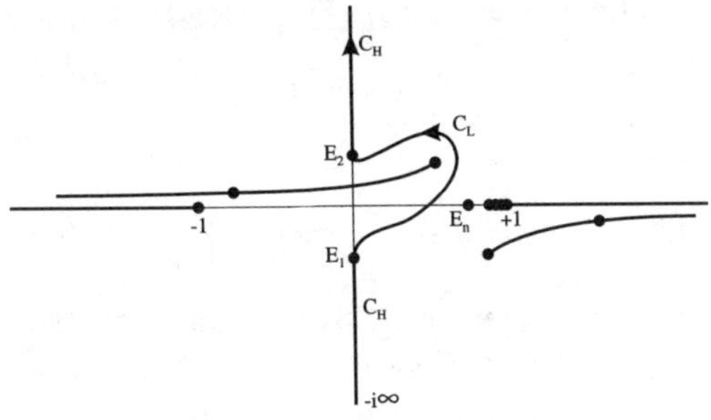

Fig. 3. The new contour in the complex energy plane which allows for the separate evaluation of the low-energy part (integration along C_L) and the high-energy part (integration along C_H), respectively

The low-energy part $\Delta E_{\rm L}$ contains no contribution from the Λ-dependent regulator term and is given by the first term of (41) integrated along the contour $C_{\rm L}$. One can check from Figs. 2 and 3 that this closed contour encloses no singularities in that case. The contour integral around the branch point at E_n can be converted to a line integral to facilitate the numerical integration. It is advantageous to work in the Coulomb gauge. Although the low-energy part itself is not gauge invariant, it can be transformed into the Coulomb gauge form of the expression plus a simple correction term [12]. Since the low-energy part is finite there it is no need to carry out a renormalization in order to evaluate it numerically. However, a finite part of the renormalization is included, and the physical level shift is smaller than the finite renormalization by a factor of order $(Z\alpha)^4$. The physical part of the low-energy part can be isolated in terms of a function $f_L(Z\alpha, R)$ defined by [13]

$$\Delta E_{\rm L} = \frac{\alpha}{\pi}\left[\frac{5}{6}E_n + \frac{2}{3}\langle\beta\rangle + \frac{7}{6}\langle V_{\rm ex}\rangle + \frac{(Z\alpha)^4}{n^3} f_L(Z\alpha, R)\right]. \tag{47}$$

The numerically significant part of the low-energy part of the self energy is calculated by evaluating the complete expression for $\Delta E_{\rm L}$ and then solving (47) numerically to obtain the function $f_{\rm L}(Z\alpha, R)$.

The complete high-energy part $\Delta E_{\rm H}$ is evaluated by integration of (41) along the contour $C_{\rm H}$ involving the decomposition of the Green's function (resolvent operator) G as the sum of the asymptotic part $G_{\rm a}$ and a remainder part $(G - G_{\rm a})$ as indicated in (42). $\Delta E_{\rm H}$ is conveniently calculated as the sum of three subtraction terms:

$$\begin{aligned}
\Delta E_{\rm H} \;&=\; \lim_{\Lambda\to\infty}\left\{ {\rm i}\alpha \int {\rm d}^3x_2\,{\rm d}^3x_1 \int_{C_{\rm H}} \frac{{\rm d}E}{2\pi}\, \phi_n^\dagger(\boldsymbol{x}_2)\alpha_\mu \right. \\
&\times\Big(\langle\boldsymbol{x}_2|F(E)|\boldsymbol{x}_1\rangle\alpha^\mu\left[\phi_n(\boldsymbol{x}_2) + (\boldsymbol{x}_1-\boldsymbol{x}_2)\cdot\boldsymbol{\nabla}_2\,\phi_n(\boldsymbol{x}_2)\right] \\
&-V_{\rm ex}(\boldsymbol{x}_2)\int {\rm d}^3x_3\langle\boldsymbol{x}_2|\left[F(E)^2\right]_{\rm D}|\boldsymbol{x}_1\rangle\alpha^\mu\phi_n(\boldsymbol{x}_2)\Big)\left(\frac{{\rm e}^{-bx_{21}}}{x_{21}} - \frac{{\rm e}^{-b'x_{21}}}{x_{21}}\right) \\
&\left.-\delta m(\Lambda)\int {\rm d}^3x\;\phi_n^\dagger(\boldsymbol{x})\beta\phi_n(\boldsymbol{x})\right\} + {\rm i}\alpha\int {\rm d}^3x_2\,{\rm d}^3x_1 \int_{C_{\rm H}}\frac{{\rm d}E}{2\pi}\,\phi_n^\dagger(\boldsymbol{x}_2)\alpha_\mu \\
&\times\Big\{\Big(\langle\boldsymbol{x}_2|F(E)|\boldsymbol{x}_1\rangle\alpha^\mu\left[\phi_n(\boldsymbol{x}_1) - \phi_n(\boldsymbol{x}_2) - (\boldsymbol{x}_1-\boldsymbol{x}_2)\cdot\boldsymbol{\nabla}_2\,\phi_n(\boldsymbol{x}_2)\right] \\
&-V_{\rm ex}(\boldsymbol{x}_2)\langle\boldsymbol{x}_2|\left[F(E)^2\right]_{\rm D}|\boldsymbol{x}_1\rangle\alpha^\mu\left[\phi_n(\boldsymbol{x}_1)-\phi_n(\boldsymbol{x}_2)\right]\Big)\frac{{\rm e}^{-bx_{21}}}{x_{21}} \\
&+\Big(\langle\boldsymbol{x}_2|\big(G(E) - F(E) + V_{\rm ex}\left[F(E)\right]_{\rm D}^2\big)|\boldsymbol{x}_1\rangle\alpha^\mu\phi_n(\boldsymbol{x}_1)\Big)\frac{{\rm e}^{-bx_{21}}}{x_{21}}\Big\}.
\end{aligned} \tag{48}$$

The first two terms arise from non-local replacements [13] of the wave function ϕ_n in the asymptotic term $G_{\rm a}$. Although the designation $|_{\rm D}$ affects the local values of the integrand functions this has no influence on the contributions to the level shift since the corresponding off-diagonal terms vanish after integrating over $\boldsymbol{x}_1$. The regulated term in wavy brackets, $\Delta E_{\mathrm{H}}^{(\mathrm{a})}$, of (48) involves the singular part of the high-energy part and can be evaluated analytically in terms of simple matrix elements [13]. Also, as an artifact of the decomposition employed the high-energy part contains finite (constant) contributions of the order of α^0 arising from renormalization. Again the physically relevant contribution to the energy shift is at least of the order of $(Z\alpha)^4$, and the numerical result can be expressed in terms of a function $f_{\mathrm{H}}^{(\mathrm{a})}$. Analogous scalings are obtained for the intermediate term $\Delta E_{\mathrm{H}}^{(\mathrm{b})}$ (second term) and for the (finite) remainder term $\Delta E_{\mathrm{H}}^{(\mathrm{c})}$ (third term). Details of the rather involved derivations of expressions can be found in [13]. As a result one obtains the exact self energy of the order of α as a sum of all these contributions:

$$\Delta E_n^{\mathrm{SE}} = \frac{\alpha}{\pi} \frac{(Z\alpha)^4}{n^3} F^{\mathrm{SE}}(Z\alpha, R)\,, \quad F^{\mathrm{SE}} = f_{\mathrm{L}} + f_{\mathrm{H}}^{(\mathrm{a})} + f_{\mathrm{H}}^{(\mathrm{b})} + f_{\mathrm{H}}^{(\mathrm{c})}\,. \tag{49}$$

The dependence of the self energy on the nuclear radius indicates that for heavy systems ($Z > 50$) also the influence of the finite nuclear size on the self-energy correction has to be taken into account [13].

4. Vacuum Polarization

In Fig. 1(b) we have drawn the Feynman diagram corresponding to the vacuum polarization correction to lowest order in α. In this section we consider a technique to evaluate this QED correction to all orders in $Z\alpha$ involving exact electron orbitals and propagators in external Coulomb potentials.

For a given state ϕ_A the unrenormalized energy shift (30) can be cast into a first-order perturbation theory expression:

$$\Delta E_A^{\mathrm{VP}} = \int \mathrm{d}^3x\, \phi_A^\dagger(\boldsymbol{x})\, \mathcal{U}^{\mathrm{VP}}(\boldsymbol{x})\, \phi_A(\boldsymbol{x})\,. \tag{50}$$

The exact first-order vacuum polarization potential induced by the static external Coulomb field of a nucleus reads (in Feynman gauge)

$$\mathcal{U}^{\mathrm{VP}}(\boldsymbol{x}) \quad = \quad e \int \mathrm{d}^3x_1\, \frac{\rho^{\mathrm{VP}}(\boldsymbol{x}_1)}{|\boldsymbol{x}-\boldsymbol{x}_1|}\,, \tag{51}$$

where the total induced vacuum polarization charge density ρ^{VP} follows as an energy integral over the trace of the exact electron propagator G along the contour C_{F}. Utilizing the integral equation (42) for the Green's function G one obtains the vacuum polarization density ρ^{VP} as a sum of the Uehling-part (one-potential term)

$$\rho_{\mathrm{Ueh}}^{\mathrm{VP}}(\boldsymbol{x}) = -\mathrm{i}e \int_{C_{\mathrm{F}}} \frac{\mathrm{d}E}{2\pi} \int \mathrm{d}^3x'\, \mathrm{Tr}\big[F(\boldsymbol{x},\boldsymbol{x}',E) V_{\mathrm{ex}}(\boldsymbol{x}') F(\boldsymbol{x}',\boldsymbol{x},E)\big] \tag{52}$$

and of the Wichmann–Kroll part summarizing all multiple-potential terms of higher orders in $Z\alpha$.

$$\begin{aligned}\rho_{\mathrm{WK}}^{\mathrm{VP}}(\boldsymbol{x}) &= -\mathrm{i}e \int_{C_{\mathrm{F}}} \frac{\mathrm{d}E}{2\pi} \Big\{ \mathrm{Tr}\big[G(\boldsymbol{x},\boldsymbol{x},E)\big] \\ &\quad - \int \mathrm{d}^3x'\,\mathrm{Tr}\big[F(\boldsymbol{x},\boldsymbol{x}',E)V_{\mathrm{ex}}(\boldsymbol{x}')F(\boldsymbol{x}',\boldsymbol{x},E)\big]\Big\}. \end{aligned} \tag{53}$$

The zero-potential term vanishes due to Furry's theorem [31]. The representation above implies a subtraction scheme for calculating the finite part of the Wichmann–Kroll potential, respectively the vacuum polarization charge density $\rho_{\mathrm{WK}}^{\mathrm{VP}}$. It is most suitably employed in connection with the angular momentum decompositions of the electron propagators F and G.

4.1 Uehling Potential

The Uehling potential represents the dominant vacuum polarization correction of the order of $\alpha(Z\alpha)$ to the one-photon exchange potential between the nuclear charge distribution and the bound electron. The virtual photon mediating the interaction between the bound electron and the nucleus creates – up to this order – a free, virtual electron-positron pair (loop) giving rise to a modification of the external Coulomb potential which causes the energy shift. It is well known that this lowest-order correction requires charge renormalization. The (quadratic) divergence may be regulated and isolated by employing the Pauli–Villars scheme. Details of the renormalization may be found in standard textbooks (e.g. [32, 33]). Finally one derives the analytically known renormalized polarization function in terms of an integral representation. It allows for the evaluation of the Uehling potential induced by a given extended nuclear charge distribution ρ_{ex}. For spherically symmetric external charge distributions one obtains the following expression [34]:

$$\begin{aligned}\mathcal{U}_{\mathrm{Ueh}}^{\mathrm{VP}}(r) &= -\frac{2e}{3\pi}\frac{Z\alpha}{r}\int_0^\infty \mathrm{d}r'\,r'\,\rho_{\mathrm{ex}}(r')\left[\chi_2(2|r-r'|)-\chi_2(2(r+r'))\right], \\ \chi_n(z) &= \int_1^\infty \mathrm{d}t\left(1-\frac{1}{t^2}\right)^{1/2}\left(1+\frac{1}{2t^2}\right)\frac{\mathrm{e}^{-zt^2}}{t^n}. \end{aligned} \tag{54}$$

The energy shift is obtained according to

$$\Delta E_A^{VP}(\mathrm{Ueh}) = \langle A|\mathcal{U}_{\mathrm{Ueh}}^{\mathrm{VP}}|A\rangle \quad . \tag{55}$$

4.2 Wichmann–Kroll Contribution

The remaining vacuum-polarization part (53) in lowest order of the fine-structure constant α but to all orders in $(Z\alpha)^n$ with $n \geq 3$ has been considered first by *Wichmann* and *Kroll* [35]. According to Furry's theorem, there are no contributions in $(Z\alpha)^n$ with even powers of n. The evaluation of this contribution for a high-Z nucleus of finite extent avoids a perturbative calculation by making use of explicit solutions of the Dirac equation in a Coulomb potential together with the partial wave decomposition of the Green's function. A detailed discussion of this type of higher-order vacuum polarization effects is presented in Ref. [36]. A special application of the computed vacuum polarization potential to muonic atoms has been presented in Ref. [37]. The numerical approach developed in [36] utilizes the decomposition of the exact Green's function into partial waves and derives the radial vacuum polarization charge density as

$$\rho^{\mathrm{VP}}(r) = -\mathrm{ie}\int \frac{\mathrm{d}u}{2\pi}\,|\kappa| \sum_{|\kappa|=1}^{\infty} \mathrm{Re}\left\{\sum_{i=1}^{2} G_{\kappa}^{ii}(r,r',\mathrm{i}u)\right\}, \tag{56}$$

where the integration contour has been Wick-rotated. This formal expression, however, is divergent. The potential expansion of the total vacuum polarization reveals that all singularities are contained in the Uehling term of order $\alpha(Z\alpha)$ and that all higher-order terms ($\alpha(Z\alpha)^n$ with $n \geq 3$) are finite. One obvious renormalization procedure for the total vacuum polarization charge density therefore is provided by subtracting off the corresponding partial wave decomposition of the Uehling contribution which leads to the finite Wichmann–Kroll part. The effect of the Uehling potential is then calculated separately based on the analytic expression (54). Denoting the partial wave components of the free Green's function with F_{κ}^{ij} the final expression for the vacuum polarization charge density of the order of $\alpha(Z\alpha)^n$ with $n \geq 3$ reads

$$\begin{aligned}\rho_{\mathrm{WK}}(r) &= \frac{e}{2}\int_0^{\infty} \frac{\mathrm{d}u}{2\pi} \sum_{\kappa=\pm 1}^{\pm\infty} |\kappa|\,\mathrm{Re}\left\{\sum_{i=1}^{2} G_{\kappa}^{ii}(r,r,\mathrm{i}u)\right.\\ &\quad\left. + \int_0^{\infty} \mathrm{d}r'\,r'^2\,V_{\mathrm{ex}}(r') \sum_{m,n=1}^{2} [F_{\kappa}^{mn}(r,r',\mathrm{i}u)]^2\right\}.\end{aligned} \tag{57}$$

With the Wichmann–Kroll density one calculates first the contribution to the vacuum polarization potential. Starting from numerical solutions of the radial Dirac equation the Wichmann–Kroll vacuum polarization charge density has been computed in Ref. [38] to all orders in $Z\alpha$ with the restriction that only angular momentum components up to $|\kappa| = 3$ were evaluated explicitly. The energy correction associated with the Wichmann–Kroll potential caused by the density (57) is usually expressed in terms of a function H_{WK}. For a bound state with principal quantum number n we write similarly to (49)

$$\Delta E_n^{\mathrm{VP}}(\mathrm{WK}) = \frac{\alpha}{\pi}\frac{(Z\alpha)^4}{n^3} H_{\mathrm{WK}}(Z\alpha, R). \tag{58}$$

The renormalization scheme for the Wichmann–Kroll part has also been employed by *Persson* et al. [18] to generate more precise results. The finite size of the nucleus has also been taken into account, and the results are in fair agreement with the corresponding data in Ref. [36]. The calculations performed in [18, 36] underline the importance of including finite-size effects in any state-of-the-art calculations of the vacuum polarization correction for high-Z systems. As in the case of the self energy it is important to examine the influence of finite nuclear size on the vacuum polarization [19] before calculating radiative effects of higher order in α.

5. Ground-State Lamb Shift in Hydrogen-Like Ions

One major task of bound state QED is to provide accurate predictions of spectroscopic data, e.g. for the Lamb shift in few-electron systems, which can be compared with high-precision measurements. Historically, the experimental observation of the (traditional) level splitting between the $2s_{1/2}$- and the $2p_{1/2}$-state in hydrogen [39] marked the first crucial test of QED in the sense that it demonstrated the physical reality of (virtual) radiative effects. Besides the idea of particles and antiparticles this it the most fundamental prediction to come from the new concept of the field-theoretical vacuum. In particular, the investigation of the Lamb shift in heavy, hydrogen-like ions allows for sensitive tests of QED in strong external Coulomb fields [3].

The Lamb shift of an individual energy level can be defined heuristically as the difference between the experimentally deduced binding energy $E_A^{\text{bind}}(\text{exp})$ and the corresponding Dirac energy (39) in the Coulomb field of a point nucleus, i.e.,

$$\Delta E_A^{\text{Lamb}} = E_A^{\text{bind}}(\text{exp}) - E_A^{\text{bind}}(\text{point}) = \Delta E_A^{\text{bind}}(\text{theory}) - E_0. \qquad (59)$$

We note that by convention [40] the Lamb shift for a single energy level accounts for all corrections ΔE_A^{bind} beyond the Dirac value for a point nucleus which contribute to the traditional $2p_{1/2} - 2s_{1/2}$-Lamb shift. The constant E_0 introduced above may contain all level-independent corrections. There is a variety of different energy corrections that add up to the theoretical Lamb shift. One may distinguish between pure QED corrections and Nuclear Structure Effects (NSE). The latter class includes energy shifts due to static nuclear properties (size and shape of the nuclear charge density distribution, finite nuclear mass corrections) and due to internal nuclear dynamics (nuclear polarization effect). All of the yet measured Lamb shifts for the $1s$-ground state in heavy systems have been obtained with a relative precision up of about 10^{-4}. These data can be explained on the basis of the dominant QED-radiative corrections of the order of α but exact in the interaction with the external field and of finite nuclear size effects, i.e.,

$$\Delta E_A^{\text{Lamb}}(\text{theory}) \approx \Delta E_A^{\text{SE}} + \Delta E_A^{\text{VP}} + \Delta E_A^{\text{FS}}. \qquad (60)$$

For a comparison between theoretical predictions for the $1s$-Lamb shift and the corresponding experimental data we first present an analysis of the calculations of the energy shifts based on the effects mentioned above. Figure 4

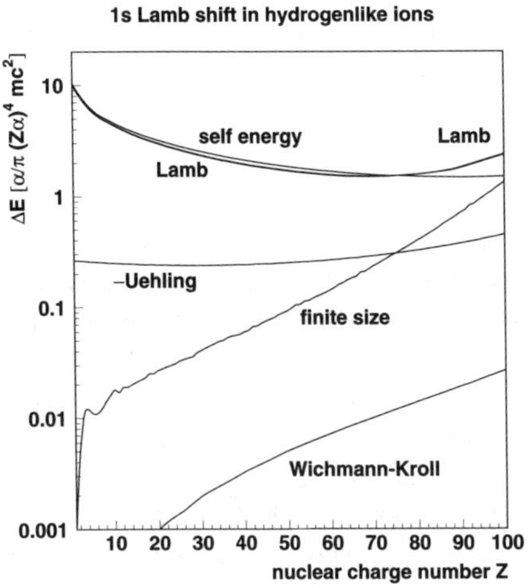

Fig. 4. Contributions of the self energy, vacuum polarization and finite nuclear size to the $1s_{1/2}$ Lamb shift in hydrogen-like ions versus the nuclear charge number Z. The energy shift ΔE is presented in units of $(\alpha/\pi)(Z\alpha)^4/n^3\, mc^2$. Lamb indicates the sum of all contributions

indicates how the individual contribution in (60) add up to the total Lamb shift as a function of the nuclear charge number Z [41] . Energies are given in units of the dimensionless quantity $(\alpha/\pi)(Z\alpha)^4/n^3$. The dominant term is provided by the self-energy shift. The Uehling contribution and the Wichmann part of the vacuum polarization are shown separately. For heavy systems the finite size correction becomes as important as the dominant self energy.

Experimental data to compare with are available for quite a number of different systems. For a review of the experimental methods and of the refined techniques applied as well as the current experimental status of Lamb shift measurements we refer to Chap. VII by *D. Liesen* and *H.F. Beyer* and references cited there. Figure 16 of Chap. VII summarizes the present situation of measurements of the ground-state Lamb shift in hydrogen-like systems. An excellent overall agreement between experimental data and the predictions based on the dominant corrections included in (60) can be stated. This is certainly true at the current experimental accuracy of about 10^{-4}. Aiming at utmost precision will require more refined calculational and experimental efforts. On one hand QED corrections can, in principle, be evaluated to any desired order in α provided all problems encountered with renormalization can be solved. On the other hand, however, one should be aware of basic, natural limitations for testing QED in atomic systems. Limits are finally set by nuclear polarization as well as by uncertainties of nuclear parameters. Estimates indicate that their influence becomes relevant at the level of relative precision of about 10^{-6}. A consistent description of spectroscopic data up to this ultimate level of accuracy requires the exact evaluation of the complete set of QED-radiative corrections of the order of α^2. To provide predictions for the Lamb shift of the ground state that meet with the ambitious standard of about 0.1 eV, e.g., in the lead or uranium system, is currently under investigation.

6. Nuclear Polarization

Nuclear polarization – as a background effect it is always present – represents a natural limitation for any high-precision tests of QED that crucially depend on the exact knowledge of the spectrum of heavy atoms. Accordingly, one needs to determine the level of accuracy at which one is leaving the framework of pure QED and at which one is just probing the influence of internal nuclear structure on the atomic spectrum. So far the nucleus has been considered as a source of the classical external Coulomb field described by a classical charge density distribution. This prescription, however, becomes inadequate when approaching the level of accuracy mentioned above. The interaction between the electron–positron field and the nucleus due to intrinsic nuclear dynamics can no longer be neglected. Due to the exchange of virtual photons the nucleus can undergo virtual electromagnetic transitions too, giving rise to additional contributions to the energy shifts of bound electrons.

The field-theoretical treatment of nuclear polarization as developed by *Plunien* et al. [42, 43] starts with the decomposition for the quantized parts of the electromagnetic source currents (2) and fields (6) introducing the concept of effective photon propagators with nuclear polarization insertions [42, 43]. When performing the perturbation expansion based on (23) and (24) now including the total radiation field $A_\mu^{\rm rad}$ one is led to Feynman diagrams with dressed internal photon lines (contractions involving $A_\mu^{\rm fluc}$), which are represented by the effective photon propagator

$$\begin{aligned} \mathcal{D}_{\mu\nu}(x,x') &= -\mathrm{i}\langle\mathcal{O}|\,\mathrm{T}\,A_\mu^{\rm fluc}(x)\;A_\nu^{\rm fluc}(x')\,|\mathcal{O}\rangle = D_{\mu\nu}(x-x') \\ &\quad + \int \mathrm{d}^4x_1\,\mathrm{d}^4x_2\,D_{\mu\alpha}(x-x_1)\;\Pi^{\alpha\beta}(x_1,x_2)\,D_{\beta\nu}(x_2-x'). \end{aligned} \tag{61}$$

The vacuum state $|\mathcal{O}\rangle$ implies that the nucleus is considered to be in its ground state. This Dyson equation defines the modification $\tilde{\mathcal{D}}^{\mu\nu}$ (second term) in terms of the nuclear polarization tensor $\Pi^{\alpha\beta}$. The latter is given by the vacuum expectation value of the nuclear transition current density

$$\mathrm{i}\,\Pi^{\alpha\beta}(x_1,x_2) = \langle\mathcal{O}|\,\mathrm{T}\,j^\alpha_{\rm fluc}(x_1)\;j^\beta_{\rm fluc}(x_2)\,|\mathcal{O}\rangle. \tag{62}$$

The transition current $j^\alpha_{\rm fluc}$ and thus the effective interaction $\tilde{\mathcal{D}}_{\mu\nu}$ can be specified by employing particular nuclear models. Practicable evaluations require further approximations: a) neglect of possible distortions of the nuclear excitation spectrum due to the presence of the electron; b) the assumption that the time evolution of the transition current $j^\alpha_{\rm fluc}$ is governed by the nuclear model Hamiltonian $H_{\rm nuc}$ only; c) neglect of the contribution of the vector current $\boldsymbol{j}_{\rm fluc}$ since the intrinsic nuclear dynamics is mainly nonrelativistic. The latter approximation implies a restriction to electric nuclear transitions only. Thus one is left with the component Π^{00} of the polarization tensor and with the longitudinal part $\tilde{\mathcal{D}}_{00}$ of the effective propagator. It can be derived for a given transition density ρ_{fluc}. Under rather general conditions, the transition density $\hat{\rho}_{\text{fluc}}$ can be decomposed in terms of electric multipoles $\hat{Q}^{(i)}_{LM}$ referring to nuclear transitions, which are labeled by angular momentum L and projection M (and perhaps additional relevant quantum numbers abbreviated by i):

$$\hat{\rho}_{\text{fluc}}(\boldsymbol{x}) = \sum_{LMi} \mathcal{R}^i_L(r)\, Y_{LM}(\theta,\varphi)\, \hat{Q}^{(i)}_{LM}. \tag{63}$$

An expansion of the charge density fluctuation according to (63) holds for the class of collective nuclear excitations (surface vibrations, rotations and giant resonances) treated within collective nuclear models [43] as well as for an extreme single-particle approach [44]. The result for the modified propagator $\tilde{\mathcal{D}}_{00}$ is obtained in the rather general form

$$\begin{aligned}\tilde{\mathcal{D}}_{00}(\boldsymbol{x},\boldsymbol{x}',E) &= \sum_{LM} B(EL; L\to 0)\, \frac{2E_L}{E^2 - E_L^2 + \mathrm{i}\eta} F_L(r) F_L(r') \\ &\quad \times Y_{LM}(\theta,\varphi)\, Y^*_{LM}(\theta,\varphi)\,, \end{aligned} \tag{64}$$

which basically depends on phenomenological parameters, namely, the excitation energies E_L and the reduced electric transition probabilities $B(EL; L \to 0)$ referring to a collective multipole transition L. The radial dependence F_L differs only slightly for the excitations under consideration. In the case of collective nuclear vibrations or low-lying rotational modes values for these parameters are tabulated in the literature. To estimate the effect due to dominant giant resonances [44, 45] one may utilize expressions for the $B(EL)$-values and excitation energies obtained from semi-empirical sum rules [46, 47].

Performing perturbation theory with H_{rad} (9) one can systematically derive effective QED corrections containing effective photon lines with nuclear polarization insertions. In this general sense this treatment enables a systematic treatment of the influence of nuclear polarization to any atomic processes. In the case of one-electron systems dominant lowest-order contributions of nuclear polarization involving one effective propagator $\tilde{\mathcal{D}}_{\mu\nu}$ (α^2 order) appear as effective self energy (NPSE) and vacuum polarization (NPVP). The corresponding diagrams are depicted in Figs. 5 and 6.

Explicit expressions for the dominant effective self-energy ΔE_A^{NPSE} and vacuum-polarization term ΔE_A^{NPVP} of electron bound states ϕ_A due to virtual nuclear excitations are obtained from (29) and (30) by the replacement of the free photon propagators by the modified propagators $\tilde{\mathcal{D}}_{00}$.

Insertion of the representation (64) for the modified propagator $\tilde{\mathcal{D}}_{00}$ and an eigenfunction expansion for the electron propagator G into the expression for the self-energy shift yields for a bound electron state $\phi_{n\kappa\mu}$ [42, 43]:

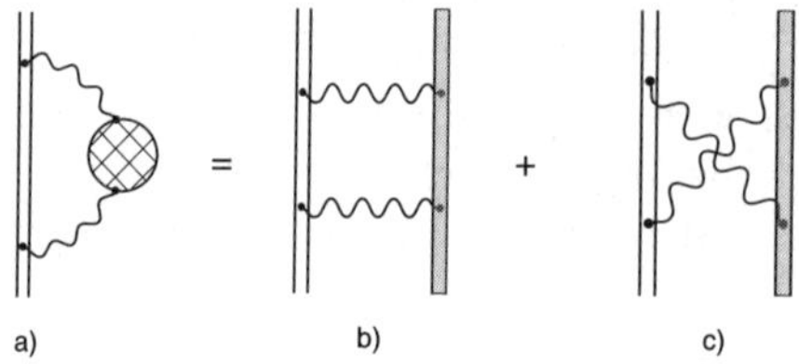

Fig. 5. Effective self-energy diagram (NPSE) corresponding to the ordinary nuclear polarization (NP) effect: a) Goldstone diagram with modified photon propagator due to NP insertion indicated by the bubble and b) the Feynman diagram. The double lines denote electron bound states (Furry picture), and the heavy lines describe the internal nuclear degrees of freedom

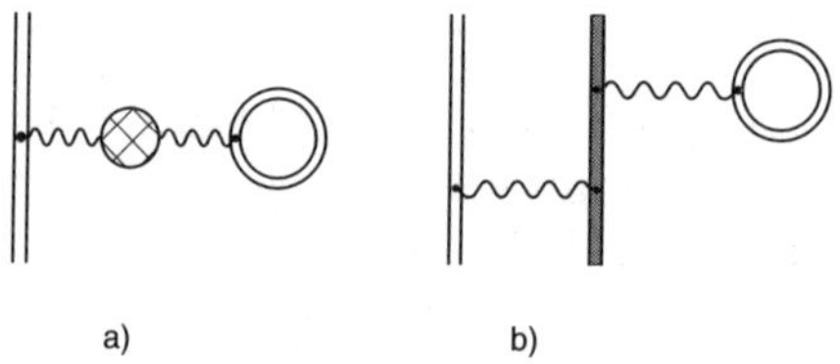

Fig. 6. Combined nuclear-polarization–vacuum-polarization (NPVP) diagram. The notation is the same as in Fig. 5

$$
\begin{aligned}
\Delta E_{n\kappa\mu}^{\mathrm{NPSE}} &= -\alpha \sum_{L\kappa_1} B(EL; L \to 0)|\kappa_1| \left(jj'L \left| \frac{1}{2} \; -\frac{1}{2} \; 0 \right. \right)^2 \\
&\times \left\{ \int_1^{\infty} \frac{\mathrm{d}E_1}{2\pi} \frac{|\langle n\kappa|F_L|E_1\kappa\rangle|^2}{E_1 - E_{n\kappa} + E_L} + \sum_{n_1} \frac{|\langle n\kappa|F_L|n_1\kappa_1\rangle|^2}{E_{n_1\kappa_1} - E_{n\kappa} + E_L + \mathrm{i}\eta} \right. \\
&\left. + \int_{-\infty}^{-1} \frac{\mathrm{d}E_1}{2\pi} \frac{|\langle n\kappa|F_L|E_1\kappa_1\rangle|^2}{E_1 - E_{n\kappa} - E_L} \right\} . \qquad (65)
\end{aligned}
$$

This contribution of nuclear polarization to the total Lamb shift carries an overall minus sign, i.e. $\Delta E_{n\kappa\mu}^{\mathrm{NPSE}}$ thus tends to increase the binding energy of electronic bound states. The effective self-energy shift depends parametrically on the excitation energies E_L, the corresponding $B(EL)$-values and, in principle, on the particular radial-dependence F_L referring to the collective nuclear multipole transitions under consideration. The first two terms in wavy brackets describe transitions of the electron in the state $|n\kappa\mu\rangle$ into higher unoccupied electron states due to virtual nuclear excitation. This intermediate state is either a positive energy continuum state $|E_1\kappa_1\rangle$ (first term) or a bound state $|n_1\kappa_1\rangle$ (second term). These two terms are equivalent to the usual expressions derived in second-order perturbation theory. The third term, however, represents a pure vacuum effect as a consequence of the field-theoretical treatment, which accounts for the different time-orderings (indicated in parts (b) and (c) of Fig. 5). It describes the interaction of the Dirac vacuum with virtual nuclear excitations where an electron from the (occupied) Dirac sea $|E_1\kappa_1\rangle$ can be excited into the occupied bound state $|n\kappa\mu\rangle$ at different space-time points, while the present electron and the positron (hole) will annihilate. This "exchange" process was first taken into account in Refs. [42, 43]. It should be mentioned, that this vacuum contribution is not included in earlier calculations of the effect of nuclear polarization in muonic atoms. However, there this vacuum contribution turns out to be negligible compared with that corresponding to the first two parts, while it contributes essentially in the case of electronic atoms.

The mixed energy contribution $\Delta E_A^{\rm NPVP}$ may be viewed as the interaction between the bound electron and the intrinsic nuclear dynamics mediated via $A_\mu^{\rm fluc}$ but screened by the vacuum polarization cloud $\rho^{\rm VP}$ induced by the external Coulomb field. The dominant contribution of this effect arises due to the interaction of virtual nuclear monopole excitations (breathing mode) and the Uehling part of the (spherically symmetric) vacuum polarization located in the interior of the nucleus [48]. For spherical nuclei the virtual excitation of this compression mode is connected with a nuclear compression energy shift $\delta E_A^{\rm comp}$ associated with fluctuations $\delta\langle r^2\rangle$ of the nuclear rms-radius. The mixed NPVP correction is of the form [48]

$$\Delta E_{n\kappa\mu}^{\rm NPVP} \;=\; \eta^{\rm VP}(R_0)\,\delta E_{n\kappa\mu}^{\rm comp}. \tag{66}$$

In Uehling approximation the prefactor $\eta^{\rm VP}$ is dominated by a logarithmic dependence on the nuclear radius R_0, i.e.

$$\Delta E_{n\kappa\mu}^{\rm NPVP} \;\simeq\; -\frac{4\,(Z\alpha)^2\,\ln(mR_0)}{25\pi\,(E_0R_0)}\,\delta E_{n\kappa\mu}^{\rm comp}. \tag{67}$$

This feature allows one to distinguish the combined NPVP effect from ordinary finite size corrections. The energy shift $\Delta E_{n\kappa\mu}^{\rm NPVP}$ carries the opposite sign to the ordinary NPSE correction. In the lead system nuclear monopole excitations contribute and the sum of both nuclear polarization corrections almost cancel completely [48]. This may indicate that it is a most preferable system for tests of QED in strong fields.

7. QED Corrections of the Order of α^2

The recent experimental progress made in the measurements of the Lamb shift in highly charged few-electron ions may indicate that calculations of higher-order QED corrections become relevant. In this regard we point out that recent measurements of the $1s_{1/2}$-ground-state Lamb shift in hydrogen-like uranium have increased the precision from 520 ± 130 eV [49] to 429 ± 63 eV, which has been achieved by *Stöhlker* et al. [50], respectively, 470 ± 16 eV measured by *Beyer* et al. [51, 52] (see also Chap. VII). For lithium-like uranium the splitting between the $2s_{1/2}$ and the $2p_{1/2}$ levels was determined to be 280.59 ± 0.09 eV corresponding to a precision of 3×10^{-4} [53].

The Feynman diagrams of the various α^2-corrections for one-electron atoms are depicted in Fig. 7. At first glance we may distinguish between diagrams involving closed fermion loops and those representing the two-photon self energy SESE. The first subset embraces the second-order vacuum polarization contributions VPVPa,b,c, the mixed self energy–vacuum polarization

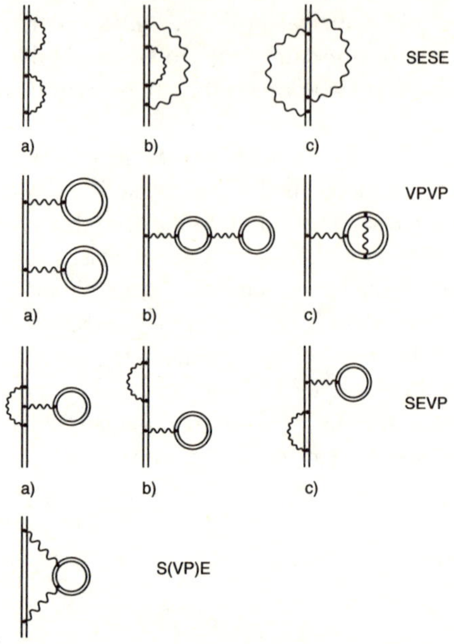

Fig. 7. Feynman diagrams corresponding to QED corrections of the order of α^2 in hydrogen-like ions. Notation is the same as in Fig. 1

SEVPa,b,c and the effective self energy–vacuum polarization contribution S(VP)E. Explicit expressions for the energy shifts in coordinate space can be found in Ref. [3]. Progress has been made towards the evaluation of all second-order corrections given above (see, e.g., [3] for a review and [54] for recent numerical results). While the calculation of the VPVP, SEVP and S(VP)E contributions are almost completed the two-photon self energy is far more elaborate to obtain and remains the subject of current research. Instead of trying to evaluate the individual energy shifts as they stand most of the diagrams involving closed fermion loops are easier to compute by employing the modified potential approach. The general idea behind this is to generate first the complete spectrum of orbitals $\widetilde{\phi}_A$ and corresponding energy eigenvalues $\widetilde{E}_A$ by solving the modified Dirac eigenvalue problem

$$\big(-\mathrm{i}\boldsymbol{\alpha}\cdot\boldsymbol{\nabla}+V_{\mathrm{Coul}}(r)+\mathcal{U}^{\mathrm{VP}}(r)+\beta m\big)\widetilde{\phi}_A(\boldsymbol{x})=\widetilde{E}_A\widetilde{\phi}_A(\boldsymbol{x})\,, \tag{68}$$

including the exact one-loop vacuum polarization potential $\mathcal{U}^{\mathrm{VP}}$ induced by the external nuclear Coulomb potential. The modified wave functions $\widetilde{\phi}_A$ and shifted energy eigenvalues $\widetilde{E}_A$ account for an infinite sequence of VP loops. Both are used for generating a partial wave representation of the modified electron propagator $\widetilde{G}$.

The idea of extracting the energy shifts that correspond to the diagram VPVPa is to calculate the difference $(\widetilde{E}_A - E_A)$ between the Dirac eigenvalues with and without the additional VP-potential. Assuming that the contributions to the energy shift of the order of $\alpha^n, n \geq 3$, are almost negligibly small, the energy shifts $\Delta E_A^{\mathrm{VPVPa}}$ can be identified after subtracting off the dominant first-order VP correction ΔE_A^{VP} from this energy difference [18, 55].

The partial wave renormalization approach is sufficiently flexible to take $\widetilde{G}$ as input for the calculation of an effective self energy as well as an effective vacuum-polarization correction of the order of α as discussed in previous sections. The latter approach allows one to reduce the calculation of the exact two-loop vacuum polarization VPVPb to that of a modified one-loop problem. The appropriate renormalization scheme for deducing the energy shifts $\Delta E_A^{\mathrm{VPVPb}}$ has been developed recently [56].

Until now, the exact Källén–Sabry correction VPVPc is known only in the Uehling approximation [54, 57]. The contributions of higher-order in $Z\alpha$ remain to be evaluated.

The three diagrams of the mixed self energy vacuum polarization SEVP can be deduced simultaneously by calculating an effective self-energy correction $\Delta \widetilde{E}_A^{\mathrm{SE}}$. The expression for the VP-corrected, effective self energy possesses a partial wave representation analogous to (46). Again one argues that higher-order vacuum polarization corrections ($\alpha^n, n \geq 3$) included become negligible. The total energy shifts $\Delta E_A^{\mathrm{SEVP}}$ due to all three SEVP diagrams can be deduced from the effective self-energy correction $\Delta \widetilde{E}_A^{\mathrm{SE}}$ after subtracting off the ordinary first-order energy shift ΔE_A^{SE} [26, 19]. This approach also has the advantage that the reference-state contribution is taken into account automatically. The numerical evaluation of the energy shifts proceeds in the same way as in the case of the ordinary first-order self-energy shifts.

The effective self-energy correction S(VP)E appears as a first-order self energy involving a modified photon propagator with a one-loop vacuum-polarization insertion. In the Uehling approximation the modified photon propagator is known analytically. Accordingly, the energy shifts $\Delta E_A^{\mathrm{S(VP)E}}$ can be obtained quite directly within the PWR-approach [19] as well as based on the angular-momentum decomposition of the electron propagator [58]. To account for the Wichmann–Kroll part requires much more elaborate work.

Until now the major theoretical uncertainty concerning the evaluation of the complete α^2-corrections arises from missing contributions of the two-photon self energy SESE. Renormalization schemes have been developed recently [59–62]. Although progress has been made in numerical evaluation [63] its final completion remains a challenge.

8. Lamb-Shift Predictions

All previous sections so far have served to provide a brief overview of all relevant QED and nuclear corrections involved in current predictions of the Lamb shift in hydrogen-like heavy systems aiming at the limiting relative precision of 10^{-6}. A summary of all presently known contributions to the Lamb shift of K- and L-shell electrons in various one-electron ions has been given in Ref.

Table 1. One-electron Lamb-shift contributions (eV) for the $1s_{1/2}$-ground state in lead and uranium ions. The nuclear shape uncertainties $\delta E^{\text{shape}}_{1s_{1/2}}$ are estimated to be 0.36 eV for uranium and 0.14 eV for lead

Ion	$^{208}Pb^{81+}$	$^{238}U^{91+}$
Binding energy E_B (point nucleus)	−101581.37	−132279.96
Correction:		
Finite nuclear size	67.25	198.82
Self energy (order α)	226.33	355.05
VP: Uehling contribution	−50.70	−93.58
VP: Wichmann–Kroll contribution	2.29	4.99
Total vacuum polarization (order α)	−48.41	−88.60
SESE a) b) c)	uncalculated	
VPVP a) (ladder diagrams)	−0.09	−0.22
VPVP b) (Källén–Sabry cont. + h.o.)	−0.07	−0.15
VPVP c) (Källén–Sabry cont.)	−0.34	−0.60
SEVP a) b) c)	0.53	1.14
S(VP)E	0.07	0.13
Reduced mass	0.27	0.30
Relativistic recoil	0.09	0.16
Total recoil	0.36	0.46
Nuclear polarization	0.00	-0.19
Sum of corrections	245.63	465.84
Resulting total binding energy	−101335.74	−131814.11
Lamb Shift (theory)	245.36	465.54

[54]. In Table 1 the current status may be exemplified here for the Lamb shift of the $1s_{1/2}$-ground state of two prominent systems: lead $^{208}_{82}Pb^{81+}$ and uranium $^{238}_{92}U^{91+}$. Similar results for the $1s$ state of hydrogen-like uranium have been given by *Mohr* [64]. The numbers presented here are slightly different both due to new calculations for the two-loop vacuum polarization diagram [56] as well as improved numerical results for the nuclear polarization [45, 48]. Since by convention [40] the Lamb shift for a single energy level accounts for all corrections beyond the point-nucleus Dirac eigenvalue which contribute to the traditional $2p_{1/2} - 2s_{1/2}$-Lamb shift, the nonrelativistic reduced mass correction has to be subtracted to obtain the proper Lamb shift. A rigorous description of nuclear recoil effects within the framework of QED has been developed by *Shabaev* et al. [65–67]. The updated numbers for the total recoil and for the relativistic recoil respectively involve also corrections due to the

finite nuclear size [68]. The predicted value for the Lamb shift of the $1s_{1/2}$-ground state in hydrogen-like uranium amounts to 465.84(36) eV, which has to be compared with the experimental value of about 470 eV. The zero value presented for the nuclear polarization in lead arises due to a different aspect of nuclear polarization, where the nucleus also interacts with a virtual electron–positron pair. This NPVP effect has been estimated recently [48], for lead, where it practically cancels that of the ordinary nuclear polarization due to existing collective monopole vibrations of this particular nucleus. Therefore the nuclear polarization effects which otherwise limit very precise Lamb shift predictions are almost completely negligible for ${}^{208}_{82}\mathrm{Pb}^{81+}$, making this ion especially suitable for the most precise theoretical predictions. Apart from the missing contribution of the two-photon self energy, currently all QED corrections to the Dirac binding energy value are expected to be known to a precision of about 0.01 eV or better. The numerical values suggest that any further improvement in computing these QED contributions may only be meaningful, if better estimates for the nuclear effects that also enter the Lamb shift predictions could be provided. From this point of view, the Pb ion is the one for which the most precise predictions can be presented. As soon as the missing two-photon self-energy values become available the goal of providing Lamb-shift predictions for all of the systems under consideration with an accuracy of about 0.1 eV will be achieved.

References

1. T. Kinoshita: *Quantum Electrodynamics* (World Scientific, Singapore 1990)
2. L. N. Labzowsky, G. Klimchitskaya, Yu. Dmitriev: *Relativistic Effects in the Spectra of Atomic Systems* (IOP Publishing, Bristol 1993)
3. P. Mohr, G. Plunien, G. Soff: Phys. Rep. **293**, 227 (1998)
4. P. G. Reinhard, W. Greiner, H. Arenhövel: Nucl. Phys. A **166**, 173 (1971)
5. J. Rafelski, B. Müller, W. Greiner: Nucl. Phys. B **68**, 585 (1974)
6. M. Gell-Mann, F. Low: Phys. Rev. **54**, 350 (1951).
7. J. Sucher: Phys. Rev. **107**, 1448 (1957).
8. J. Goldstone: Proc. Roy. Soc. London, A **239**, 267 (1957)
9. G. C. Wick: Phys. Rev. **80**, 268 (1950)
10. M. E. Rose: *Relativistic Electron Theory* (Wiley, New York, 1961)
11. T. Franosch, G. Soff: Z. Phys. D **18**, 219 (1991)
12. P. J. Mohr: Ann. Phys. (NY) **88**, 26 (1974)
13. P. Indelicato, P. J. Mohr: Phys. Rev. A **46**, 172 (1992)
14. P. J. Mohr, G. Soff: Phys. Rev. Lett. **70**, 158 (1993)
15. H. M. Quiney, I. P. Grant: Phys. Scr. **T46**, 132 (1993)
16. H. Persson, I. Lindgren, S. Salomonson: Phys. Scr. **T46**, 125 (1993)
17. H. M. Quiney, I. P. Grant: J. Phys. B **27**, L299 (1994)
18. H. Persson, I. Lindgren, S. Salomonson, P. Sunnergren: Phys. Rev. A **48**, 2772 (1993)
19. H. Persson, I. Lindgren, L. N. Labzowsky, G. Plunien, T. Beier, G. Soff: Phys. Rev. A **54**, 2805 (1996)

20. W. Pauli, F. Villars: Rev. Mod. Phys. **21**, 434 (1949)
21. G. E. Brown, J. S. Langer, G. W. Schaefer: Proc. Roy. Soc. (London) A **251**, 92 (1959)
22. A. M. Desiderio, W. R. Johnson: Phys. Rev. A **3**, 1267 (1971)
23. S. A. Blundell, N. J. Snyderman: Phys. Rev. A **44**, 1427 (1991)
24. I. Lindgren, H. Persson, S. Salomonson, A.Ynnermann: Phys. Rev. A **47**, 4555 (1993)
25. H. Persson, S. M. Schneider, W. Greiner, G. Soff, I. Lindgren: Phys. Rev. Lett. **76**, 204 (1996)
26. I. Lindgren, H. Persson, S. Salomonson, V. Karasiev, L. Labzowsky, A. Mitrushenkov, M. Tokman: J. Phys. B **26**, L503 (1993)
27. P. J. Mohr: Ann. Phys. (NY) **88**, 52 (1974)
28. P. J. Mohr: Phys. Rev. Lett. **34**, 1050 (1975)
29. P. J. Mohr: Phys. Rev. A **46**, 4421 (1992)
30. P. J. Mohr: Y.-K. Kim, Phys. Rev. A **45**, 2727 (1992)
31. W. H. Furry: Phys. Rev. **81**, 115 (1951).
32. J. D. Bjorken, S. D. Drell: *Relativistic Quantum Mechanics* (McGraw-Hill, New York 1968)
33. C. Itzykson, J.-B. Zuber: *Quantum Field Theory* (McGraw-Hill, New York 1980)
34. S. Klarsfeld: Phys. Lett. B **66**, 86 (1977)
35. E. H. Wichmann, N. M. Kroll: Phys. Rev. **101**, 843 (1956)
36. G. Soff, P. Mohr: Phys. Rev. A **38**, 5066 (1988)
37. J. M. Schmidt, G. Soff, P. J. Mohr: Phys. Rev. A **40**, 2176 (1989)
38. G. Soff: Z. Phys. D **11**, 29 (1989)
39. W. E. Lamb Jr., R. C. Retherford: Phys. Rev. **72**, 241 (1947)
40. J. R. Sapirstein, D. R. Yennie: in *Quantum Electrodynamics*, T. Kinoshita, ed. (World Scientific, Singapore 1990), p. 560
41. W. R. Johnson, G. Soff: At. Data Nucl. Data Tables **33**, 405 (1985)
42. G. Plunien, B. Müller, W. Greiner, G. Soff: Phys. Rev. A **39**, 5428 (1989)
43. G. Plunien, B. Müller, W. Greiner, G. Soff: Phys. Rev. A **43**, 5853 (1991)
44. G. Plunien, G. Soff: Phys. Rev. A **51**, 1119 (1995); G. Plunien, G. Soff: Phys. Rev. A **53**, 4614 (1996)
45. A. V. Nefiodov, L. N. Labzowsky, G. Plunien, G. Soff: Phys. Lett. A **222**, 227 (1996)
46. B. Hofmann, G. Baur, J. Speth: Z. Phys. **315**, 57 (1984)
47. P. Ring and P. Schuck: *The Nuclear Many Body Problem*, (Springer, New York 1980)
48. L. N. Labzowsky, A. V. Nefiodov, G. Plunien, T. Beier, G. Soff: J. Phys. B **29**, 3841 (1996)
49. J. P. Briand, P. Chevallier, P. Indelicato, K. P. Ziock, D. D. Dietrich: Phys. Rev. Lett. **65**, 2761 (1990)
50. Th. Stöhlker, P.H. Mokler, K. Beckert, F. Bosch, H. Eickhoff, B. Franzke, M. Jung, T. Kandler, O. Kleppner, C. Kuzhuharov, R. Moshammer: F. Nolden, H. Reich, P. Rymuza, P. Spädtke, M. Steck: Phys. Rev. Lett. **71**, 2184 (1993)
51. H. F. Beyer: IEEE Trans. Instrum. Meas. **44**, 510 (1995)
52. H. F. Beyer, G. Menzel, D. Liesen, A. Gallus, F. Bosch, R. Deslattes, P. Indelicato, Th. Stöhlker, O. Klepper, R. Moshammer, F. Nolden, H. Eickhoff, B. Franzke, M. Steck: Z. Phys. D **35**, 169 (1995)
53. J. Schweppe, A. Belkacem, L. Blumenfield, N. Claytor, B. Feinberg, H. Gould, V.E. Koster, L. Levy, S. Misawa, J.R. Mowat, M.H. Prior: Phys. Rev. Lett. **66**, 1434 (1991); R.L. Mills, N.M. Kroll: Phys. Rev. **98**, 1489 (1955)

54. T. Beier, P. Mohr, H. Persson, G. Plunien, M. Greiner, G. Soff: Phys. Lett. A **236**, 329 (1997)
55. T. Beier, G. Plunien, M. Greiner, G. Soff: J. Phys. **B28**, 2761 (1997)
56. G. Plunien, T. Beier, G. Soff, H. Persson: Europhys. J. D **1**, 177 (1998)
57. S.M. Schneider, W. Greiner, G. Soff: J. Phys. B **26**, L529 (1993)
58. S. Mallampalli, J. Sapirstein: Phys. Rev. A **54**, 2714 (1996)
59. L. N. Labzowsky, A. O. Mitrushenkov: Phys. Lett. A **198**, 333 (1995)
60. A. O. Mitrushenkov, L. N. Labzowsky, I. Lindgren, H. Persson, S. Salomonson: Phys. Lett. A **200**, 51 (1995)
61. L. N. Labzowsky, A. O. Mitrushenkov: Phys. Rev. A **53**, 3029 (1996)
62. I. Lindgren, H. Persson, S. Salomonson, P. Sunnergren: Phys. Rev. A **58**, 1001 (1998)
63. S. Mallampalli, J. Sapirstein: Phys. Rev. A **57**, 1548 (1998)
64. P.J. Mohr: Nucl. Instrum. Methods. B **87** (1994) 232
65. A.N. Artemyev, V.M. Shabaev, V.A. Yerokhin: Phys. Rev. A **52**, 1884 (1995)
66. A.N. Artemyev, V.M. Shabaev, V.A. Yerokhin: J. Phys. B **28**, 5201 (1995)
67. V. M. Shabaev: Phys. Rev. A **57**, 59 (1998)
68. V.M. Shabaev, T. Beier, G. Plunien, V.A. Yerokhin, G. Soff: Phys. Ref. A **57**, 4235 (1998)

IV Electron Correlation Effects in Few-Electron Atoms

Paul Indelicato

Summary. The physics of the relativistic many-body problem in both simple and complex few-electron systems is reviewed. Its connection to QED and its limitations are discussed. Examples of application to both simple (2 and 3-electron ions) and complex (Be-like to Ne-like ions) systems are presented. The correlation effects on transition energies and transition rates, as well as their interplay with pure QED effects are described in detail. The case of neutral helium, for which theory lags well behind experiment, is also discussed, as an example of the tighter integration that must be realized between QED and many-body methods, to achieve reliable complete calculations.

1. Introduction

In addition to the study of Quantum Electrodynamics (QED) effects in one-electron heavy systems, there has been a great deal of attention paid to the relativistic many-body problem in heavy ions in the last 15 years. This interest originates from several directions. First, there is the need to get a fundamental understanding of the relativistic many-body problem, which is important for many applications, e.g., for the analysis of parity or time-reversal conservation violation in atoms. Second, several experiments have provided high-precision results for few-electron systems at high Z, which can actually provide very high precision test of strong-field QED, if one is able to disentangle one and many-body effects. The determination of one-electron QED contributions to the $2s$ Lamb shift deduced from energy measurements in Li-like ions could provide a more accurate test of strong field QED than the one deduced from the $1s$ Lamb shift of H-like ions, when all two- and three-body QED contributions are calculated.

It is impossible to separate completely electronic correlation from QED and relativistic effects, as these effects are more and more entangled, when going to higher orders of the theory. Indeed, there are correlation contributions to radiative corrections, and nonradiative QED corrections are really relativistic correlation contributions. This distinction is thus interesting only if it has an operational value. This is the case because there are methods that allow one to get high accuracy on correlation, when QED would be too difficult to use. It is particularly important to distinguish between correlation and QED contributions for light systems, for which correlation effects are much larger than QED effects. In helium a direct QED calculation would have a difficult time reproducing the experimental results because of the large size and slow convergence of high-order contributions to the correlation.

Because of the broad nature of phenomena in which electron correlations are important, this chapter will be limited to the aspects of electron correlation concerning *singly excited*, few-electron ions. Only correlation effects on energies and lifetimes will be described.

In this chapter, we first describe the fundamental aspects of the problem (Sect. 2). Then we devote Sect. 3 to the different methods used to solve the relativistic many-body problem. Relativistic Many-Body Perturbation Techniques (RMBPT) and derived all-order techniques are presented in Sect. 3.1 and variational methods like the Multi-Configuration Dirac–Fock (MCDF) are discussed in Sect. 3.2. Results concerning transition energies, with comparison with experimental data are presented in Sect. 4, while results concerning one-electron operators are presented in Sect. 5.

2. The Relativistic Many-Body Problem

2.1 The Atomic Many-Body Problem from the Point of View of Bound-State QED

Standard bound-state QED (BSQED) provides a complete description of any atomic system and should enable us to calculate its energy to any desired accuracy. It is described in detail in Chap. III and in Refs. [1, 2]. The unperturbed Dirac Hamiltonian h_0 contains the Coulomb field of the nucleus V_{N}

$$h_0 = c\boldsymbol{\alpha} \cdot \boldsymbol{p} + \beta mc^2 + V_{\mathrm{N}}(r) . \tag{1}$$

All effects of this Coulomb field are thus included to all orders in all the quantities involved in the calculation. Here $\boldsymbol{\alpha}$ and β are Dirac matrices.

BSQED is a perturbation theory. The perturbing Hamiltonian is the one-photon, electron–electron interaction, which leads to an expansion in α, the fine-structure constant. Each extra photon line, connected to two vertices, adds an extra power of α. Each contribution to a given order can be represented as a Feynman diagram.

Giving a general and operational definition of correlation is a challenge from the QED point of view. In the nonrelativistic limit, correlation is defined as the difference between the Hartree–Fock solution of the many-electron Hamiltonian and the exact solution. The Hartree–Fock solution is the best possible independent-particle approximation, and thus the most legitimate reference. In QED one could define correlation as the difference between a calculation which includes all nonradiative corrections (i.e., all corrections without self-energy or vacuum-polarization loops) with any number of electron–electron photon lines and a relativistic equivalent of the Hartree–Fock solution. Such a solution would be represented by the diagrams of Fig. 1.

Although complete and perfectly founded from a field-theoretic point of view, the formalism of BSQED is however mostly inapplicable except for very

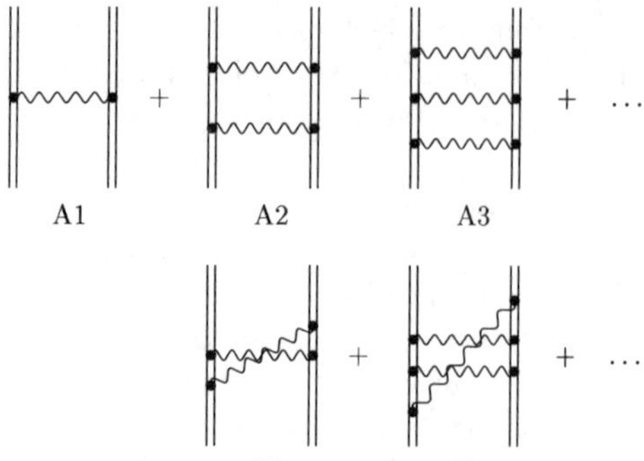

Fig. 1. Ladder (A1, A2, A3) and crossed ladder (B1, B2) approximation diagrams

simple heavy systems. This is because a diagram belonging to the ladder approximation (Fig. 1) with $n+1$ photon lines is only of order $1/Z$ compared to a diagram of the same structure with n photon lines. Thus unless Z is very large, the direct use of the QED formalism for evaluating energy to the high order required for light atoms or many-electron systems is impractical. Only a few attempts have been made to use QED directly, and this only for heavy two-electron ions [3].

For a long time, people have thus tried to use well-established methods from nonrelativistic atomic physics to tackle this problem. It can be proven that all-order nonrelativistic methods (like the Hylleraas method, the Configuration Interaction method (CI), pair-equation or the multiconfiguration Hartree–Fock method) amount to a re-summation of the Coulomb ladder approximation. However, naive approaches to this problem lead to difficulties when dealing with the relativistic problem.

2.2 Role of the Negative-Energy Continuum and the Relativistic Many-Electron Hamiltonian

One of the most fascinating consequences of special relativity is the possibility of particle creation due to the equivalence between mass and energy. Any attempt to establish a relativistic Schrödinger equation leads necessarily to a negative-energy continuum. The spectrum of the Dirac equation, which is suitable for the description of fermions, has two continua, one of energy $E > mc^2$ and one of energy $E < -mc^2$. The existence of the $E < -mc^2$ continuum led Dirac to postulate the existence of the positron (the anti-electron) before it was actually discovered. Because of that any relativistic quantum theory is inherently a theory with an infinite number of particles. To avoid the decay of a one-electron state to the negative energy continuum Dirac had to postulate that all these levels where filled with electrons, and that positrons would manifest themselves when a hole is created in this *Dirac sea* leading to the appearance of a real electron–positron pair. For a bound electron pair it is possible to find an infinite number of pairs of states, one in the upper continuum, and one in the lower continuum, the energy of which is degenerate with the bound electrons' total energy. Any electron–electron interaction will couple the bound pair with this continuum, leading to the disintegration of the atom (this was noticed as early as 1951 by *Brown* and *Ravenhall* [4]). For a long time it was wrongly thought that only the Breit interaction would lead to such a disintegration. The prescription was thus to use it only in first-order perturbation theory. In fact, only BSQED can avoid these difficulties, because it re-interprets the negative energy states as positive energy positrons (propagating backward in time). Yet since the use of BSQED is impractical in the many-electron problem, an Hamiltonian equation has been derived from BSQED [4, 5], in which consequences of the existence of the negative energy states are included correctly. This Hamiltonian is called the *no-pair* Hamiltonian and gives precise results as long as the energy of the electrons is small compared to the pair creation energy $2mc^2$. The no-pair Hamiltonian is written

$$\mathcal{H}^{\text{no pair}} = \sum_{i=1}^{m} h_0(r_i) + \sum_{i<j} \mathcal{V}\left(|\boldsymbol{r}_i - \boldsymbol{r}_j|\right), \tag{2}$$

where $h_0$is a one-electron Dirac operator (1) and $\mathcal{V}$ is an operator representing the two-body interaction. This operator has the expression

$$\mathcal{V}_{ij} = \Lambda_{ij}^{++} V_{ij} \Lambda_{ij}^{++}. \tag{3}$$

Here $\Lambda_{ij}^{++} = \Lambda_i^{+} \Lambda_j^{+}$ is an operator projecting onto the one-electron positive energy states. Any effects from virtual electron–positron pairs must be included afterwards, as it is not possible to include pair creation in a Hamiltonian formalism, in which the norm of the wave function, and thus the number of particle, is conserved. The principle and difficulties associated with the no-pair Hamiltonian have been the subject of much works in recent years [6–9]. With the appearance of the projection operators, the effect of *any* electron–electron interaction can safely be evaluated to all orders of perturbation theory. However all QED corrections that explicitly involve negative energy states cannot be accounted for in this manner and can be evaluated only with a QED calculation. Up to now this has only been done in the case of the ground state of He-like ions [10, 11] and for Li-like ions [12]. Solving exactly the Hamiltonian of (2) would sum all the no-pair contributions of diagrams A2, A3, ... (obtained by replacing the electron–electron interaction in all expressions by the projected interaction). Nonradiative QED corrections thus include both the contribution of these diagrams beyond the no-pair approximation and all diagrams with one or more photon crossings, like those of type B.

2.2.1 The electron–electron interaction. The next difficulty in the establishment of a many-electron relativistic formalism is the determination of the electron–electron interaction operator. In the relativistic problem each particle must have its own proper time. In contrast, the Hamiltonian formalism is a one time formalism. The interaction operator is deduced from the diagram A1 in Fig. 1. In the Coulomb gauge, the operator is

$$\begin{aligned} V_{ij} &= \frac{1}{r_{ij}} - \frac{\boldsymbol{\alpha}_i \cdot \boldsymbol{\alpha}_j}{r_{ij}} - \frac{\boldsymbol{\alpha}_i \cdot \boldsymbol{\alpha}_j}{r_{ij}} \left(\cos\left(\omega_{ij} r_{ij}\right) - 1\right) \\ &\quad + \left(\boldsymbol{\alpha}_i \cdot \boldsymbol{\nabla}_i\right)\left(\boldsymbol{\alpha}_j \cdot \boldsymbol{\nabla}_j\right) \frac{\cos\left(\omega_{ij} r_{ij}\right) - 1}{\omega_{ij}^2 r_{ij}}, \end{aligned} \tag{4}$$

where $r_{ij} = |\boldsymbol{r}_i - \boldsymbol{r}_j|$ is the inter-electronic distance, ω_{ij} is the energy of the photon exchanged between the electrons, and $\boldsymbol{\alpha}_i$ are Dirac matrices ($\boldsymbol{\nabla}$ operators act on the r_{ij} and not on the following wave functions). The remnants of the two-time nature of the interaction are explicit through the presence of the ω_{ij}. The first term in (4) is the Coulomb interaction, the second is the magnetic interaction that reduces to spin–spin, spin–other orbit and orbit–orbit interaction in the Pauli approximation. The rest of the operator is called the retardation term. The operator including all magnetic and retardation terms up to retardation of the second order in ω_{ij} is called the *Breit operator.* Retardation beyond Breit is called *higher-order retardation.* Much literature has been devoted to this operator, its effects and the problems that it poses [4, 9, 13–16].

Because of the dependence of V_{ij} on the ω_{ij}, the Hamiltonian (2) is not Hermitian. Also in a many-body context the ω_{ij} are difficult to define properly. Finally, the gauge dependence of the operator is a problem when dealing with approximate solutions. The gauge invariance is preserved as long as the individual electronic wave functions are solutions of local differential equations. When non-local potentials are used, as in the MCDF method, the gauge dependence is strong [14, 15]. In that case it has been shown that the Coulomb gauge gives the best answer: the Lorentz gauge introduces spurious terms that can be canceled only when adding all crossed ladder diagrams (Fig. 1B) [16].

3. Methods for the Relativistic Many-Body Problem

3.1 Many-Body Perturbation Theory

3.1.1 Perturbation techniques. These techniques, usually called Relativistic Many-Body Perturbation Technique (RMBPT), find their roots in the work of *Goldstone* [17] who applied Rayleigh–Schrödinger perturbation theory to many-body systems (for more details see for example [18]). They were first brought to nonrelativistic atomic physics in the pioneering work of *Kelly* [19]. Their extension to the relativistic problem was started by *Johnson* and collaborators [20–22], and by *Lindgren* and collaborators [23]. The difficulty in using perturbation theory for orders larger than one lies in the infinite sums over states to which it is associated. Nowadays, finite basis sets are used to solve this difficulty [20]. Other techniques that involve turning the sum into a differential equation will be considered in Sect. 3.1.2.

RMBPT is a theory very similar to BSQED. The main difference is that the reference state is no longer the vacuum, but an atomic state (usually a Hartree–Fock closed-core state) and that the perturbing potential is the difference between the *no-pair, one-photon* electron–electron interaction and the Hartree–Fock potential.

Modern approaches to RMBPT use basis sets. The atom is enclosed in a spherical box. A complete, finite set of solutions of the one-electron Dirac equation can thus be obtained, which possesses a pseudo-spectrum made of a few bound states, and of a discrete set of positive and negative energy states. All the sums of perturbation theory can thus be done by discrete finite summations. Two types of basis set have been used. The first one has developed by *Johnson* and collaborators and has been applied to a wide variety of heavy ions [20–22] (see [24, 25] for recent reviews). This method describes wave functions as linear combinations of basis (or B) splines. The B splines provide a complete set of nonorthogonal functions over a finite interval.

The other technique which has been used is discretization [26, 27]. Again the atom is enclosed in a box, and the wave function is described on a mesh. The differential operator is then transformed into a finite difference operator, and the Dirac equation then turns into an eigensystem. The eigenvalues of this system give the pseudo-spectrum, and the eigenvector the value of the wave functions on each point of the mesh. In both methods projection operators are easily implemented by restricting all sums to positive energy eigenstates.

3.1.2 All-order equations based on MBPT; pair-equation. All-order methods are employed to fix the difficulties of going to very high-order in perturbation theory that would be required to get very high accuracy in light or complex atoms or ions. Re-summation of large classes of diagrams are automatically done by these methods, which can lead to very high accuracy. There are basically two classes of methods: methods that use a *variational principle* (described in Sect. 3.2), and methods that use a differential equation built from RMBPT that is solved iteratively and is described here. These kinds of methods allow one to sum certain subclasses of diagrams to all orders. However it is not possible in this case to sum all classes to all orders. Only variational methods allow one to re-sum all no-pair contributions to the ladder approximation to all orders, at the price of more complexity.

The pair equation and coupled-cluster approaches were developed in the nonrelativistic regime during the 1970s (see, e.g., [28]). The relativistic formalism was developed first in the Breit–Pauli approximation [23] and then for the general case using numerical [29] or basis set methods [30–32]. It allows one to sum up all contributions from pair excitation, and is thus exact *only* for two-electron atoms. For many-electron atoms triple or quadruple excitations can be added to add new classes of diagrams to all orders.

In the simple case of two electrons the method is the following. The wave function is separated into two parts, an independent-particle wave function and a remainder

$$\Psi(1,2) = \{\psi_a(1)\psi_b(2) + \rho(1,2)\}\,, \tag{5}$$

where the curly brackets denote antisymmetrization and a and b are the quantum numbers of the state being studied. The remainder $\rho(1,2)$ is required to be orthogonal to $\psi_a(1)\psi_b(2)$. The Hamiltonian (2) is separated in the same way, including for example the Dirac–Fock potential in the one-electron Dirac equation (1).

The Hamiltonian equation (2) is then rewritten

$$\mathcal{H}\Psi(1,2) = (\epsilon_a + \epsilon_b + \Delta E)\{\psi_a(1)\psi_b(2) + \rho(1,2)\}\,, \tag{6}$$

with

$$\Delta E = \langle\{\psi_a(1)\psi_b(2)\}|\,\mathcal{V}(1,2)\,|\{\psi_a(1)\psi_b(2) + \rho(1,2)\}\rangle\,. \tag{7}$$

This finally leads to a differential equation for $\rho(1,2)$. This equation can then be solved iteratively, replacing the terms containing $\mathcal{V}(1,2)$ in the right-hand side of (6) by 0 to get a first approximation of $\rho(1,2)$, which is then re-injected in the RHS of (6) to provide a better approximation. The procedure is then repeated until consistency is reached.

3.1.3 $1/Z$ expansion of correlation energy. The dependence of the ladder approximation in $1/Z$ as described in Sect. 2.1 was recognized early without reference to its QED origin. This is due to the scale invariance of the Shrödinger equation under a variable change $\rho = r/Z$. The electron–electron interaction takes the form $1/(\rho Z)$ and is the only place where Z appears. One can then do a perturbation expansion, using zeroth-order wave functions made of products of hydrogen-like orbitals. This method has been used, e.g., by *Safronova* and collaborators in calculations of few-electron heavy-ion energies [33, 34]. In the nonrelativistic domain the total energy behaves as

$$E_{\text{Tot}} = E_0 Z^2 + E_1 Z + E_2 + E_3\left(\frac{1}{Z}\right) + E_4\left(\frac{1}{Z}\right)^2 + \cdots\,. \tag{8}$$

The first coefficient E_0 is just the sum of the hydrogen energy over all involved orbitals. It can be shown that the Hartree–Fock approximation reproduces exactly E_0 and E_1, and contains pieces of E_2 and E_3. E_n correspond to the contribution of an n-photon ladder-approximation diagram. The nonrelativistic correlation energy is written

$$E^{\mathrm{c}} = E_0^{\mathrm{c}} + E_1^{\mathrm{c}}\left(\frac{1}{Z}\right) + \cdots\,. \tag{9}$$

In the relativistic domain this formalism provides

$$E_{\text{Tot}} = \sum_{\substack{i=1\\j=0}}^{\infty} E_{i,j} \left(Z\alpha\right)^{2i} \left(\frac{1}{Z}\right)^{j} . \tag{10}$$

Here the sum of all terms with $j = 0$ gives back the sum of all hydrogen Dirac energies over all involved orbitals.

3.2 Variational All-Orders Methods

Variational methods are all-orders methods in which the total wave function of the atom is represented by a trial function, which is built to cover most of the functional space of the solution of the Hamiltonian. Depending upon the form of the trial functions, variational methods can be divided into three groups. In the first group, the trial function is build out of linear combination of Slater determinants, and depends only on single electron coordinates $|\boldsymbol{r}_i|$ but not on relative coordinates $r_{ij} = |\boldsymbol{r}_i - \boldsymbol{r}_j|$. One then minimize the total energy. If the only variational parameters are the mixing coefficients, the method is called Relativistic Configuration Interaction (RCI). If both the one-electron wave functions (called one-electron orbitals) and the mixing coefficient are optimized, the method is called Multiconfiguration Dirac–Fock. If the trial function explicitly involves r_{ij} or other types of coordinates (hyperspheric ...), the techniques is often referred to as the Hylleraas technique. As it cannot be directly applied to the relativistic problem it will not be described here.

3.2.1 The multiconfiguration Dirac–Fock and the relativistic interaction of configuration methods. In the MCDF and RCI methods the total wave function is calculated with the help for the variational principle. The total energy of the ion is written as [7, 8, 35]

$$\mathcal{H}^{\text{no pair}} \Psi_{\Pi,J,M}\left(r_1, \ldots, r_m\right) = E_{\Pi,J,M} \Psi_{\Pi,J,M}\left(r_1, \ldots, r_m\right), \tag{11}$$

where Π is the parity, J is the total angular momentum eigenvalue, and M is the eigenvalue of its projection on the z axis. The MCDF method is defined by the particular choice of a trial function to solve (11) as a linear combination of configuration state functions (CSF):

$$| \Psi_{\Pi,J,M}\rangle = \sum_{\nu=1}^{n} c_\nu \mid \nu \Pi J M\rangle . \tag{12}$$

The CSF are eigenfunctions of the parity Π, the total angular momentum $\boldsymbol{J}^2$ and its projection J_z. The CSF are antisymmetric products of one-electron wave functions expressed as a linear combination of Slater determinants of Dirac 4-spinors.

$$| \nu \Pi J M\rangle = \sum_{i=1}^{N_\nu} d_i \begin{vmatrix} \psi_1^i(r_1) & \cdots & \psi_m^i(r_1) \\ \vdots & \ddots & \vdots \\ \psi_1^i(r_m) & \cdots & \psi_m^i(r_m) \end{vmatrix} . \tag{13}$$

The d_i are obtained by requiring that the CSF are eigenstates of $\boldsymbol{J}^2$ and J_z. In the RCI method one-electron orbitals are calculated in a model potential and used to build the Hamiltonian matrix that provides the mixing coefficients c_ν and level energy. The product of c_ν^2 and the number of electrons in an orbitals is called the effective occupation. Orbitals that have small effective occupation numbers are called correlation orbitals.

In the MCDF method the variational principle also provides the integro-differential equations to determine the radial wave functions. The MCDF method thus starts by a RCI calculation which provides starting wave functions, potential and mixing coefficients. The direct and exchange potential that appears in the integro-differential equations, which represents the interaction with the other electrons, are then evaluated and new wave functions evaluated. The process is repeated until all potentials and wave functions are stationary. This iterative process is called the self-consistent field (SCF) process. The Hamiltonian matrix is regularly diagonalized during the process. The MCDF method can reach high accuracy with limited numbers of CSF because the wave functions are optimized. On the other hand the SCF process is highly non-linear, and the exchange potential can be very large for highly excited correlation orbitals. In that case convergence of the iterative process can be very difficult or impossible [7, 36, 37]. RCI methods needs huge numbers of CSF to be accurate, and are thus limited to systems with few electrons. For a large number of electrons, a combination of MCDF for the core orbitals and intra-shell correlation and of RCI for core-valence and valence correlation can be very efficient (see, e.g., [38]).

The RCI method has been mostly developed on finite basis sets [39, 40]. The projection operators required to implement the no-pair Hamiltonian are obtained by restricting the basis set to positive energy eigenstates. In the MCDF method in contrast, which uses wave functions obtained by numerical integration of differential equations, the continua do not appear explicitly. Projection operators have thus been implemented only recently [7]. The severe numerical trouble encountered in early MCDF calculations, particularly when the magnetic interaction was included in the SCF process, were often spurious effects due to the lack of projection operators [7]. Even when the calculation converges, the energy cannot be correct. However such an incorrect procedure is still the most widely used. This is because in practice the deviation from correct energy is bound to be small for the convergence to occur. The MCDF method is actually well protected from pair creations which cannot occur in first order [6] and by the fact as analyzed by Mittleman [41] that the projection operators are stationary around the Dirac–Fock potential. It should also be noted that MCDF code using analytic basis functions (as, e.g., [42]) are immune to this problem because again positive and negative energy states are easily identified.

4. Transition Energies and Comparison with Experiment

4.1 Theoretical Evaluation of Transition Energies

In order to make a significant comparison with experiment, theoretical transition energies should contain all necessary contributions to a consistent accuracy. However, it is not yet possible to calculate transition energies to a given order in α. Actually it is not required to achieve a consistent level in α as some contributions belonging to the nonrelativistic ladder approximation can be much larger at low Z than radiative corrections with a lower formal order. This is because all radiative corrections diagrams (i.e., with at least one closed photon or electron loop) with a given number of electron–electron interactions are multiplied by at least a factor of $(Z\alpha)^4$ compared to a ladder approximation diagram with the same number of electron–electron interactions.

On the other hand the magnitude of the ladder-diagram contribution decreases only as $1/Z$, which is very slow at low Z. The techniques used at low Z must then include correlations to all orders to yield high accuracy, while at high Z QED corrections to order α^2 must be taken into account. Another important parameter to take into account for heavy ions is the nuclear size. All the energies must be calculated using experimental nuclear charge distribution for the nucleus when available, as well as nuclear shapes for $Z \geq 90$ as was shown in the study of $K\alpha$ transitions in neutral atoms ionized in inner shells [43] and in lithium-like uranium [44]. For low Z the strategy as used by *Drake* in his *unified method* is to calculate highly accurate nonrelativistic correlations using Hylleraas type coordinates and then to evaluate relativistic and QED corrections perturbatively [45, 46].

In all methods, exact one-electron radiative corrections are added afterwards (see Chap. VII). Calculation of the two-electron self-energy, which occurs at the α^2 level, is still in its infancy. It was only in 1991 that a method was devised to do such a calculation, in an approximation in which the screening electron is represented as a perturbing potential [47]. A limited number of calculations have been performed in the same spirit in which the self-energy was calculated using a Hartree–Fock potential in the zeroth-order Hamiltonian (1) used in the BSQED [40, 48–52]. Complete calculations with exchange have been performed only for the ground state of He-like ions [53, 54] and for Li-like uranium [52]. For the screened vacuum polarization, most of the effect can be obtained by using the Uehling potential perturbatively with any of the many-body methods described in the previous sections. However, a few specific exact calculations have been performed for two- and three-electron ions [52, 53, 55].

For most complex ions the only screening correction has been obtained by either an effective nuclear charge deduced from a comparison of the single-orbital wave-function mean radius to the radius of a hydrogen-like orbital of the same quantum numbers. Because self-energy is mostly sensitive to what happens in a region of one Compton wavelength around the nucleus, this is not satisfactory. A better method, which works very well despite its simplicity is to use the Welton model of the self-energy to calculate an electronic-density correction near the origin [37, 56, 57].

4.2 Neutral Helium

Helium has obviously been the subject of many theoretical studies and experiments. Since the advent of laser spectroscopy, many accurate measurements have been performed to test correlation and radiative corrections. Because Z is only 2 for helium, correlation plays a central role, even at the level of radiative corrections. Hylleraas-type methods are the best ones, although all-order RMBPT (which can provide 7 significant figures [31]) and even MCDF [58] have been used. A complete summary of methods and results can be found in [46]. Because of the high precision of the measurements and of the small mass of the nucleus, nuclear recoil effects have to be included precisely, taking into account correlation effects. The default of the present calculations is that because operators are expanded in powers of $1/c$ it is very difficult to keep track of all the contributions, and it is difficult to systematically improve the results. Present experimental accuracy for transitions between excited states [59, 60] is sometimes two orders of magnitude more precise than state-of-the-art theory. Combining several experiments and assuming theoretical values for highly excited states, it is possible to make a very detailed comparison between theory and experiment. The results of such a comparison are shown in Table 1.

Such a detailed comparison shows that our understanding of the correlation problem is at least adequate to the level of the accuracy of the radiative corrections. Most of the time the experimental accuracy is much better than the theoretical one. It is worth remembering that the present limitation to the accuracy of the evaluation of radiative corrections is mostly due to correlation effects. For example the agreement between experiment and theory for the 2^1S_0 in Table 1 is only due to the evaluation of the two-electron Bethe logarithm to all orders in $1/Z$ [64]. Using only the first few terms in $1/Z$ as evaluated in Ref. [65, 66] gives a 93 MHz disagreement. The two-electron Bethe logarithm is the main, nonrelativistic contribution to the self energy correction to the electron–electron interaction. It is really a self-energy correction to all ladder diagrams of Fig. 1.

4.3 Highly Charged Ions

With the use of Tokamaks, laser produced plasmas, heavy-ion accelerators and EBIT (Electron-Beam Ion Traps), a sizable corpus of measurement of transition energies in few-electron ions has become available. Two and three-electron ions have now been studied up to unranium. Many other isoelectronic

Table 1. Theoretical and Experimental Lamb shift in helium (MHz). The experimental values are obtained after subtraction of nonrelativistic energy, first relativistic corrections and finite nuclear size corrections [46]. For the theoretical uncertainty the revised, somewhat larger values from Ref. [46] are used. Note that there is a misprint in Ref. [46], Table 11.7 on $n = 4$, S states (*G. Drake*, private communication (1998))

Level	Ref.	Experiment	Theory [46]
$3\,^3P_1$	[60]	−357.3±1.5	−359.7±2.4
$3\,^3P_2$	[60]	−356.6±1.5	−359.7±2.4
$3\,^3P_0$	[60]	−359.51±0.20	−359.7±2.4
$3\,^1P_1$	[61]	15.6±1.3	14.2±2.4
$4\,^3D_1$	[60]	−10.1±1.8	−7.89±0.01
$5\,^3D_1$	[60]	−5.4±2.4	−4.224±0.006
$2\,^3P$	[60]	−1253.9±0.1	−1259.5±8
$2\,^3S_1$	[60]	4057.28±0.06	4062.3±8.
$2\,^1S_0$	[61]	2809.92±0.15	2808.48±8
$2\,^1S_0$	[62]	2810.57±0.21	2808.48±8
$1\,^1S_0$	[63]	41224.±45	41233.±180
$4\,^3S_1$	[60]	−398.7±2.4	394.1±16
$5\,^3S_1$	[60]	196.7±1.8	187.3±8
$6\,^3S_1$	[60]	122.7±27.0	101.5±4

sequences have been studied in detail. From the point of view of theory one can divide these ions into several categories. The easiest one to study is the one with one electron outside a closed core. Li-like ions are the prototype of this category, but Na-like or Cu-like ions have also been studied [67, 52]. The second category deals with excited states of closed core systems. In that second category we find He-like or Ne-like ions, but also Be-like or Mg-like ions. There are, finally, complex ions like Ti-like ions, with a $3d^4$ shell structure (only titanium is $3d^2 4s^2$), but little is known still about methods to evaluate the correlation with high accuracy in those cases [68–70].

4.3.1 Two-electron ions. As in the case of neutral helium, this is the best test system for relativistic many-body theory. If nonrelativistic correlation can be best tested in He, relativistic correlation is better studied on the ground state of He-like heavy ions. It is on this system that the strong influence of the magnetic interaction on the correlation was first noticed by *Desclaux* [71]. Detailed studies of correlation with the MCDF method then confirmed this influence [7, 36, 72] and showed that correlation due to the magnetic and retardation part of the Breit interaction become much stronger than the Coulomb contribution at high Z [7]. The correlation energy for the ground state of two-electron ions is shown in Fig. 2, with different approximations. While the nonrelativistic Coulomb correlation is around −1.2 eV and varies very slowly at high Z, relativistic corrections have a strong Z-dependence. It is an important result of MCDF [7], confirmed by direct QED

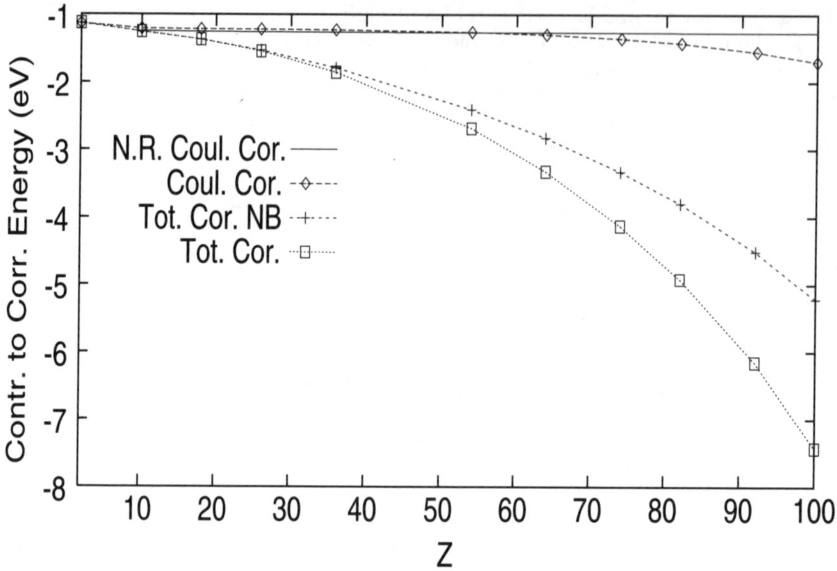

Fig. 2. Correlation energy of the ground state of two-electron ions. N.R. Coul. Cor.: Exact nonrelativistic contribution. Coul. Corr.: MCDF Coulomb correlation. Tot. Cor. NB: Total Breit (Coul.+ Mag. + Ret.) correlation, magnetic interaction in perturbation. Tot. Cor.: Total Breit correlation, magnetic interaction self-consistent [7]

calculation [10, 11], RMBPT calculations [32], and RCI calculations [39, 50], that the magnetic interaction plays a central role. Figure 2 shows that in the MCDF method the magnetic interaction must be evaluated self-consistently. It is not enough to include the magnetic interaction in the mixing coefficient, but it must also be included in the wave function differential equation. At $Z = 92$ magnetic and retardation correlations are of the same order of magnitude as α^2 one and two-body QED corrections (Fig. 3). Recently several measurements of the two-body energy of heavy He-like ions were made at the Livermore Super-EBIT [73]. A comparison of these experimental results and theory is shown in Table 2. This measurement provides a direct experimental value for the two-body contribution to the ground state, but they are not yet precise enough to show the interplay of correlation and QED. Many transition energies in two-electron ions have been measured that provide more precise values for low to medium Z. Both $\Delta n = 1$ $(2p - 1s)$ and $\Delta n = 0$ $(2p - 2s)$ transitions have been measured over a wide range of Z. The most studied $\Delta n = 1$ transition is the diagram line $2\,^1P_1 \to 1\,^1S_0$ transition. This is a transition very sensitive to ground state correlation and QED corrections. Observed–calculated energy differences for this transition are presented in Fig. 4. The overall agreement is good, although a few experimental points lay several standard deviations away from both theoretical calcula-

Table 2. Two body effects for the ground-state of helium-like ions

Z	32	54	66	74	83	92
MCDF+Welton [37]	562.1	1028.3	1336.7	1574.0	1881.3	2245.2
RCI [39]	561.6	1027.1				
RMBPT [74]	562.0	1028.5	1337.0	1575.5		2245.9
Drake [45]	562.1	1028.8	1338.2	1576.6	1886.3	
RMBPT [53]	562.0	1028.2	1336.6	1573.9	1881.5	2246.0
Experiment [73]	562.5	1027.2	1341.6	1568.0	1876.0	
Accuracy	± 1.6	± 3.5	± 4.3	± 15.0	± 14.0	

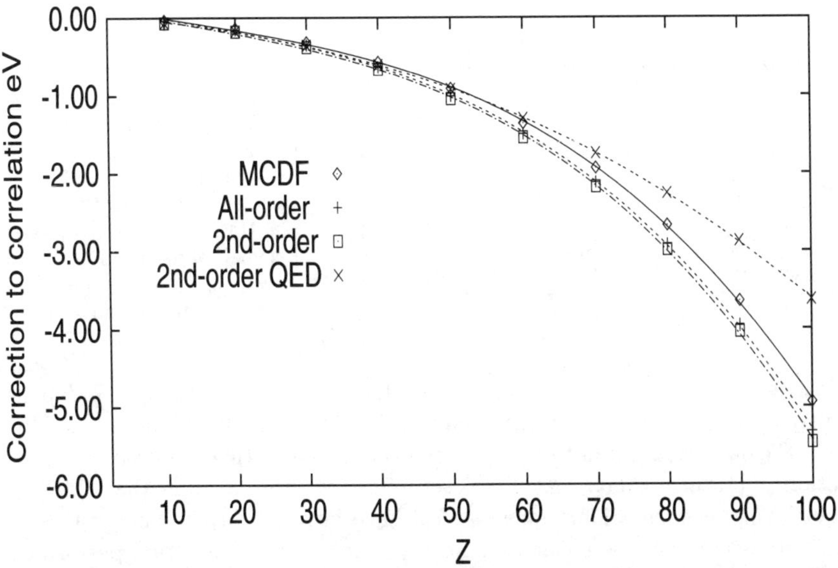

Fig. 3. Comparison of relativistic and QED contributions to correlation energy. MCDF : MCDF with magnetic and retardation correlation [7]. All-order : All-order RMBPT [74]. 2nd order : 2nd order RMBPT [10]. 2nd-order QED : exact calculation of the sum of the contributions of diagrams A2 and B1 of Fig. 1 [10]

tions. Both the MCDF calculation (which uses the Welton approximation for two-electron QED) and the RMBPT calculation (which uses Drake's two-electron QED) are in good agreement. These results, which are dominated by ground-state correlation and QED effects, thus confirm the conclusions from Table 2. Again more precise measurements at high Z would be needed to get more detailed information.

4.3.2 Three-electron ions. Li-like ions are very good candidates for accurate tests of relativistic correlation and QED effects. They are simple enough that very large numbers of configurations can be included for the MCDF or RCI methods; and because they are composed of one electron outside of a very simple core, they are also very easy to study by RMBPT. The first

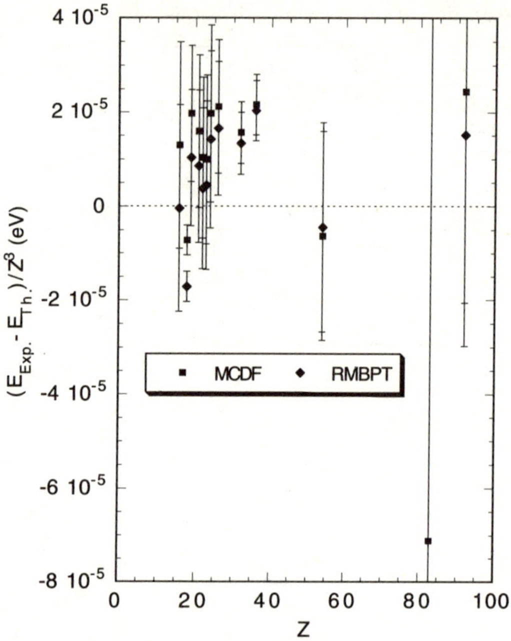

Fig. 4. Comparison between theory and experiment for the $2\,^1P_1 \rightarrow 1\,^1S_0$ transition in helium-like ions. MCDF: Refs. [37, 56, 7], *Plante* et al.; RMBPT: Ref. [74]. Experimental data can be found in Ref. [74]. Error bars are from experiment only

relativistic calculation of the three lower levels of Li-like ions were independently performed using both the MCDF [57] and RMBPT [22] methods. This MCDF calculation did not include higher-order magnetic correlation, or projection operators, which makes it less precise at large Z than the RMBPT result. Very precise calculations [44, 75] have been performed since then for uranium because of the measurement of the $2p_{1/2} \rightarrow 2s$ transition performed at Berkeley in 1991 [76]. More recently the Super-EBIT at Livermore was used to produce spectra of 3–10 electron $2p_{3/2} \rightarrow 2s$ transitions with $Z = 83$ [77], 90 [78] and 92 [79], which have been measured with high accuracy. These transitions in Li-like ions have been the object of MCDF [80], RMBPT [75, 51] and RCI [40] calculations (although the last two methods have been used only for U), and are thus an excellent test case. Here we focus on the lithium-like ions and the other ones will be considered in Sect. 4.3.3. Table 3 presents the most precise calculation for the $2p_{1/2} \rightarrow 2s$ transition of Li-like uranium. It gives a good idea of the current status of such calculations. Less detailed calculations (in particular without the exchange correction to two-electron QED corrections) by three different methods are presented in Table 4. For $2p_{1/2} \rightarrow 2s$ transitions the size of the sum of all uncalculated terms can be as small as a few 0.01 eV out of 280 eV. It can be as small as a few 0.1 eV on $2p_{3/2} \rightarrow 2s$ transitions, out of 2 to 4 keV.

4.3.3 Complex few-electron ions. The situation for ions with more than three electrons is very different. The MCDF method has great difficulties

Table 3. Contributions to the $2p_{1/2} \to 2s$ transition energies for Li-like uranium (eV)

Contribution	Value	Ref.	Other res.	
Dirac–Fock	323.00			
Correlation	−0.59	RMBPT [44]	−0.67	RMBPT [81]
Nonrad. QED	0.02	[12]		
Nuclear Recoil	−0.08	[44]		
Nuclear Polarization	0.03	[82]		
Radiat. QED	−41.66	[12, 44]		
Total	280.72		280.54	[12]
Theoretical Uncertainty	0.15		0.15	
Experiment	280.59	[76]	280.59	
Exp. Precision	0.09		0.09	
Difference	−0.13		0.05	

in treating ions with more than 8–10 electrons, if very precise results are needed. It is very inefficient, in particular, when excitations from the core are needed (see, e.g., [80]). On the other hand it is very efficient to get intra-shell correlations like in the ground state of Be-like ions (strong interaction between $2s^2$ and $2p^2$). The RCI method may work well also if not too many electrons need to be excited from the core. An efficient method may combine MCDF for strongly interacting electrons and RCI to get the rest of the correlation. *Froese Fischer* and collaborators [83, 38, 84] have developed a very elaborate strategy in that direction. RMBPT on the other hand has been proven to be very efficient for calculations involving alkali ions or more generally for cases involving a single valence electron outside of a closed core [52, 67, 85]. It also works well in the case of excited states of noble gases configurations [86, 87]. A very illustrative example is again found in the 4–10-electron ions with a hole in the $2s$ shell. Experiment–theory results for the $2p_{3/2} \to 2s$ are shown in Table 4. The large scale RCI calculations are the more efficient for 3–9-electron ions, but cannot work with 10-electron ions, because of the size of the calculation involved. RMBPT has great difficulties with, or cannot do, complex outer-shell structure like Be-like or B-like, but is very efficient for Li-like (one electron) or F-like (one hole).

5. Correlation Effects on Lifetimes

Correlation also plays a large or sometimes crucial role in quantities other than energy. For example, it is well known that all phenomena that involve simultaneous changes of state for two electrons are completely forbidden if correlation is not taken into account. In this section we describe some examples of particular significance, because relativity and correlation are particularly involved. The interest of such phenomena from a fundamental point of

Table 4. Energy differences between measurement and calculation for 3- to 10-electron ions. Experimental values for Bi are from Ref. [77], for Th from Ref. [78] and for U from Ref. [79]. Key: As more than one transition has been identified for some ions, the line is represented by the ion isoelectronic sequence symbol and a number, following the original paper. Unc.: experimental uncertainty. DF: MCDF values [80]. MB A and MB B: RMBPT calculations [51] using two different potentials to generate zeroth-order wave function. MB2: RMBPT values [80]. RCI: RCI calculation [40]

Key	DF	Unc.	DF	MB A	MB B	MB 2	RCI	Unc.
	Bi							
Li	0.326	0.018						
	Th		U					
Li	0.62	0.15	0.63	−0.12	0.24		0.06	0.35
Be	−0.25	0.13	−0.18	−1.16	−2.56		0.19	0.27
B1	1.00	0.50	0.70				0.09	0.60
B2	0.84	0.50	0.98				−0.14	0.60
C	1.64	0.13	2.24	−0.07	0.37		0.02	0.20
N1	−0.38	0.14	−0.25				0.08	0.20
N2	0.57	0.28	0.39				−0.46	0.58
N3	0.85	0.23	0.85				0.09	0.26
O1	1.17	0.20	0.84				0.39	0.25
O2	2.17	0.17	2.07				−0.18	0.48
O3	0.99	0.14	1.36				−0.02	0.21
F	0.99	0.16	1.00	−0.48	−0.19	−0.55	0.16	0.17
Ne	2.90	0.45	4.14	−1.92	−1.77			0.33

view lies in the fact that they enable us to test properties of the wave function other than energy. In variational methods in particular, the energy is made stationary with respect to wave function changes, and thus is less sensitive to it than lifetimes or hyperfine structure. Also for one-electron operators in first order, it make sense to include the negative energy continuum in the calculation, and then the negative energy continuum plays a very different role than in energy calculations.

Radiative lifetimes offer some of the most interesting examples of the effect of correlation. In this section, after considering some special aspects of the evaluation of one-body operators with the MCDF method, we will review some of the most striking cases of relativistic correlation effects in multiply-charged ions. More details of RMBPT evaluations of lifetimes can be found in, e.g., [88].

There are some special problems involved when computing mean values of nondiagonal matrix elements of one-electron operators with the MCDF method (nondiagonal means between two different states). The full potential of the MCDF methods for correlation can only be achieved if the wave function of each state involved in the calculation is optimized separately (Optimized Levels, or OL method). The OL method has however an important drawback: the spin orbitals of identical symmetry in the initial and final state are not orthogonal. One has then to develop a formalism to take care of the non-orthogonality. *Froese-Fischer* [89] and collaborators have developed and implemented such a formalism, based on bi-orthogonal transformations, which works for large classes of configurations. Other implementations use Racah algebra [90] or direct evaluation with determinants [91–93], following the formalism of *Löwdin* [94, 95], and are completely general.

In this section we will consider three examples. The first one will be the so-called *relativistic* M1 transition in He-like ions. The second will be the spin forbidden transitions in light Be-like ions. The last one will be a special case of two-electron one-photon transition in Be-like ions. In all three of them correlation and relativity play a fundamental role.

5.1 The Relativistic $1s2s\,^3S_1 \rightarrow 1s^2\,^1S_0$ M1 Transition in Two-Electron Ions. Role of the Negative Energy Continuum

The $1s2s\,^3S_1 \rightarrow 1s^2\,^1S_0$ transition is called the *relativistic* M1 transition because its probability is exactly zero in the nonrelativistic regime. This happens because the nonrelativistic transition operator acts only on spatial coordinates, and because the initial and final states are orthogonal spin eigenstates. This transition has been intensely studied and measurements have been made for $2 \leq Z \leq 54$. In particular very precise measurements were performed for carbon [96] and for Nb ($Z = 41$) [97]. The latter measurement is in clear disagreement with Dirac–Fock theory (i.e., without correlation).

Several calculations have been performed for this transition. The most precise *ab initio* relativistic calculations were done by RMBPT, which was used for Ar [98], and MCDF [92]. Here we will focus on the RMBPT and MCDF calculations. The MCDF calculation is performed with several kinds of OL wave functions, and uses exact expressions for non-orthogonal states. The wave function was made self-consistent with or without projection operators in the interaction (4), and with or without the magnetic interaction.

A comparison between calculations and experiment is presented in Fig. 5. The effect of correlation on the radial part of the matrix element ranges from 47% at $Z = 2$ to 0.5% at $Z = 92$. As can be seen in Fig. 5 the $0.96 \pm 0.35\%$ disagreement between the Dirac–Fock calculation and experiment for Nb becomes 0.1% when correlation to the matrix element is included using MCDF wave functions evaluated with projection operators. For $Z = 6$ the difference decreases from $15 \pm 0.2\%$ to 7.8% but is not canceled. In the same way, at $Z = 18$, the difference between the RMBPT value of [98] and the MCDF value decreases from $4.1 \pm 0.2\%$ to 2.3%. Most of the remaining difference is recovered when removing the projection operators from the wave function used to evaluate the $M1$ matrix element (obviously the energies must still be computed with projection operators). For the same elements, the latter calculation gives $0.64 \pm 0.35\%$ (Nb), $2.4 \pm 0.2\%$ (C) and $0.3 \pm 0.2\%$ (Ar) respectively, which significantly reduces the difference between calculation and the most precise experiments to date ($Z = 6$, 10, 41, 47). At that stage it should

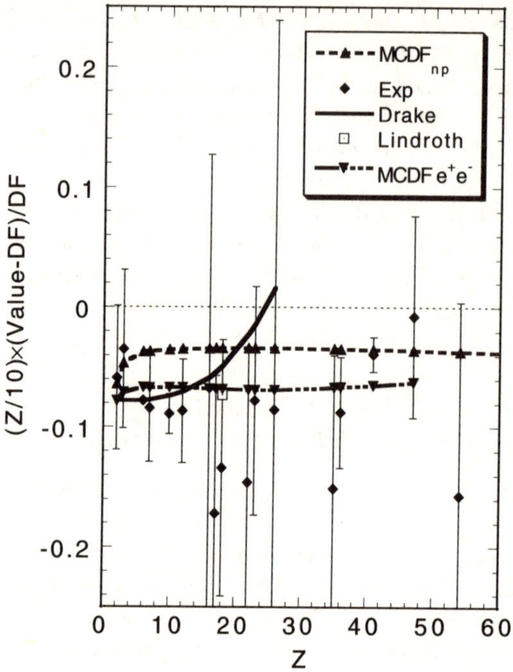

Fig. 5. Effect of correlation on the relativistic M1 transition in He-like ions. All rates are normalized to the DF value. MCDF$_{\rm np}$: Highly correlated Dirac–Fock calculation using no-pair Hamiltonian [92]. MCDF e^+e^-: same without the projection operator [92]. References to other calculations values are given in [92]

be noted that if the magnetic part of the interaction (4) is not included in the evaluation of the wave function, the transition rate changes by almost 5% at Z=6, thus destroying the agreement between theory and experiment. This has, however, little effect for large Z.

Here we thus have a very peculiar case in which magnetic interaction and negative continuum play a very strong role at low Z while having a negligible effect compared to correlation at high Z (the relative contribution of correlation is much larger at low Z). The role of the negative continuum is also very dependent on the method. In the RMBPT calculation the sum of all contributions containing interactions between positive and negative energy states cancels out to a very high accuracy. This shows that the role of the negative energy continuum in this case depends very strongly on the nature of the method used.

5.2 Correlation Effects on Spin-Forbidden Transitions

Spin-forbidden transitions at low Z like the $1s^22s2p\,^3P_1 \rightarrow 1s^22s^2\,^1S_0$ intercombination line in Be-like ions when obtained from a relativistic calculation has almost zero probability because the total radiative matrix element is obtained as the difference of two large, nearly equal, matrix elements (one between $2s$ and $2p_{1/2}$, and the other between $2s$ and $2p_{3/2}$), which add up when computing the $1s^22s2p\,^1P_1 \rightarrow 1s^22s^2\,^1S_0$ diagram line. Such a small difference, which is several orders of magnitude smaller than each matrix element taken separately, is easily perturbed, and depends strongly on correlation.

Recently precise lifetime measurements of $1s^2 2s2p\,^3P_1 \rightarrow 1s^2 2s^2\,^1S_0$ in Be-like carbon were performed [99, 100] using respectively RF-traps and storage ring. These two measurements strongly disagree (see Table 5). The status of the theory is not much better. A summary of recent theoretical and experimental results is presented in Table 5. There is a lot of effort to get an accurate value for this transition rate as it is of great astrophysical importance.

Table 5. Comparison between theory and experiment for the $1s^2 2s2p\,^3P_1 \rightarrow 1s^2 2s^2\,^1S_0$ transition rate in Be-like C^{++}

Method	Ref.	Decay rate (s^{-1})
Theory:		
CI	[101]	104.±4.
MCHF	[102]	103.±3.
MCRRPA	[103]	118
MCDF/CI	[38]	100.3±4.0
Z expansion	[104]	120
MCDF/OL/NO	[105]	102.1±1.
Experiment:		
RF ion trap	[99]	121.±7.
Ion storage ring	[100]	102.94±0.14

The CI calculation of Table 5 is done in the Breit–Pauli approximation, and includes core-polarization by two different methods. The MCHF procedure was done following the method of Ref. [90] to account for non-orthogonality in a partial way. The MCDF/CI uses GRASP-2 [106] to build a large set of configurations. Orbitals with small principal quantum number are made fully self-consistent, and more excited orbitals are treated in a CI model. The change of the rate with successive additions of configurations is quite erratic, and the final value may not be very reliable. There is finally an accurate MCDF/OL/NO calculation, in which full account of non-orthogonality is done, as well as a complete, independent optimization of initial and final state. The magnetic interaction is made fully self-consistent and orbitals with $n \leq 6$ are included. During this calculation it was found that using a relativistic method introduced severe problems. Such problems have been identified a long time ago for fine structure [107], but never for transition rates. They come from the fact that $j = l - 1/2$ and $j = l + 1/2$ orbitals are optimized separately, while in the nonrelativistic limit their radial part should be identical. The $c \rightarrow \infty$ limit of the rate should be exactly zero. That means that all mixing coefficients for singlet states in triplet wave functions must cancel. Yet when adding some configurations like $1s^2 3d4f$ to the $1s^2 2s2p$ and $1s^2 2p3d$ there is a residual mixing. It was found that in fact this residual mixing disappears if one introduces non-orthogonal orbitals inside a given state (see, e.g., [108]). Thus here the set of configurations will include the $1s^2 2s2p$ and $1s^2 2p'3d$ and $1s^2 3d'4f$. In this case the nonrelativistic offset vanishes [105] and one gets an accurate result.

5.3 Two-Electron One-Photon Transitions

Two-electron one-photon rates are exactly 0 if there is no correlation and if initial and final state orbitals are orthogonal. If an OL calculation is done, and non-orthogonality accounted for, the transition probability becomes non-zero (shake-down effect). However only a full account of correlation (which implies full account of non-orthogonality) can lead to a meaningful result. Here we describe relativistic and correlation effects in the decay of the $1s^2 2s3p\,^3P_0$ state for very high-Z ions.

In order to investigate the lifetime of the $1s^2 2s3p\,^3P_0$ state one needs to know the structure of the lower levels of a highly charged Be-like ion. The first group of levels consists of 6 levels of even parity $1s^2 2s^2$ $J = 0$, $1s^2 2p^2$ $J = 0$, $J = 1$ and $J = 2$, and of 4 levels of odd parity $1s^2 2s2p$ $J = 0$, $J = 1$, $J = 2$. The second group of levels consists of all the $1s^2 2s3s$, $1s^2 2s3p$, $1s^2 2s3d$ $1s^2 2p3s$ and $1s^2 2p3d$. The $1s^2 2s3p\,^3P_0$ level can decay to only one level of the second group, the $1s^2 2s3s\,^3S_1$ level, by an $E1$ transition. Because of the small transition energy the transition probability is small. There are several possible transitions to the $6 \times J \neq 0$ levels of the first group. There are one $E1$, two $M1$, one $E2$ and two $M2$ transitions possible. One should note that the $E1$ transition to the $1s^2 2p^2\,^3P_1$ is a two-electron one-photon transition.

One can calculate the rates for all these transitions in the independent particle approximation. From this calculation it appears that the $E2$ transition dominates all the others for $Z \geq 54$. The two $M2$ transitions are too weak to be of interest. Of the two $M1$, only the $1s^2 2s3p\,^3P_0 \rightarrow 1s^2 2s2p\,^1P_1$ is significant (a few percent of the $E2$ rate). The *normal* $E1$ to the $1s^2 2s3s\,^3S_1$ is only a few percent of the $E2$ transition. Because of the non-orthogonal nature of the spin orbitals, the two-electron one-photon rate is non-zero, but its rate is weak, in the order of 0.1% of the $E2$ rate. Correlation is then added [93]. The results of such a calculation are shown in Table 6.

Correlation makes the two-electron one-photon rate almost 20 times larger for $Z = 54$. In fact it is shown in Ref. [93] that the rate of this $E2$ transition is very dependent on Z, because it is dominated by two matrix elements the difference of which cancels twice between $Z = 18$ and $Z = 92$.

6. Conclusion and Perspectives

In this chapter we have described the connection between bound-state QED and the relativistic many-body problem. We have also given a survey of

Table 6. Correlation effects in a two-electron one photon transition. Rates (in s^{-1}) for transitions originating from the $1s^2 2s3p\,^3P_0$ level [93]. † indicates that the value includes correlation. When such a value is set in the table the corresponding uncorrelated value is displayed just below in italic

initial	final		$Z = 26$	$Z = 54$	$Z = 92$
$2s3p\,^3P_0$					
→	$2s3s\,^3S_1$	$E1$	7.374×10^{8}	2.428×10^{9}†	8.318×10^{9}
				2.461×10⁹	
→	$2p^2\,^1D_2$	$M2$	7.552×10^{3}	2.526×10^{6}	1.002×10^{7}
→	$2p^2\,^3P_2$	$M2$	2.067×10^{3}	1.054×10^{4}	8.419×10^{3}
→	$2p^2\,^3P_1$	$E1$	3.155×10^{10}†	9.495×10^{9}†	1.885×10^{8}†
			1.675×10⁸	*4.888×10⁸*	*1.125×10⁸*
→	$2s2p\,^3P_1$	$M1$	6.698×10^{4}	1.310×10^{7}†	3.152×10^{10}
				2.097×10⁷	
→	$2s2p\,^1P_1$	$M1$	7.473×10^{5}	2.135×10^{9}†	6.489×10^{11}
				2.117×10⁹	
→	$2s2p\,^3P_2$	$E2$	5.240×10^{9}†	4.938×10^{11}†	1.181×10^{13}†
			4.307×10⁹	*4.644×10¹¹*	*1.166×10¹³*
Total			3.753×10^{10}	5.078×10^{11}	1.250×10^{13}
			5.291×10⁹	*4.695×10¹¹*	*1.236×10¹³*
Lifetime (s)			2.66×10^{-11}	1.97×10^{-12}	8.00×10^{-14}

current state-of-the-art methods for dealing with the many-body relativistic problem, and described some examples of current interest to illustrate these methods. For the evaluation of transition energies in simple atoms, we showed that the limit towards high accuracy lies in correlation correction to the self-energy. Nonradiative QED corrections are small except in the ground state of helium-like ions. It should also be noted that there are new methods being developed that bring the hope of direct QED evaluation of correlation in complex atoms [109, 110], and that could lead to a better approximation of the correlation contribution on the screening of radiative corrections.

For complex atoms, the question of how to develop a general method to provide correlation energy to high accuracy, with reasonable effort is still open. *Kaldor* and *Ishikawa* [42, 111] have developed combined MCDF coupled-clusters methods that look promising. In general, coupling the MCDF method to get intra-shell correlation easily, and improve the result by RMBPT calculations, seems a very promising direction.

For transition probabilities the situation is even more difficult, as the wave function is optimized on the energy. Relativistic methods can lead to unreliable results if one does not take care of non-orthogonality between orbitals in MCDF calculations, and if nonrelativistic limits are not correct.

References

1. P.J. Mohr: in *Physics of Highly-Ionized Atoms*, R. Marrus, ed. (Plenum, New York, 1989), pp. 111–141
2. P.J. Mohr, G. Plunien, G. Soff: Phys. Rep. **293**, 227 (1998)
3. P.J. Mohr: Phys. Rev. A **32**, 1949 (1985)
4. G.E. Brown, D.E. Ravenhall: Proc. Roy. Soc. (London), A **208**, 552 (1951)
5. J. Sucher: Phys. Rev. A **22**, 348 (1980)
6. J.L. Heully, I. Lindgren, E. Lindroth, A.M. Mårtensson-Pendrill: Phys. Rev. A **33**, 4426 (1986)
7. P. Indelicato: Phys. Rev. A **51**, 1132 (1995)
8. I.P. Grant, H.M. Quiney: Adv. At. Mol. Phys. **23**, 37 (1988)
9. I.P. Grant: J. Phys. B **20**, L735 (1987)
10. S.A. Blundell, P.J. Mohr, W.R. Johnson, J. Sapirstein: Phys. Rev. A **48**, 2615 (1993)
11. I. Lindgren, H. Persson, S. Salomonson, L. Labzowsky: Phys. Rev. A **51**, 1167 (1995)
12. H. Persson, I. Lindgren, L. Labzowsky, G. Plunien, T. Beier, G. Soff: Phys. Rev. A **54**, 2805 (1996)
13. J. Sucher: J. Phys. B **21**, L585 (1988)
14. O. Gorceix, P. Indelicato: Phys. Rev. A **37**, 1087 (1988)
15. E. Lindroth, A.-M. Mårtensson-Pendrill: Phys. Rev. A **39**, 3794 (1989)
16. I. Lindgren: J. Phys. B **23**, 1085 (1990)
17. J. Goldstone: Proc. Roy. Soc. (London), Ser A **239**, 267 (1957)
18. A.L. Fetter, J.D. Walecka: *Quantum Theory of Many-Particle Systems* (McGraw-Hill, New York 1971), p. 601.
19. H.P. Kelly: Phys. Rev. **131**, 684 (1963)
20. W.R. Johnson, J. Sapirstein: Phys. Rev. Lett. **57**, 1126 (1986)
21. W.R. Johnson, S. Blundell, J. Sapirstein: Phys. Rev. A **37**, 307 (1988)
22. W.R. Johnson, S. Blundell, J. Sapirstein: Phys. Rev. A **37**, 2764 (1988)
23. J.L. Heully, I. Lindgren, E. Lindroth, S. Lundqvist, A.M. Mårtensson-Pendrill: J. Phys. B **19**, 2799 (1986)
24. J. Sapirstein, W.R. Johnson: J. Phys. B **29**, 5213 (1996); *ibid*, **29**, 5213 (1996)
25. J. Sapirstein: Rev. Mod. Phys. **70**, 55 (1998)
26. S. Salomonson, P. Öster: Phys. Rev. A **40**, 5548 (1989)
27. S. Salomonson, P. Öster: Phys. Rev. A **40**, 5559 (1989)
28. I. Lindgren, J. Morrison: *Atomic Many-Body Theory* (Springer, Berlin 1982)
29. E. Lindroth: Phys. Rev. A **37**, 316 (1988)
30. S.A. Blundell, W.R. Johnson, Z.W. Liu, J. Sapirstein: Phys. Rev. A **40**, 2233 (1989)
31. S.A. Blundell, W.R. Johnson, Z.W. Liu, J. Sapirstein: Phys. Rev. A **39**, 3768 (1989)
32. W.R. Johnson, J. Sapirstein: Phys. Rev. A **46**, R2197 (1992)
33. E.P. Ivanova, U.I. Safronova: J. Phys. B **8**, 1591 (1975)
34. E.V. Aglitsky, P.S. Antsiferov, S.L. Mandelstam, A.M. Panin, U.I. Safronova, S.A. Ulitin: Phys. Scr. **38**, 136 (1988)
35. J.P. Desclaux: *Methods and Techniques in Computational Chemistry* (STEF, Cagliary 1993), Vol. A.
36. O. Gorceix, P. Indelicato, J. P. Desclaux: J. Phys. B **20**, 639 (1987)
37. P. Indelicato, O. Gorceix, J.P. Desclaux: J. Phys. B **20**, 651 (1987)
38. A. Ynnerman, C. Froese-Fischer: Phys. Rev. A **51**, 2020 (1995)

39. K.T. Cheng, M.H. Chen, W.R. Johnson, J. Sapirstein: Phys. Rev. A **50**, 247 (1994)
40. K.T. Cheng, M.H. Chen: Phys. Rev. A **53**, 2206 (1996)
41. M.H. Mittleman: Phys. Rev. A **24**, 1167 (1981)
42. Y. Ishikawa, K. Koc: Phys. Rev. A **50**, 4733 (1994)
43. P. Indelicato, E. Lindroth: Phys. Rev. A **46**, 2426 (1992)
44. S.A. Blundell, W.R. Johnson, J. Sapirstein: Phys. Rev. A **41**, 1698 (1990)
45. G.W.F. Drake: Can. J. Phys. **66**, 586 (1988)
46. G.W.F. Drake: in *Atomic, Molecular, and Optical Physics Handbook*, G. W. F. Drake, ed. (AIP Press, New York 1996)
47. P. Indelicato, P.J. Mohr: Theor. Chem. Acta. **80**, 207 (1991)
48. K.T. Cheng, W.R. Johnson, J. Sapirstein: Phys. Rev. Lett. **66**, 2960 (1991)
49. S.A. Blundell: Phys. Rev. A **46**, 3762 (1992)
50. M.H. Chen, K.T. Cheng, W.R. Johnson: Phys. Rev. A **47**, 3692 (1993)
51. W.R. Johnson, J. Sapirstein, K.T. Cheng: Phys. Rev. A **51**, 297 (1995)
52. S.A. Blundell: Phys. Rev. A **47**, 1790 (1993)
53. H. Persson, S. Salomonson, P. Sunnergren, I. Lindgren: Phys. Rev. Lett. **76**, 204 (1996)
54. V.A. Yerokhin, V.M. Shabaev: Phys. Lett. A **207**, 274 (1995)
55. A.N. Artemyev, V.M. Shabaev, V.A. Yerokhin: Phys. Rev. A **56**, 3529 (1997)
56. P. Indelicato: Nucl. Instrum. Methods. B **31**, 14 (1988)
57. P. Indelicato, J.P. Desclaux: Phys. Rev. A **42**, 5139 (1990)
58. F.A. Parpia, I.P. Grant: J. Phys. B **23**, 211 (1990)
59. F. Marin, F. Minardi, F.S. Pavone, M. Inguscio: Phys. Rev. A **49**, R1523 (1994)
60. C. Dorrer, F. Nez, B. de Bauvoir, L. Julien, F. Biraben: Phys. Rev. Lett. **78**, 3658 (1997)
61. W. Lichten, D. Shiner, Z.-X. Zhou: Phys. Rev. A **43**, 1663 (1991)
62. C.J. Sansonetti, J.D. Gillaspy, C.L. Cromer: Phys. Rev. Lett. **65**, 2539 (1990)
63. K.S. Eikema, W. Ubachs, W. Vassen, W. Hogervorst: Phys. Rev. A **55**, 1866 (1997)
64. J.D. Baker: Bull. Am. Phys. Soc. **38**, 1127 (1993)
65. S.P. Goldman, G.W.F. Drake: J. Phys. B **16**, L183 (1983)
66. G.W.F. Drake, R.A. Swainson: Phys. Rev. A **40**, 1243 (1990)
67. W.R. Johnson, S.A. Blundell, J. Sapirstein: Phys. Rev. A **42**, 1087 (1990)
68. F. Parente, J.P. Marques, P. Indelicato: Europhys. Lett. **26**, 437 (1994)
69. F.G. Serpa, E.S. Meyer, C.A. Morgan, J.D. Gillaspy, J. Sugar, J.R. Roberts, C.M. Brown, U. Feldman: Phys. Rev. A **53**, 2220 (1996)
70. D.R. Beck: Phys. Rev. A **56**, 2428 (1997)
71. J.P. Briand, M. Tavernier, R. Marrus, J.P. Desclaux: Phys. Rev. A **29**, 3143 (1984)
72. P. Indelicato, J.P. Desclaux: Phys. Scr. **T46**, 110 (1993)
73. R.E. Marrs, S.R. Elliot, T. Stöhlker: Phys. Rev. A **52**, 3577 (1995)
74. D.R. Plante, W.R. Johnson, J. Sapirstein: Phys. Rev. A **49**, 3519 (1994)
75. A. Ynnerman, J. James, I. Lindgren, H. Persson, S. Salomonson: Phys. Rev. A **50**, 4671 (1994)
76. J. Schweppe, A. Belkacem, L. Blumenfeld, N. Claytor, B. Feynberg, H. Gould, V. Kostroun, L. Levy, S. Misawa, R. Mowat, M. Prior: Phys. Rev. Lett. **66**, 1434 (1991)
77. P. Beiersdorfer, A. Osterheld, J. Scofield, J. Crespo-López-Urrutia, K. Widmann: Preprint UCRL-JC-127437 (1997)
78. P. Beiersdorfer, A. Osterheld, S.R. Elliott, M.H. Chen, D. Knapp, K. Reed: Phys. Rev. A **52**, 2693 (1995)

79. P. Beiersdorfer, D. Knapp, R.E. Marrs, S.R. Elliot, M.H. Chen: Phys. Rev. Lett. **71**, 3939 (1993)
80. J. P. Santos, J.P. Marques, F. Parente, E. Lindroth, S. Boucard, P. Indelicato: Europhys. Phys. D **2**, 1 (1998)
81. I. Lindgren, H. Persson, S. Salomonson, A. Ynnermann: Phys. Rev. A **47**, 4555 (1993)
82. G. Plunien, B. Muller, W. Greiner, G. Soff: Phys. Rev. A **43**, 5853 (1991)
83. T. Brage, D.S. Leckrone, C. Froese-Fischer: Phys. Rev. A **53**, 192 (1996)
84. J. Bieron, P. Jönsson, C. Froese-Fischer: Phys. Rev. A **53**, 1 (1996)
85. S.A. Blundell: Phys. Scr. **T46**, 144 (1993)
86. E. Avgoustoglou, W.R. Johnson, Z.W. Liu, J. Sapirstein: Phys. Rev. A **51**, 1196 (1995)
87. E. Avgoustoglou, W.R. Johnson, D.R. Plante, J. Sapirstein, S. Sheinerman, S.A. Blundell: Phys. Rev. A **46**, 5478 (1992)
88. W.R. Johnson, D.R. Plante, J. Sapristein: in *Advances in Atomic, Molecular and Optical Physics*, B. Bederson, H. Walters, eds. (Addison-Wesley, New York 1995), Vol. 35, pp. 255–329.
89. J. Olsen, M. Godefroid, P. Jönsson, P.A. Malmqvist, C.Froese-Fischer: Phys. Rev. E **52**, 4499 (1995)
90. A. Hibbert, C. Froese-Fischer, M.R. Godefroid: Comput. Phys. Commun. **51**, 285 (1988)
91. K.T. Cheng, W.R. Johnson: Phys. Rev. A **16**, 263 (1977)
92. P. Indelicato: Phys. Rev. Lett. **77**, 3323 (1996)
93. P. Indelicato: Hyp. Int. **108**, 39 (1997)
94. P.O. Löwdin: Phys. Rev. **97**, 1474 (1955)
95. J.C. Slater: *Quantum Theory of Molecules and Solids* (McGraw-Hill, New York 1963), Vol. 1, p. 485.
96. H.T. Schmidt, P. Forck, M. Grieser, D. Habs, J. Kenntner, G. Miersch, R. Repnow, U. Schramm, T. Schüssler, D. Schwalm, A. Wolf: Phys. Rev. Lett. **72**, 1616 (1994)
97. A. Simionovici, B.B. Birkett, R. Marrus, P. Charles, P. Indelicato, D.D. Dietrich, K. Finlayson: Phys. Rev. A **49**, 3553 (1994)
98. E. Lindroth, S. Salomonson: Phys. Rev. A **41**, 4659 (1990)
99. V.H.S. Kwong, Z. Fang, T.T. Gibbons, W.H. Parkinson, P.L. Smith: Astrophys. J. **411**, 431 (1993)
100. J. Doefert, E. Träbert, A. Wolf, D. Schwalm, O. Uwira: Phys. Rev. Lett. **78**, 4355 (1997)
101. J. Fleming, A. Hibbert, R.P. Stafford: Phys. Scr. **49**, 316 (1994)
102. C. Froese-Fischer: Phys. Scr. **49**, 323 (1994)
103. H.S. Chou, H.C. Chi, K.-N. Huang: Chin. J. of Phys. **32**, 261 (1994)
104. Y.V. Ral'chenko L.A. Vainshtein: Phys. Rev. A **52**, 2449 (1995)
105. Y.K. Kim, J. Marques, F. Parente, P. Indelicato, J.P. Desclaux: submitted to Phys. Rev. A (1998)
106. K.G. Dyall, I.P. Grant, C.T. Johnson, F.A. Parpia, E.P. Plummer: Comput. Phys. Commmun. **55**, 425 (1989)
107. K.N. Huang, Y.-K. Kim, K.T. Cheng, J.P. Desclaux: Phys. Rev. Lett. **48**, 1245 (1982)
108. C. Froese-Fischer: *The Hartree–Fock Method for Atoms* (Wiley, New York, 1977)
109. H. Warston, I. Lindgren, S. Salomonson: Phys. Rev. A **55**, 2757 (1997)
110. D.J. Hylton, N.J. Snyderman: Phys. Rev. A **55**, 2651 (1997)
111. E. Eliav, U. Kaldor, Y. Ishikawa: Phys. Rev. Lett. **74**, 1079 (1995)

V Relativistic Effects in Spectra of One- and Two-Electron Ions

Vitalij G. Pal'chikov

Summary. A Quantum-Electrodynamical (QED) perturbation theory is developed for one- and two-electron systems with interaction between bound electrons and the radiative field being treated as the lowest-order perturbation. In calculations of the energy terms for the $1s3\ell$-configurations of helium-like ions, the effects of the gauge dependence are considered. An analytical relativistic basis-set representation for the Coulomb Green's function is investigated for the high-order QED calculations. The Green's function method provides a basis to obtain the second-order correlation energy for the ground state of the He-like systems in the limits of both small and large values of the nuclear charge Z. Another application is presented to determine electric dipole polarizabilities and hyperpolarizabilities for one- and two-electron ions.

1. Introduction

Experimental studies of spectra in multicharged ions have stimulated the development of relativistic methods for calculating the spectral characteristics of these atomic systems [1]. Besides, progress in increasing the accuracy of measurements has become a steady trend during the last few years [2, 3]. Thus, a rigorous theory of multicharged ions has to account for the high-order effects since their contribution to the observed spectral characteristics becomes extremely important in the analysis of precise measurements.

At present, it is possible to delineate two basic approaches in the relativistic theory of multicharged ions:

1. the use of different kinds of generalization in the relativistic Hartree–Fock method or other methods followed from the theory of many particles, in particular, the random-phase approximation [4, 5], relativistic all-order many-body calculations [6–8], the configuration-interaction (CI) method [9, 10], multiconfiguration Dirac–Fock (MCDF) [11, 12], relativistic variational methods [13, 14], etc.,
2. the use of the adiabatic S-matrix formalism [15] or the evolution operator method [16] equivalent to the theory of perturbation on interaction between electrons. This approach is the most advisable for systems with a few electrons and an arbitrary nuclear charge.

It is worth noting that from a theoretical viewpoint, the second method is more systematic since it permits, besides correlation effects, the calculation of the contribution of the vacuum QED corrections to the spectral characteristics of multiply charged ions. Although in the last few years the relativistic version of the general theory of the adiabatic approach has been used in a fairly detailed manner, its application to specific calculations is very difficult, mainly because of the lack of effective methods of calculation that could permit one to perform a summation over the intermediate states of the S-matrix elements.

In our previous publications [17–19], a method for calculating the correlation diagrams has been proposed using the Green's Coulomb function of the Dirac equation for summation over virtual states, which gives exact results in all orders on $(\alpha Z)^2$, where α is the fine-structure constant. In particular, it was shown that regarding the interaction between electrons, the indicated difficulties are eliminated only in the first orders when one applies the single-electron relativistic Green's functions of the Dirac equation with a Coulomb potential.

The present chapter presents a continuation of this method and a generalization of the approach developed for calculation of relativistic and correlation effects in the higher orders of perturbation theory. The proposed method permits us on the one hand, to get rid of projections on the states with positive and negative energies arising in the higher orders of relativistic perturbation theory [11, 20] and to represent the results in terms of integrals over single-electron Green's function of the Dirac equation and, on the other hand, to express the results in the form of an integral over well-convergent series suitable for practical calculations.

The plan of this chapter is as follows. In Sect. 2 the evolution operator formalism is described. In Sect. 3 we present the analytical form of the relativistic Coulomb Green's function. The goal of Sect. 4 is a theoretical analysis of the gauge dependence in QED-calculations of the correlation energy of helium-like ions for all energy terms of the $1s3\ell$-configurations. The use of the Green's function method is illustrated for the calculation of the second-order correlation energy of the ground state in He-like ions in the limits of both small and large values of the nuclear charge Z. In Sect. 5 the influence of the high-order Stark effect on radiative characteristics of multicharged ions is discussed.

2. Formulation of the QED Bound-Interaction Representation

The field theory applied to hydrogen- and helium-like ions is a generalization of the quantum-electrodynamical approach to systems with a discrete spectrum based on the Feynman diagrams for the bound electron corresponding to the lowest-order terms in the expansion of the S-matrix. Here, a field theoretical basis of atomic structure including relativistic and quantum electrodynamics (QED) effects is described within the framework of the Furry bound-interaction picture of QED [21].

The electron–positron field operators $\psi_{\rm F}(x)$ and $\overline{\psi_{\rm F}}(x)$ are expanded in terms of electron creation (annihilation) operators a_n^+, a_n and positron creation (annihilation) operators b_n^+, b_n [1]:

$$\left.\begin{aligned} \psi_{\rm F}(x) &= \sum\nolimits_{n(+)} a_n \psi_n^{(+)}(x) + \sum\nolimits_{n(-)} b_n^+ \psi_n^{(-)}(x), \\ \overline{\psi_{\rm F}}(x) &= \sum\nolimits_{n(+)} a_n^+ \overline{\psi_n^{(+)}}(x) + \sum\nolimits_{n(-)} b_n \overline{\psi_n^{(-)}}(x), \end{aligned}\right\} \tag{1}$$

where $\psi_n^{\pm}(x) = \psi_n^{\pm}(\boldsymbol{r}) \exp\left(-{\rm i}E_n^{(\pm)} t\right)$ are time-dependent solutions of the Dirac equation in an external four-dimensional potential

$$A_\mu(x) = \{A_0(\boldsymbol{r}) - \boldsymbol{A}(\boldsymbol{r})\}\,.$$

The states with $n(+)$ are interpreted as those of electrons and those with $n(-)$ as the negative energy states or positrons.

In the evolution-operator formalism, the change in the energy ΔE_n of an isolated nondegenerate state $|n\rangle$ is given by [16]

$$\Delta E_{\rm n} = i\frac{\partial}{\partial t} \ln\langle n|U(t,t_0)|n\rangle, \tag{2}$$

where $U(t,t_0)$ is the evolution operator

$$U(t,t_0) = T\left\{\exp\left(-\frac{\rm i}{\hbar}\int_{t_0}^{t} {\rm d}t' H_I(t')\right)\right\}, \tag{3}$$

which is introduced as the time displacement operator, i.e., $|t\rangle = U(t,t_0)|t_0\rangle$, and allows us to obtain the state vector $|t\rangle$ at time t from a known state vector $|t_0\rangle$ at the initial time t_0. Here T is the Wick chronological operator. The limiting transition $\tau = t - t_0 \to \infty$ is taken in the final stage of the calculations.

The interaction between the electron–positron field and the photon field is represented by the interaction Hamiltonian

$$H_{\rm I}(x) = j^\mu(x) A_\mu(x), \tag{4}$$

where j^μ is the current operator

$$j^\mu(x) = -\frac{e}{2}\left[\overline{\psi_{\rm F}}(x)\gamma^\mu, \psi_{\rm F}(x)\right] \tag{5}$$

where $[a,b]$ denotes the commutator of operators a and b, and γ^μ are Dirac matrices. In what follows we will not include the mass renormalization operator $\delta M(x)$ in $H_{\rm I}(x)$ nor the radiation correction and we will limit ourselves to consideration of only processes without participation of real photons.

In calculating the correlation energy, it is convenient to represent $U(t,t_0)$ in the form [22]

[1] All equations are written in atomic units. We insert these quantities explicitly if it helps to bring out the dimensionality.

$$U(t,t_0) = T\exp\left[-\frac{\alpha}{2}\int\limits_{t_0}^{t}\int\limits_{t_0}^{t} \mathrm{d}^4x_1\mathrm{d}^4x_2 j^{\mu}(x_1)D(x_1,x_2)j_{\mu}(x_2)\right]. \tag{6}$$

We note that in this case the Wick chronogical operator acts only on the operators of the electron–positron field. The photon Green's function $D(x_1, x_2)$ depends only on the difference $x_1 - x_2 \equiv x$ and satisfies the equation

$$\langle 0|T\left\{A_{\mu}(x_1)A_{\nu}(x_2)\right\}|\rangle 0 = D(x_1 - x_2)g_{\mu\nu}, \tag{7}$$

where $|0\rangle$ is the vacuum state, and $g_{\mu\nu}$ is the metric tensor [16]. The operator $U(t, t_0)$ in (6) has the perturbation expansion

$$U(t,t_0) = \sum_k U^{(k)}(t,t_0) \tag{8}$$

and, accordingly, the corresponding expansion for the energy shift is given by [19]

$$\begin{aligned} \Delta E_n &= \sum_k \Delta E_n^{(k)}, \\ \Delta E_n^{(1)} &= i\frac{\partial}{\partial t}\langle n|U^{(1)}(t,t_0)|n\rangle, \\ \Delta E_n^{(2)} &= i\frac{\partial}{\partial t}\left[\langle n|U^{(2)}(t,t_0)|n\rangle - \frac{1}{2}\left(\langle n|U^{(1)}(t,t_0)|n\rangle\right)^2\right], \\ &\dots \end{aligned} \tag{9}$$

The expression in (6) can be written as

$$\begin{aligned} &U^{(k)}(t,t_0) = \\ &\quad \frac{1}{n!}\left(-\frac{\alpha}{2}\right)^k \int \mathrm{d}^4x_1 \dots \mathrm{d}^4x_{2k} D(x_1 - x_2)\dots D(x_{2k-1} - x_{2k}) \\ &\quad \times T\left\{j^{\mu}(x_1)j_{\mu}(x_2)\dots j^{\nu}(x_{2k-1})j_{\nu}(x_{2k})\right\}. \end{aligned} \tag{10}$$

The details of the QED calculations of the correlation energy for the first and second orders are described elsewhere [18, 23–25] and are omitted here for brevity.

3. Analytical Basis Set for High-Z QED Calculations

For high-Z ions, the main difficulty in the computation of the amplitudes of $N \geq 2$ orders is related to the presence of infinite sums extended to the whole atomic spectrum. All possible intermediate states are incorporated into the Green's function sum over all bound states, all positive energy continuum electron states in the strong Coulomb field from the nucleus, and all negative energy continuum states. Therefore, for practical computer calculations it is necessary to have the most convenient approximation for the Green's function.

We have used the Green's-function method and have investigated the analytical Laguerre polynomial series expansion for the relativistic Coulomb Green's function [19, 26, 27]. This method is similar to an analytical Sturmian basis-set representation [28–31]. The applicability of the proposed method is illustrated in the following sections.

The perturbation expansion of the evolution operator $U(t,t_0)$ leads to the Feynman diagrams in which the electron–positron lines correspond to the bound-electron propagation function [22]

$$\begin{aligned} S^{(e)}(x,x') &= \langle 0|T\left\{\psi_F(x)\overline{\psi_F}(x')\right\}|0\rangle, \\ &= \Theta(t-t')\sum_k \psi_k^{(+)}(x)\overline{\psi_k^{(+)}}(x') \\ &\quad -\Theta(t-t')\sum_k \psi_k^{(-)}(x)\overline{\psi_k^{(-)}}(x'), \\ &= \frac{1}{2\pi i}\int\limits_{-\infty}^{\infty} \mathrm{d}E\exp[-\mathrm{i}E(t-t')]G(E;\boldsymbol{r},\boldsymbol{r}'), \end{aligned} \tag{11}$$

where $\Theta(t-t')$ is the Heaviside step function which is unity if $t>t'$ and zero otherwise; $G(E;\boldsymbol{r},\boldsymbol{r}')$ is the Green's function in the coordinate representation

$$G(E;\boldsymbol{r},\boldsymbol{r}') = \sum_{n(\pm)} \frac{\psi_n(\boldsymbol{r})\overline{\psi_n}(\boldsymbol{r}')}{E_n(1-\mathrm{i}0)-E}. \tag{12}$$

In many applications it is convenient to present the Green's function in the form of a partial expansion over the spherical spinors $\Omega_{JLM}(\boldsymbol{n})$ with a definite set of quantum numbers J, L and M [26, 27]

$$G(E;\boldsymbol{r},\boldsymbol{r}') = \sum_{JLM}\begin{pmatrix} G_{JL}^{(1)}(r,r')\Upsilon_{L,L}(\boldsymbol{n},\boldsymbol{n}'); & \mathrm{i}G_{JL}^{(2)}(r,r')\Upsilon_{L,2J-L}(\boldsymbol{n},\boldsymbol{n}') \\ G_{JL}^{(3)}(r,r')\Upsilon_{2J-L,L}(\boldsymbol{n},\boldsymbol{n}'); & -G_{JL}^{(4)}(r,r')\Upsilon_{2J-L,2J-L}(\boldsymbol{n},\boldsymbol{n}') \end{pmatrix} \tag{13}$$

where $\Upsilon_{L,L}(\boldsymbol{n},\boldsymbol{n}') = \Omega_{JLM}(\boldsymbol{n})\Omega_{JLM}^{(+)}(\boldsymbol{n}')$, J is the total electron momentum, L is the orbital momentum, and M is the projection of the total momentum. The radial parts $G_{JL}^{(i)}(r,r')$ may be presented in the form of a polynomial series expansion [19, 32]

$$\begin{aligned} G_{JL}^{(1,4)}(r,r') &= -\frac{2(1\pm\varepsilon)}{\nu}\Big\{(Z\nu+\kappa)S_J^{(1)} \\ &\quad +\left[(Z\nu-\kappa)\mp 2(\lambda_N+\eta)\right]S_J^{(0)} \pm S_J^{(+)}\Big\}, \end{aligned} \tag{14}$$

$$\begin{aligned} G_{JL}^{(2,3)}(r,r') &= -\frac{2\sqrt{1-\varepsilon^2}}{\nu}\Big\{(Z\nu+\kappa)S_J^{(1)} - (Z\nu-\kappa)S_J^{(0)} \\ &\quad \mp S_J^{(-)}\Big\}, \end{aligned} \tag{15}$$

$$S_J^{(i)}(E;r,r') = \Phi(rr')\sum_{m=0}^{\infty}\frac{m!L_m^{\mu}(x)L_m^{\mu}(x')}{\Gamma(m+\mu+1)(m+\mu/2+i-\eta)}, \tag{16}$$

$$S_J^{(\pm)}(E;r,r') = \Phi(rr')\sum_{n=0}^{\infty}\frac{m!\left[L_m^{\mu-1}(x)L_m^{\mu}(x') \pm L_m^{\mu}(x)L_m^{\mu-1}(x')\right]}{\Gamma(m+\mu)(m+\mu/2-\eta)} \tag{17}$$

where

$$\Phi(rr') = (xx')^{\mu/2-1}\exp\left(-\frac{x+x'}{2}\right).$$

Here, L_m^a is the Laguerre polynomial and the coefficients $\lambda_N, \eta, \nu, \kappa, \mu$ and variables x, x' are defined by: $x = 2r/\nu$, $x' = 2r'/\nu$, $\nu = \alpha/\sqrt{1-\varepsilon^2}$, $\varepsilon = E$, $\eta = \varepsilon Z\nu$, $\lambda_N = \left((J+1/2)^2 - (\alpha Z)^2\right)^{1/2}$, $\eta\pm = (\eta \pm 1/2, \lambda_N)$, $\mu = 2\lambda_N$, $\kappa = (J+1/2)^{J-1/2}$.

We note that the series in (16) and (17) converge for all r, r' (excluding the point $r = r' = 0$) when $E < 0$. The advantage of this representation is that it constitutes an eigenfunctions representation in which the effective wave functions for intermediate states are polynomials times exponentials, similar to Dirac bound-state wave functions, although such an analytical representation describes the full Green's function including both positive and negative energy continuum states.

The nonrelativistic approach of the function (13) can be expressed in terms of the spherical harmonics [33]

$$G(E;r,r') = \sum_{lm} g_l(E;r,r')Y_{lm}(\boldsymbol{n})Y_{lm}^*(\boldsymbol{n}'), \tag{18}$$

where

$$g_l(E;r,r') = -\frac{4}{\bar{\nu}}(xx')^l \exp\left(-\frac{x+x'}{2}\right) \times \sum_{m=0}^{\infty}\frac{m!L_m^{2l+1}(x)L_m^{2l+1}(x')}{(m+2l+1)!(m+l+1-Z\bar{\nu})}, \tag{19}$$

and $x = 2r/\nu; \bar{\nu} = 1/\sqrt{-2E}$. The poles of the radial Green's function $g_l(E;r,r')$ correspond to the hydrogenic energies $E_n = -Z^2/2n^2$ and the residues at these poles equal the products of the corresponding eigenfunctions. The other various representations for the nonrelativistic Coulomb Green's function are presented in [19, 34, 35]. The generalization of the Sturmian basis set for the $E > 0$ energy range is considered in [36, 37].

The reduced Coulomb Green's function corresponding to the discrete energy level E_{nj} is defined by the limit

$$G^R(E_{nj};\boldsymbol{r},\boldsymbol{r}') = \frac{\partial}{\partial E}\left[(E-E_{nj})G(E;\boldsymbol{r},\boldsymbol{r}')\right], \ E \to E_{nj}. \tag{20}$$

The radial parts of $G^R(E_{nj};\boldsymbol{r},\boldsymbol{r}')$ are presented in [19]. The expressions in (13) are particularly convenient when the contribution to the energy shift or the amplitude of the radiation process under consideration is given by a finite number of partial terms (for example, because of the *selection rules*). The Laguerre polynomial series expansion for the Green's function may also be derived using the Sturm-function expansion [28–31].

4. Leading Corrections in Perturbation Theory of Helium-Like Ions

The expansion of the spectral characteristics in inverse powers of the nuclear charge Z provides a valuable analytical technique for investigating the accuracy of atomic characteristics calculated on the lowest orders of perturbation theory [19, 38, 39]. For very highly ionized atoms, where the number of electrons is smaller than the nuclear charge, the Z^{-1}-expansion converges rapidly, though αZ is not necessarily a good expansion parameter, and so all orders in αZ must be included. Below, such an approach is illustrated by examples of calculation of the first- and second-order correlation energies for the ground and excited states in He-like ions.

4.1 Gauge Dependence of the First-Order Correlation Energy

High-precision measurements of transition energies in two-electron ions, when compared with ab initio theoretical calculations, provide hints that previously neglected effects have to be included due to their influence on the spectral characteristics [40, 41].

In this section, the relativistic electron–electron interaction in He-like ions is discussed. We show that the relativistic Breit operator in the Coulomb and Feynman gauges gives identical results in the case of a local potential, but different values for the nondiagonal matrix elements.

The photon Green's function $D(x_1 - x_2)$ defined in (7) has the form

$$D(x_1 - x_2) = \frac{\mathrm{i}c_\mathrm{o}}{4\pi^3} \int \frac{\mathrm{d}^4 k \exp(-ikx)}{k^2 + i0} \tag{21}$$

$$= \frac{\mathrm{i}c_\mathrm{o}}{4\pi^3} \int \frac{\exp(\mathrm{i}\boldsymbol{kr} - \mathrm{i}\omega t)}{\omega^2/c_\mathrm{o}^2 - k^2 + \mathrm{i}0)} \mathrm{d}\omega \mathrm{d}\boldsymbol{k}, \tag{22}$$

where $k^\mu \equiv (\omega/c_\mathrm{o}, \boldsymbol{k})$ and c_o is the speed of light in vacuum. Carrying out the integral with respect to $\boldsymbol{k}$ in (22), we obtain the following one-dimensional integral for $D(x)$

$$D(x) = -\frac{\mathrm{i}}{2\pi r_{12}} \int \exp\left(-\mathrm{i}\omega t + \frac{\mathrm{i}|\omega| r_{12}}{c_\mathrm{o}}\right) \mathrm{d}\omega. \tag{23}$$

This expression is convenient in analyzing the effect of the interelectronic interaction since it can be evaluated directly by the limit $c_\mathrm{o} \to \infty$ corresponding to exclusion of the retardation effects.

Equation (23) is written in the Feynman gauge for the electromagnetic potentials. By changing the gauge, we can represent the right-hand side of (23) in the following form

$$D(x_1 - x_2)g_{\mu\nu} \leftrightarrow (g_{\mu\nu} + \lambda\partial_\mu\partial_\nu\hat{D}^{-2})D(x_1 - x_2), \tag{24}$$

where $\hat{D} \equiv \partial/\partial x_\mu \cdot \partial/\partial x^\mu$, and λ is an arbitrary parameter. In particular, the case $\lambda = -1$ corresponds to the transverse gauge (Landau gauge), while $\lambda = 2$ corresponds to the Fried–Yennie gauge, which has actively been used in recent years for calculating the radiative corrections. To perform relativistic calculations in a Coulomb gauge it is convenient to separate the part $D^{\mathrm{C}}(x)$ of the $D(x)$ function corresponding to the purely Coulomb interaction,

$$DC(x) = -\frac{\mathrm{i}}{r_{12}}\delta_{\mu,0}\delta_{\nu,0}\delta(t), \tag{25}$$

and the part $D^t_{lm}(x)$ (l and m are indices running from 0 to 3) corresponding to a transverse photon. The explicit expression for $D^t_{jm}(x)$ has the form [19]

$$D^t_{jm}(x) = g_{\mu\nu}D(x) - \int \nabla_{1j}\nabla_{2m}F(x,\omega)\frac{\mathrm{d}\omega}{\omega^2}, \tag{26}$$

where $D(x)$ is defined in (23), ∇_j denotes the jth component of the gradient, and $F(x,\omega)$ is given by

$$F(x,\omega) = -\frac{\mathrm{i}\alpha}{2\pi r_{12}}\exp(\mathrm{i}\omega t)\left[\exp\left(\mathrm{i}\frac{|\omega| r_{12}}{c_o}\right) - 1\right]. \tag{27}$$

The correction to the Coulomb interaction for the electrons, $V_{\mathrm{C}} = 1/r_{12}$, owing to exchange of a transverse photon, is usually referred to as the Breit interaction V_{B}. Using (22) and (26) it is easy to show that V_{B} in the Coulomb gauge is given by [42]

$$\begin{aligned} V_{\mathrm{B}}^{\mathrm{C}} &= -\frac{\boldsymbol{\alpha}_1 \cdot \boldsymbol{\alpha}_2 \cos(\omega r_{12}/c_{\mathrm{o}})}{r_{12}} \\ &\quad + (\boldsymbol{\alpha}_1 \cdot \boldsymbol{\nabla}_{r_{12}})(\boldsymbol{\alpha}_2 \cdot \boldsymbol{\nabla}_{r_{12}})\frac{\cos(\omega r_{12}/c_{\mathrm{o}}) - 1}{\omega^2 r_{12} c_{\mathrm{o}}^2}. \end{aligned} \tag{28}$$

Here $\boldsymbol{\alpha}$ is the usual Dirac matrix [22]. In (27) the imaginary part, corresponding to the radiative width of the value, has been omitted. As (27) implies, the operator V_{B}^{C} depends explicitly on the rate of exchange for transverse photons of frequency ω. The expression for V_{B} in the Feynman gauge has the form

$$V_{\mathrm{B}}^{\mathrm{F}} = -\frac{1}{r_{12}}\left[\boldsymbol{\alpha}_1 \cdot \boldsymbol{\alpha}_2 \cos\frac{\omega r_{12}}{c_{\mathrm{o}}} + 1 - \cos\frac{\omega r_{12}}{c_{\mathrm{o}}}\right]. \tag{29}$$

Despite the rather substantial difference between these two representations for V_{B}, their use in relativistic calculations leads to identical results when basic wave functions, which are solutions of the Dirac equation with a local potential, are employed [42]. This situation arises, for examples, when perturbation theory on the Z^{-1} parameter is used (i.e., when the electron–electron interaction is small compared to the electron–nuclear interaction) or in the case where a local approach is applied to the electron–electron interactions in self-consistent field methods (the Dirac–Hartree–Fock–Slater method). In this case, the equivalence of the representations (28) and (29) follows from the properties of the commutation relation

$$\left[H_1, \left[H_2, \frac{\exp(\mathrm{i}\omega r_{12}/c_\mathrm{o}) - 1}{\omega^2 r_{12}}\right]\right], \tag{30}$$

used to represent the second term in (28). In (30) H_i are the one-particle Dirac Hamiltonians [22]. Using the formula for a double commutator and applying it to the second term of (28), we obtain the term:

$$-\frac{\exp(\mathrm{i}\omega r_{12}/c_\mathrm{o}) - 1}{r_{12}}. \tag{31}$$

Thus, (28) and (29) are completely identical in all orders of the $\mathrm{i}\omega r_{12}/c_\mathrm{o}$ parameter.

When the basis functions are used with a nonlocal potential (e.g., in self-consistent field methods [40]), the commutation relation (30) is valid and a gauge dependence begins to show up, even in the lowest orders of the expansions (28) and (29) on the retardation parameter $i\omega r_{12}/c_\mathrm{o}$

$$\Delta V_\mathrm{B} = V_\mathrm{B}^\mathrm{F} - V_\mathrm{C}^\mathrm{F} = -\frac{\omega^2 r_{12}}{2} - \frac{\boldsymbol{\alpha}_1 \cdot \boldsymbol{\alpha}_2}{2r_{12}} + \frac{(\boldsymbol{\alpha}_1 \cdot \mathbf{r}_{12})(\boldsymbol{\alpha}_2 \cdot \mathbf{r}_{12})}{2r_{12}^3}. \tag{32}$$

For example, it has been shown [40] that the average value for $\Delta V_\mathrm{B} \approx 20$ cm^{-1} is of the same order of magnitude as the error in the measurement of the energy of the $2\,^3P_2 - 2^3S_1$ transition for helium-like argon (32 cm^{-1}). Note that the use of wave functions with a local potential, applying (30) to the first term of (32), again yields $\Delta V_\mathrm{B} = 0$.

A gauge dependence also appears in relativistic calculations of the nondiagonal matrix elements in the case where an intermediate scheme for coupling of the angular momenta is used to determine the state vector of an ion. For helium-like systems, the simplest prototypes of these states are two excited $1s2p$ states with the total angular momentum $J = 1$, since in the intermediate coupling scheme these states are written in terms of linear combinations of the $1s_{1/2}2p_{1/2}$ and $1s_{1/2}2p_{3/2}$ states. The energies of the one-particle $2p_{1/2}$ and $2p_{3/2}$ states are separated by the fine-structure interval. Thus, the Breit operator in the forms (28) and (29) evidently cannot be used in relativistic calculations of the nondiagonal matrix elements since, according to (6), the energy conservation law must hold for arbitrary initial and final states after integrating with respect to the time variable in (6). In order to eliminate this contradiction, a gauge dependent generalized Breit operator [43]

$$\tilde{V}_\mathrm{B} = \frac{1}{2}[V_\mathrm{B}(\omega_\mathrm{AC}) + V_\mathrm{B}(\omega_\mathrm{BD})] \tag{33}$$

has been introduced which corresponds to the scattering amplitude between arbitrary initial $|AB\rangle$ and final $|CD\rangle$ two-particle states in two representations:

$$M^{\mathrm{F}}_{\mathrm{CD,AB}} = \langle CD|\frac{(1-\boldsymbol{\alpha}_1\cdot\boldsymbol{\alpha}_2)}{r_{12}}G(r_{12})|AB\rangle, \tag{34}$$

and

$$M^{\mathrm{F}}_{\mathrm{CD,AB}} = \langle CD|\frac{F(r_{12})-\boldsymbol{\alpha}_1\cdot\boldsymbol{\alpha}_2 G(r_{12})}{r_{12}}|AB\rangle. \tag{35}$$

The functions F and G are given in [42]. By using the explicit form of these functions, it is easy to show that for the diagonal matrix elements with a local potential, (34) and (35) transform identically one into another, but as for the nondiagonal matrix elements, (34) and (35) are no longer identical because of (32).

In this section we present numerical calculations for the nondiagonal matrix elements corresponding to singlet–triplet mixing of the $1s3p\ ^{1,3}P$ and $1s3d\ ^{1,3}D$ states:

$$|E_1\rangle = \cos\frac{\theta}{2}|1s3\ell\ ^3L_J\rangle + \sin\frac{\theta}{2}|1s3\ell\ ^1L_J\rangle, \tag{36}$$

and

$$|E_2\rangle = -\sin\frac{\theta}{2}|1s3\ell\ ^3L_J\rangle + \cos\frac{\theta}{2}|1s3\ell\ ^1L_J\rangle. \tag{37}$$

Here $E_\lambda(\lambda = 1,2)$ are the eigenvectors of the Hamiltonian of an ion and θ is the singlet–triplet *mixing angle* [13]. The nondiagonal matrix elements, owing to the spin–spin interaction between the $1s3s\,^3S_1$ and $1s3d\ ^3D_1$ states, have been examined in detail in [24]. For the relativistic calculations of the nondiagonal matrix elements (34) and (35), we have used the basis Dirac wave functions in the $\boldsymbol{jj}$-coupling scheme for the angular momenta. The details of the calculations of the angular and radial integrals have been described in [24]. We note only that the use of analytic methods for calculating the radial integrals makes it possible, in principle, to calculate the matrix elements (34) and (35) with an arbitrary accuracy.

The results of the calculations for the nondiagonal matrix elements are given in Table 1, which lists the differences of the functions $Q(\alpha Z)$ and $P(\alpha Z)$: $\Delta Q = Q^{\mathrm{F}}(\alpha Z) - Q^{\mathrm{C}}(\alpha Z)$, $\Delta P = P^{\mathrm{F}}(\alpha Z) - P^{\mathrm{C}}(\alpha Z)$. These functions determine the relativistic first-order correlation energy according to the formulas

$$M\left(3\,^3P_1 - 3\,^1P_1\right) = ZQ(\alpha Z), \tag{38}$$

and

$$M\left(3\,^3D_2 - 3\,^1D_2\right) = ZP(\alpha Z). \tag{39}$$

The functions $Q(\alpha Z)$ and $P(\alpha Z)$ are dimensionless and can be written in the form of expansions in powers of the relativistic parameter $(\alpha Z)^2$

$$Q(\alpha Z) = q^{(0)} + q^{(2)}(\alpha Z)^2 + q^{(4)}(\alpha Z)^4 + \cdots, \tag{40}$$

and

$$P(\alpha Z) = p^{(0)} + p^{(2)}(\alpha Z)^2 + p^{(4)}(\alpha Z)^4 + \cdots. \tag{41}$$

Table 1. Gauge dependence of the nondiagonal matrix elements

Z	ΔQ	ΔP	Z	ΔQ	ΔP
2	0.5204×10^{-17}	0.2288×10^{-19}	60	0.3129×10^{-8}	0.2110×10^{-10}
10	0.8304×10^{-13}	0.3662×10^{-15}	70	0.7066×10^{-8}	0.4980×10^{-10}
20	0.5236×10^{-11}	0.2464×10^{-13}	80	0.1339×10^{-7}	0.9489×10^{-10}
30	0.5812×10^{-10}	0.2997×10^{-12}	90	0.2108×10^{-7}	0.1426×10^{-9}
40	0.3140×10^{-9}	0.1791×10^{-11}	100	0.2503×10^{-7}	0.1505×10^{-9}
50	0.1133×10^{-8}	0.7096×10^{-19}	110	0.2902×10^{-7}	0.1600×10^{-9}

An analysis of the numerical data obtained in [41] leads to the following conclusions:

1. In relativistic calculations of the correlation energy of helium-like ions there is first-order dependence of the correlation energy on the choice of gauge for the Breit operator used to account for the electron–electron interaction effects. The relative contribution of the gauge dependence for $3\,^3P_1 - 3\,^1P_1$ transition is 1.5×10^{-10} and 8×10^{-8}, respectively, for $Z = 30$ and $Z = 100$ (Table 1). For the $3\,^3D_1 - 3\,^1D_1$ transition, this contribution is 1.6×10^{-11}. and 4×10^{-9}, respectively. For $Z > 20$ the difference $Y_i = M_i^{\mathrm{F}} - M_i^{\mathrm{C}}$ of the amplitudes (34) and (35) is approximated, with quite high accuracy, by the following simple formulas:
$$Y\left(3\,^3P_1 - 3\,^1P_1\right) = Z^{5.548} \times 3.424 \times 10^{-19}, \tag{42}$$
$$Y\left(3\,^3D_2 - 3\,^1D_2\right) = Z^{5.723} \times 1.040 \times 10^{-21}. \tag{43}$$
2. The nondiagonal matrix elements, which contain the generalized Breit operator depending on the nuclear charge Z, change sign at $Z > 70$ for the $3\,^3P_1 - 3\,^1P_1$ transition and at $Z > 60$ for the $3\,^3D_1 - 3\,^1D_1$ transition. The sign change is caused by a contribution from the retardation effect and by higher-order relativistic effects which are important for high nuclear charges Z .
3. The expansion coefficients in (40) and (41) calculated with a Coulomb gauge are given in Table 2. The effects of the gauge dependence begin to show up only in coefficients of sixth and higher orders and, as an example, for the $3\,^3P_1 - 3\,^1P_1$ transition the differences in the values of $q^{(6)}$ and $q^{(8)}$ are equal to 0.55297×10^{-6} and -0.56201×10^{-5}, respectively (for the coefficients $p^{(6)}$ and $p^{(8)}$ similar calculations yield 0.23165×10^{-8} and -0.41291×10^{-8}. These results are consistent with estimates for singlet–triplet mixing of the $2\,^1P_1$ and $2\,^3P_1$ states of helium-like ions.

Table 2. Leading terms in the expansions of the functions P and Q in terms of the relativistic parameter $(\alpha Z)^2$, (40) and (41)

Coefficient	Value	Coefficient	Value
$q^{(0)}$	-0.42726×10^{-2}	$p^{(0)}$	-0.24219×10^{-3}
$q^{(2)}$	0.11676×10^{-1}	$p^{(2)}$	0.11663×10^{-2}
$q^{(4)}$	0.28273×10^{-2}	$p^{(4)}$	0.18216×10^{-3}
$q^{(6)}$	0.20279×10^{-2}	$p^{(6)}$	-0.42390×10^{-3}
$q^{(8)}$	-0.11513×10^{-3}	$p^{(8)}$	-0.62353×10^{-5}

4. Numerically, the contribution of the gauge dependence is extremely small even for high Z. A similar conclusion follows also from relativistic calculations of the second-order correlation energy [44]. This circumstance justifies a critical attitude towards self-consistent field methods in relativistic variants employing the exact Breit operator [40] (the Dirac–Hartree–Fock method). Contradictions of this kind do not arise when a local approach to the theory of the self-consistent field is applied (the Dirac–Hartree–Fock–Slater method).

4.2 Relativistic Calculations of the Second-Order Correlation Effects

Starting from the Dirac equation, a relativistic Z^{-1}-expansion theory applied to calculations of the second-order correlation energy $\Delta E^{(2)}$, leads to sums over intermediate states that are difficult to evaluate accurately. The main reason for the difficulty follows from the fact that the spectrum of the Dirac equation consists of an infinite set of combinations of the bound and electron–positron continuum states [11, 18, 20]. In this section we shall consider the second-order correlation energy for the ground state of He-like ions. After QED calculation of $\Delta E^{(2)}$, we obtain [18]

$$\begin{aligned}\Delta E^{(2)} &= \sum_{n,j}{}' \frac{\langle 1s1s|U(k_{jm})\Lambda_1|jm\rangle\langle jm|U(k_{jm})|1s1s\rangle}{\varepsilon_j+\varepsilon_m-2\varepsilon_{1s}} \\ &\quad + \sum_{j,m}\frac{\langle 1sm|U(k_{1sj})\Lambda_2|j1s\rangle\langle 1sj|U(k_{1sm})|m1s\rangle}{\varepsilon_j-\varepsilon_m}, \end{aligned} \tag{44}$$

where

$$U(k)=(1-\boldsymbol{\alpha}_1\times\boldsymbol{\alpha}_2)\exp(ikr_{12})/r_{12}, \qquad k_{jm}=\alpha|\varepsilon_j-\varepsilon_m|,$$

$$\Lambda_1=\Lambda_j^{(-)}\Lambda_j^{(-)}-\Lambda_j^{(+)}\Lambda_j^{(+)}, \qquad \Lambda_2=\Lambda_j^{(+)}\Lambda_j^{(-)}-\Lambda_j^{(-)}\Lambda_j^{(+)}.$$

The operators $\Lambda_k^{(\pm)}$ are the projections on the single-particle states with positive and negative energies.

For calculation of the relativistic corrections of the order of $(\alpha Z)^2$ (with respect to the nonrelativistic value of $\Delta E^{(2)}$ which is designated below as $\Delta \bar{E}^{(2)}$), it is convenient to use a Coulomb gauge for the photon propagator in the matrix elements to separate the retarded and Coulomb interactions. The basic contribution to $\Delta E^{(2)}$ occurs from the matrix element corresponding to the pure Coulomb interaction

$$\Delta E^{(2)} = \sum_{j,m}{}' \frac{\langle 1s1s|1/r_{12}\Lambda_1|jm\rangle\langle nm|1/r_{34}|1s1s\rangle}{\varepsilon_j + \varepsilon_m - 2\varepsilon_{1s}}. \tag{45}$$

In calculating $\Delta E^{(2)}$ in (45), the problem of calculating a spectral sum of the following form arises (we introduce the designation S_1 for it):

$$S_1 = \sum_{j,m}{}' \frac{\Lambda_1|jm\rangle\langle jm|}{\varepsilon_j + \varepsilon_m - 2\varepsilon_{1s}}. \tag{46}$$

We rewrite it in an equivalent form, separating the summation according to electron and positron states [18]:

$$\begin{aligned} S_1 &= 2\psi_{1s_{1/2}}(\boldsymbol{r}_1)\bar{\psi}_{1s_{1/2}}(\boldsymbol{r}_2)\sum_m{}' \frac{\Lambda_m^{(+)}|m\rangle\langle m|}{\varepsilon_m - \varepsilon_{1s}} + \sum_{j,m} \frac{\Lambda_{jm}^{(++)}|jm\rangle\langle jm|}{\varepsilon_j + \varepsilon_m - 2\varepsilon_{1s}} \\ &\quad - \sum_{j,m} \frac{\Lambda_{jm}^{(--)}|jm\rangle\langle jm|}{\varepsilon_j + \varepsilon_m - 2\varepsilon_{1s}}, \qquad \varepsilon_j > \varepsilon_{1s},\ \varepsilon_m > \varepsilon_{1s}. \end{aligned} \tag{47}$$

The notation of this expression may be simplified if one uses the following integral equation [18]:

$$\frac{1}{2\pi}\int_{-\infty}^{+\infty} \frac{\mathrm{d}\omega}{(\varepsilon_j - \varepsilon_{1s} + \mathrm{i}\omega)(\varepsilon_m - \varepsilon_{1s} - \mathrm{i}\omega)} = \pm\frac{1}{\varepsilon_j + \varepsilon_m - 2\varepsilon_{1s}}. \tag{48}$$

The proof of this expression is based on Cauchy's residue theorem. Actually, if $\varepsilon_m - \varepsilon_{1s}$ and $\varepsilon_j - \varepsilon_{1s}$ have different signs, the polar peculiarities will be presented in one half-plane of the variable ω, which leads to a zero value of (48), in agreement with Cauchy's theorem. If one simultaneously changes the signs of $\varepsilon_m - \varepsilon_{1s}$ and $\varepsilon_j - \varepsilon_{1s}$, the poles in the upper and lower half-planes exchange places, which reverses the sign of the integral. An integral equation of type (48) was first used for calculation of the constant of dispersive interaction between two atoms in [45] (the Casimir–Polder formula). Using the definition of the single-particle Coulomb Green's function and (48), we represent (46) in the form

$$\begin{aligned} S_1 &= -2\psi_{1s}(\boldsymbol{r}_1)\bar{\psi}_{1s}(\boldsymbol{r}_3)\sum_m \frac{\Lambda_m^{(-)}|m\rangle\langle m|}{\varepsilon_m - \varepsilon_{1s}} \\ &\quad +2\psi_{1s}(\boldsymbol{r}_1)\bar{\psi}_{1s}(\boldsymbol{r}_2)G^R(\varepsilon_{1s};\boldsymbol{r}_2,\boldsymbol{r}_4) \\ &\quad +\frac{1}{2\pi}\int_{-\infty}^{+\infty}\left[G(\varepsilon_{1s} - \mathrm{i}\omega;\boldsymbol{r}_1,\boldsymbol{r}_3) - \frac{\psi_{1s}(\boldsymbol{r}_1)\bar{\psi}_{1s}(\boldsymbol{r}_3)}{\mathrm{i}\omega}\right] \end{aligned}$$

$$\times \left[G(\varepsilon_{1s} + \mathrm{i}\omega; \boldsymbol{r}_2, \boldsymbol{r}_4) + \frac{\psi_{1s}(\boldsymbol{r}_2)\bar{\psi}_{1s}(\boldsymbol{r}_4)}{\mathrm{i}\omega} \right] \mathrm{d}\omega. \tag{49}$$

We note that the order of contribution of the first term in (49) to $\Delta E^{(2)}$ is $(\alpha Z)^3$, and in our approximation $(\alpha Z)^2$ may be omitted. $G^R_{\varepsilon_{1s}}$ is the reduced Coulomb Green's function of the Dirac equation [19]. In this way, (48) permits introducing the single-particle Green's function on the basis of (49) into (45), which allows us to use the analytical methods of calculation developed in [17, 18, 24]. Then the result of the integration over the radial variables is represented in the form of absolutely converging series of a combination of the hypergeometric Gaussian function ${}_2F_1$. Integration over ω was performed numerically, using the Gauss integration formula. In the vicinity of $\omega = 0$, an expansion of the integrand was performed in a series on ω, with mutual compensation of the terms $1/\omega$ occurring.

Finally, the numerical value of $\Delta E^{(2)}$ with an accuracy of $(\alpha Z)^2$ terms including the remaining part of the Breit operator (one Breit + one Coulomb) is equal to $\Delta E^{(2)} = -0.15767[1 + 0.631(\alpha Z)^2]$. We can compare this result for the relativistic correction with values inferred from a Z^{-1}-expansion fit to variational calculations [46] $(-0.637(\alpha Z)^2)$, and with corresponding results obtained by using B-spline approximation [47]: $-0.634(\alpha Z)^2$. Including corrections obtained for the energy of the ground state of the He-like ion, we have the following relativistic $(\alpha Z)^2$ generalization of the Hyllearaas expansion, containing only the terms that do not decrease with increasing nuclear charge [18, 19]:

$$E = \left[-Z^2 + \frac{5}{8}Z - 0.15767 + (\alpha Z)^2 \left(-\frac{Z^2}{4} + 0.480147Z - 0.631 \right) \right]. \tag{50}$$

Since the nonrelativistic value of the third-order correlation energy is $\Delta \bar{E}^{(3)} = 0.008699/Z$ [13], it follows from (50) that the relativistic corrections to $\Delta \bar{E}^{(2)}$ exceed $\Delta \bar{E}^{(3)}$ at $Z \geq 7$.

As discussed in the introduction, for high-Z ions, besides correlation effects, it is necessary to account for the radiation corrections. However, the matrix elements determining the radiation corrections (in any case to the fourth order of interaction of electrons with the field) contain only one-electron Green's functions, and, therefore, the difficulties associated with the presence of two-electron Green's functions are absent. Then in the full relativistic matrix elements, it is necessary to use extremely complex methods [44, 47–49]. In this connection, an approximate method of calculation of correlation diagrams for large Z is proposed below.

In both the relativistic and nonrelativistic cases, the contribution to $\Delta E^{(2)}$ is given by the intermediate terms of three types, corresponding to the discrete–discrete $\Delta E^{(2)}_{\mathrm{DD}}$, the discrete-continuous $\Delta E^{(2)}_{\mathrm{DC}}$, and continuous-continuous $\Delta E^{(2)}_{\mathrm{CC}}$ states in (45). In the nonrelativistic case, the contribution of $\Delta E^{(2)}_{\mathrm{DD}}$ is 20%, and it is evident that allowance for relativistic effects cannot substantially change this value. Since at large Z the contribution of relativistic effects becomes important, one may neglect $\Delta E^{(2)}_{\mathrm{CC}}$, which gives an accuracy of accounting for correlation effects of order 80%. In this approach, the summation over m in (45) is performed only over the first excited states of the discrete spectrum.

The summation over the intermediate states is extended to electron states and to positron states. The combined matrix elements include a summation over virtual electron–positron states, including the continuous spectrum which immediately permits us to use the method of the hydrogen-like Green's function. For example, if one takes into account only the single intermediate state $1s$, we have for $\Delta E^{(2)}_{1s}$

$$\Delta E^{(2)}_{1s} = 2\langle 1s1s|\bar{\psi}_{1s}(\boldsymbol{r}_1)U(0)G^R(\varepsilon_{1s};\boldsymbol{r}_2,\boldsymbol{r}_4)U(0)\psi_{1s}(\boldsymbol{r}_3)|1s1s\rangle . \tag{51}$$

The calculation using the nonrelativistic Coulomb Green's function gives

$$\Delta \bar{E}^{(2)}_{1s} = \frac{1}{2^7}\left[\frac{104}{27} - 36\ln\left(\frac{3}{4}\right)\right], \tag{52}$$

which coincides with results of the Hartree–Fock calculation [50]. Results of an analytical calculation of $\Delta\bar{E}^{(2)}_n$ including the states with $n = 2$ and $n = 3$ are given in Table 3. Taking into account the most exact value of

Table 3. Contribution of discrete states to $\Delta\bar{E}^{(2)}$ (nonrelativistic calculation)

States, n	$\Delta\bar{E}^{(2)}_n$
$1s$	−0.111003170
$n = 2$	−0.011540126
$n = 3$	−0.002515011

$\Delta\bar{E}^{(2)} = -0.1576664295$ [51], it is evident that the combined contribution of $n = 1$ and $n = 2$ is 79%. The contribution $\Delta\bar{E}^{(2)}_3 = -0.002515011$ of the states with $n = 3(3s, 3p, 3d)$ gives 1.5% of the contribution. In this way, the contribution of the double excited states with $m > 2$ is not substantial and in our approximation may be neglected.

In the nonrelativistic limit, the inclusion in the sum over m of only the $1s_{1/2}, 2s_{1/2}$, and $2p_j$ states gives

$$\Delta\bar{E}^{(2)}_{1,2} = -0.122543296. \tag{53}$$

The combined result of the relativistic calculation of $\Delta E^{(2)}_{1s_{1/2}}$ in (51) and of similar expressions for $\Delta E^{(2)}_{2s_{1/2},2p_j}$ is conveniently represented in the form

$$\Delta E^{(2)}_{1,2} = \Delta\bar{E}^{(2)}_{1,2}W(Z), \tag{54}$$

where the $W(Z)$ function determines the contribution of relativistic effects depending on Z [$W(0) = 1$]. The numerical values of $W(Z)$ are given in

Table 4. Contribution of relativistic effects to $\Delta\bar{E}^{(2)}$ for large Z

Z	$\Delta E^{(2)}$	$W(Z)$	Z	$\Delta E^{(2)}$	$W(Z)$
20	−0.13849	1.1074	85	−0.19165	1.5325
25	−0.13992	1.1188	90	−0.20119	1.6088
30	−0.14167	1.1328	95	−0.21269	1.7007
35	−0.14370	1.1491	100	−0.22664	1.8123
40	−0.14616	1.1687	105	−0.24757	1.9796
45	−0.14899	1.1924	110	−0.26025	2.0811
50	−0.15222	1.2172	115	−0.27573	2.2064
55	−0.15376	1.2295	120	−0.33759	2.6995
60	−0.16016	1.2807	125	−0.42169	3.3720
65	−0.16484	1.3181	130	−0.77082	6.1637
70	−0.17007	1.3599	135	−1.3298	9.9139
80	−0.18225	1.4573	137	−4.5329	36.246

Table 4. From Table 4 it follows that the contribution of relativistic effects is 20% for $Z = 45$. Thus, for $Z \geq 45$ the relativistic effects exceed the contribution of the continuous–continuous spectrum $\Delta E^{(2)}_{\mathrm{CC}}$, which justifies the approximation made above for ions of a high degree of ionization.

5. High-Order Stark Effect

The dipole moment induced in an ion by a uniform electric field of strength F is expressed as a linear function of F, but higher terms on F may become important when the field strength is sufficiently large. The induced dipole moment μ is given by

$$\mu = \beta F + \frac{1}{6}\gamma F^3 + \ldots, \tag{55}$$

where β is the electric dipole *polarizability*, and γ represents the deviation of μ from a linear law. The parameter γ is often termed the electric dipole *hyperpolarizability*. The energy associated with the induction of moment μ leads to an expression for the Stark shift ΔE as an asymptotic power series of the field strength [52]

$$\Delta E = E - E_0 = -\beta F^2/2! - \gamma F^4/4! - \ldots, \tag{56}$$

where E_0 is the field-free energy.

Accurate *ab initio* calculations of β and γ are not routine because both the relativistic and electron correlation effects can be very important in the case of multicharged ions. Besides being of purely theoretical interest, the relativistic generalization of these quantities can be applied for different applications, e.g. for the determination of the Stark shift of x-ray lines in the inner-shells of heavy atoms.

Below, we report relativistic computations of β and γ for the ground states of H- and He-like multicharged ions by using a Coulomb Green's function method on second- and fourth-order approximations of the perturbation theory.

5.1 Hydrogen-Like Ions

The dipole polarizability β is known exactly in the nonrelativistic limit, and analytically up to the order of $(\alpha Z)^4$ in the relativistic case [53]

$$\beta_{\mathrm{H}} \simeq \left[\frac{9}{2} - \frac{14}{3}(\alpha Z)^2 + \frac{1}{48}\left(\pi^2 + \frac{31}{2}\right)(\alpha Z)^4\right]. \tag{57}$$

Taking into account the perturbation potential $\hat{d}(r) = -(\boldsymbol{rF})$ according to the Dirac theory, the relativistic polarizability of the H-like ion can be presented as

$$\beta_{\mathrm{H}} = 2\langle 1s_{1/2}(\boldsymbol{r_1}|\hat{d}(r_1)G(\epsilon_{1s_{1/2}};\boldsymbol{r_1},\boldsymbol{r_2})\hat{d}(r_2)|1s_{1/2}\rangle. \tag{58}$$

Performing the integration over angular variables, we obtain

$$\begin{aligned}\beta_{\mathrm{H}} &= \tfrac{2}{9}\left[R^{(1)}_{1/2} - \xi_1 R^{(2)}_{1/2} - \xi_1 R^{(3)}_{1/2} - \xi_1^2 R^{(4)}_{1/2}\right. \\ &\quad \left. +2\left(R^{(1)}_{3/2} - \xi_1 R^{(2)}_{3/2} - \xi_1 R^{(3)}_{3/2} - \xi_1^2 R^{(4)}_{3/2}\right)\right],\end{aligned} \tag{59}$$

where

$$R^{(i)}_{JL} = \int_0^\infty \int_0^\infty r_1^3 r_2^3 \psi_{1s_{1/2}}(r_1)\psi_{1s_{1/2}}(r_2) G^{(i)}_{JL}(\epsilon_{1s_{1/2}}; r_1, r_2) \mathrm{d}r_1 \mathrm{d}r_2, \tag{60}$$

and $G^{(i)}_{jl}$ are radial components of the Green's function, $\xi_1 = \alpha Z/(1 + \sqrt{1-(\alpha Z)^2})$.

Three different analytical representations for the relativistic polarizability β_H are presented in [28, 53, 54]. Computed values of β_{H} [54] in the $Z \leq 110$ range are given in Table 5. The numerical results are in very good agreement with those calculated in [28, 53]. At small Z, the relativistic corrections to $\beta^{nr}_{\mathrm{H}} = 9/(2Z^4)$ are not significant, but with Z increasing their contribution becomes quite important and reach about 50% at $Z = 100$. Using the full relativistic calculations of β_{H} and (57), the leading corrections to β^{nr}_{H} can be rewritten as [54]

$$\begin{aligned}\beta_{\mathrm{H}} &= \beta^{nr}_{\mathrm{H}}[1 - 28(\alpha Z)^2/27 + (2\pi^2 + 31)(\alpha Z)^4/432 \\ &\quad +0.7496060 \times 10^{-2}(\alpha Z)^6 - 0.11175273 \times 10^{-1}(\alpha Z)^8 \\ &\quad +0.2425135 \times 10^{1}(\alpha Z)^{10}].\end{aligned} \tag{61}$$

The dipole hyperpolarizability γ_H is defined by

$$\begin{aligned}\gamma_{\mathrm{H}} &= \langle 1s_{1/2}(\boldsymbol{r}_1)|\hat{d}(r_1)G(\epsilon_{1s_{1/2}};\boldsymbol{r}_1,\boldsymbol{r}_2)\hat{d}(r_2)G(\epsilon_{1s_{1/2}};\boldsymbol{r}_2,\boldsymbol{r}_3)\hat{d}(r_3) \\ &\quad \times G(\epsilon_{1s_{1/2}};\boldsymbol{r}_3,\boldsymbol{r}_4)\hat{d}(r_4)|1s_{1/2}(\boldsymbol{r}_4) > -\frac{1}{2}\alpha_H S_{-3},\end{aligned} \tag{62}$$

Table 5. The ratio $\beta_{\mathrm{H}}/\beta_{\mathrm{H}}^{\mathrm{nr}}$ ($\beta_{\mathrm{H}}^{\mathrm{nr}} = 9/2Z^4$) and hyperpolarizability γ_{H} in H-like ions in the ground state [54]

Z	$\beta_{\mathrm{H}}/\beta_{\mathrm{H}}^{\mathrm{nr}}$	γ_{H}
1	0.99994477	$1.3329687\times10^{+3}$
10	0.99313253	1.3175691×10^{-7}
20	0.97796390	1.2417310×10^{-10}
30	0.95056932	2.0269375×10^{-12}
40	0.91249938	1.0459400×10^{-13}
50	0.86403977	9.9836241×10^{-15}
60	0.80556203	1.3814085×10^{-15}
70	0.73752563	2.4456595×10^{-16}
80	0.66047816	5.0590889×10^{-17}
90	0.57504883	1.1392008×10^{-17}
100	0.48192161	2.6734156×10^{-18}
110	0.38174026	5.7759208×10^{-19}

where

$$S_{-3} = \langle 1s_{1/2}(\boldsymbol{r}_1)|\hat{d}(r_1)G(\epsilon_{1s_{1/2}};\boldsymbol{r}_1,\boldsymbol{r}_2)G(\epsilon_{1s_{1/2}};\boldsymbol{r}_2,\boldsymbol{r}_3)\hat{d}(r_3)|1s_{1/2}(\boldsymbol{r}_3)\rangle . \tag{63}$$

The relativistic values of γ_{H} have been obtained using exact analytical methods developed in [54] and also presented in Table 5.

5.2 Helium-Like Ions

The Stark effect for the ground state of He-like ions is similar to that of H-like ions because the ground state is nondegenerate in zero-order perturbation theory. Taking into account the interelectronic interaction, the polarizability of the He-like ion can be presented as

$$\beta_{\mathrm{He}} = \beta_{\mathrm{He}}^{(0)}\left(1 + \sum_{j=1}^{\infty}\frac{a_j}{Z^j}\right), \tag{64}$$

where $\beta_{\mathrm{He}}^{(0)}$ is the polarizability in the H-like approximation

$$\beta_{\mathrm{He}}^{(0)} = 2\beta_{\mathrm{H}}. \tag{65}$$

The exact nonrelativistic value of the coefficient a_1 has been obtained in [55]:

$$a_1^{nr} = 207/144. \tag{66}$$

In the "screened-charge" approximation, β_{He}^{nr} can be written in the form [56]:

$$\beta_{\mathrm{He}}^{nr} = \left(Z - \frac{197}{500}\right)^{-4}\left[9 - 1.24489\left(Z - \frac{197}{500}\right)^{-1} + 1.47425\left(Z - \frac{197}{500}\right)^{-2}\right]. \tag{67}$$

The relativistic values of the coefficient a_1^{rel} in (64) have been obtained in [19, 57]. The most accurate numerical results obtained using the relativistic CI method have been obtained in [10]. Extensive numerical results obtained using the relativistic random-phase method have been tabulated in [5]. Exact variational calculations are presented in [58].

To illustrate our method of calculation, we give in Tables 6 and 7 the numerical results for $\beta_{\rm He}$. The second column in Table 6 gives the correction function a_1^{rel}/a_1^{nr} , the third one gives $\beta_{\rm He}$ in the H-like approximation, the fourth is given the value of $\beta_{\rm He}$ obtained in the "screened-charge" approximation [59], i.e.,

$$\bar{\beta}_{\rm He} = 9/(Z+b)^4 - 28\beta_{\rm He}^{(0)nr}(\beta Z)^2/27, \tag{68}$$

where $b = -a_1^{rel}/4$. For comparison we also give the relativistic results presented in [5, 10]. For small Z, the interelectronic corrections to $\beta_{\rm He}^{(0)nr}$ are significant (60% for $Z = 2$), but these corrections decrease with increasing Z (2% for $Z = 10$). On the other hand, the relativistic effects rapidly increase with Z increasing (90% for $Z = 100$).

Similar estimations of $\gamma_{\rm He}$ are much more complicated due to the complexity of calculations of the sums over intermediate states including three two-particle Green's function. Nevertheless, taking into account the electron–electron interaction as the first-order perturbation potential, we can express the fourth-order energy shift by means of the one-particle Coulomb Green's function [59]. The two leading terms of the order of Z^{-10} and Z^{-11} in the ground state case are [59]:

Table 6. Relativistic polarizability $\beta_{\rm He}$ of He-like ions in the ground state (low Z)

Z	$a_1^{\rm rel}/a_1^{\rm nr}$	$\beta_{\rm He}^{(0)}$	$\hat{\beta}_{\rm He}$	$\beta_{\rm He}$ [56]	$\beta_{\rm He}$ [10]
2	1.0002	5.6250×10^{-1}	$1.2421 \times 10^{+0}$	$1.3222 \times 10^{+0}$	$1.383123 \times 10^{+0}$
3	1.0005	1.1111×10^{-1}	1.8505×10^{-1}	1.8947×10^{-1}	1.924088×10^{-1}
4	1.0009	3.5156×10^{-2}	5.1201×10^{-2}	5.1857×10^{-2}	5.224179×10^{-2}
6	1.0021	6.9444×10^{-3}	8.8769×10^{-3}	8.9353×10^{-3}	5.224179×10^{-3}
8	1.0038	2.1973×10^{-3}	2.6330×10^{-3}	2.6479×10^{-3}	2.645215×10^{-3}
10	1.0059	8.9999×10^{-4}	1.0369×10^{-3}	1.0437×10^{-3}	1.040062×10^{-3}
20	1.0239	5.6250×10^{-5}	5.9239×10^{-5}		

Table 7. Relativistic polarizability $\beta_{\rm He}$ of He-like ions in the ground state (high Z)

Z	$a_1^{\rm rel}/a_1^{\rm nr}$	$\hat{\beta}_{\rm He}$	$\beta_{\rm He}$ [5]
30	1.0552	1.1113×10^{-5}	1.111×10^{-5}
40	1.1019	3.3380×10^{-6}	3.331×10^{-6}
50	1.1666	1.2867×10^{-6}	1.283×10^{-6}
80	1.5088	1.4909×10^{-7}	1.486×10^{-1}
100	1.9137	4.4572×10^{-8}	

$$\begin{aligned} \gamma_{\rm He} &= \gamma_{\rm He}^{(0)} + \gamma_{\rm He}^{(1)}/Z, \\ \gamma_{\rm He}^{(0)} &= 10665/(4Z^{10}), \\ \gamma_{\rm He}^{(1)} &= 11347.8/Z^{10}. \end{aligned} \tag{69}$$

The "screened-charge" approximation for $\gamma_{\rm He}$ similar to (68) has the form

$$\hat{\gamma}_{\rm He} = \frac{10665}{4(Z-4.2561)^{10}}. \tag{70}$$

The numerical results for $\gamma_{\rm He}$ are listed in Table 8. For comparison we give here the results of multiconfiguration Hartree–Fock calculations [56] (the fourth column in Table 8).

Table 8. Static dipole hyperpolarizability of He-like ions in the ground state [59]

Z	$\gamma_{\rm He}^{(0)}$	$\hat{\gamma}_{\rm He}$	$\gamma_{\rm He}$ [56]
2	$2.6038 \times 10^{+0}$	$2.8496 \times 10^{+1}$	$2.8496 \times 10^{+1}$
3	4.1553×10^{-2}	2.0852×10^{-1}	2.2740×10^{-1}
4	2.5427×10^{-3}	7.8322×10^{-3}	8.1910×10^{-3}
5	2.7302×10^{-4}	6.6461×10^{-4}	6.8308×10^{-4}
6	4.4095×10^{-5}	9.2030×10^{-5}	9.3740×10^{-5}
7	9.4389×10^{-6}	1.7674×10^{-5}	1.7910×10^{-5}
8	2.4831×10^{-6}	4.2897×10^{-5}	4.3325×10^{-6}
9	7.6467×10^{-7}	1.2413×10^{-6}	1.2510×10^{-6}
10	2.6662×10^{-7}	4.1190×10^{-7}	4.1447×10^{-7}

6. Outlook and Conclusions

In this chapter, we have presented analytical and numerical results concerning various high-order relativistic effects on the spectra of multicharged ions with one and two electrons. More precisely, we described these effects within a common framework based on the Coulomb Green's function method. Several general conclusions can be made from the analysis presented here.

1. We derived analytical expressions for the relativistic Coulomb Green's function of a hydrogen-like atom. These analytical representations allowed us to get very accurate numerical values of the considered atomic characteristics.
2. We also studied the gauge dependence in relativistic calculations of the correlation energy of helium-like atoms. It was shown that the relativistic Breit operator in the Coulomb and Feynman gauges yields identical results for the diagonal matrix elements in a local potential, but different values for the nondiagonal matrix elements.

3. An approach was formulated that permits us to use analytical methods in calculations of correlation effects of the high-orders of perturbation theory for computing the correlation energy of multicharged ions. As an important application, relativistic corrections to the second-order correlation energy in He-like ions were calculated.

Acknowledgement. Financial support was provided in part by the Russian Foundation for Basic Research (grant RFBR No 96-02-00257) and by Deutsche Forschungsgemeinschaft (grant No 436 RUS 113/ 164/ O(R,S)).

References

1. H.F. Beyer, H.-J. Kluge, V.P. Shevelko: *X-ray Radiation of Highly Charged Ions* (Springer, Berlin, Heidelberg 1997)
2. K. Widmann, P. Beiersdorfer, V. Declaux, M. Bitter: Phys. Rev. A **53**, 2200 (1996)
3. A. Bartnik, E. Biemont, V.M. Dyakin, A.Ya. Faenov, H. Fiedorowicz, W.H. Goldstein, R. Jarocki, J. Kostecki, A.L. Osterheld, V.G. Pal'chikov, Yu-bo Qiu, I.Yu. Skobelev, M. Szezurek, Yu Zou: J. Phys. B **30**, 4453 (1997)
4. W.R. Johnson, C.D. Lin: Phys. Rev. A **14**, 565 (1976)
5. W.R. Johnson, D. Kolb, K.N. Huang: At. Data Nucl. Data Tables **28**, 333 (1983)
6. S. Salomonson, P. Öster: Phys. Rev. A **40**, 5548 (1989)
7. E. Lindroth, A.M. Märtensson-Pendrill: Phys. Rev. A **39**, 3794 (1989)
8. I. Lindgren: Int. J. Quantum. Chem. **57**, 683 (1996)
9. K.T. Cheng, M.H. Chen, W.R. Johnson, J. Sapirstein: Phys. Rev. A **50**, 247 (1994)
10. W.R. Johnson, K.T. Cheng: Phys. Rev. A **53**, 1375 (1996)
11. P. Indelicato: Phys. Rev. A **51**, 1132 (1995)
12. F.A. Parpia, Ch. Froese-Fischer, I.P. Grant: Comput. Phys. Commun. **94**, 249 (1996)
13. G.W.F. Drake: Can. J. Phys. **66**, 586 (1988)
14. W. Cencek, W. Kutselnigg: J. Chem. Phys. **105**, 5878 (1996)
15. M. Gell-Mann, F. Low: Phys. Rev. **84**, 350 (1951)
16. S.S. Schweber: *An Introduction to Relativistic Quantum Field Theory* (Row, Peterson and Co., Evanston, New York 1961)
17. S.A. Zapryagaev, N.L. Manakov, V.G. Pal'chikov: Opt. Spectrosc. **46**, 119 (1979)
18. S.A. Zapryagaev, N.L. Manakov, V.G. Pal'chikov: Opt. Spectrosc. **52**, 248 (1982)
19. S.A. Zapryagaev, N.L. Manakov, V.G. Pal'chikov: *The Theory of Multicharged Ions with One and Two Electrons* (Energoatomizdat, Moscow 1985) (in Russian)
20. I. Lindgren: Physica Scripta **T59**, 179 (1995)
21. W.H. Furry: Phys. Rev. **81**, 115 (1951)
22. A.I. Akhieser, V.B. Berestetskii: *Quantum Electrodynamics* (John Wiley, New York 1965)
23. U.I. Safronova, M.S. Safronova, N.J. Snyderman, V.G. Pal'chikov: Phys. Scr. **50**, 29 (1994)
24. V.G. Pal'chikov, G.von Oppen: Phys. Scr. **52**, 366 (1995)

25. P.J. Mohr: Phys. Rev. A **32**, 1949 (1985)
26. N.L. Manakov, S.A. Zapryagaev: Phys. Lett. A **58**, 23 (1976)
27. N.L. Manakov, S.A. Zapryagaev: Sov. Phys.- Nucl. Phys. **23**, 917 (1976)
28. R. Czmytkowski: J. Phys. B **30**, 825 (1997)
29. R.A. Swainson, G.W.F. Drake: J. Phys. B **24**, 79 (1991); *ibid*, 95(1991); *ibid*, 1801 (1991)
30. G.W.F. Drake, S.P. Goldman: Adv. At. Mol. Phys. **25**, 393 (1988)
31. I.P. Grant: in: *The Effects of Relativity in Atoms, Molecules, and the Solid State.* eds. S. Wilson, I.P. Grant, B.L. Gyorffy (Plenum, New York 1991) pp. 17–43
32. A.V. Borovskii, S.A. Zapryagaev, O.I. Zatsarinnii, N.L. Manakov: *Plasma of Multicharged Ions* (Khimiya, St.Petersburg 1995) (in Russian)
33. L. Hostler: J. Math. Phys. **11**, 2966 (1970)
34. A. Maquet: Phys. Rev. A **15**, 1088 (1977)
35. N.L. Manakov, V.D. Ovsiannikov, L.P. Rapoport: Phys. Rep. **141**, 319 (1986)
36. R. Shakeshaft: Phys. Rev. A **34**, 5119 (1986)
37. N.L. Manakov, S.I. Marmo, A.G. Fainshtein: Theor. Math. Phys. **59**, 351 (1984)
38. V.A. Boyko, V.G. Pal'chikov, I.Yu. Skobelev, A.Ya. Faenov: *Spectra of Atoms with One and Two Electrons* (CRC Press, Boca Raton 1994)
39. V.G. Pal'chikov, V.P. Shevelko: *Reference Data on Multicharged Ions* (Springer, Berlin, Heidelberg 1995)
40. O. Gorceix, P. Indelicato: Phys. Rev. A **37**, 1087 (1988)
41. V.G. Pal'chikov, G.von Oppen: Sov. Ohys. - JETP **85**, 885 (1996)
42. J. Hata, I.P. Grant: J. Phys. B **17**, L107 (1984)
43. M. Mittleman: Phys. Rev. A **4**, 897 (1971); *ibid* A **5**, 2395 (1972)
44. I. Lindgren, H. Persson, S. Salomonson, L. Labzowsky: Phys. Rev. A **51**, 1167 (1995)
45. H.B.G. Casimir, D. Polder: Phys. Rev. **73**, 360 (1948)
46. Y. Accad, C.L. Pekeris, B. Schiff: Phys. Rev. A **4**, 516 (1972)
47. S.A. Blundell, P.J. Mohr, W.R. Johnson, J. Sapirstein: Phys. Rev. A **48**, 2615 (1993)
48. V.M. Shabaev: J. Phys. B **26**, 4703 (1993)
49. V.M. Shabaev, I.G. Fokeeva: Phys. Rev. A **49**, 4489 (1994)
50. N.L. Rahman: J. Chem. Phys. **67**, 1684 (1977)
51. J.D. Baker, D.E. Freund, R.N. Hill, J.D. Morgan III: Phys. Rev. A **41**, 1247 (1990)
52. H.A. Bethe, E.E. Salpeter: *Quantum Mechanics of One- and Two-Electron Atoms* (Plenum, New York 1977)
53. B.A. Zon, N.L. Manakov, L.P. Rapoport: Sov. Phys.- Nucl. Phys. **15**, 282 (1972)
54. V.G. Pal'chikov: *Theory of Atoms with One and Two Electrons* (Doctoral Thesis, Moscow 1990) (unpublished)
55. A. Dalgarno: Adv. Phys. **11**, 281 (1962)
56. H.K. Mc Dowell, R.N. Porter: J. Chem. Phys. **66**, 472 (1977)
57. N.L. Manakov, V.G. Pal'chikov: Opt. Spectrosc. **46**, 688 (1979)
58. M.K. Harbola, A. Banerjce: Phys. Rev. A **54**, 283 (1996)
59. V.G. Pal'chikov, A.N. Tkachev: Opt. Spectrosc. **67**, 995 (1989)

VI Hyperfine Structure of Highly Charged Ions

Vladimir M. Shabaev

Summary. The theory of hyperfine-structure splitting in highly charged ions is reviewed. The available theoretical results for the ground-state hyperfine splitting in hydrogen-like and lithium-like ions are presented. It is shown that the Bohr–Weisskopf effect, which mainly defines the uncertainty of the theoretical values, can be eliminated in a combination of the hyperfine splitting values of hydrogen-like and lithium-like ions. It allows one to make high precision predictions for the ground-state hyperfine splitting in lithium-like ions if the ground-state hyperfine splitting of the corresponding hydrogen-like ions is known from experiment. The transition probabilities between the hyperfine components are also presented.

1. Introduction

Interaction of atomic electrons with the magnetic dipole or electric quadrupole moment of the nucleus causes splitting of atomic level into components with each of them corresponding to a fixed value of the total moment of the atomic system (electrons plus nucleus). This splitting, which is called the *hyperfine splitting*, can be easily understood if we take into account that the total electronic moment $\boldsymbol{J}$ and the nuclear moment $\boldsymbol{I}$ are no longer integrals of the motion but the total atomic moment $\boldsymbol{F} = \boldsymbol{I} + \boldsymbol{J}$ is conserved.

Calculations of the hyperfine splitting in multicharged ions were initiated several years ago by *Sunyaev* and *Churazov* [1]. They suggested to use astronomical search of the radio lines in the millimeter range, which correspond to the transition between the hyperfine structure components of multicharged ions in the $Z < 30$ nuclear charge range, for the investigation of the chemical composition of hot astrophysical plasmas. Accurate calculations of the ground-state hyperfine splitting in these hydrogen-like and lithium-like ions have been done in [2, 3]. These results are also reviewed in [4].

Renewed interest in the calculations of the hyperfine splitting in highly charged ions has been triggered by high precision measurements of the ground state hyperfine splitting in hydrogen-like Bi [5] and Ho ions [6]. Recently [7], the first experimental value of the ground-state hyperfine splitting in lithium-like bismuth was reported. The status of the theory of the hyperfine structure splitting of these ions as well as of other ions, which seem to be candidates for ESR and Super-EBIT experiments, is discussed in the present chapter.

2. Hydrogen-Like Ions

2.1 Magnetic-Dipole Hyperfine Splitting

The magnetic-dipole hyperfine splitting in hydrogen-like ions can be written in the form [8]

$$\begin{aligned} \Delta E_\mu &= \frac{\alpha(\alpha Z)^3}{n^3}\frac{\mu}{\mu_N}\frac{m_\mathrm{e}}{m_\mathrm{p}}\frac{F(F+1)-I(I+1)-j(j+1)}{2Ij(j+1)(2l+1)}m_\mathrm{e}c_\mathrm{o}^2 \\ &\quad \times\{A(\alpha Z)(1-\delta)(1-\varepsilon)+x_\mathrm{rad}\}\,. \end{aligned} \tag{1}$$

Here α is the fine-structure constant, Z is the nuclear charge, m_e is the electron rest mass, m_p is the proton mass, c_o is the speed of light in vacuum, μ is the nuclear magnetic moment, μ_N is the nuclear magneton, I is the nuclear spin, j is the total electron moment, l is the orbital electron moment, F is the total atomic moment, and n is the principal quantum number. $A(\alpha Z)$ denotes the relativistic factor [9–11]

$$A(\alpha Z)=\frac{n^3(2l+1)\kappa[2\kappa(\gamma+n_r)-N]}{N^4\gamma(4\gamma^2-1)}\,, \tag{2}$$

where $\kappa=(-1)^{j+l+\frac{1}{2}}(j+\frac{1}{2})$, $\gamma=\sqrt{\kappa^2-(\alpha Z)^2}$, $N=\sqrt{n_r^2+2n_r\gamma+\kappa^2}$, and n_r is the radial quantum number $n=n_r+|\kappa|$; δ is the nuclear charge distribution correction, ε is the nuclear magnetization distribution correction (the Bohr–Weisskopf effect), and x_rad is the QED correction. We neglect here the nuclear recoil correction because it is small for high nuclear charge Z. In the non-relativistic approximation this correction is determined by the factor $(M/(M+m))^3$, where M is the nuclear mass.

2.2 Nuclear Charge Distribution Correction

With accuracy of the order of ZR/a_0, where R is the nuclear charge radius and a_0 is the Bohr radius, the nuclear charge distribution correction is calculated analytically in [8]. There, the correction δ is tabulated for s and p states in the case of uniformly charged nuclei. More accurate numerical results for specific nuclei, based on the two-parameter Fermi model for the nuclear charge distribution

$$\rho(r)=\frac{\rho_0}{1+\exp\left[(r-r_0)/z\right]}\,, \tag{3}$$

are considered in [12–17]. Here, r_0 is the half-density radius and z is related to the skin thickness t by $t=(4\log 3)z$, defined as the distance over which the charge density falls from 90% to 10% of its maximum value.

Table 1 displays δ-values for the $1s$ and $2s$ states [14, 16]. We note also that in [17], in addition to the uniformly charged nucleus and the Fermi model, calculations based on the potential derived from a Fourier–Bessel expansion of the nuclear charge distribution have been performed. The results of these calculations are very close to the two-parameter Fermi model results.

Table 1. The nuclear charge distribution correction δ for the $1s$ and $2s$ states [14, 16]. The root-mean-square nuclear charge radii used in calculations are taken from [18–20]

Ion	$\langle r^2\rangle^{1/2}$, fm	$\delta^{(1s)}$	$\delta^{(2s)}$
$^{113}In^{48+}$	4.598	0.0170	0.0175
$^{121}Sb^{50+}$	4.681	0.0191	0.0197
$^{123}Sb^{50+}$	4.689	0.0191	0.0197
$^{127}I^{52+}$	4.749	0.0213	0.0220
$^{133}Cs^{54+}$	4.8 04	0.0237	0.0245
$^{139}La^{56+}$	4.85	0.0263	0.0273
$^{141}Pr^{58+}$	4.892	0.0292	0.0304
$^{151}Eu^{62+}$	5.044	0.0365	0.0382
$^{159}Tb^{64+}$	5.099	0.0407	0.0427
$^{165}Ho^{66+}$	5.19	0.0456	0.0480
$^{175}Lu^{70+}$	5.37	0.0575	0.0607
$^{181}Ta^{72+}$	5.48	0.0645	0.0683
$^{185}Re^{74+}$	5.351	0.0706	0.0749
$^{203}Tl^{80+}$	5.463	0.0988	0.1054
$^{205}Tl^{80+}$	5.470	0.0989	0.1055
$^{207}Pb^{81+}$	5.497	0.1047	0.1118
$^{209}Bi^{82+}$	5.533	0.1111	0.1187

2.3 Nuclear Magnetization Distribution Correction

Calculations of the Bohr–Weisskopf (BW) correction within the single particle nuclear model have been performed by many authors. In this model, the nuclear magnetization is ascribed to the odd nucleon with an effective g_S factor chosen to yield the observed nuclear magnetic moment (the g_L factor is commonly set to 1 and 0 for a proton and a neutron, respectively). In [14] the nucleon wave function is calculated by the Schrödinger equation with the Woods–Saxon potential

$$U(r) = V(r) + V_{\rm SO}(r) + V_{\rm Coul}(r)\,, \tag{4}$$

where

$$V(r) = -V_0 f(r)\,,$$

$$V_{\rm SO}(r) = \phi_{\rm SO}(r)(\boldsymbol{s}\cdot\boldsymbol{l})\,,$$

$$V_{\rm Coul}(r) = \begin{cases} \alpha(Z-1)(3-r^2/R_0^2)/2R_0 & r \le R_0 \\ \alpha(Z-1)/r & r \ge R_0 \end{cases}\,,$$

$$\phi_{\rm SO}(r) = \frac{\lambda}{2}\left(\frac{\hbar}{m_{\rm p}c_{\rm o}}\right)^2 \frac{V_0}{r}\frac{{\rm d}f_{\rm SO}(r)}{{\rm d}r}\,,$$

$$f(r) = \Big[1 + \exp\left[(r-R_0)/a\right]\Big]^{-1}\,,$$

$$f_{\rm SO}(r) = \Big[1 + \exp\left[(r-R_{\rm SO})/a\right]\Big]^{-1}\,.$$

Here, V_0 is the depth of the central nuclear potential, R_0 is its radius, and a is its diffusivity; λ is a positive dimensionless parameter of the nuclear spin–orbit interaction and $R_{\rm SO}$ is its radius. The diffusivity of the spin–orbit interaction is taken to be the same as for the central nuclear potential. In the neutron case, the term $V_{\rm Coul}$ must be omitted.

The potential parameters employed in the calculation are chosen to yield the nuclear binding energies in the lead region [21]. The phenomenological spin–orbit interaction ($V_{\rm SO}$) is larger by an order of magnitude than that followed from the Dirac equation for the nucleon [22]. Inclusion of this interaction in the Woods–Saxon potential results in an additional term in the nuclear magnetic moment [23] and in the BW effect [14, 15]. Taking $g_L = 0$ in the neutron case and $g_L = 1$ in the proton case, g_S is chosen to yield the experimental value of the magnetic moment within the single particle approximation

$$\frac{\mu}{\mu_N} = \frac{1}{2} g_S + \Big[I - \frac{1}{2} + \frac{2I+1}{4(I+1)} \frac{m_{\rm p}}{\hbar^2} \langle \phi_{\rm SO} r^2 \rangle \Big] g_L \tag{5}$$

for $I = L + 1/2$, and

$$\frac{\mu}{\mu_N} = -\frac{I}{2(I+1)} g_S + \Big[\frac{I(2I+3)}{2(I+1)} - \frac{2I+1}{4(I+1)} \frac{m_{\rm p}}{\hbar^2} \langle \phi_{\rm SO} r^2 \rangle \Big] g_L \tag{6}$$

for $I = L - 1/2$. The Bohr–Weisskopf correction within the single particle approximation is given by

$$\begin{aligned} \varepsilon &= \frac{g_S}{g_I} \Big[\frac{1}{2I} \langle K_S \rangle + \frac{(2I-1)}{8I(I+1)} \langle K_S - K_L \rangle \Big] \\ &\quad + \frac{g_L}{g_I} \Big[\frac{(2I-1)}{2I} \langle K_L \rangle + \frac{(2I+1)}{4I(I+1)} \frac{m_{\rm p}}{\hbar^2} \langle \phi_{\rm SO} r^2 K_L \rangle \Big] \end{aligned} \tag{7}$$

for $I = L + 1/2$, and

$$\begin{aligned} \varepsilon &= \frac{g_S}{g_I} \Big[-\frac{1}{2(I+1)} \langle K_S \rangle - \frac{(2I+3)}{8I(I+1)} \langle K_S - K_L \rangle \Big] \\ &\quad + \frac{g_L}{g_I} \Big[\frac{(2I+3)}{2(I+1)} \langle K_L \rangle - \frac{(2I+1)}{4I(I+1)} \frac{m_{\rm p}}{\hbar^2} \langle \phi_{\rm SO} r^2 K_L \rangle \Big] \end{aligned} \tag{8}$$

for $I = L - 1/2$. Here

$$\langle K_S \rangle = \int_0^\infty K_S(r) |u(r)|^2 r^2 \, {\rm d}r \,,$$

$$\langle K_L \rangle = \int_0^\infty K_L(r) |u(r)|^2 r^2 \, {\rm d}r \,,$$

$$\langle \phi_{\rm SO} r^2 K_L \rangle = \int_0^\infty \phi_{\rm SO}(r) r^2 K_L(r) |u(r)|^2 r^2 \, {\rm d}r \,,$$

$$K_S(r) = \frac{\int_0^r gf\,\mathrm{d}r'}{\int_0^\infty gf\,\mathrm{d}r'}\,,$$

$$K_L(r) = \frac{\int_0^r \left(1-\frac{r'^3}{r^3}\right) gf\,\mathrm{d}r'}{\int_0^\infty gf\,\mathrm{d}r'}\,,$$

g and f are the radial parts of the Dirac electron wave function defined by

$$\psi_{n\kappa m}(\boldsymbol{r}) = \begin{pmatrix} g_{n\kappa}(r)\Omega_{\kappa m}(\boldsymbol{n}) \\ \mathrm{i} f_{n\kappa}(r)\Omega_{-\kappa m}(\boldsymbol{n}) \end{pmatrix}.$$

Here $u(r)$ is the radial part of the wave function of the odd nucleon. Neglecting the spin–orbit terms in (7)–(8) yields the original Bohr–Weisskopf formulae [24–26]. For s and p electron states, the functions $K_S(r)$ and $K_L(r)$, with a reasonable precision, can be calculated by using the following approximate formulae [8]

$$K_S(r) = b\left[\frac{a_1}{2}\left(\frac{r}{R}\right)^2 + \frac{a_2}{4}\left(\frac{r}{R}\right)^4 + \frac{a_3}{6}\left(\frac{r}{R}\right)^6\right], \tag{9}$$

$$K_L(r) = 3b\left[\frac{a_1}{10}\left(\frac{r}{R}\right)^2 + \frac{a_2}{28}\left(\frac{r}{R}\right)^4 + \frac{a_3}{54}\left(\frac{r}{R}\right)^6\right], \tag{10}$$

where R is the radius of the uniformly charged nucleus. The coefficients a_1, a_2, and a_3 are given by

$$a_1^{(s)} = \frac{1}{2}\alpha Z\,, \qquad a_1^{(p_{\frac{1}{2}})} = \frac{1}{2}\alpha Z + \frac{2}{3}\frac{Rm_e c_o}{\hbar}\,,$$

$$a_2^{(s)} = a_2^{(p_{1/2})} = -\frac{1}{10}\alpha Z - \frac{3}{10}(\alpha Z)^3\,,$$

$$a_3^{(s)} = a_3^{(p_{1/2})} = \frac{19}{140}(\alpha Z)^3 + \frac{81}{1120}(\alpha Z)^5\,.$$

The coefficient b is calculated analytically with an accuracy of the order of ZR/a_0 in [8] and is tabulated there in the range of $Z = 5$–100. Using the approximate formulae (9)–(10) results in a few percent error in the Bohr–Weisskopf correction. Generally, it is much smaller than the error due to a deviation from the single particle nuclear model.

Table 2 gives the Bohr–Weisskopf corrections found within the single particle nuclear model [14, 15]. Unfortunately, there is no well-defined method to estimate the uncertainty of the Bohr–Weisskopf correction. Taking into account that the single particle model with the effective g_S-factor usually gives reasonable agreement with experiment for neutral atoms [27, 28], the corresponding errors for ε are assumed in [14]. For ions, where the spin and orbital parts in (7)–(8) are of the same sign (it results in relatively large values of ε), the uncertainty of ε is expected to be about 30%. For ions, where the spin and orbital parts in Eqs. (7)–(8) are of opposite sign (it results in relatively small values of ε), the uncertainty of ε is expected to be about 20% of $\langle K_S\rangle$ ($0.2\langle K_S\rangle \sim 0.03\alpha Z b$). In the case of Pb, instead of the approximate formulae (9)–(10), the exact numerical calculations of $K_S(r)$ and $K_L(r)$ are employed and the uncertainty of ε is assumed to be about 10%. The last estimate looks reasonably if we take into account that in case of lead, first, the single particle model well predicts the nuclear energy levels and, second, using the free g-factors for the neutron ($g_S^{(n)} = -3.8263$ and $g_L^{(n)} = 0$) yields the magnetic moment which deviates from the observed magnetic moment by less than 10%. It follows that in case of lead, the core polarization effects, which are omitted in the single particle model, should contribute to the magnetic moment not more than 10 %. In this connection, it is natural to expect that the contribution of the core polarization effects to the Bohr–Weisskopf correction is also about 10%. We stress, however, that the uncertainties of ε found this way should be considered only as the order of magnitude of the expected error. More accurate calculations of the Bohr–Weisskopf effect must be based on many-particle nuclear models and include a more complete procedure for the determination of the uncertainty. In the case of Bi, a calculation of the BW correction including many-particle nuclear effects has been performed in [29].

Table 2. The Bohr–Weisskopf correction within the single particle nuclear model for the $1s$ and $2s$ states [14, 15]. g_S is the effective spin g-factor which is chosen to yield the experimental value of the magnetic moment

Ion	Nucleon state	g_S	$\varepsilon^{(1s)}$	$\varepsilon^{(2s)}$
$^{113}In^{48+}$	$1g_{9/2}$	3.67	0.0047	0.0048
$^{121}Sb^{50+}$	$2d_{5/2}$	3.04	0.0052	0.0053
$^{123}Sb^{50+}$	$1g_{7/2}$	4.21	0.0014	0.0014
$^{127}I^{52+}$	$2d_{5/2}$	1.95	0.0052	0.0054
$^{133}Cs^{54+}$	$1g_{7/2}$	4.14	0.0017	0.0018
$^{139}La^{56+}$	$1g_{7/2}$	3.64	0.0025	0.0026
$^{141}Pr^{58+}$	$2d_{5/2}$	4.88	0.0075	0.0078
$^{151}Eu^{62+}$	$2d_{5/2}$	3.27	0.0080	0.0084
$^{159}Tb^{64+}$	$2d_{3/2}$	−0.203	0.0069	0.0073
$^{165}Ho^{66+}$	$1f_{7/2}$	2.90	0.0089	0.0094
$^{175}Lu^{70+}$	$1g_{7/2}$	5.10	0.0006	0.0006
$^{181}Ta^{72+}$	$1g_{7/2}$	4.76	0.0017	0.0018
$^{185}Re^{74+}$	$2d_{5/2}$	2.71	0.0122	0.0130
$^{203}Tl^{80+}$	$3s_{1/2}$	3.47	0.0179	0.0193
$^{205}Tl^{80+}$	$3s_{1/2}$	3.50	0.0179	0.0193
$^{207}Pb^{81+}$	$3p_{1/2}$	−3.56	0.0419	0.0451
$^{209}Bi^{82+}$	$1h_{9/2}$	2.80	0.0118	0.0127

2.4 QED Correction

The radiative correction is the sum of the vacuum-polarization (VP) and self-energy (SE) contributions. These contributions are defined by the Feynman diagrams presented in Figs. 1 and 2, respectively, where the dashed lines denote the hyperfine interaction.

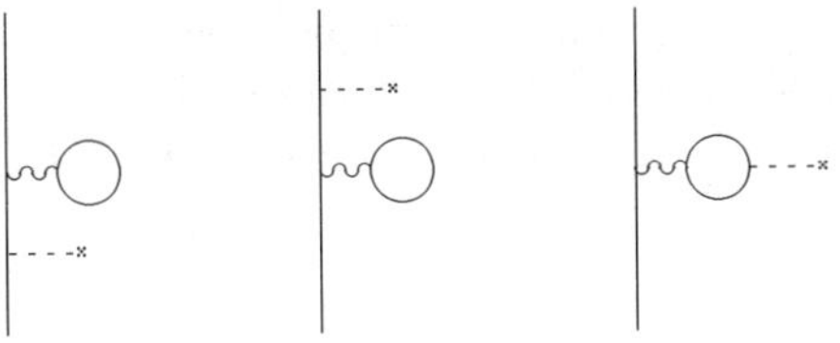

Fig. 1. Vacuum polarization – hyperfine splitting diagrams

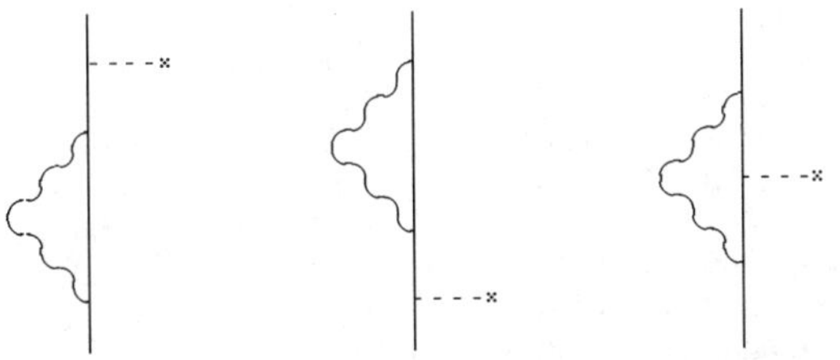

Fig. 2. Self energy – hyperfine splitting diagrams

These diagrams correspond to possible insertions of the hyperfine interaction into the electron lines of the first order vacuum polarization and self-energy diagrams. At present, the QED corrections are calculated for the $1s$ state [14, 30–34] and for the $2s$ state [15, 35, 36] only. The calculation of the VP correction includes completely the Uehling contribution. As to the Wichman–Kroll (WK) contribution, only the part which describes the electric loop corrections to the electron wave function has been calculated. The WK magnetic loop contribution, which involves a renormalization of the nuclear magnetic moment [37, 38], has not been calculated so far. However, calculation of the corresponding term to the VP screening diagrams for two-electron ions [39] allows one to expect that this term is quite small. The values of the QED corrections for the $1s$ and $2s$ states taken from [14–16] are listed in Table 3.

2.5 Ground-State Hyperfine Splitting

The energy difference between the ground-state hyperfine splitting components is given by

$$\Delta E^{(1s)} = \frac{4}{3}\alpha(\alpha Z)^3 \frac{\mu}{\mu_N}\frac{m_{\rm e}}{m_{\rm p}}\frac{2I+1}{2I} m_{\rm e} c_{\rm o}^2$$

$$\times\{A^{(1s)}(\alpha Z)(1-\delta^{(1s)})(1-\varepsilon^{(1s)})+x_{\rm rad}^{(1s)}\}\,, \tag{11}$$

where the relativistic factor $A^{(1s)}(\alpha Z)$ is

$$A^{(1s)}(\alpha Z)=\frac{1}{\gamma(2\gamma-1)}=1+\frac{3}{2}(\alpha Z)^2+\frac{17}{8}(\alpha Z)^4+\cdots\,, \tag{12}$$

$$\gamma=\sqrt{1-(\alpha Z)^2}\,.$$

The energies and the wavelengths of the transition between the ground-state hyperfine splitting components are listed in Table 4. The uncertainties given in the table should be considered only as the order of magnitude of the expected error.

Table 3. The QED corrections to the hyperfine splitting for the $1s$ and $2s$ states in terms of x defined by (1) [14–16]. $x_{\rm VP}$ is the VP contribution without the WK magnetic loop part, $x_{\rm SE}$ is the SE contribution, and $x_{\rm rad}$ is the total QED correction

Z	$x_{\rm VP}^{(1s)}$	$x_{\rm SE}^{(1s)}$	$x_{\rm rad}^{(1s)}$	$x_{\rm VP}^{(2s)}$	$x_{\rm SE}^{(2s)}$	$x_{\rm rad}^{(2s)}$
49	0.0031	−0.0074	−0.0043	0.0034	−0.0079	−0.0045
53	0.0036	−0.0084	−0.0048	0.0040	−0.0092	−0.0052
57	0.0043	−0.0096	−0.0053	0.0048	−0.0107	−0.0059
63	0.0054	−0.0116	−0.0062	0.0063	−0.0134	−0.0071
67	0.0063	−0.0132	−0.0069	0.0075	−0.0156	−0.0081
71	0.0074	−0.0150	−0.0076	0.0091	−0.0183	−0.0092
75	0.0088	−0.0171	−0.0083	0.0111	−0.0216	−0.0104
82	0.0119	−0.0218	−0.0099	0.0160	−0.0291	−0.0131
83	0.0125	−0.0226	−0.0101	0.0169	−0.0304	−0.0135

Table 5 shows the individual contributions to the ground-state hyperfine splitting in hydrogen-like Ho and Bi ions. As one can see from this table, in both cases the theoretical results are in reasonable agreement with experiment. The theoretical errors, which are mainly defined by the uncertainty of the BW correction, are comparable with the total QED corrections. It restricts possibilities for testing the QED effects by the direct comparison of theory and experiment for the ground-state hyperfine splitting in high-Z hydrogen-like ions.

2.6 Electric Quadrupole Splitting

For the states with $j\neq 1/2$, in addition to the magnetic dipole interaction, the electric quadrupole interaction has to be taken into account. Relativistic calculation of the electric quadrupole splitting for a hydrogen-like ion leads to [8, 11]

$$\Delta E_Q = \Delta + BC(C+1)\,, \tag{13}$$

where

Table 4. The energies (ΔE) and the wavelengths (λ) for the transitions between the ground-state hyperfine structure components in hydrogen-like ions [14]. $A^{(1s)}$ is the relativistic factor, $\delta^{(1s)}$ is the nuclear charge distribution correction, $\varepsilon^{(1s)}$ is the nuclear magnetization distribution correction (the Bohr–Weisskopf correction), and $x_{\rm rad}^{(1s)}$ is the radiative correction (see equation (1)). The nuclear magnetic moments μ are taken from [40–42]

Ion	μ/μ_N	$A^{(1s)}$	$\delta^{(1s)}$	$\varepsilon^{(1s)}$	$x_{\rm rad}^{(1s)}$	ΔE (eV)	λ (nm)
$^{113}In^{48+}$	5.5289(2)	1.2340	0.0170	0.0047	−0.0043	0.9148(13)	1355.3(1.9)
$^{121}Sb^{50+}$	3.3634(3)	1.2582	0.0191	0.0052	−0.0045	0.6891(11)	1799.3(2.8)
$^{123}Sb^{50+}$	2.5498(2)	1.2582	0.0191	0.0014	−0.0045	0.4994(7)	2482.5(3.5)
$^{127}I^{52+}$	2.81327(8)	1.2843	0.0213	0.0052	−0.0048	0.6587(10)	1882.2(3.0)
$^{133}Cs^{54+}$	2.58202	1.3125	0.0237	0.0017	−0.0051	0.6582(11)	1883.6(3.0)
$^{139}La^{56+}$	2.78305	1.3430	0.0263	0.0025	−0.0053	0.8052(14)	1539.9(2.6)
$^{141}Pr^{58+}$	4.2754(5)	1.3761	0.0292	0.0075	−0.0056	1.464(3)	847.0(1.9)
$^{151}Eu^{62+}$	3.4717(6)	1.4509	0.0365	0.0080	−0.0062	1.513(4)	819.4(2.0)
$^{159}Tb^{64+}$	2.014(4)	1.4933	0.0407	0.0069	−0.0065	1.099(3)	1128.(3.)
$^{165}Ho^{66+}$	4.177(5)	1.5395	0.0456	0.0089	−0.0069	2.189(7)	566.3(1.7)
$^{175}Lu^{70+}$	2.2327(11)	1.6453	0.0575	0.0006	−0.0076	1.482(4)	836.6(2.4)
$^{181}Ta^{72+}$	2.3705(7)	1.7061	0.0645	0.0017	−0.0080	1.758(5)	705.3(2.2)
$^{185}Re^{74+}$	3.1871(3)	1.7731	0.0706	0.0122	−0.0083	2.749(10)	451.0(1.7)
$^{203}Tl^{80+}$	1.62226	2.0217	0.0988	0.0179	−0.0096	3.229(18)	384.0(2.1)
$^{205}Tl^{80+}$	1.63821	2.0217	0.0989	0.0179	−0.0096	3.261(18)	380.2(2.1)
$^{207}Pb^{81+}$	0.592583(9)	2.0718	0.1047	0.0419	−0.0099	1.215(5)	1020.3(4.5)
$^{209}Bi^{82+}$	4.1106(2)	2.1250	0.1111	0.0118	−0.0101	5.101(27)	243.0(1.3)

Table 5. The individual contributions (in eV) to the ground-state hyperfine splitting in $^{209}Bi^{82+}$ ($\mu = 4.1106(2)\mu_N$) and in $^{165}Ho^{66+}$ ($\mu = 4.177(5)\mu_N$) [14]

Contribution	$^{209}Bi^{82+}$	$^{165}Ho^{66+}$
Nonrelativistic value	2.7479(1)	1.5108(18)
Relativistic value (point nucleus)	5.8393(3)	2.3257(28)
Nuclear size effect	−0.6487(20)	−0.1062(7)
Bohr–Weisskopf effect	−0.061(27)	−0.0198(59)
Vacuum polarization	0.0343	0.0095
Self energy	−0.0621	−0.0199
Total QED	−0.0278	−0.0104
Total theoretical value	5.101(27)	2.1894(66)
Experiment [5, 6]	5.0840(8)	2.1645(6)

$$
\begin{aligned}
C &= F(F+1) - I(I+1) - j(j+1)\,, \\
\Delta &= -\frac{4}{3} I(I+1) j(j+1) B\,, \\
B_{j\neq\frac{1}{2}} &= \frac{3Q\alpha(\alpha Z)^3[N^2(1+2\gamma^2) - 3\kappa N(\gamma+n_r) + 3(\alpha Z)^2(N^2-\kappa^2)]}{8I(2I-1)j(j+1)N^5(\gamma^2-1)\gamma(4\gamma^2-1)} \\
&\quad\times \frac{m_{\rm e}^3 c_{\rm o}^4}{\hbar^2}\,, \\
B_{j=\frac{1}{2}} &= 0\,.
\end{aligned}
$$

Here Q is the electric quadrupole moment of the nucleus. The nuclear size corrections to ΔE_Q are very small since the wave functions of the states with $j \neq \frac{1}{2}$ equal zero at $r = 0$.

3. Lithium-Like Ions

The energy difference between the ground-state hyperfine splitting components of a high-Z lithium-like ion can be written in the form

$$\begin{aligned}\Delta E^{(2s)} &= \frac{1}{6}\alpha(\alpha Z)^3 \frac{m_\mathrm{e}}{m_\mathrm{p}} \frac{\mu}{\mu_N} \frac{2I+1}{2I} m_\mathrm{e} c_\mathrm{o}^2 \\ &\quad \times \Big\{ [A^{(2s)}(\alpha Z)(1-\delta^{(2s)})(1-\varepsilon^{(2s)}) + x_\mathrm{rad}^{(2s)}] \\ &\quad + \frac{1}{Z} B(\alpha Z) + \frac{1}{Z^2} C(\alpha Z) + \cdots \Big\} . \end{aligned} \tag{14}$$

Here $A^{(2s)}(\alpha Z)$ denotes the one-electron relativistic factor for the $2s$ state

$$\begin{aligned} A^{(2s)}(\alpha Z) &= \frac{2[2(1+\gamma) + \sqrt{2(1+\gamma)}]}{(1+\gamma)^2 \gamma (4\gamma^2 - 1)} \\ &= 1 + \frac{17}{8}(\alpha Z)^2 + \frac{449}{128}(\alpha Z)^4 + \cdots , \end{aligned} \tag{15}$$

$\delta^{(2s)}$ is the one-electron nuclear charge distribution correction, $\varepsilon^{(2s)}$ is the one-electron nuclear-magnetization distribution correction, and $x_\mathrm{rad}^{(2s)}$ is the one-electron radiative correction. The terms $B(\alpha Z)/Z$ and $C(\alpha Z)/Z^2$ correspond to the interelectronic interaction corrections of the first and second orders on the parameter $1/Z$, respectively.

3.1 One-Electron Contribution

The one-electron contribution is enclosed in square brackets in (14). We denote it by $a^{(2s)}$:

$$a^{(2s)} = A^{(2s)}(\alpha Z)(1-\delta^{(2s)})(1-\varepsilon^{(2s)}) + x_\mathrm{rad}^{(2s)} . \tag{16}$$

The values of the one-electron corrections $\delta^{(2s)}$, $\varepsilon^{(2s)}$, and $x_\mathrm{rad}^{(2s)}$ have already been listed in Tables 1, 2, and 3, respectively.

3.2 Interelectronic Interaction Corrections

The function $B(\alpha Z)$ is defined by the Feynman diagrams shown in Fig. 3. These diagrams correspond to possible insertions of the hyperfine interaction into the electron lines of the one-photon-exchange diagram. Calculations of these diagrams have been performed for point nuclei [43] as well as for extended nuclei [15, 16]. As in the case of the one-electron contribution, it is convenient to represent the function $B(\alpha Z)$ in the form

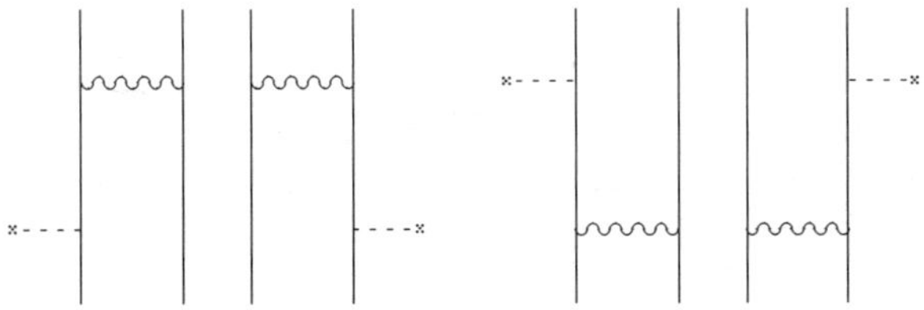

Fig. 3. Interelectronic interaction – hyperfine splitting diagrams

$$B(\alpha Z) = B_0(\alpha Z)(1 - \delta_B)(1 - \varepsilon_B)\,, \tag{17}$$

where $B_0(\alpha Z)$ is the point nucleus approximation for $B(\alpha Z)$, δ_B is the nuclear charge distribution correction to this function, and ε_B is the nuclear magnetization distribution correction. To the lowest orders on αZ the function $B_0(\alpha Z)$ is given by [43]

$$B(\alpha Z) = -2.6557 - 6.2138(\alpha Z)^2 + \cdots\,. \tag{18}$$

The values B_0 and δ_B, taken from [15, 16], are given in the second and third columns of Table 6. Comparing Tables 1 and 6, one can see that the correction δ_B is very close to the related one-electron correction $\delta^{(2s)}$. In Ref. [15] it is assumed that the nuclear magnetization correction ε_B is also close to the related one-electron value $\varepsilon^{(2s)}$. This assumption is based on the fact that the interaction of the $2s$ electron with the closed $1s$ shell, which gives the dominant part of the interelectronic interaction contribution (see Ref. [43]

Table 6. The function $B(\alpha Z)$ defined by (14) [16]. $B_0(\alpha Z)$ is the point-nucleus value, δ_B is the nuclear charge distribution correction, $B_{\rm NS}(\alpha Z) = B_0(\alpha Z)(1-\delta_B)$, and $B_{\rm NS,BW}(\alpha Z) = B_0(\alpha Z)(1-\delta_B)(1-\varepsilon)$

Ion	$B_0(\alpha Z)$	δ_B	$B_{NS}(\alpha Z)$	$B_{NS,BW}(\alpha Z)$
$^{113}In^{46+}$	−3.677	0.018	−3.611	−3.594
$^{121}Sb^{48+}$	−3.788	0.020	−3.712	−3.692
$^{123}Sb^{48+}$	−3.788	0.020	−3.712	−3.707
$^{127}I^{50+}$	−3.909	0.023	−3.821	−3.800
$^{133}Cs^{52+}$	−4.042	0.025	−3.939	−3.932
$^{139}La^{54+}$	−4.186	0.028	−4.068	−4.057
$^{141}Pr^{56+}$	−4.344	0.032	−4.207	−4.174
$^{151}Eu^{60+}$	−4.707	0.040	−4.520	−4.482
$^{159}Tb^{62+}$	−4.916	0.045	−4.697	−4.663
$^{165}Ho^{64+}$	−5.147	0.050	−4.889	−4.843
$^{175}Lu^{68+}$	−5.687	0.064	−5.324	−5.321
$^{181}Ta^{70+}$	−6.003	0.072	−5.572	−5.561
$^{185}Re^{72+}$	−6.357	0.079	−5.855	−5.779
$^{203}Tl^{78+}$	−7.711	0.112	−6.849	−6.717
$^{205}Tl^{78+}$	−7.711	0.112	−6.849	−6.717
$^{207}Pb^{79+}$	−7.992	0.119	−7.043	−6.726
$^{209}Bi^{80+}$	−8.292	0.126	−7.247	−7.154

for details), in the nuclear region, changes mainly the normalization factor of the 2s wave function but not its behavior. So, in the last column of Table 6 the values $B(\alpha Z)$ corrected by the factor $(1-\varepsilon^{(2s)})$ are listed.

The term $C(\alpha Z)/Z^2$, which is small enough for high Z, is estimated in the non-relativistic approximation as

$$\frac{C(\alpha Z)}{Z^2} \approx \frac{C(0)}{Z^2} . \tag{19}$$

The coefficient $C(0)$ is found from the configuration-interaction Hartree–Fock (CI-HF) calculation of Ref. [2] to be $C(0) = 0.87 \pm 0.05$.

3.3 Total Theoretical Values

In Table 7, the total theoretical values of the energies and wavelengths are given for the transition between the ground-state hyperfine splitting components of high-Z lithium-like ions [15, 16].

Table 7. The energies ΔE and the wavelengths λ for the transitions between the ground-state hyperfine structure components in lithium-like ions [16]. The Bohr–Weisskopf effect is calculated within the single particle nuclear model. $A^{(2s)}(\alpha Z)$ is the relativistic one-electron factor defined by (15) and $a^{(2s)}$ is the total one-electron contribution defined by (16). $B(\alpha Z)/Z$ and $C(0)/Z^2$ are the interelectronic interaction contributions defined by (14). $\eta = 8\Delta E^{(2s)}/\Delta E^{(1s)}$. The nuclear magnetic moments used are the same as in Table 4

Ion	$A^{(2s)}(\alpha Z)$	$a^{(2s)}$	$B(\alpha Z)/Z$	$C(0)/Z^2$	ΔE (eV)	$\lambda(\mu m)$	η
$^{113}In^{46+}$	1.3425	1.3081	−0.0733	0.0004	0.11742(18)	10.56(2)	1.0268(2)
$^{121}Sb^{48+}$	1.3791	1.3400	−0.0724	0.0003	0.08929(15)	13.89(2)	1.0366(2)
$^{123}Sb^{48+}$	1.3791	1.3452	−0.0727	0.0003	0.06472(9)	19.16(3)	1.0366(2)
$^{127}I^{50+}$	1.4188	1.3749	−0.0717	0.0003	0.08617(15)	14.39(2)	1.0465(2)
$^{133}Cs^{52+}$	1.4620	1.4180	−0.0715	0.0003	0.08697(14)	14.26(2)	1.0570(2)
$^{139}La^{54+}$	1.5089	1.4580	−0.0712	0.0003	0.10746(20)	11.54(2)	1.0677(3)
$^{141}Pr^{56+}$	1.5601	1.4945	−0.0707	0.0002	0.1974(5)	6.282(16)	1.0786(3)
$^{151}Eu^{60+}$	1.6770	1.5922	−0.0711	0.0002	0.2084(6)	5.948(16)	1.1021(3)
$^{159}Tb^{62+}$	1.7440	1.6497	−0.0717	0.0002	0.1531(5)	8.10(3)	1.1145(3)
$^{165}Ho^{64+}$	1.8175	1.7059	−0.0723	0.0002	0.3085(10)	4.018(13)	1.1274(3)
$^{175}Lu^{68+}$	1.9879	1.8569	−0.0749	0.0002	0.2141(7)	5.79(2)	1.1556(3)
$^{181}Ta^{70+}$	2.0871	1.9312	−0.0762	0.0002	0.2572(9)	4.82(2)	1.1703(3)
$^{185}Re^{72+}$	2.1973	1.9958	−0.0770	0.0002	0.407(2)	3.045(13)	1.1850(4)
$^{203}Tl^{78+}$	2.6141	2.2808	−0.0829	0.0001	0.498(3)	2.487(15)	1.2350(6)
$^{205}Tl^{78+}$	2.6141	2.2806	−0.0829	0.0001	0.503(3)	2.463(15)	1.2350(6)
$^{207}Pb^{79+}$	2.6994	2.2763	−0.0820	0.0001	0.1886(9)	6.57(3)	1.2417(5)
$^{209}Bi^{80+}$	2.7904	2.4144	−0.0862	0.0001	0.800(4)	1.550(9)	1.2542(5)

The uncertainties are mainly defined by errors of the Bohr–Weisskopf correction discussed in the previous section. As one can see from Tables 1 and 2 (see also Tables 1 and 2 in [8]), the nuclear corrections ε and δ are functions weakly dependent on the principal quantum number n for s states and, therefore, cancel significantly in the ratio of the $2s$ and $1s$ hyperfine splitting values. It means that, if the value ε is calculated in the same nuclear model for the $1s$ and $2s$ states, the ratio has a higher precision than the individual hyperfine splitting values. In this connection, in the last column of Table 7 the value η defined by

$$\eta = \frac{8\Delta E^{(2s)}}{\Delta E^{(1s)}} = \frac{A^{(2s)}(1-\delta^{(2s)})(1-\varepsilon^{(2s)}) + x_{\rm rad}^{(2s)} + \frac{B(\alpha Z)}{Z} + \frac{C(0)}{Z^2}}{A^{(1s)}(1-\delta^{(1s)})(1-\varepsilon^{(1s)}) + x_{\rm rad}^{(1s)}}\,, \quad (20)$$

is given. This value may be used for comparison of the experimental hyperfine splitting values in hydrogen-like and lithium-like ions. Since the one-electron QED corrections are also significantly canceled in the ratio (20), the η value is mainly defined by the functions $A(\alpha Z)$ and $B(\alpha Z)$.

3.4 Bohr–Weisskopf Correction Derived from the $1s$ Hyperfine Splitting

As in the hydrogen-like case, the accuracy of the theoretical calculations of the hyperfine splitting in lithium-like ions is mainly defined by the uncertainty of the Bohr–Weisskopf correction. However, this uncertainty can be significantly reduced if the $1s$ hyperfine splitting value is known from experiment. For an arbitrary nuclear model in which the residual nucleon-nucleon interaction does not depend on the nucleon momenta, the Bohr–Weisskopf correction is given by

$$\varepsilon = \frac{\langle IM_I|\Delta Q_S^z + \Delta Q_L^z + \Delta Q_{\rm SO}^z|IM_I\rangle}{\langle IM_I|Q_\mu^z|IM_I\rangle}\,, \quad (21)$$

with

$$\Delta Q_S^z = \sum_{i=1}^{\rm A} g_{si}[s_{zi}K_S(r_i) + \sqrt{\frac{\pi}{2}}[Y_{2i}\otimes\sigma_i]_z^1(K_S(r_i) - K_L(r_i))]\,, \quad (22)$$

$$\Delta Q_L^z = \sum_{i=1}^{\rm A} g_{li}l_{zi}K_L(r_i)\,, \quad (23)$$

$$\Delta Q_{\rm SO}^z = \sum_{i=1}^{\rm A} g_{li}\frac{2m_p}{3\hbar^2}[s_{zi} + \sqrt{\frac{\pi}{2}}[Y_{2i}\otimes\sigma_i]_z^1]\phi(r_i)r_i^2K_L(r_i)\,, \quad (24)$$

$$Q_\mu^z = \sum_{i=1}^{\rm A}[g_{si}s_{zi} + g_{li}l_{zi} + g_{li}\frac{2m_p}{3\hbar^2}(s_{zi} + \sqrt{\frac{\pi}{2}}[Y_{2i}\otimes\sigma_i]_z^1)\phi(r_i)r_i^2]\,, \quad (25)$$

where A is the number of nucleons. The functions $K_S(r)$ and $K_L(r)$ are defined in the previous section. As one can see from (21)–(25), the BW correction depends on the electronic structure through the functions $K_S(r)$ and $K_L(r)$. It is important that this dependence is linear. From the expressions (9)–(10), it follows that the functions $K_S(r)$ and $K_L(r)$ for the 2s state differ from those for the 1s state only by the overall factor b. With a good precision, this is also confirmed by accurate numerical calculations. So, in the case of Z=83 the exact numerical calculation for the uniformly charged nucleus gives

$$\frac{K_S^{(2s)}(r)}{K_S^{(1s)}(r)} = \frac{K_L^{(2s)}(r)}{K_L^{(1s)}(r)} = 1.0782 \pm 0.0003$$

in the nuclear region (by the nuclear region we imply here an interval of r, e.g., $0 \leq r \leq 30$ fm, where the wave functions of the nucleons may differ from zero). This fact can easily be understood if we take into account that the binding energy of the electron ($W = E - m_e c_0^2$) is small compared to the nuclear potential $V(r)$ in the nuclear region. So, the binding energy gives a very small correction to the behavior of the functions $g(r)$ and $f(r)$ in the nuclear region. It follows that the Bohr–Weisskopf effect for the 2s state can be found by using ε for the 1s state and the values of the overall factors tabulated in [8] (while the relative precision of b in [8] is of the order of ZR/a_0, the precision of the ratio $b^{(2s)}/b^{(1s)}$ is higher by an order of magnitude).

If the 1s hyperfine splitting is known from experiment, the BW effect for the 1s state is derived from the equation

$$\varepsilon^{(1s)} = \frac{\Delta E_{\rm NS}^{(1s)} + \Delta E_{\rm QED}^{(1s)} - \Delta E_{\rm exp}^{(1s)}}{\Delta E_{\rm NS}^{(1s)}}, \tag{26}$$

where $\Delta E_{\rm NS}^{(1s)}$ is the theoretical hyperfine splitting value incorporating the relativistic and nuclear charge distribution effects, $\Delta E_{\rm QED}^{(1s)}$ is the QED contribution for the 1s state, and $\Delta E_{\rm exp}^{(1s)}$ is the experimental value of the 1s hyperfine splitting. The Bohr–Weisskopf effect for the 2s state is calculated by

$$\varepsilon^{(2s)} = \frac{b^{(2s)}}{b^{(1s)}} \varepsilon^{(1s)}$$

which in the case $Z = 83$ gives $\varepsilon^{(2s)} = 1.078\,\varepsilon^{(1s)}$. Using the experimental and theoretical (without the BW correction) values of the 1s hyperfine splitting from Table 5, one has $\varepsilon^{(1s)} = 0.0152(3)$, $\varepsilon^{(2s)} = 0.0164(4)$ for $^{209}\mathrm{Bi}^{82+}$ and $\varepsilon^{(1s)} = 0.0202(12)$, $\varepsilon^{(2s)}$=0.0213(13) for $^{165}\mathrm{Ho}^{66+}$.

For comparison, the direct calculation, based on the single particle nuclear model, gives $\varepsilon^{(1s)} = 0.0118(49)$, $\varepsilon^{(2s)} = 0.0127(53)$ for $^{209}\mathrm{Bi}^{82+}$ and $\varepsilon^{(1s)} = 0.0089(27)$, $\varepsilon^{(2s)} = 0.0094(28)$ for $^{165}\mathrm{Ho}^{66+}$.

The results of the calculation of the ground-state hyperfine splitting in $^{209}\mathrm{Bi}^{80+}$ and $^{165}\mathrm{Ho}^{64+}$, based on using the 1s experimental value, are presented in Table 8. This table includes also a rough estimate of the combined interelectronic interaction - QED correction (see [15, 16] for details)

$$\Delta E^{(2s)}_{\rm int,QED} \sim \frac{B(\alpha Z)}{Z} \frac{\Delta E^{(2s)}_{\rm QED}}{A^{(2s)}(\alpha Z)} \,. \tag{27}$$

The accuracy of this estimate is assumed to be 50%.

Table 8. The individual contributions (in eV) to the ground-state hyperfine splitting in $^{209}Bi^{80+}$ ($\mu = 4.1106(2)\mu_{\rm N}$) and in $^{165}Ho^{64+}$ ($\mu = 4.177(5)\mu_{\rm N}$) [16]. The Bohr–Weisskopf effect is found by using the experimental values of the $1s$ hyperfine splitting, $\Delta E^{(1s)}_{\rm exp} = 5.0840(8)$ eV for Bi [5] and $\Delta E^{(1s)}_{\rm exp} = 2.1645(6)$ eV for Ho ions [6]

Contribution	$^{209}Bi^{80+}$	$^{165}Ho^{64+}$
Nonrelativistic one-electron value	0.34349(2)	0.1888(2)
Relativistic one-electron value	0.95850(5)	0.3432(4)
Nuclear size	−0.1138(3)	−0.0165(1)
Bohr–Weisskopf	−0.0139(3)	−0.0070(4)
One-electron QED	−0.0046	−0.0015
Interelectronic interaction	−0.02945(4)	−0.01345(4)
Interelectronic-interaction QED	0.00016(8)	0.00006(3)
Total theoretical value	0.7969(2)	0.3049(1)
Experiment [7]	0.820(26)	

It should be stressed here that the uncertainty of the total hyperfine splitting values given in Table 8 is not equal to the sum of the uncertainties of the individual contributions. It is due to the fact that the total hyperfine splitting value found in this way is stable enough with respect to possible variations of the nuclear charge radius and the magnetic moment. To prove this fact, let us represent the $2s$ hyperfine splitting value in the form

$$\begin{aligned} \Delta E^{(2s)} &= \Delta E^{(2s)}_{\rm NS} + \Delta E^{(2s)}_{\rm int,NS} + \nu(\Delta E^{(1s)}_{\rm exp} - \Delta E^{(1s)}_{\rm NS}) \\ &\quad + \Delta E^{(2s)}_{\rm QED} + \Delta E^{(2s)}_{\rm int,QED} - \nu \Delta E^{(1s)}_{\rm QED} \,, \end{aligned} \tag{28}$$

where

$$\nu = \frac{b^{(2s)}}{b^{(1s)}} \frac{\Delta E^{(2s)}_{\rm NS} + \Delta E^{(2s)}_{\rm int,NS}}{\Delta E^{(1s)}_{\rm NS}} \,. \tag{29}$$

Here $\Delta E^{(2s)}_{\rm int,NS}$ is the interelectronic interaction contribution for an extended nucleus. (We note here that the theoretical value of the Bohr–Weisskopf effect is eliminated completely in (28).) Taking, for example, a small variation of the magnetic moment $\delta\mu$ we get

$$\begin{aligned} \delta(\Delta E^{(2s)}) &= \frac{\delta\mu}{\mu} \Big[(\Delta E^{(2s)}_{\rm NS} + \Delta E^{(2s)}_{\rm int}) \Big(1 - \frac{b^{(2s)}}{b^{(1s)}}\Big) \\ &\quad + \Delta E^{(2s)}_{\rm QED} + \Delta E^{(2s)}_{\rm int,QED} - \nu \Delta E^{(1s)}_{\rm QED} \Big] \,. \end{aligned} \tag{30}$$

Because the factor $(1 - b^{(2s)}/b^{(1s)})$ is small (it equals -0.078 for Z=83), the ratio $\delta(\Delta E^{(2s)})/\Delta E^{(2s)}$ is smaller at least by an order of magnitude than $\delta\mu/\mu$. Considering in the same way a small variation of the nuclear charge radius and taking into account that $\delta^{(1s)}$, $\delta^{(2s)}$, and δ_B are significantly canceled in (29) (e.g., in the case $Z = 83$, $(\delta^{(2s)} - \delta^{(1s)})/\delta^{(1s)}$=0.069 and $(\delta_B - \delta^{(1s)})/\delta^{(1s)}$=0.14), we get a similar result. As to a small variation of the 1s experimental hyperfine splitting value, we find $\delta(\Delta E^{(2s)})/\Delta E^{(2s)}$= $\delta(\Delta E^{(1s)}_{\rm exp})/\Delta E^{(1s)}_{\rm exp}$. So, the uncertainty of the total hyperfine splitting value is mainly defined by $\delta(\Delta E^{(1s)}_{\rm exp})$ and the combined interelectronic-interaction QED term estimated by (27). In the case of $^{209}Bi^{80+}$, the total theoretical value constitutes 0.7969(2) eV and agrees with the experimental value 0.820(26) eV [7], although it is close to the limit of experimental error. For comparison, a calculation, based on using the single-particle nuclear model for the Bohr–Weisskopf effect (see Table 7), gives 0.800(4) eV.

3.5 Testing QED Effects

One of the main objects of the investigations of the hyperfine splitting of highly charged ions consists in testing QED effects in the strong electric and magnetic fields of heavy nuclei. As one can see from Table 8, the QED contributions for the 2s state are larger than the uncertainties of the hyperfine splitting values with the Bohr–Weisskopf correction found from the 1s hyperfine splitting. However, since the calculation of the Bohr–Weisskopf effect includes the QED correction for the 1s state, it is natural to consider a value derived from QED as the sum of the last three terms in (28). We find that the value derived from QED is 0.0002(1) eV for ^{209}Bi^{80+} and 0.0001 eV for ^{165}Ho^{64+}. Comparing these values with the uncertainty of the complete theoretical values discussed above allows us to expect that high precision measurements of the ground-state hyperfine splitting in hydrogen-like and lithium-like ions can provide a possibility for testing QED effects in the hyperfine-splitting investigations.

4. Transition Probabilities

4.1 Hydrogen-Like Ions

The transition probability between the hyperfine-splitting components of an atomic level is defined by the $M1$ transition. For a hydrogen-like atom it is given by [15]

$$w_{F\to F'} = \alpha \frac{\omega^3}{c_o^2} \frac{(2F'+1)(2j+1)^3}{3j(j+1)} \begin{Bmatrix} j & F' & I \\ F & j & 1 \end{Bmatrix}^2 \left[\int_0^\infty gfr^3 \mathrm{d}r\right]^2 , \qquad (31)$$

where ω is the transition frequency, j is the electron moment, F and F' are the total atomic moments in the initial and final states, respectively, and

$g(r)$ and $f(r)$ are the upper and lower radial components of the hydrogen-like Dirac wave function defined above. For a point nucleus, using formulae from [44] one simply finds

$$\int_0^\infty gfr^3\mathrm{d}r = \frac{2\kappa\epsilon - m_\mathrm{e}c_\mathrm{o}^2}{4m_\mathrm{e}c_\mathrm{o}^2}\frac{\hbar}{m_\mathrm{e}c_\mathrm{o}}, \tag{32}$$

where

$$\epsilon = \frac{m_\mathrm{e}c_\mathrm{o}^2}{\sqrt{1+\left(\frac{\alpha Z}{\gamma+n_r}\right)^2}}$$

is the one-electron Dirac–Coulomb energy. For s electrons we get

$$w_{F\to F'} = \alpha\omega^3\frac{\hbar^2}{m_\mathrm{e}^2c_\mathrm{o}^4}\frac{4}{27}\frac{I}{2I+1}\left[\frac{2\epsilon}{m_\mathrm{e}c_\mathrm{o}^2}+1\right]^2. \tag{33}$$

In the case of the $^{209}\mathrm{Bi}^{82+}$ ground state, this formula coincides with the corresponding formula in [45] (see also [12]). Because the integrand in (31) is a strongly decreasing function of r at $r \to 0$, the finite nuclear size corrections to (32) and (33) can be neglected. Table 9 displays the transition probabilities and the lifetimes ($\tau = 1/w$) calculated by the formula (33) employing the transition energies from Table 4. According to (31)–(33) the uncertainty of the transition probability is three times larger than the uncertainty of the transition energy.

Calculation of the lifetime for the Bi and Ho ions, based on employing the experimental values of the $1s$ hyperfine splitting in (33), gives $\tau = 0.400(1)$ ms for $^{209}\mathrm{Bi}^{82+}$ and $\tau = 4.76(1)$ ms for $^{165}\mathrm{Ho}^{66+}$. The uncertainties are chosen to be of the order of the uncalculated QED corrections. In the bismuth case, the theoretical result disagrees with the experimental value $\tau_\mathrm{exp} = 0.351(16)$ ms [5] by about 14%.

4.2 Lithium-Like Ions

In the case of lithium-like ions, the interelectronic interaction correction of the order of $1/Z$ has to be added to the one-electron contribution. Such calculations have been performed in Ref. [15]. There, it is found that this correction is small enough. So, it increases w only by 0.23% for $^{209}\mathrm{Bi}^{80+}$ and by 0.17% for $^{165}\mathrm{Ho}^{64+}$. Calculation of the transition probability for a many electron atom, including an approximate treatment of the $1/Z$ term, can also be done by the formula

$$w_{F\to F'} = \alpha\frac{\omega^3}{3c_o^2}(2F'+1)J(J+1)(2J+1)\begin{Bmatrix} J & F' & I \\ F & J & 1 \end{Bmatrix}^2\gamma^2(J), \tag{34}$$

where

$$\gamma(J) = \frac{\langle JM_J|\sum_i[\boldsymbol{r}_i\times\boldsymbol{\alpha}_i]_z|JM_J\rangle}{M_J},$$

Table 9. Transition probabilities (w) and lifetimes ($\tau = 1/w$) for the ground-state hyperfine splitting of hydrogen-like ions calculated with the transition energies from Table 4

Ion	w (s^{-1})	τ (ms)
$^{113}In^{48+}$	17.82(8)	56.11(24)
$^{121}Sb^{50+}$	7.00(3)	142.9(7)
$^{123}Sb^{50+}$	2.796(12)	357.6(1.5)
$^{127}I^{52+}$	6.06(3)	165.0(8)
$^{133}Cs^{54+}$	6.29(3)	158.9(8)
$^{139}La^{56+}$	11.41(6)	87.6(5)
$^{141}Pr^{58+}$	64.7(4)	15.45(10)
$^{151}Eu^{62+}$	70.0(6)	14.29(11)
$^{159}Tb^{64+}$	23.86(20)	41.9(3)
$^{165}Ho^{66+}$	217.6(2.0)	4.60(5)
$^{175}Lu^{70+}$	65.8(5)	15.19(12)
$^{181}Ta^{72+}$	108.4(9)	9.23(8)
$^{185}Re^{74+}$	389.(4.)	2.569(28)
$^{203}Tl^{80+}$	362.(6.)	2.77(5)
$^{205}Tl^{80+}$	372.(6.)	2.69(4)
$^{207}Pb^{81+}$	19.11(25)	52.3(7)
$^{209}Bi^{82+}$	2523.(40.)	0.396(6)

and J and M_J are the total electronic moment and its projection, respectively. Such a calculation [15], based on the CI-HF method [2], confirms the exact (in $1/Z$) results. The transition probabilities and the lifetimes for lithium-like ions, calculated using the transition energies from Table 7, are presented in Table 10.

More accurate calculation of the transition probability for $^{209}Bi^{80+}$ and $^{165}Ho^{64+}$, based on the transition energies from Table 8, gives $w = 12.03(2)$ s^{-1} ($\tau = 0.0831(1)$ s) for $^{209}Bi^{80+}$ and $w = 0.673(1)$ s^{-1} ($\tau = 1.486(2)$ s) for $^{165}Ho^{64+}$.

5. Conclusion

In this chapter we have reviewed the present status of theory of the hyperfine splitting in high-Z hydrogen-like and lithium-like ions. For such systems, the αZ expansion is no longer applicable, and only the complete αZ-dependence calculations are valid.

The ground-state hyperfine splitting in hydrogen-like Bi and Ho ions have been measured very accurately [5, 6], and a reasonable agreement between theory and experiment is obtained by considering relativistic, nuclear, and QED effects. The uncertainty of the theoretical results, which is mainly defined by the error of the Bohr–Weisskopf correction, is comparable to the

Table 10. The transition probabilities w and the lifetimes $\tau = 1/w$ for the ground-state hyperfine splitting of high-Z lithium-like ions calculated with the transition energies from Table 7 [16]

Ion	w (s^{-1})	τ (s)
$^{113}In^{46+}$	0.0404(2)	24.77(11)
$^{121}Sb^{48+}$	0.01641(8)	61.0(3)
$^{123}Sb^{48+}$	0.00656(3)	152.4(6)
$^{127}I^{50+}$	0.01472(8)	67.9(4)
$^{133}Cs^{52+}$	0.01585(8)	63.1(3)
$^{139}La^{54+}$	0.02985(16)	33.5(2)
$^{141}Pr^{56+}$	0.1757(13)	5.69(4)
$^{151}Eu^{60+}$	0.206(2)	4.86(4)
$^{159}Tb^{62+}$	0.0732(7)	13.66(13)
$^{165}Ho^{64+}$	0.698(7)	1.434(14)
$^{175}Lu^{68+}$	0.232(2)	4.32(4)
$^{181}Ta^{70+}$	0.400(4)	2.50(3)
$^{185}Re^{72+}$	1.51(2)	0.663(8)
$^{203}Tl^{78+}$	1.64(3)	0.609(11)
$^{205}Tl^{78+}$	1.69(3)	0.591(11)
$^{207}Pb^{79+}$	0.0888(13)	11.26(17)
$^{209}Bi^{80+}$	12.2(2)	0.0822(14)

QED contribution. It strongly restricts testing the QED contributions by direct comparison of theory and experiment. More accurate calculations of the Bohr–Weisskopf effects, based on many particle nuclear models, are required.

The ground-state hyperfine splitting in lithium-like ^{209}Bi ions has been measured in [7] to be $\Delta E_{\rm exp} = 0.820(26)$ eV. It agrees with the theoretical value found within the single particle nuclear model ($\Delta E = 0.800(4)$ eV [15]), although it is close to the limit of experimental error. It is found that the Bohr–Weisskopf effect for the lithium-like ions can be calculated with a high precision by using the experimental value of the ground-state hyperfine splitting in hydrogen-like ions. Since the Bohr–Weisskopf effect gives a dominant contribution to the uncertainty of the theoretical value, this method allows significant improvement of the theoretical predictions. So, using the experimental result for the $1s$ hyperfine splitting in ^{209}Bi, the ground-state hyperfine splitting in lithium-like ^{209}Bi is calculated to be $\Delta E = 0.7969(2)$eV. The uncertainty of this value is by a factor of 20 smaller than the error of the theoretical value calculated using the single particle nuclear model.

Acknowledgement. The author wants to express his thanks to A.N. Artemyev, Th. Kühl, M.B. Shabaeva, M. Tomaselli, I.I. Tupitsyn, and V.A. Yerokhin for stimulating collaboration. Valuable conversations with T. Beier, P. Beiersdorfer, J.R. Crespo Lopez-Urrutia, S.G. Karshenboim, G. Plunien, S.M. Schneider, P. Seelig, and G. Soff are gratefully acknowledged.

References

1. R.A. Sunyaev, E.M. Churazov: Sov. Phys. - Pis'ma Astron. Zh. **10**, 483 (1984)
2. V.M. Shabaev, M.B. Shabaeva, I.I. Tupitsyn: Phys. Rev. A **52**, 3686 (1995)
3. V.M. Shabaev, M.B. Shabaeva, I.I. Tupitsyn: Astron. Astrophys. Trans. **12**, 243 (1997)
4. H.F. Beyer, H.-J. Kluge, V.P. Shevelko: *X-ray Radiation of Highly Charged Ions* (Springer, Berlin 1997)
5. I. Klaft, S. Borneis, T. Engel, B. Fricke, R. Grieser, G. Huber, T. Kühl, D. Marx, R. Neumann, S. Schröder, P. Seelig, L. Völker: Phys. Rev. Lett. **73**, 2425 (1994)
6. J.R. Crespo Lopez-Urrutia, P. Beiersdorfer, D. Savin, K. Widman: Phys. Rev. Lett. **77**, 826 (1996)
7. P. Beiersdorfer, A. Osterheld, J. Scofield, J. Crespo Lopez-Urrutia, K. Widmann: Phys. Rev. Lett. **80**, 3022
8. V.M. Shabaev: J. Phys. B **27**, 5825 (1994)
9. P. Pyykko, E. Pajanne, M. Inokuti: Int. J. Quantum Chem. **7**, 785 (1973)
10. S.A. Zapryagaev: Opt. Spectrosc. **47**, 9 (1979)
11. V.M. Shabaev: Opt. Spectrosc. **56**, 244 (1984)
12. M. Finkbeiner, B. Fricke, T. Kühl: Phys. Lett. A **176**, 113 (1993)
13. S.M. Schneider, J. Schaffner, W. Greiner, G. Soff: J. Phys. B **26**, L581 (1993)
14. V.M. Shabaev, M. Tomaselli, T. Kühl, A.N. Artemyev, V.A. Yerokhin: Phys. Rev. A **56**, 252 (1997)
15. V.M. Shabaev, M.B. Shabaeva, I.I. Tupitsyn, V.A.Yerokhin, A.N. Artemyev, T. Kühl, M. Tomaselli, O.M. Zherebtsov: Phys. Rev. A **57**, 149 (1998); ibid **58,** 1610 (1998)
16. V.M. Shabaev, M.B. Shabaeva, I.I. Tupitsyn, V.A. Yerokhin, A.N. Artemyev, T. Kühl, M. Tomaselli, O.M. Zherebtsov: E-print archive, physics/9707006 (http://xxx.lanl.gov)
17. M.G.H. Gustavsson, A.-M. Mårtensson-Pendrill: *Modern Trends in Atomic Physics*, D. Hanstorp, H. Persson, eds., Advances in Quantum Chemistry **30**, 125 (1998)
18. G. Fricke, C. Bernhardt, K. Heilig, L.A. Schaller, L. Schellenberg, E.B. Shera, C.W. de Jager: At. Data Nucl. Data Tables **60**, 177 (1995)
19. H. de Vries, C.W. de Jager, C. de Vries: At. Data Nucl. Data Tables **36**, 495 (1987)
20. W.R. Johnson, G. Soff: At. Data Nucl. Data Tables **33**, 405 (1985)
21. E. Rost: Phys. Lett. B **26**, 184 (1968)
22. L.N. Labzowsky, W.R. Johnson, G. Soff, S.M. Schneider: Phys. Rev. A **51**, 4597 (1995)
23. J.H.D. Jensen, M.G. Mayer: Phys. Rev. **85**, 1040 (1952)
24. A. Bohr, V.F. Weisskopf: Phys. Rev. **77**, 94 (1950)
25. A. Bohr: Phys. Rev. **81**, 331 (1950)
26. M. Le Bellac: Nucl. Phys. **40**, 645 (1963)
27. J.T. Eisinger, B. Bederson, B.T. Feld: Phys.Rev. **86**, 73 (1952)
28. H. Kopfermann: *Kernmomente* (Akademische Verlagsgesellschaft M.B.H., Frankfurt am Main, 1956)
29. M. Tomaselli, S.M. Schneider, E. Kankeleit, T. Kühl: Phys. Rev. C **51**, 2989 (1995)
30. S.M. Schneider, W. Greiner, G. Soff: Phys. Rev. A **50**, 118 (1994)
31. S.M. Schneider: *Die Hyperfinestrukturaufspaltung von Einelektronenatomen*, Ph. D. Thesis, Johann Wolfgang Goethe-Universität (Frankfurt am Main 1995)

32. H. Persson, S.M. Schneider, W. Greiner, G. Soff, I. Lindgren: Phys. Rev. Lett. **76**, 1433 (1996)
33. S.A. Blundell, K.T. Cheng, J. Sapirstein: Phys. Rev. A **55**, 1857 (1997)
34. V.A. Yerokhin, V.M. Shabaev, A.N. Artemyev: Sov. Phys. - Pis'ma v ZhETF **66**, 19 (1997) [JETP Lett. **66**, 18 (1997)]
35. V.A. Yerokhin, V.M. Shabaev, A.N. Artemyev: E-print archive, physics/9705029 (http://xxx.lanl.gov).
36. L.N. Labzowsky, A.V. Nefiodov, G. Plunien, G. Soff, P. Pyykko: Phys. Rev. A **56**, 4508 (1997)
37. A.I. Milstein, A.S. Yelkhovsky: Phys. Lett. B **233**, 11 (1989)
38. A.I. Milstein, A.S. Yelkhovsky: Sov. Phys. - JETP **72**, 592 (1991)
39. A.N. Artemyev, V.M. Shabaev, V.A. Yerokhin: Phys. Rev. A **56**, 3529 (1997)
40. P. Raghavan: At. Data Nucl. Data Tables **42**, 189 (1989)
41. G. Nachtsheim: *Präzisionsmessung der Hyperfeinstruktur-Wechselwirkung von* ^{165}*Ho im Grundzustand*, Ph.D. Thesis (Bonn 1980)
42. M.G.H. Gustavsson, A.-M. Mårtensson-Pendrill: Phys. Rev. A **58**, November (1998) (in press)
43. M.B. Shabaeva, V.M. Shabaev: Phys. Rev. A **52**, 2811 (1995)
44. V.M. Shabaev: J. Phys. B **24**, 4479 (1991)
45. S.M. Schneider, W. Greiner, G. Soff: Z. Phys. D **31**, 143 (1994)

VII Ground-State Lamb Shift in Heavy Ions

Dieter Liesen and Heinrich F. Beyer

Summary. Radiative corrections to the binding energies of electrons in the field of a nucleus have been the subject of experimental and theoretical investigations since the discovery of the Lamb shift more than 50 years ago. In order to disentangle radiative corrections and many-body effects, most of the efforts have been concentrated on hydrogen- and helium-like ions. Ions of high nuclear charge Z are of particular interest because the radiative corrections approximately increase as $\sim Z^4/n^3$, whereas the binding energies grow only as $\sim Z^2/n^2$. Therefore, QED effects are the stronger the larger the nuclear charge Z is. New experimental possibilities for the high-Z region have opened up with a new generation of heavy-ion accelerators and especially of heavy-ion storage rings like the ESR at the GSI Darmstadt. We will present the basic ideas of the experiments performed so far and of the experiments which are starting at the present time. Finally, we will discuss the present experimental results and compare them to theoretical calculations.

1. Introduction

In 1928 Dirac introduced the concepts of the theory of special relativity into quantum mechanics and derived a relativistic equation for the electron [1]. This so-called *Dirac equation*, which takes account of the spin of the electron, can be solved analytically for a single electron bound to a point-like nucleus, thus for hydrogen and hydrogen-like ions. The solutions describe the fine structure of the spectrum correctly and predict that levels with the same value of the principal quantum number n and of the total angular quantum number j are energetically degenerate. About 50 years ago, *Lamb* and *Retherford* [2] proved in a series of famous experiments on hydrogen, using microwave techniques, that the binding energy of the $2s_{1/2}$ state is shifted towards smaller values relative to the $2p_{1/2}$ state. According to the Dirac theory, both states should be degenerate because both have the same values $n = 2$ and $j = 1/2$. In hydrogen, the $2s_{1/2}$–$2p_{1/2}$ energy splitting, the so-called *Lamb shift*, is only a very tiny effect amounting to 4.4×10^{-6} eV.

Despite the small size this observation had enormous consequences. It stimulated the development and formulation of modern quantum electrodynamics (QED), which today is considered to be the most precise theory in physics. QED extends the Dirac theory in that it provides a complete description of the interaction between a charged particle and the quantized electromagnetic field. A bound electron interacts with the external central field of the nucleus and with its own (virtual) radiation field. The latter interaction, which is missing in the Dirac theory, leads to radiative corrections responsible for the $2s_{1/2}$–$2p_{1/2}$ splitting, and more generally, for the shift of all atomic levels. Additional level shifts are caused by the finite size of the nucleus, relativistic recoil corrections and related higher-order effects. The sum of all the shifts of an atomic energy level with respect to the eigen energy of the Dirac equation for a point-like nucleus is usually called the Lamb shift of the considered level. The main radiative corrections to the binding energy are the *self energy* and the *vacuum polarization*. More details are given by *G. Plunien* and *G. Soff* in Chap. III.

The size of the radiative corrections strongly increases with increasing strength of the external field or, in other words, with decreasing distance between electron and nucleus and with increasing nuclear charge. Therefore, the level shifts are largest for wave functions with high radial densities at the origin such as s states. For a given n, the Lamb shift for the s states is much larger than the shift for the p and higher ℓ states. A scheme of the low-lying levels and of the radiative transitions in a one-electron ion is presented in Fig. 1. There, the numerical values for the binding energies are those for hydrogen-like uranium, U^{91+}. Note the strong shift of the $1s_{1/2}$ ground state of 469 eV for uranium with the nuclear charge $Z = 92$ compared to 3.4×10^{-5} eV for hydrogen with $Z = 1$.

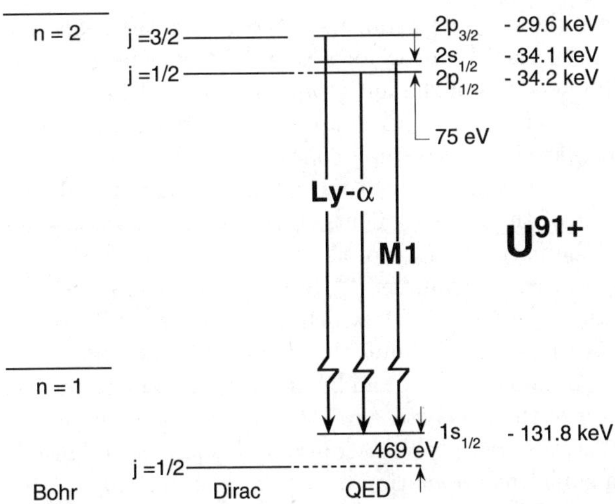

Fig. 1. Level scheme of hydrogen-like uranium according to the various atomic models. Approximate binding energies are indicated

It can be shown [3] that the Lamb shift S of a level with the principal quantum number n in the field of an extended nucleus can be written as

$$S = \frac{\alpha}{\pi} \frac{(Z\alpha)^4}{n^3} F(\alpha Z) \, m_e c_o^2 \, , \tag{1}$$

where α is the fine-structure constant and $m_e c_o^2$ is the rest energy of the electron. $F(\alpha Z)$ is a slowly varying function of Z which comprises all the QED corrections including the effects of the finite nuclear size. Equation (1) clearly reflects the fact that the Lamb shift is largest for ground-state electrons in the field of high-Z nuclei.

In the following, we shall concentrate on measurements of the ground-state Lamb shift in hydrogen-like ions of high nuclear charge ($Z \geq 18$), which are ideal objects because only one electron is bound to the nucleus and the many-body problem is circumvented. For completeness, we want to mention, that the ground-state Lamb shift in the lightest element, hydrogen, has been measured with the extraordinary relative accuracy of 5.6×10^{-6} using Doppler-free, two-photon laser spectroscopy [4]. The result of 8172.798(46) MHz is in agreement with QED calculations taking the larger of two existing measured values of the proton radius. Laser-spectroscopic methods cannot be applied in measurements for the highly charged ions considered here, because the energies of transitions into the ground state lie in the x-ray region. The characteristic Lyman-α lines in U^{91+}, for example, have energies around 100 keV (see Fig. 1). Therefore, the experimentalist is faced with the problem of measuring the energy of hard x rays [5, 6] with a relative accuracy of the order of 10^{-5} or even better, which corresponds to the typical uncertainty of 1 eV of the theoretical calculations.

The difference between the measured transition energy and the transition energy calculated from the Dirac equation for a point-like nucleus gives the difference between the Lamb shifts of the initial np and the final $1s_{1/2}$ state. This difference is mainly determined by the Lamb shift of the ground state because the shifts of the p levels is very small compared to the ground-state shift. We want to mention that the accuracy to which the ground-state Lamb shift can be extracted out of the measurements of the prompt Lyman lines is principally limited by the natural width of the initial p levels. This width increases drastically with the atomic number [7]. In U^{91+} the natural width of the $2p$ level amounts to approximately 30 eV which is already about 7% of the $1s_{1/2}$ Lamb shift. In order to defeat this limit, the energy of the *forbidden* M1 transition from the $2s_{1/2}$ to the $1s_{1/2}$ state can be measured as an alternative, because the natural width of the $2s_{1/2}$ state in U^{91+} amounts to only about 70 meV. However, in this case the problem consists of selecting a suitable process for generating a sufficient population of the $2s_{1/2}$ state.

In the following, we will discuss some of the aspects of how to deal with the problems connected with the x-ray spectroscopy of heavy ions. A first, but basic condition is the availability of hydrogen-like ions in excited states. Whereas ions of intermediate Z are available for instance in laser-produced plasmas, at tokamaks, Van-de-Graaff accelerators, cyclotrons, etc., there are only two places providing also the very heavy ions like U^{91+} or U^{92+} for spectroscopic experiments. One is the *Experimentier-Speicher-Ring* (ESR) at the GSI Darmstadt and the other is the *Electron Beam Ion Trap* at the LLNL

Livermore. The principles of operation of both the machines are illustrated in Chaps. I and II by *A. Wolf* and *D. Schneider*, respectively. A second point concerns the experimental techniques to achieve a precision as high as possible.

2. Methods

2.1 Hydrogen-Like Ions

Although QED corrections to the binding energies in neutral or weakly ionized heavy atoms may be large, they cannot easily be separated in a multi-electron atom. Therefore, *one*-electron ions have to be prepared in the laboratory. For the experimental determination of transition energies in heavy one-electron systems, it is important *how* these ions are generated.

Limitations arise from the simultaneous population of unwanted excited states in neighboring charge states if they cannot be discriminated experimentally. The presence of one or more additional *spectator* electrons in shells with high principal quantum numbers n leads to a small shift of the transition energy relative to the case of a pure one-electron system. It is very difficult to correctly interpret and analyze spectra from these environments. Problems of this kind prevail when the radiation is observed from ions inside a solid or a dense plasma. Consequently, the radiative decay in vacuum or dilute gas is generally preferred. When the ions acquire a high velocity, serious problems from Doppler corrections may arise. The magnitude of the velocity and of the relative angle of observation are the important quantities needed to be controlled in the spectroscopy of fast ions.

2.2 Collisional Population of States

To produce highly charged ions from the neutral species, successive ionization by electron bombardment can be employed. This is the main physical process upon which the operation of powerful ion sources is based [5]. Prominent examples of the latter are the *Electron Cyclotron Resonance* (ECR) Ion source [8], the *Electron Beam Ion Source* (EBIS) [9] and the *Electron Beam Ion Trap* (EBIT) [10].

The attainable degree of ionization for a given atomic species is governed by the balance of reaction rates for ionization and recombination. It strongly depends on the densities and on the kinetic energy or temperature of the electrons. For a quantitative description of the stripping processes in ion sources the various rate coefficients for ionization and recombination, treated by *A. Müller* in Chap. XII, have to be known. Generally, the high recombination and capture rates for very high charge states demand low densities of atoms and electrons thereby limiting the absolute number of ions in high charge states.

Similar processes as in the ion sources are responsible for the stripping of fast ion beams traversing stripper foils ar gases. Once the ion beam has a sufficiently high energy, fully stripped ions may constitute the prevailing charge state. There are some advantages in starting with completely stripped ions to which later single electrons are added. For the heaviest ions, this requires accelerator facilities for the formation and manipulation of ion beams, in particular, storage rings as explained by *A. Wolf* in Chap. I. The underlying

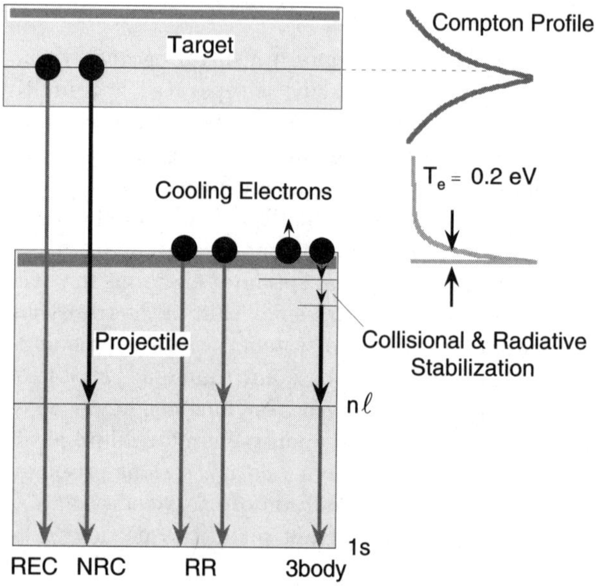

Fig. 2. Various modes of the transfer of an electron into a highly stripped heavy ion. The electron transferred may initially be bound to a target nucleus or it may be taken from the beam of cooling electrons from an electron cooler

physical processes for one-electron transfer are sketched in Fig. 2. Initially, the electron may be bound to a target nucleus with some momentum distribution as characterized by the corresponding Compton profile. There may also be a large offset in energy between the target and projectile represented by the kinetic energy $T_{\mathrm{r}} = (\gamma - 1)m_{\mathrm{e}}c_{\mathrm{o}}^2$ of an electron traveling with the projectile velocity βc_{o}. The two most important reactions are Radiative (REC) and Nonradiative (NRC) Electron Capture.

Population of the states of interest can also occur in an electron cooler. There, the free electrons are moving with the same average velocity as the ions with only a very narrow velocity distribution mainly characterized by the electron-beam temperature. For bare ions the only recombination process is *Radiative Recombination* (RR), which is the inverse of the photoelectric effect. If there are one or more electrons bound to the projectile, three-body recombination becomes possible, but this process becomes important only at high electron density and very low electron temperature.

Radiative recombination in the electron cooler has gained some importance [11]. For the very low relative kinetic energies involved in the electron cooler, RR may be described by the non-relativistic theory of *Stobbe* [12] or, for high n, approximated by the *Kramers formula* [13]. The rate coefficient $\alpha(n)$ falls off with the principal quantum number approximately as $1/n$. The maximum of the rate coefficient shifts to small angular momentum ℓ as the principal quantum number n increases. The main radiative transitions, useful for spectroscopy, are the direct radiative transition to the $1s$ state and the $2p_{1/2,3/2} \rightarrow 1s_{1/2}$ transitions mainly induced by a strong cascade to the $2p_{1/2,3/2}$ states. Detailed calculations [11] show that the $2s_{1/2}$ is only weekly populated, confirmed also by the nearly statistical intensity ratio of the observed Lyman-$\alpha_{1,2}$ doublet as will be shown in Sect. 3.2.

The situation for radiative electron capture from an initially *bound* state is characterized by the dramatically broadened spectral profile as illustrated in Fig. 3. There, the calculated and measured profiles of the K REC are shown for collisions of bare Dy^{66+} ions with argon atoms [14]. In the inertial system of the projectile the double-differential cross section for REC may be written as [15]

$$\begin{aligned} \frac{\partial^2 \sigma_{\mathrm{REC}}}{\partial \Omega'_x \partial E'_x} &= \sum_{n\ell} g_{n\ell} \int \mathrm{d}^3 \boldsymbol{p}_{\mathrm{e}} \, |\Phi_{n\ell}(\boldsymbol{p}_{\mathrm{e}})|^2 \\ &\quad \times \frac{\partial \sigma_{\mathrm{RR}}(\boldsymbol{p}'_{\mathrm{e}})}{\partial \Omega'_x} \, \delta[E'_x - (E'_i - E'_f)], \end{aligned} \tag{2}$$

where the primes indicate the projectile system, $\Phi_{n\ell}(\boldsymbol{p}_{\mathrm{e}})$ denote the target wave functions in momentum space, $E_{i,f}$ the initial and final binding energies, E_x the x-ray energy and the summation runs over the target $n\ell$ substates

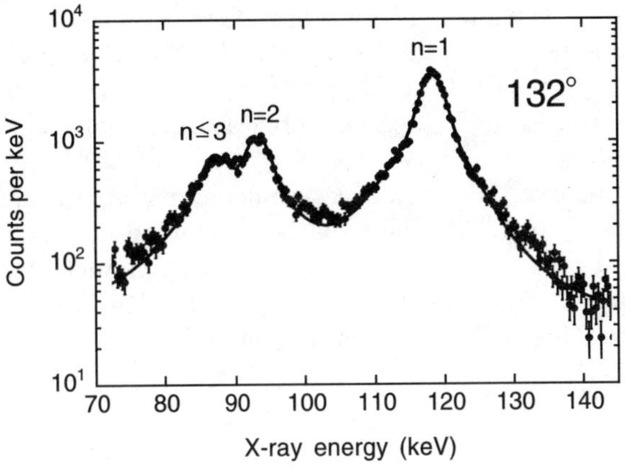

Fig. 3. Experimental REC spectrum (data points) and theoretical profile (full line) for Dy^{66+}–Ar collisions at a beam energy of 292 MeV/u, [14]

accounting for their statistical weights $g_{n\ell}$. The cross sections σ_{RR} inserted are those for radiative recombination with free electrons.

The large width of the REC line together with an asymmetric lineshape and a systematic centroid shift make it difficult to extract binding energies accurately and thus QED corrections from the measured line profile. On the other hand, the REC line can be a large fraction of the emitted photon spectrum. This fraction mainly depends on the collision energy and on the atomic number of the target gas. If only the characteristic x-ray spectrum is used for binding-energy determination the parameters can be chosen such that the dominant electron transfer occurs by nonradiative electron capture. Angular distributions of REC have been shown to be sensitive to relativistic effects [16] also covered in Chap. XI by *J. Eichler* and *Th. Stöhlker.*

2.3 X-Ray Spectrometry

The measurement process for the relatively weak x-ray sources, represented by the interaction of heavy ions with gas atoms or free electrons, is divided into two steps [6]. In the first step, suitable wavelength references are established and in the second step the x-ray line under study is compared with the reference. Strong x-ray and gamma-ray sources can be used in combination with precision instruments with low throughput to achieve the desired secondary gamma-ray standards linked to the optical region of the electromagnetic spectrum [17]. For the second part, an efficient wavelength comparator is needed that allows a relative measurement between the wavelengths of the gamma-ray standard and the x-ray line. Ideally, the wavelength difference between the reference line and the line to be measured is very small, relaxing the requirements on the comparators.

Crystal spectrometers and solid-state germanium detectors have been used as comparators. The stability and linearity of crystal spectrometers are well known. Solid-state detectors usually need a large number of calibration lines if they are used over an extended energy range. This is caused by an undefined offset and a non-linearity of the energy-response function of the detector and the electronics attached to it. Despite those difficulties, state-of-the-art germanium detectors have the capability of measuring small energy differences with high accuracy.

Figure 4 shows an example of the use of this technique for the measurement of the 1s binding energy of U^{91+}. The radiative-recombination line of U^{91+} was observed at the ESR electron cooler at an observation angle near zero degrees and the ion-beam energy was tuned to overlap the RR line with the 295.956 keV gamma-ray line of ^{192}Ta used as a standard.

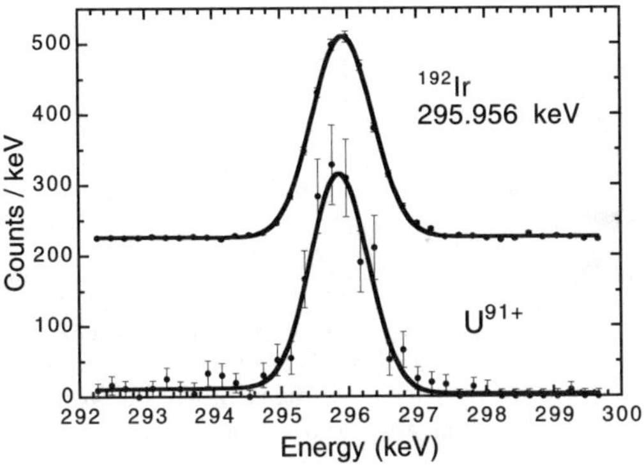

Fig. 4. The radiative recombination line of U^{91+} measured in relation to the 295.956 keV gamma-ray line of ^{192}Ir used as a secondary standard. The experiment was performed at the electron cooler of the ESR [18, 19]

2.4 Fast Ion Beams and the Doppler Effect

Photons emitted from fast particles are subject to the Lorentz transformation

$$E_{\mathrm{o}} = E_{\mathrm{x}}\,\gamma(1 - \beta\cos\vartheta), \tag{3}$$

where E_{o} and E_{x} denote the x-ray energies in the emitter and in the laboratory frame of reference, respectively. For an ion beam with $\beta = 0.5$, for instance, the forward-backward energy ratio amounts to $E_{\mathrm{x}}(0°)/E_{\mathrm{x}}(180°) = 3$. The large Doppler shifts can be used to displace the spectral line of interest into the neighborhood of a calibration line as demonstrated in the example of Fig. 4. The influence from uncertainties of the beam velocity β and of the observation angle ϑ can be estimated by inspecting the first derivatives of (3) with respect to ϑ and β resulting in a relative uncertainty of the x-ray energy in the emitter frame amounting to

$$\begin{aligned}\left[\frac{\Delta E_{\mathrm{o}}}{E_{\mathrm{o}}}\right]^2 &= \left[\frac{\beta\sin\vartheta}{1-\beta\cos\vartheta}\,\Delta\vartheta\right]^2 \\ &\quad + \left[\left(\beta\gamma^2 - \frac{\cos\vartheta}{1-\beta\cos\vartheta}\right)\Delta\beta\right]^2 + \left[\frac{\Delta E_{\mathrm{x}}}{E_{\mathrm{x}}}\right]^2 .\end{aligned} \tag{4}$$

This is graphically illustrated in Fig. 5 as a polar diagram. The closed curves labeled *angle* and *velocity* are the multipliers of $\Delta\vartheta$ and $\Delta\beta$ in (4).

An experiment may be optimized by carefully arranging one or more spectrometers relative to the fast-beam source depending on the actual magnitudes of the angular and velocity uncertainties. There are two extreme geometries where either the angular or the velocity uncertainties are aggravated or diminished. In a velocity-sensitive geometry the detector is placed at 0° or 180° making angular uncertainties unimportant. An angular-sensitive geometry is ideally realized for the *magic angle* at which $\cos\vartheta = \beta$. There, influences from velocity uncertainties vanish.

The velocity-sensitive geometry is the only practical one at the electron cooler. Therefore, great effort has been put into accurate measurements of

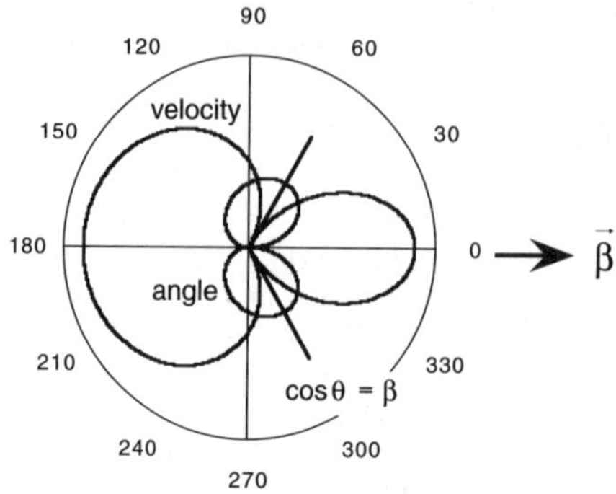

Fig. 5. The sensitivity towards angular ($\Delta\vartheta$) and velocity ($\Delta\beta$) uncertainties or widths, plotted as a polar diagram for $\beta = 0.5$

the beam velocity [18, 19]. The methods are based on the measurement of the terminal voltage of the electron cooler, on a purely spectroscopic method or on the measurement of the revolution frequency in the storage ring. Whereas the three methods give consistent results it turned out that the measurement of the voltage yielded the most accurate velocities. The spectroscopic results were gained by a comparison of the measured and the theoretical energy difference between the $1s$ RR line and the Lyman-α_1 line. In this difference the $1s$ binding energy, with its QED contributions, is canceled out. A spectroscopic measurement of the beam velocity can also be performed with collinear laser spectroscopy (see Chap. VIII).

2.5 Absorptiometry

Absorptiometric x-ray measurements date back to the early days of x-ray physics. The measurements, however, were restricted to the fixed energies of the emission lines and absorption edges of the elements. With the tuning possibilities of a fast-beam source, discussed in the previous section, this limitation can be overcome [20].

The Doppler-shifted emission line can be matched with a known absorption feature, like a K edge, of a thin filter inserted between the fast-beam source and the detector. The energy in the emitter frame of reference can be calculated if the beam velocity β, the observation angle ϑ at which the absorption feature is detected and the absorption profile is known. The method is directly based on the Doppler formula (3) and makes use of the variable Doppler dispersion (4) as a function of the observation angle ϑ and the ion velocity β.

In practice, the absorption structure can be mapped by a variation of the observation angle at a fixed beam velocity (angle-tuned spectroscopy) or by scanning the velocity at a fixed observation angle (velocity-tuned spectroscopy). Sophisticated combinations of both, with several, position-sensitive, detectors, might be planned for maximum accuracy. Figure 6 shows the relevant parameters for velocity-tuned spectroscopy at the electron cooler.

The RR line of U^{91+} can be scanned across the Bi-K edge near $\beta \approx 0.36$ and $\vartheta = 180°$. There, the line appears beyond the short-wavelength limit of the cooler-Bremsstrahlung background.

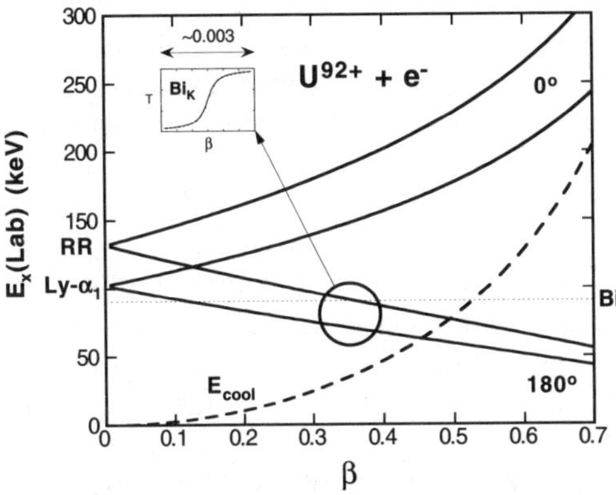

Fig. 6. Variation of the laboratory energy of the RR and of the Lyman-α_1 line of U^{91+} with the emitter velocity β at a fixed observation angle $\vartheta = 0°$ or $180°$, respectively. At $\beta \approx 0.36$ and $\vartheta = 180°$, the RR line matches the Bi-K edge. The line appears above the kinetic energy $E_{\rm cool}$ of the cooling electrons

3. Measurements

3.1 Ions with Intermediate Nuclear Charge: $26 \leq Z \leq 36$

The first experiment on a hydrogen-like ion in the region of intermediate nuclear charges yielding a meaningful result for the 1s Lamb shift was performed at the SuperHILAC in Berkeley by *Briand* et al. [21]. Hydrogen-like Fe^{25+} ions in excited states were produced at an energy of 8.5 MeV/u in a thin carbon foil. Lyman-α lines emitted in the foil were observed at an observation angle of 90° by a flat-crystal spectrometer (Si 220) which reflected the x rays at the Bragg angle $\theta_{\rm B}$ into a position-sensitive detector of the backgammon type. The spectrometer was calibrated by an in-situ measurement of the energy of Co $K\alpha$ x rays. The absolute energy of the Lyman-α lines was determined with an accuracy of 90 ppm which was mainly limited by the uncertainty of the Doppler parameters, ion velocity and emission angle. From the energies of the Lyman-$\alpha_{1,2}$ lines values of 3.4(6) eV and 4.13(70) eV for the Lamb shift of the 1s ground state were deduced, which corresponds to an accuracy of about 17%. A comparison of the experimental and theoretical results will be presented in Sect. 4.

In a similar setup, the energies of the the Lyman-α_1 line of hydrogen-like krypton has been measured at GANIL in Caen, France, with an accuracy of 36 ppm, from which the ground-state Lamb shift could be determined with an uncertainty of 4% to 11.95(50) eV [22]. The essential improvement in this experiment as compared to the former one consisted in observing the

Bragg reflection of the Lyman-α line and the $K\alpha$ calibration x rays of Cr in the position-sensitive detector in two arrangements, symmetric with respect to the main axis of the spectrometer. This $(\theta, -\theta)$ *method* compensates for possible misalignments of the ion beam. The absolute velocity of the 35 MeV/u Kr^{36+} ions was measured with the aid of a magnetic spectrometer with a relative uncertainty of 4×10^{-4} which results in a contribution of only 4 ppm to the total uncertainty. The dominant contribution (26 ppm) originates from the inaccurately known energy loss of the ions in the stripper and the excitation foil.

This drawback has been overcome in an experiment conducted at the UNILAC in Darmstadt, where bare nickel ions passed through a differentially pumped gas cell filled with argon at a pressure of about 0.2 mbar [23]. Under this condition, the energy loss of the projectiles is negligibly small. The excited $2p$ states are populated by one-electron capture processes mainly into higher n states followed by subsequent cascading. In order to control the uncertainties introduced by the Doppler effect, x-ray spectra were taken at four different projectile energies (3.1, 5.2, 6.7, and 8.4 MeV/u). The velocities of the ions could be determined to $\Delta\beta/\beta = 4 \times 10^{-4}$ by a time-of-flight measurement. The experimental configuration is shown in Fig. 7. The x rays with

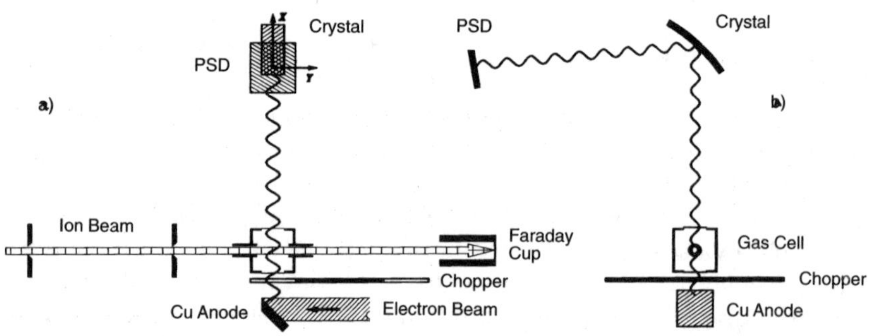

Fig. 7. Schematic of the apparatus used in the nickel experiment at the UNILAC

an energy of about 8 keV are reflected in second order by a curved Ge-220 crystal, used in Johann geometry, under a 90° observation angle and detected in a two-dimensional position-sensitive proportional counter. A position resolution of about 150 μm in the direction of dispersion (x direction) is obtained from a charge division of the two signals from the backgammon-type cathode. Twelve anode wires which are separated by 2.032 mm and oriented along the x direction provide the position of the ionizing event in the direction perpendicular to the dispersion (y direction). An on-line calibration was established by the observation of the Cu $K\alpha$ spectrum chopped in such a way that only during the *off* periods of the pulsed ion beam could x rays from the Cu anode reach the detector.

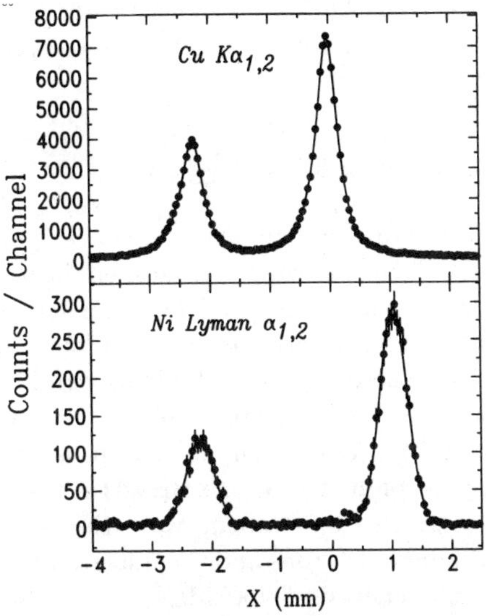

Fig. 8. The Lyman-$\alpha_{1,2}$ doublet of hydrogen-like Ni^{27+} calibrated with the Cu $K\alpha_{1,2}$ lines

A sample Lyman-α spectrum at an energy of 5.2 MeV/u is shown in Fig. 8 together with the Cu calibration spectrum taken from an anode wire close to the centre of the detector corresponding to $y = 0$. All the spectra obtained from the 12 anode wires at the four different projectile energies and the calibration spectra were fitted with Voigt and Gaussian profiles, respectively, in order to find the line positions. Due to the Doppler effect, the positions of the Lyman-$\alpha_{1,2}$ lines move along the x direction for the different wires corresponding to different values of y. The position of the calibration lines is unshifted, as can be seen by the points and the dashed lines in the two-dimensional plot in Fig. 9. The x-y dependence of the line positions can be calculated analytically [24] and was fitted to all the data points in Fig. 9. Inserting the velocity of the ion beam and the wavelength of the Cu calibration line, the only free parameters besides a trivial geometrical factor are the wavelengths of the Lyman-$\alpha_{1,2}$ lines. The result of the fit is given by the solid lines in Fig. 9 from which the wavelength of the Lyman-α_1 line emerges with an uncertainty of 13 ppm dominated by nonuniformities in the bending of the cylindrical crystal. The resulting 1s Lamb shift in hydrogen-like nickel amounts to 5.07(10) eV.

The final accuracy achieved in these experiments with intermediate Z ions is probably not the most important aspect. Rather the step-by-step improvements of the experimental configurations yielded the method for an accurate spectroscopy of very high Z, hydrogen-like ions. The method entails the use of bare projectiles which capture a single electron in a thin gas target in order to make the energy loss negligibly small and to avoid the problem of spectator electrons. Measurements of coincidences between singly down-charged projectiles and x rays will reduce the background drastically. Low projectile energies are favorable because the electron capture rate increases, while the Doppler corrections decrease.

3.2 Ions with Very High Nuclear Charges: $54 \leq Z \leq 92$

An exploratory study of the spectrum of hydrogen-like heavy ions has been performed at GANIL by Briand and coworkers [25] using conventional beam-foil techniques for the preparation of hydrogen-like xenon and a Ge(i) detector for the measurement of the Lyman and Balmer spectra. From a comparison between the Lyman and the Balmer lines, the energies of the hydrogen-like lines could be determined with an uncertainty of 320 ppm yielding a value of 54(10) eV for the $1s$ Lamb shift in Xe^{53+}. With a similar setup, a first experiment on the heaviest ion, U^{91+}, was conducted at the BEVALAC [26] in Berkeley. The most severe problems of this experiment were the large background of bremsstrahlung from secondary electrons and the uncertainties of the Doppler corrections at a projectile energy of 400 MeV/u. A value of 520(130) eV for the $1s$ Lamb shift of hydrogen-like uranium could be extracted from the data.

All the experiments of the following generation have been performed at the electron cooler and the internal gas target of the heavy-ion storage ring ESR in Darmstadt. The combination of thin gas targets, fast variation of

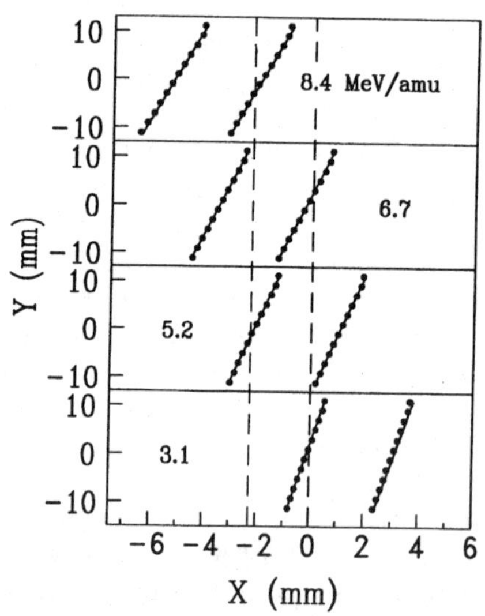

Fig. 9. The two-dimensional image of the Lyman-$\alpha_{1,2}$ doublet of hydrogen-like Ni^{27+} at four different beam energies. The two dashed lines indicate the positions of the Cu $K\alpha_{1,2}$ lines

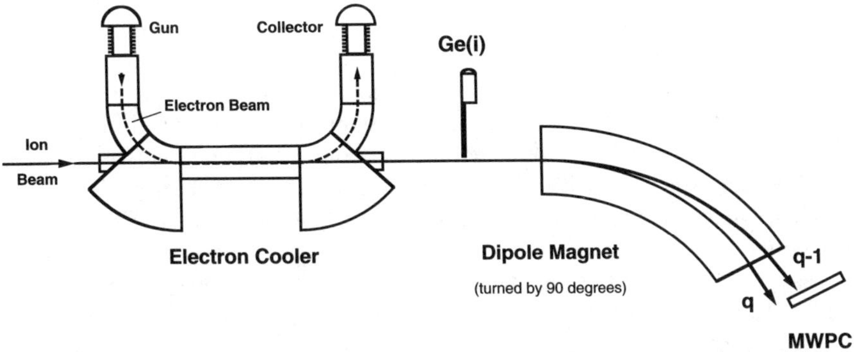

Fig. 10. Scheme of the experimental arrangement for x-ray spectroscopy at the electron cooler of the ESR, from [27]

the projectile velocity, and recycling of the stored bare ions fulfills almost all the conditions for a precision spectroscopy of very heavy ions. Figure 10 schematically shows the experimental configuration at the electron cooler. Bare ions stored in the ESR which have captured one electron in the cooler are deflected from the central orbit in the next following dipole magnet and detected either in a position-sensitive proportional counter or in a scintillator. The emitted x rays are measured in a Ge(i) detector in coincidence with the down-charged particles. The coincidence condition strongly reduces the background and from the position of the detected peaks of down-charged ions single-electron capture is guaranteed. The angle of observation of the Ge(i) detector can be made as small as 0.6°, so that angular uncertainties are unimportant for this velocity-sensitive geometry, as was shown in Sect. 2.4.

An x-ray spectrum of hydrogen-like gold taken at an energy of 266 MeV/u is shown in Fig. 11 [27]. The spectrum is dominated by the $2p_{1/2,3/2} \rightarrow 1s$ Lyman-α lines and the $\infty \rightarrow 1s$ K-RR line (see the upper part of Fig. 11), while the Lyman-β and -γ lines are only weakly visible. It has been shown [11] that the relative intensities observed in Fig. 11 are characteristic for the population of states via RR and subsequent cascading, and that the $2s_{1/2}$ population amounts to less than 5% of the total $n = 2$ population. A projectile energy of 266 MeV/u was chosen for shifting the dominant lines by the Doppler effect very close to calibrated gamma-ray lines from ^{182}Ta and ^{192}Ir sources. The energies of the dominant lines could be extracted with an accuracy corresponding to 85 ppm yielding a value of 202.3(7.9) eV for the Lamb shift of the ground state of Au^{78+} which nicely agrees with the most recent theoretical result of 205.73 eV [28]. The experimental uncertainty of ±7.9 eV originates from a ±3.4 eV statistical and a ±4.5 eV systematic uncertainty caused by the beam velocity.

In a subsequent experiment, the x-ray spectrum of hydrogen-like uranium has been measured [19]. Figure 12 shows the same region of the spectrum as Fig. 11 for gold. A striking feature of the Lyman lines from uranium is observed. The characteristic lines show distinctive tails on the low-energy side and the RR-line does not. A similar, but much less pronounced effect may be observed in Fig. 11. The reason for the low-energy tails is a pure geometric effect. The RR line comes from photons emitted within the cooler promptly with the capture of a cooling electron. These photons are observed by the x-ray detector with a small angular width. The characteristic lines originate from a direct population of the 2ℓ states within the cooler and a feeding by fast and delayed cascades. The delayed cascades can lead to emission of Lyman photons on the approximately 3-m-long route from the end of the cooler to the x-ray detector. These photons can hit the detector under angles between 0.8° and 9°, which gives rise to substantial Doppler

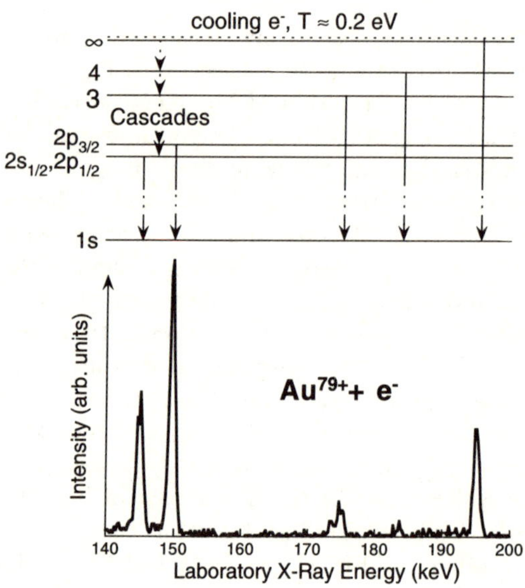

Fig. 11. X-ray spectrum generated via the interaction of completely stripped Au^{79+} ions with free cooling electrons. The lines are attributed to radiative recombination into the K shell or to recombination into higher $n\ell$ states with subsequent cascades as indicated in the corresponding level scheme

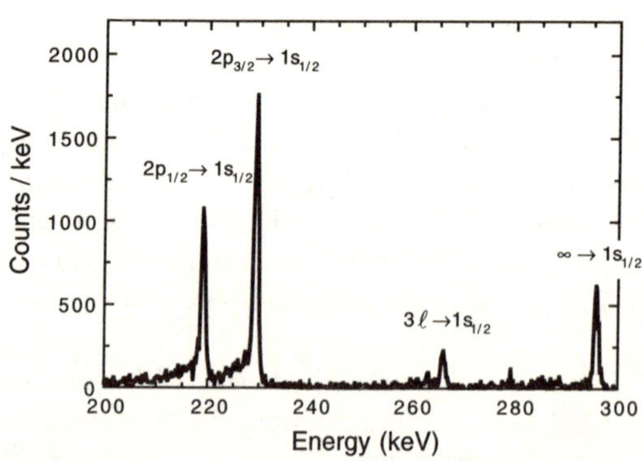

Fig. 12. X-ray spectrum from the interaction of bare U^{92+} ions with free cooling electrons, from [19]

shifts towards lower x-ray energies. The resulting line shape can be computed by considering the time evolution of each single cascade into the $2p$ states starting from an initial $n\ell$ level and by summing up over all initial states. The numerical result can accurately be parameterized as a Gauß function on top of a single-sided linear slope, where the only free parameters are the absolute height and the position of the peak. Details of the calculation can be found in [18, 19]. Figure 13 shows a fit of the calculated to the experimental data for the Lyman-α_1 and the Lyman-α_2 lines of U^{91+}. A remarkably good agreement between the measured and the calculated line profile is observed. From this fit of the characteristic lines and from a fit of a Gaussian to the RR-line, the corresponding energies could be determined with an uncertainty in the order of 150 ppm, which gives a $1s$ Lamb shift of 470(16) eV for hydrogen-like uranium. The uncertainty of ± 16 eV results from a statistical uncertainty of ± 9.1 eV, linearly added to ± 6.9 eV from an uncertainty of 4.5×10^{-5} for the ion velocity. For comparison, we quote the most recent theoretical prediction of 465.59 eV [28].

At the internal gas target of the ESR, an angular-sensitive setup was realized for geometrical reasons. In order to gain control over the Doppler corrections, usually several detectors are mounted around the target with one detector as close as possible to the magic angle, and the measurements are performed at different projectile velocities, preferentially at low velocities. The combination of several detectors and projectile velocities provides a highly redundant set of spectra for an internally consistent analysis. Figure 14 shows such an arrangement of detectors which was used for the measurement of the Lyman-α radiation of hydrogen-like uranium [29]. Four detectors observe the x rays emitted from the overlap region between the ion beam and the gas target which is formed by a jet of N_2 molecules. The detectors are mounted at angles of 48°, 90°, and 132° with respect to the predicted ion-optical axis and were very precisely adjusted by laser assisted trigonometry. For large projectile energies, the forward angle is close to the magic angle which shifts towards 90° with decreasing velocity. One of the two for-

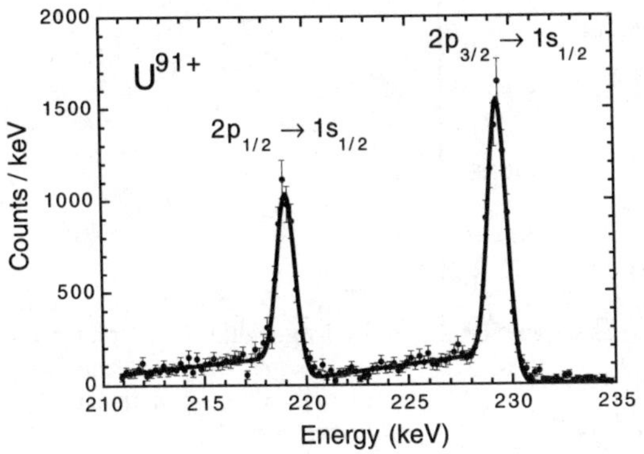

Fig. 13. The Lyman-α doublet of U^{91+} induced by radiative recombination. The sloping intensity observed at the low-energy side of the peaks is a consequence of a delayed cascade population, from [18]

ward detectors is built up of seven precisely measured, equidistant, parallel segments which can be read out separately. Thereby, a sum spectrum with a small Doppler broadening can be generated from the known internal geometry of each of the segments. The energy calibration of the detectors was achieved using gamma-ray lines from ^{179}Yb and ^{182}Ta sources. Like in the cooler experiments, coincidences between x rays in all of the detectors and down-charged particles were measured.

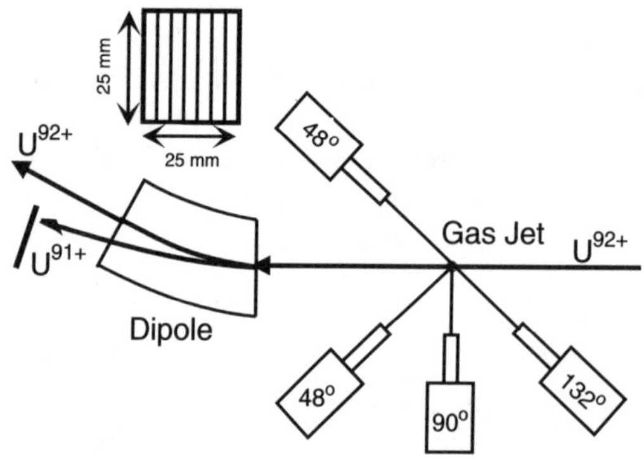

Fig. 14. Experimental arrangement at the ESR gas target with several germanium detectors

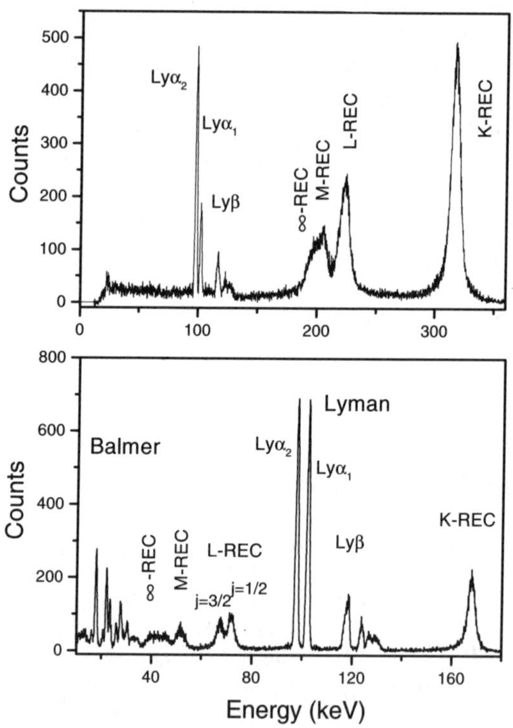

Fig. 15. X-ray spectra of hydrogen-like U^{91+} observed at the gas target of the ESR at two different projectile energies, from [29]

A coincident x-ray spectrum from collisions between bare uranium ions and N_2 molecules at a projectile energy of 358 MeV/u is shown in the upper part of Fig. 15 [29]. Note that the energy scale has already been transformed into the emitting frame of reference. The main features of the spectrum are the characteristic Lyman-α lines and the x rays emitted with the capture of an electron from a target atom which are labeled as K-REC, L-REC and so on. In contrast to the RR lines, the REC lines are broadened due to the momentum distribution of the initial bound target states, the Compton profile. This point and the problem of accurately extracting binding energies is discussed in Sect. 2.2.

A drastic change of the spectrum is found, when the projectile velocity is reduced by decelerating the ions in the ring. The lower part of Fig. 15 shows the spectrum at a projectile energy of 68 MeV/u. At such low energies, the REC lines are suppressed and the spectrum is strongly dominated by a variety of characteristic Balmer $n = 3, 4, \ldots, \infty \rightarrow n = 2$ and Lyman transitions. The increase of the intensity of the characteristic lines and the decrease of the REC lines is caused by the fact that at low beam velocities the target electrons are mainly captured nonradiatively into high lying projectile states, and subsequent cascades lead to the emission of Balmer and Lyman lines. It is obvious from Fig. 15 that experiments with low energy ions have an enormous potential for high precision spectroscopy, since also the uncertainties of the Doppler corrections reduce with decreasing projectile velocity. The large amount of uranium data revealed that *systematic* uncertainties like angular and calibration uncertainties and the limited resolving power of the Ge detectors, and not the counting statistics, will be the factors limiting the achievable accuracy of the $1s$ Lamb shift [30].

In a preceding experiment on hydrogen-like uranium at the gas target only two Ge detectors were used, a segmented and a conventional one [31]. That experiment suffered from a low luminosity, so that the $1s$ Lamb shift could only be determined to 429(63) eV, where the dominant contribution to the total uncertainty comes from statistics (± 46 eV). The uncertainty of the projectile velocity and of the calibration amount to ± 32 eV and ± 30 eV, respectively; the individual uncertainties were summed up quadratically for the final result.

4. Comparison of Experiment and Theory

A general comparison between experimental and theoretical results for the whole range of the nuclear charge Z can easily be made on the basis of the scaling relation (1) (see also Chap. III) for $n = 1$ by plotting $F(\alpha Z)$ as function of Z. This function is obtained by dividing the experimental and theoretical values for the $1s$ Lamb shift by $\alpha/\pi\,(Z\alpha)^4\, m_e c_o^2$. The corresponding plot is shown in Fig. 16. The symbols with uncertainty bars give the experimental and the solid line the theoretical [3] results. The solid symbols represent the

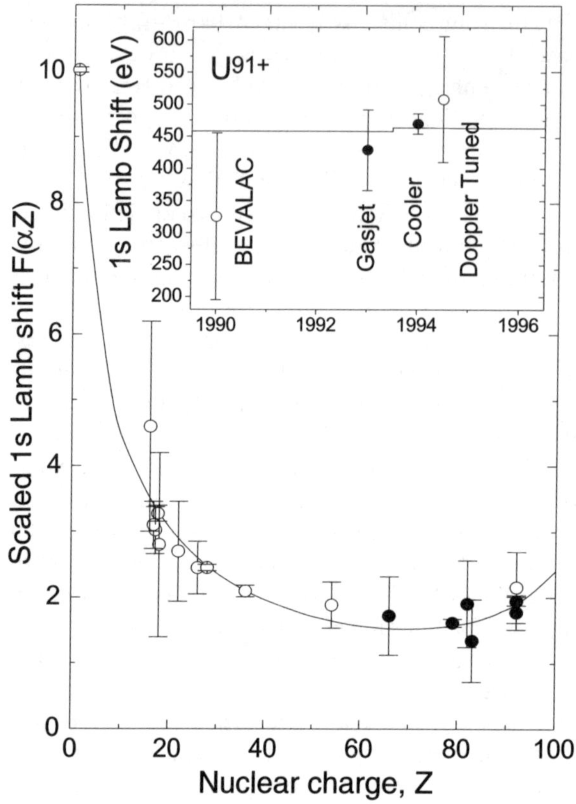

Fig. 16. The 1s Lamb shift of hydrogen and hydrogen-like ions, scaled according to relation (1), as a function of the nuclear charge Z. The experimental results are represented by the data points and the theoretical calculations [3] by the continuous curve. In the inset, the history of improved results for uranium are shown

experimental results obtained at the accelerator facilities of the GSI. An excellent agreement between experiment and theory is observed over the whole range of nuclear charges. For $Z > 50$, most of the experiments provide a test of the calculations on a 30% level; only the data from the electron cooler at the ESR have a considerably higher sensitivity corresponding to 4% for gold and 3% for uranium. The experimental efforts have been concentrated on U^{91+} as the heaviest available hydrogen-like ion. Therefore, the inset in Fig. 16 separately shows the evolution of the experimental and theoretical results within the last few years where the solid symbols represent the ESR data. The size of the error bars demonstrates the substantial improvement in precision due to the excellent experimental conditions at a storage ring as compared to the BEVALAC [26, 32]. The solid line shows the developments of theory over the years [3, 28, 33]. At present, the theoretical prediction is 465.59 eV with an estimated uncertainty of ± 2 eV [28]. This has to be compared to the most precise experimental value of 470(16) eV. The largest contributions to the theoretical uncertainty come from QED effects of second order in the fine-structure constant α, which are not yet fully calculated, while the uncertainty due to the nuclear-size effect is only ~ 0.1 eV.

An experimental test of the theoretical calculations demands an improvement of the experimental accuracy by one order of magnitude, implying a requested fractional uncertainty of 10^{-5} in the measurement of x-ray energies around 100 keV with the complication that the x rays are emitted from a fast moving source. In order to obtain such a precision, the energy resolution of the instruments have to be improved. Germanium detectors in combination with Doppler-tuned spectroscopy [20], either velocity- or angular-tuned, or a combination of both may lead to a determination of x-ray energies to a precision of about 5 eV having the advantage of large solid angles. An alternative are high resolution instruments like transmission-type crystal spectrometers with two-dimensional position-sensitive detectors [34] and calorimetric low-temperature detectors [35]. Currently, these schemes are under construction or have already been tested.

References

1. P.A.M. Dirac: Proc. R. Soc. London A **117,** 610 (1928)
2. W.E. Lamb, R.C. Retherford: Phys. Rev. **72,** 241 (1947)
3. W.R. Johnson, G. Soff: At. Data Nucl. Data Tables **33,** 405 (1988)
4. S. Bourzeix, B. de Beauvoir, F. Netz, M.D. Plimmer, F. de Tomasi, L. Julien, F. Biraben, D.N. Stacey: Phys. Rev. Lett. **76,** 384 (1996)
5. H.F. Beyer, H.-J. Kluge, V.P. Shevelko: *X-Ray Radiation of Highly Charged Ions* Springer Ser. Atoms Plasmas, Vol. 19 (Springer, Berlin, Heidelberg 1997)
6. H.F. Beyer: In *Physics with Multiply Charged Ions*, ed. by D. Liesen, NATO ASI Series B, Vol. 348 (Plenum, New York 1995)
7. M. Hillery, P.J. Mohr: Phys. Rev. A **21,** 24 (1980)
8. R. Geller: *Electron Cyclotron Resonance Ion Sources* (IOP, Bristol 1996)
9. E.D. Donets: Rev. Sci. Instrum. **67,** 873 (1996)
10. R.E. Marrs, P. Beiersdorfer, D. Schneider: Physics Today, **47** 27 (October 1994)
11. D. Liesen, H.F. Beyer, K.D. Finlayson, F. Bosch, M. Jung, O. Klepper, R. Moshammer, K. Beckert, H. Eickhoff, B. Franzke, F. Nolden, P. Spädtke, M. Steck, G. Menzel, R.D. Deslattes: Z. Phys. D **30,** 307 (1994)
12. M. Stobbe: Ann. Physik (Leipzig) **7,** 661 (1930)
13. H.A. Kramers: Philos. Mag. **46,** 836 (1923)
14. H.F. Beyer, K.D. Finlayson, D. Liesen, P. Indelicato, C.T. Chantler, R.D. Deslattes, J. Schweppe, F. Bosch, M. Jung, O. Kleppner, W. König, R. Moshammer, K. Beckert, H. Eickhoff, B. Franzke, A. Gruber, F. Nolden, P. Spädtke, M. Steck: J. Phys. B **26,** 1557 (1993)
15. M. Kleber, D.H. Jakubassa: Nucl. Phys. A **252,** 152 (1975)
16. T. Stöhlker, H. Geissel, H. Irnich, T. Kandler, C. Kozhuharov, P.H. Mokler, G. Münzenberg, F. Nickel, C. Scheidenberger, T. Suzuki, M. Kucharski, A. Warczak, P. Rymuza, Z. Statchura, A. Kriessbach, D. Dauvergne, B. Dunford, J. Eichler, A. Ichihara, T. Shirai: Phys. Rev. Lett. **73,** 3520 (1994)
17. R.D. Deslattes, E.G. Kessler, W.C. Sauder, A. Henins: Ann. Phys. (NY) **129,** 378 (1980)
18. H.F. Beyer: IEEE Trans. Instr. Meas. **44,** 510 (1995)

19. H.F. Beyer, G. Menzel, D. Liesen, A. Gallus, F. Bosch, R.D. Deslattes, P. Indelicato, T. Stöhlker, O. Klepper, R. Moshammer, F. Nolden, H. Eickhoff, B. Franzke, M. Steck: Z. Phys. D **35,** 169 (1995)
20. R.W. Schmieder, R. Marrus: Nucl. Instrum. Methods. **110,** 459 (1973)
21. J.P. Briand, M. Tavernier, P. Indelicato, R. Marrus, H. Gould: Phys. Rev. Lett. **50,** 832 (1983)
22. M. Tavernier, J.P. Briand, P. Indelicato, D. Liesen, P. Richard: J. Phys. B **18,** L327 (1985)
23. H.F. Beyer, P. Indelicato, K.D. Finlayson, D. Liesen, R.D. Deslattes: Phys. Rev. A **43,** 223 (1991)
24. H.F. Beyer, D. Liesen: Nucl. Instrum. Methods. A **272,** 895 (1988)
25. J.P. Briand, P. Indelicato, A. Simionovici, V. San Vicente, D. Liesen, D. Dietrich: Europhys. Lett. **9,** 225 (1989)
26. J.P. Briand, P. Chevallier, P. Indelicato, D. Dietrich, K. Ziock: Phys. Rev. Lett. **65,** 2761 (1990)
27. H.F. Beyer, D. Liesen, F. Bosch, K.D. Finlayson, M. Jung, O. Klepper, R. Moshammer, K. Beckert, H. Eickhoff, B. Franzke, F. Nolden, P. Spädtke, M. Steck, G. Menzel, R.D. Deslattes: Phys. Lett. A **184,** 435 (1994)
28. T. Beier, P.J. Mohr, H. Persson, G. Plunien, M. Greiner, G. Soff: Phys. Lett. A **236,** 329 (1997)
29. P.H. Mokler, T. Stöhlker, R.W. Dunford, A. Gallus, T. Kandler, G. Menzel, H.-T. Prinz, P. Rymuza, Z. Stachura, P. Swiat, A. Warczak: Z. Phys. D **35,** 77 (1995)
30. Th. Stöhlker: unpublished (1998)
31. T. Stöhlker, P.H. Mokler, K. Beckert, F. Bosch, H. Eickhoff, B. Franzke, M. Jung, T. Kandler, O. Kleppner, C. Kozhuharov, R. Moshammer, F. Nolden, H. Reich, P. Rymuza, P. Spädtke, M. Steck: Phys. Rev. Lett. **71,** 2184 (1993)
32. J.H. Lupton, D.D. Dietrich, C.J. Hailey, R.E. Stewart, K.P. Ziock: Phys. Rev. A **50,** 2150 (1994)
33. H. Persson, S. Salomonson, P. Sunnergren, I. Lindgren, M.G.H. Gustavsson: Hyp. Int. **108,** 3 (1997)
34. H.F. Beyer: Nucl. Instrum. Methods. A **400,** 137 (1997)
35. P. Egelhof, H.F. Beyer, D. McCammon, F.v. Feilitzsch, A.v. Kienlin, H.-J. Kluge, D. Liesen, J. Meier, S.H. Moseley, T. Stöhlker: Nucl. Instrum. Methods. A **370,** 26 (1996)

VIII Laser Spectroscopy

Thomas Kühl

Summary. Precision laser spectroscopy of fast heavy ions imposes stringent requirements on the quality and control of the accelerated beam. The ability of heavy ion storage rings to meet these demands was recognized early during the proposal phase, and implemented into experiments otherwise thought to be infeasible. Within the accelerator-storage ring scenario, exotic ion species can be produced and stored. Most notably, at the high-energy accelerators highly charged and radioactive ions can be formed via reactions of energetic ions with fixed targets prior to storage. Once stored, reactions with gas targets or the electrons in the electron cooler allow the preparation of specific ionic configurations. Laser spectroscopy can be applied to probe these exotic atoms with high precision, allowing unprecedented tests of fundamental properties. The two examples given here concern the quantum electrodynamics in high fields and the theory of special relativity at velocities surpassing by far the typical movements of celestial frames of reference.

1. Heavy Ions in Laser Light – Test Stand for Quantum Electrodynamics in Strong Fields

Due to its central role in the *standard model*, QED has been extensively tested. Accuracies and agreement in the 10^{-6} range have been reported for the Lamb-shift in hydrogen, and even better for the g-factor measurements of the electron and the muon in magnetic fields around one Tesla. Highly charged ions represent a very peculiar possibility to test electromagnetic interactions in exceedingly large fields. This has been exploited for a long time in x-ray spectroscopy of the Lyman-series. The K-shell electrons in a highly charged ion are exposed to an electric field Z^4 higher than the outer electrons in neutral atoms. In order to remove such an electron in the case of lead or uranium atoms, however, energies in excess of 100 keV are needed. Still the most effective method for the production of highly charged ions is therefore the *stripping* of energetic ion beams – in the energy range of some 100 MeV/u – upon passage of a foil of a few micron thickness. Alternatively electron bombardment at a few 100 keV in an EBIT (Electron Beam Ion Trap) can be used. X-ray spectroscopy experiments using highly charged ions in the storage ring are discussed in Chap. VII. Completely new and complementary is the possibility of testing QED in the combination of strong electric and magnetic fields by laser spectroscopy of the ground state hyperfine splitting in high-Z hydrogen-like systems. In all cases, where the nucleus has a spin-value larger than or equal to 1/2, and therefore exhibits a magnetic moment, the combination of electric and magnetic interaction leads to the hyperfine splitting of the spectral lines. A transition between the hyperfine structure components is possible as a pure M1-transition. The transition wavelength, observed as the 20-cm line in neutral hydrogen, is boosted into the optical regime in highly charged ions. This is due to the large magnetic field in the vicinity of the nucleus, which exceeds 100 million Tesla in the case of bismuth or lead. The ground-state hyperfine splitting of hydrogen-like ions is often given as a product of the nonrelativistic solution multiplied by correction factors:

$$\begin{aligned}\Delta E(\mu) &= \frac{4}{3}\alpha(\alpha Z)^3 \frac{\mu}{\mu_N} \frac{m}{m_p} \frac{2I+1}{2I} mc^2 \\ &\times \{A(\alpha Z)(1-\delta)(1-\varepsilon) + x_{rad}\}. \end{aligned} \quad (1)$$

Here α is the fine-structure constant, Z is the nuclear charge, m is the electron mass and m_p the proton mass, μ is the nuclear magnetic moment, μ_N is the nuclear magneton, and I is the nuclear spin. The relativistic correction $A(\alpha Z)$ is typically already included in the exact solution of the Dirac equation with a Coulomb potential. The factor $(1 - \delta)$, correcting for the finite spatial distribution of the nuclear charge (*Breit–Schawlow correction*), is also explicitly taken into account in most modern approaches. More complicated is the $(1 - \varepsilon)$ correction for the finite spatial distribution of the nuclear magnetization (the Bohr–Weisskopf correction) [1], and the QED correction x_{rad}. The governing term in the splitting energy is the Z^3 dependence, which also reflects the increase of the magnetic field strength experienced by the electron. An important consequence of the enlargement of the transition energy is the reduction of the transition lifetime from 10^7 years in atomic hydrogen into the millisecond regime.

2. Experiment on Hydrogen-Like $^{208}Bi^{82+}$ at the ESR

In principle the high velocity of ions needed for the production of highly charged ions represents a very problematic complication to high resolution laser spectroscopy. In order to provide an interaction time of only a few nanoseconds, several meters of overlap between laser and ion beam have to be provided. The Doppler effect at $\beta = v/c = 0.6$, a typical ion velocity at the ESR, shifts the laboratory wavelength of the laser by nearly a factor of two. Small deviations of the individual ion velocity from the average velocity of the beam ions lead to an accordingly large Doppler width. In this situation only a small number of laser spectroscopy experiments has ever been performed directly on accelerator beams. This has been changed drastically for experiments in the cooler-storage rings. Recirculating the ions through the laser interaction region restores the possibility for multiple excitations of each single ion despite the geometrical limitations. In conjunction with the electron cooler, which produces the required quality in the ion beam, and in addition effectively controls the mean beam velocity, it is possible to exploit the entire potential of laser spectroscopy. In fact the large Doppler shift now becomes useful, since the wavelength range accessible by tunable lasers can be extended into the UV and the infrared. Even fixed frequency lasers can be used, when the resonance is scanned via the Doppler shift by a variation of the ion velocity. As a drawback, geometrical arguments usually restrict the effective length of practical detection systems to a small fraction of the storage ring circumference. But by the choice of the observation angle the wavelength of light emitted from the ions can be shifted into the spectrum detectable by the most sensitive detectors. The experimental setup at the ESR is given in Fig. 1. Highly charged ions are produced with an efficiency of

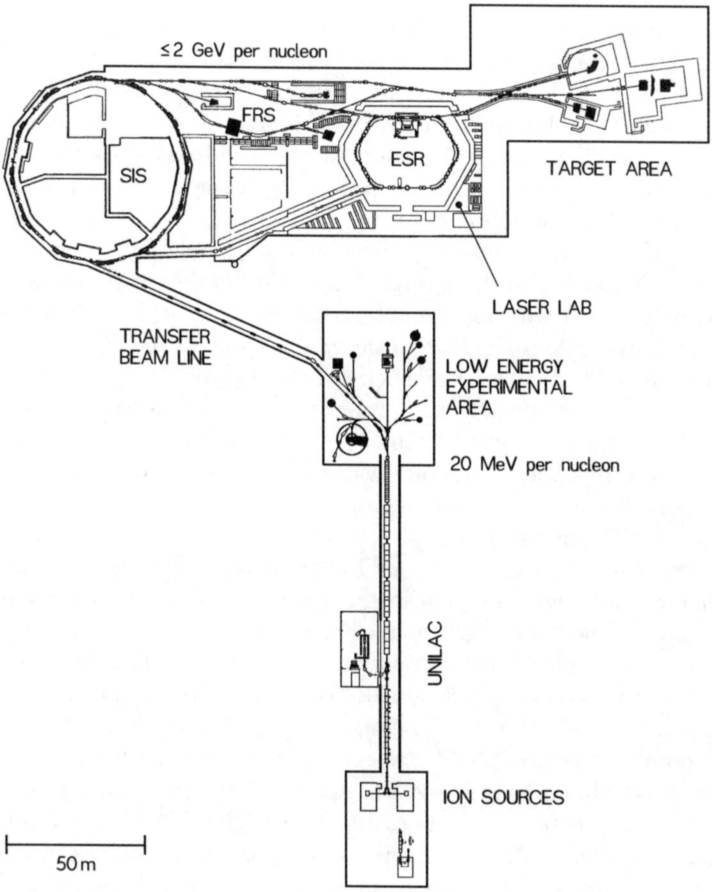

Fig. 1. Laser spectroscopy at the ESR. Heavy ions are pre-accelerated in the UNILAC linear accelerator to energies up to 20 MeV per nucleon. The final energy is limited by the bending fields of 18 Tm to, e.g., 1 GeV per nucleon for uranium. At this energy hydrogen-like ions can be prepared by additional stripping. Within the ESR the velocity of the ions is cooled by an electron cooler. Laser excitation takes place in either of the straight sections. Fluorescence light is collected by a mirror system inside the vacuum tube, and detected by photomultipliers

about 30% by in-flight stripping of lower-charged ions accelerated to about 200 MeV/nucleon in the heavy-ion synchrotron SIS. Hydrogen-like ions are then selected from the resulting charge-state distribution and injected into the ESR storage ring. The high initial velocity spread of the ions is significantly reduced by electron cooling [2]. Since the beam lifetime in the storage ring is in the order of one hour, it is possible to increase the beam intensity in the storage ring by accumulation of a number of cycles of the SIS. For the first measurement of a hyperfine splitting in the optical regime, up to 1.8 $\times$ 10^8 ions of $^{209}Bi^{82+}$ were stored at a velocity of $\beta = v/c = 0.58666(11)$. The ion beam was overlapped in a collinear, antiparallel ($\theta = 180^0$) geometry with 10 mJ laser pulses of 30 ns duration from a dye laser pumped by an excimer laser. The alignment and overlap of the ion beam (8 mm diameter) and the laser beam (10 mm diameter) in a field-free straight section of the storage ring was controlled to an accuracy of 0.33 mrad by pairs of mechanical scrapers separated by 6 meters. The Doppler effect shifted the wavelength for laser excitation in the given geometry from $\lambda_o = 250$ nm in the rest frame of the ions to $\lambda = 489$ nm in the laboratory frame at the given β. This large shift makes it extremely important to determine the ion velocity accurately. The velocity was deduced from the acceleration voltage of the electron cooler, which can be determined with an absolute accuracy of 10 V, including the effects of space charge potentials. Due to the cooling process the ion velocity can be assumed to be identical to the velocity of the electrons. Schottky analysis of the signal induced in capacitive probes showed that electron cooling reduced the relative spread in velocity to $\delta\beta/\beta = 7\times10^{-5}$ at the given beam intensity. This corresponds to a Doppler width of 40 GHz in the laboratory frame for the conditions given above.

Due to its low transition probability, the M1-decay of the $F = 5$ sublevel takes place along the entire ESR orbit of 108.25 m. To collect fluorescence from a maximum solid angle, a 60-cm long cylindrical, elliptical mirror system was installed in the ultra-high vacuum beam tube of the storage ring. This focuses 5 $\times$ 10^{-4} of the total emission in the ring through sapphire windows onto three photomultipliers outside the vacuum. The Doppler effect causes the spectral distribution of the emitted light to range from 128 nm to 490 nm. Photons were accepted between the transmission cut-off of the sapphire windows at about 200 nm and the edge of a color-glass filter at 450 nm. This filter was necessary to reject scattered laser light. The detected fluorescence exhibits a time structure because only a portion of the ions in the ring can be excited by each laser pulse. As a result, the light is modulated with the revolution frequency of the ions of 1.626 MHz corresponding to the revolution period of 615 ns. A time interval of 120 ns of signal is followed by about 495 ns during which ions not irradiated by the laser pass the detector. In the case of bismuth this pattern was maintained throughout the decay time of 350 microseconds because of the low velocity spread of the stored ions and could, therefore, be used for the discrimination of background. The

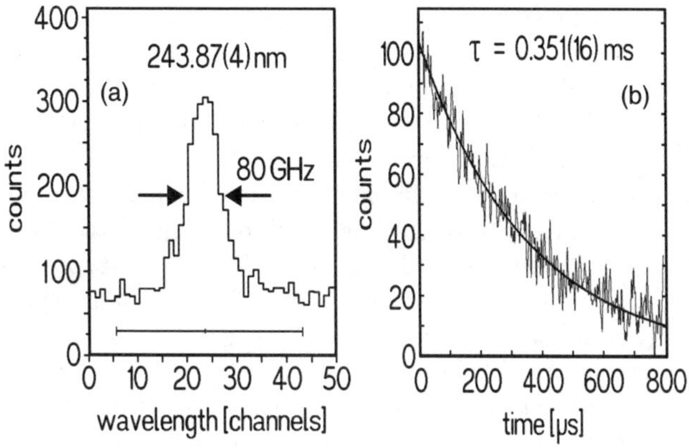

Fig. 2. Results for bismuth: (a) Resonance signal of the M1-transition in the ground state of $^{209}Bi^{82+}$. The Doppler width of the line is determined by the velocity spread of the stored ions. (b) Lifetime measurement of the $F = 5$ ground state hyperfine level of $^{209}Bi^{82+}$. The fluorescence intensity is plotted as a function of the time delay with respect to the pulsed excitation

ground-state HFS resonance in Bi^{82+}, shown in Fig. 2, was found at a laser wavelength of $\lambda = 477.794(4)$ nm. The quoted error is 10% of the Doppler width, which was 40 GHz as expected. The laboratory value λ corresponds to $\lambda_0 = 243.87(2)$ nm in the rest frame of the ions. The quoted accuracy of this number is roughly a factor of two better than given in [3] due to the improved velocity definition proved by [4].

The error is dominated by the $\Delta\beta/\beta = 1.0 \times 10^{-4}$ uncertainty of the ion velocity in the ring entering as $\Delta\lambda_0 = \lambda_0\beta\gamma^2\Delta\beta/\beta$. The effects of divergence and relative angle of the laser and ion beams are negligible compared to the uncertainty in velocity. The same is the case for Zeeman and Stark splittings.

The fluorescence decay curve acquired after excitation by the laser pulse at resonance is shown in Fig. 2b and implies a lifetime for an ion at rest of $\tau_0 = 0.351(16)$ ms. To evaluate QED effects, one must combine the prediction of the Dirac–Fock calculations [5, 6] with information on the nuclear magnetization distribution which leads to the Bohr–Weisskopf correction. For bismuth the size of this correction is estimated to be 2% of the hyperfine splitting.

Accurate assessment of the Bohr–Weisskopf contribution is closely related to the problem of nuclear magnetism itself. Note that for ^{209}Bi the shell-model prediction for the magnetic moment differs from the experimental value by nearly a factor of two. This discrepancy is reduced by core polarization calculations that introduce state-mixing via particle hole excitations. Recently, reasonable success has been obtained by calculations that include all allowed configurations, and binding energies below as well as above the Fermi level of the nucleus [7]. Summation of these contributions according to the additivity rules and their respective separation energy reveals that a number of states, which are usually not included in core polarization models, contribute sizably. These calculations have reproduced the magnetic moments of holmium, rhenium, thallium, lead and bismuth [8] very well. Since the charge distribution obtained in these calculations is also in agreement with the measured charge radius, it seems reasonable to use this model to determine the magnetic moment distribution in the nucleus. Use of the magnetic moment distribution given by this procedure in combination with the Dirac–Fock calculations results in a theoretical transition wavelength of $\lambda = 243.84(50)$ nm, which leaves little room for any contribution from QED. In the case of bismuth, an exceptional additional source of information exists: due to some peculiarity in the level scheme, a measurement of the same hyperfine transition was reported for the muonic atom [9]. There the energy difference between the spin orientations of the K-shell muon was observed. Due to the higher mass of the muon the QED contributions for the muonic levels are practically negligible. Although the muonic wavefunction is different from the wavefunction of the electron, the two cases can be very well compared for a given nuclear model. It turns out that the very large Bohr–Weisskopf contribution for the muonic hyperfine splitting is well described by the dynamic correlation model. The experimental precision of the muonic measurement is sufficient to warrant a reliability of better than 10%. On the basis of the precisions assigned to the corrections the observed discrepancy is therefore quite significant. For holmium, where the hyperfine splitting was measures at the Super EBIT, a smaller discrepancy between the calculated and experimental values was reported [10]. However, the theoretical analysis [11] did not take into account the nuclear polarization effect which is expected to contribute significantly.

3. Experiment on Hydrogen-Like $^{207}Pb^{81+}$

In order to disentangle the contributions due to the nuclear size, the QED corrections and the Bohr–Weisskopf effect, the hyperfine structure splitting of the ground states of hydrogen-like $^{207}Pb^{81+}$ was measured at the ESR [12]. This candidate is exceptional since the nuclear structure is particularly simple. The M1-transition between the ground-state hyperfine levels in hydrogen-like lead is in the infrared region at a wavelength of about 1 μm. As a consequence, the lifetime of the upper hyperfine level is $\tau = 50$ ms, about 100 times longer than in the previous experiment [3] on $^{209}Bi^{82+}$. In order to study the spectroscopy of such an extremely long-lived infrared transition, it was important to develop a new technique employing bunching of the circulating ions. This new technique has several advantages: Without bunching only a small fraction of the stored ions is illuminated by the pulsed laser. With two bunches in the ring it is possible to illuminate 50% of the ions stored in the ring. It also becomes possible to compare fluorescence from a bunch containing excited ions with the light detected from a bunch containing no excited ions. Thus bunching both improves the efficiency with which ions are excited by the laser pulses and allows background subtraction, even when the transition lifetime is longer than the ion diffusion time in the ring. The new technique therefore improves the signal quality by nearly a factor of ten. The circulating ions were compressed into two bunches by applying a radio-frequency acceleration voltage with an amplitude of 20 V. The bunch size was determined to be about 11 meters in length, i.e., about 10% of the ring circumference.

To improve the collection of fluorescence photons the array of cylindrical and elliptical mirrors centered around the ion beam was slightly tilted with respect to the beam direction to enhance the detection of light emitted into forward angles. The storage ring also allowed the use of a fixed-frequency Nd:YAG laser, doubled to 532.222(5) nm (pulse energy 250 mJ), to search for the resonance. Tuning was done by changing the beam velocity to vary the Doppler shift. The velocity of the ions was varied by changing both the acceleration voltage of the electron cooler and the buncher frequency. Although the velocity is affected mainly by the buncher frequency, a mismatch with the electron velocity would lead to a wider spread of the ion velocities. To minimize such broadening, the circulation frequency of the coasting beam was

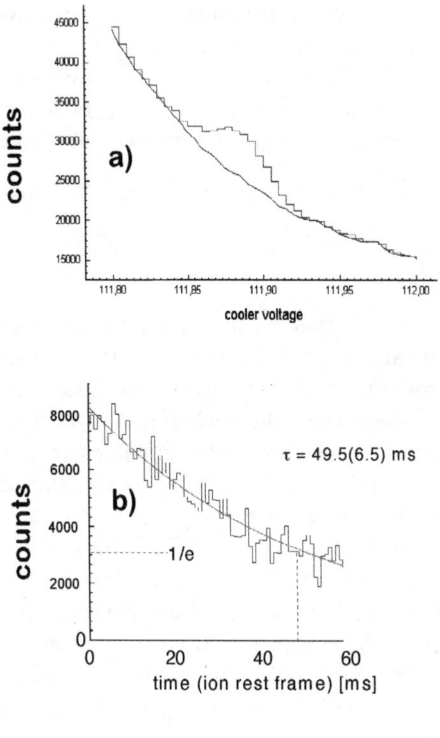

Fig. 3. Raw data for lead: a) Laser induced fluorescence spectrum, time during the measurement approximately 30 minutes; b) Decay-time measurement

precisely measured through electrostatic pickup signals at a given electron-cooler voltage. The buncher frequency was adjusted exactly to this value. As a result the precision of the determination of the ion velocity became the precision of the determination of the electron cooler's acceleration potential, as in the measurements on a coasting beam. The laser pulses were synchronized with the buncher frequency, so that only atoms in one of the two bunches were excited. The detection electronics was also synchronized with the bunches, and photons detected during the passage of the different bunches were counted into different registers. The raw data are shown in Fig. 3 where the photon counts are plotted against the acceleration voltage applied at the electron cooler. The counts of the beam-dependent background (lower line), mostly from ionization and excitation of residual gas, show a decrease as the acceleration voltage is changed, since the data were recorded for about 30 minutes after each filling of the ESR. During this time the ion-beam intensity decreased by a factor of three. The count rate in the bunch excited by the laser (upper line) shows the same behavior, but with the fluorescence signal added at resonance.

The signal width of 0.1 nm is in good agreement with the expected Doppler width $\Delta\lambda_0 = \lambda_0 \beta\gamma^2 \Delta\beta/\beta$ where $\Delta\beta/\beta = 2 \times 10^{-4}$. The centroid of the resonance peak corresponds to 1019.7(2) nm. The accuracy of the measurement is better than the width of the signal, because mismatches between the buncher frequency and the velocity of cooler electrons would show up primarily as an increased width, but the result also reflects the accuracy of the velocity determination. The lifetime was deduced from this data as $\tau =$ 48(5) ms. Within the limited statistical quality of this result a contradiction with the theoretical prediction of 52 ms is not evident. The result, however, clearly excludes drastic deexcitation processes in the ring, as needed for an explanation of the discrepancy in bismuth.

4. Implications of the Result

As shown in Table 1, the combination of the theoretically predicted nuclear and QED contributions leaves a significant gap between the experiment and the theoretical value. The presently remaining uncertainties from QED corrections (0.05 nm) and the quoted uncertainties of the nuclear quantities (0.2 nm) are seemingly too small to account for this discrepancy. The nuclear contributions are supposedly well understood in the case of $^{207}Pb^{81+}$. Its nuclear magnetic moment is given very exactly by the value of the $p_{1/2}$ neutron hole in the last shell before the doubly magic shell closure at ^{208}Pb (82 protons and 126 neutrons), so that this nucleus seems to be well described by the single-particle model, and the effects of core polarization are expected to be small. The difference between the Bohr–Weisskopf contribution in the extreme single particle calculation, where the magnetic moment is regarded as being attached exclusively to the unpaired neutron, and the dynamic mixing

Table 1. Contributions to the hyperfine splitting of the ground state of hydrogen-like lead and bismuth

Ion	$^{209}Bi^{82+}$	$^{207}Pb^{81+}$
RMS Radius	5.519 fm	5.497 fm
Magnetic Moment (corrected)	4.1106(2) μ_N [13]	0.58219(2) μ_N [14]
Point nucleus (Dirac) [5, 6, 16]	212.320(1) nm	885.76(3) nm
+ Breit–Schawlow	238.791(50) nm	989.66(10) nm
+ Bohr–Weisskopf [7]	243.91(38) nm	1019.1 (1.9) [0.4] nm
Vacuum Polarization [15]	−1.64 nm	−6.83 nm
Self energy [6, 16]	+ 2.86 nm	+ 11.9 nm
Total QED [15, 17]	+ 1.22(10) nm	+ 5.08(50) nm
Theory incl. QED	245.13 (58) nm	1024.2 (2.4) nm
Experiment	243.87(2) nm(1)	1019.7(2) nm

model is indeed very small, less than 0.25 nm. One can argue that this gives a reasonable range for the size of this contribution. Alternatively one can go back to the proven reliability of the dynamic mixing model found in muonic bismuth, which would imply at least a reliability of 1.9 nm. These two alternatives are given in the table. Using arguments given by *Bohr* [1] an additional term, due to the coupling with the residual nucleons, should be included in the (single-particle) Bohr–Weisskopf formula. Applying $g_l = Z/A$, as in the original paper, this indeed improves the already good coincidence with the dynamic mixing value. A more justifiable application of the same interaction [18], however, increases the total contribution by nearly 5 nm. Tauntingly, the deviation of an older NMR value for the magnetic moment exactly counteracts this effect, reproducing the measured value reasonably well for the combination of these two! The accuracy of the double-resonance measurement, which has been accounted for, and which was critically repeated by reason of the large deviation from the older NMR value, nonetheless excludes this alternative. The present situation urgently requires an independent verification. Theoretical investigations by several authors indicate that this is possible by the measurement of the 2s-hyperfine structure in lithium-like bismuth. The transition wavelength can be predicted with high accuracy using the result from the 1s-hyperfine splitting in the hydrogen-like case as input for an elimination of the Bohr–Weisskopf-effect [16]. The QED effects in the two states differ due to the interelectronic interaction, whereby a considerable fraction of the correction can be extracted free from the uncertainties of the nuclear parameters. With the present accuracy of the bismuth data a precision of 17% of the QED contribution is numerically possible, as soon as the lithium-like transition is measured. Moreover, the test would directly benefit from improved experimental precision, allowing for a QED test to a 1% level. This can be achieved by using the measurement in the lithium-like ion for a determination of the beam velocity in a beam of mixed charge-states. This is possible with the available equipment, since the predicted wavelength of 1553.5(3) nm can be reached in parallel and antiparallel excitation. The transition lifetime of 82.7(1) ms is similar to the situation in hydrogen-like lead, allowing for an encouraging estimate of the expected signal-to-noise ratio based on the present results.

5. Experiment at the TSR

An important consequence of the theory of special relativity is the dilatation of time in a moving system. In a spectroscopic experiment, the quantity can be tested by a precise measurement of the transverse Doppler effect. A first experimental proof had been achieved by *Ives* and *Stillwell* [19], by experiments on a fast moving hydrogen beam. In the special situation at the storage ring one can do this experiment at relativistic velocities. This provides a large increase in the measured effect in a much cleaner experimental situation. For comparison, it should be kept in mind that the Earth cannot be regarded as a reference system at rest. The anisotropy of background radiation from outer space is usually explained by assuming a velocity of 350 km/s relative to some interstellar frame of reference, as sketched in Fig. 4.

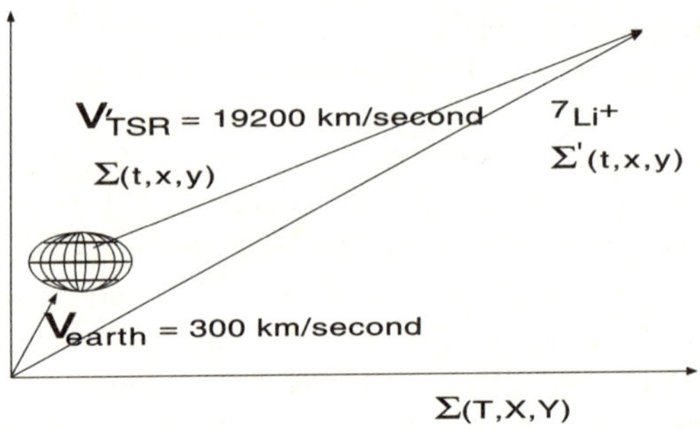

Fig. 4. Reference frames involved in the test of special relativity

At the TSR storage ring in Heidelberg, precision laser spectroscopy of fast $^7Li^+$ ions in the storage ring was used to perform a modern Ives and Stilwell-type experiment [20]. Well-known optical transition frequencies in the $^7Li^+$ ion [21, 22] were used as clocks moving at relativistic velocities. Doppler-free saturation spectroscopy of the Λ-type three level system $^3S_1 \rightarrow {}^3P_2\,(F = \frac{5}{2} \leftrightarrow F' = \frac{5}{2} \leftrightarrow F = \frac{3}{2})$ was used to determine the resonance frequency of the $^7Li^+$ ions by precisely aligned counterpropagating laser beams.

The laser sources were a frequency-stabilized Ar^+ ion laser and a scanning dye laser. By this scheme deviations

$$\gamma = \gamma_{\mathrm{SRT}}\,(1 + \delta\alpha\,\beta^2 + ...)$$

of the time dilation factor γ from the value

$$\gamma_{\mathrm{SRT}} = (1-\beta^2)^{-\frac{1}{2}}$$

given by the special theory of relativity can be observed with high precision. Besides these absolute deviations, an Ives and Stillwell experiment would also be sensitive to an additional anisotropy $A = \alpha \frac{w}{c} \cos(\Omega)$, where Ω denotes the angle of the reference frame of the experiment and a movement relative to some hypothetical "universal" frame of reference. A candidate for a universal frame of reference might be deduced from the anisotropy of the 3 K background radiation. For this case an experiment fixed to Earth would be exposed to modulations of such a deviation by the sidereal motion. The experiments at the TSR storage ring were performed on singly-charged $^7\mathrm{Li}^+$ ions accelerated to 13.3 MeV, corresponding to a velocity of $\beta = 0.064$. About 10–20% of the ions formed in the stripper foil of the TANDEM accelerator are in the metastable ^3S-state. For the experiments 10^6 $^7\mathrm{Li}^+$ ions were stored in the ring. The triplet system of helium-like $^7\mathrm{Li}^+$ shows a well resolved and precisely known fine and hyperfine structure multiplet. For the stringent test of relativity, high-resolution saturation spectroscopy was performed in the Λ system given by the transitions $^3\mathrm{S}_1$ F=3/2 $\to$ $^3\mathrm{P}_2$ F′=5/2 and $^3\mathrm{S}_1$ F=5/2 $\to$ $^3\mathrm{P}_2$ F′=5/2. The experimental setup is shown in Fig. 5.

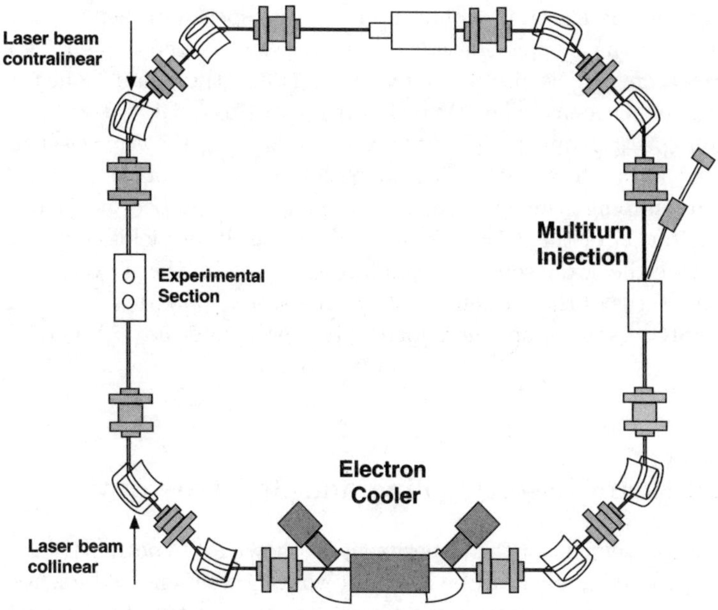

Fig. 5. Experimental setup of the relativity experiment at the TSR. Two counter-propagating single-mode lasers are used to produce a Doppler-free signal

The experiment used a single-mode Ar^+ laser stabilized on a $^{127}I_2$ saturation resonance, that is in resonance at 514 nm, a recommended frequency standard, and a tunable dye laser modified for better frequency stability. The frequency of the scanning tunable laser was determined by comparison with simultaneously recorded $^{127}I_2$ HFS resonances of the R(99) 15-1 transition, recorded simultaneously during the measurement of the transition in lithium. The frequency of the i-component of this transition was calibrated with high accuracy at the Michelson interferometer at the PTB-Braunschweig [20]. In order to assure a well-defined collinearity of the two laser beams, both beams were sent through the same polarization-preserving single-mode fiber over a distance of 20 m before entering the experimental section of the TSR. This bichromatic beam was focused on a plane mirror behind the experimental section at 15 m distance. An angular accuracy of $\Delta\phi \leq \pm 40$ μrad was achieved by adjusting the retro-reflected laser beams for maximum transmission back into the fiber. The optical resonances were detected in fluorescence by two photomultipliers equipped with filters to suppress stray light. The alignment of the laser beams and the ion beam was controlled relative to four capacitive position pickups mounted within the experimental section of the TSR. The alignment accuracy of this procedure corresponds to an angular uncertainty $\delta\theta \leq \pm 370$ μrad.

Before starting the laser spectroscopy measurement, the injected ion beam was first electron-cooled for 7 s. During this time the laser beams were blocked mechanically in order to avoid optical pumping while the longitudinal velocity changed during the electron cooling process. The spectrum displayed in trace 1 of Fig. 6 is the Doppler-broadened fluorescence of the $F = 5/2 \rightarrow F' = 7/2$ transition, and is produced by sending only the beam from the dye laser to the experiment. The HWHM corresponds to $\delta v/v = 3 \times 10^{-5}$. The saturation spectroscopy signal is obtained by keeping the Ar^+ laser at resonance with the $F = 5/2 \rightarrow F' = 5/2$ transition. Depletion of the $F = 5/2$ level by optical pumping is evident when the population is probed by the dye laser scan. Excitation of the $F = 3/2 \rightarrow F'=5/2$ transition yields a strong fluorescence, since the two lasers now simultaneously excite the Λ system.

The present status of the measurements with a stored 13.3 MeV $^7Li^+$ ion beam in the TSR sets an upper limit for the test parameter $\delta\alpha \leq 8 \times 10^{-7}$ [20].

6. Bunched-Beam Laser Cooling and Spectroscopy

An important improvement for this experiment, as shown for hyperfine transition spectroscopy cited above, can be achieved working with an RF-bunched ion beam. In this scheme, phase synchronous detection of the fluorescence light can be applied. As a result, the experiments can be performed with enhanced signal-to-noise ratio, or with a smaller number of ions. For the test

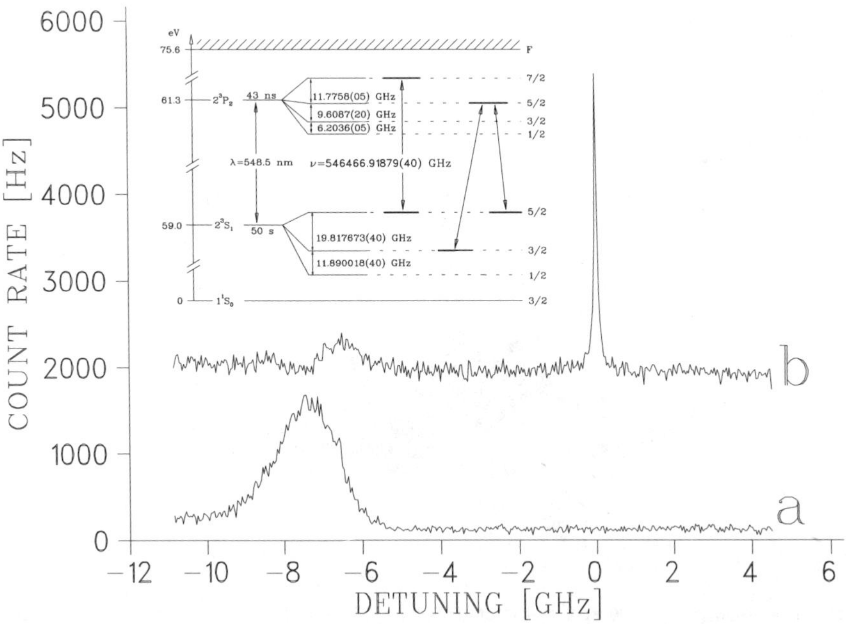

Fig. 6. Sub-Doppler laser spectroscopy at the TSR: The essential part of the level scheme is shown on top. The Doppler broadened signal from the $F = 5/2 \rightarrow F'$ $= 7/2$ transition excited by the dye laser shows a dip caused by the simultaneous excitation by the fixed-frequency Ar+ laser. The narrow resonance occurs when the dye-laser is tuned to the $F = 3/2 \rightarrow F' = 5/2$ transition

of special relativity systematic uncertainties caused by internal beam interactions have to be well controlled. An excellent possibility for this is provided by laser cooling of the bunched beam.

To test this scheme a $^7Li^+$ ion beam in the TSR was bunched on the tenth harmonic of the revolution frequency $\nu_r = 345$ kHz. After electron precooling, the ions were decelerated by the induction accelerator until resonance with the co-propagating Ar^+ ion laser beam occurred. With an additional RF Voltage of 1.1 V at 3.45 MHz, the correct frequency for the cooled ion revolution frequency, the ions were bunched, and laser cooling of the RF-bunched ion beam took place [23]. In this situation two processes occur at the same time. The high frequency field forces the ions into an oscillatory motion relative to the center position of the dynamic "bucket" created by the field gradients. The spontaneous light force of the laser leads to damping of the oscillatory ion movement depending on the amplitude of the ion motion. A strong compression of the phase space occupied by the ions in the bunch is expected. The resulting phase space structure of the ion beam was observed directly by time resolved, phase synchronous detection of the fluorescence light (spatial distribution) and by probe scans performed with a counterpropagating dye laser (momentum distribution). The time resolved data acquisition

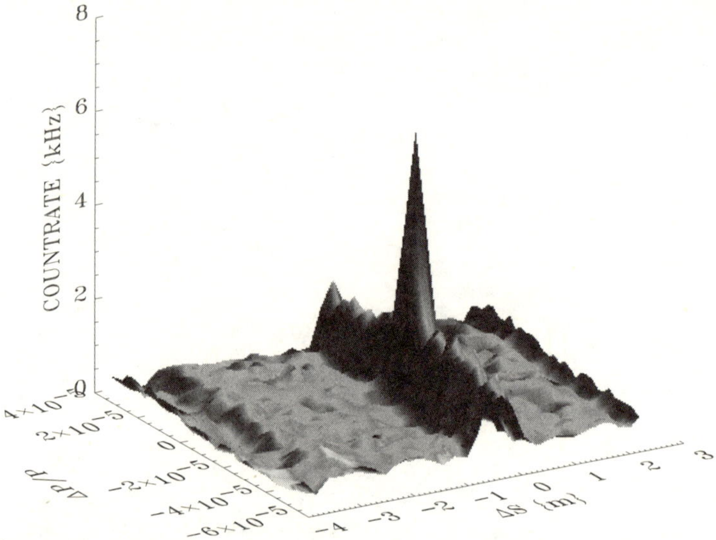

Fig. 7. Measured phase space structure of laser cooled, bunched $^7Li^+$-ions in the TSR by time resolved laser induced fluorescence spectroscopy. The extracted bunch length is $l = 0.53 \pm 0.07$ m (FWHM) while a momentum spread of $\Delta P/P < (3.1 \pm 0.2) \times 10^{-6}$ is deduced. For details see text

comprises 32 scalers, gated recurrently for 9.1 ns per scaler corresponding to a length of 17.3 cm of the ion beam. A typical measurement of the phase space of the ions in the buckets is shown in Fig. 7. In the three-dimensional spectrum the fluorescence intensity is plotted against the frequency detuning of the probing dye laser, corresponding via the Doppler-shift to the relative velocity of the excited ions, and the position within the bunch. The bucket length in this experiment was 5.5 m, i.e., a relative distance of 0.86 m corresponds to a relative spatial phase $\Delta\Phi = 1$. In accordance with the simulation in Fig. 7, laser cooling leads to a compression of the ions in the center of the bunch (0 m). Independently from the detuning of the probe laser the ions can be localized by their fluorescence signal, since they are excited by the Ar^+-laser used for the laser cooling. The measurement exhibits a bunch length $l = 0.53 \pm 0.07$ m (FWHM). The additional excitation by the scanning dye laser probes the velocity distribution of the ions. A steep and narrow velocity distribution is found for the laser cooled ions in the bunch center. From the width of this signal a momentum spread $\Delta P/P < (3.1 \pm 0.2) \times 10^{-6}$ (FWHM) is deduced.

The spatial compression of the ions towards the center of the bunch might invoke effects of intra beam scattering and space charge observed in dense, ultra-cold beams [24, 25]. Therefore it would be desirable to remove lithium ions in the ionic ground state which do not contribute to the laser induced fluorescence signal. In addition to the obvious benefit of applying laser cooling, the combination of laser cooling and bunching finally allows the separation of the ground state ions from the metastable ions. Lithium ions in the ground state are not excited by visible laser light, and not affected by the laser cooling process. This can be used to induce a small velocity difference between the metastable ions and ground-state ions. As described above, prior to bunching the ions were subjected to laser cooling. During this process ions in the metastable state were kept at the fixed velocity given by the laser wavelength, while the velocity of the ions in the ground state was continuously lowered under the influence of the induction accelerator. This separation in velocity by itself would not be sufficient to remove the ground state ions from the beam. Due to the small velocity difference, however, their capture into the buckets takes a longer time, and can be strongly suppressed by increasing the force of the induction accelerator shortly after the RF is applied. This was demonstrated in a recent experiment at the TSR. The application of laser cooled bunches, containing only metastable ions, together with further improvements of the present experimental setup, should allow at the TSR a test of the theory of special relativity with an accuracy of $\delta\alpha < 2 \times 10^{-7}$. This limit might be further reduced by one order of magnitude by performing the experiment at the much higher beam velocity at the ESR. At a velocity of $\beta = 1/3$ a sensitivity in $\delta\alpha$ close to the 10^{-9}-range would be possible.

7. Future Applications of Lasers in Storage Rings

The combination of storage ring and laser technology allows for a complete new class of experiments. At the present stage it does not seem over-optimistic to foresee a further increase in the number and quality of experiments. In addition to the experiments described here, laser-induced recombination in the electron cooler has become a tool for precision spectroscopy of highly excited Rydberg states in highly charged ions. Tests of relativity at high β will be limited only by the achievable velocity of electrons in the electron cooler. To improve the test of QED in highly-charged systems, laser spectroscopy is most likely to be used to explore nuclear effects by comparative studies of neighboring elements and isotopes. In the future, radioactive ions will also become available in the storage rings in quantities sufficient for laser applications. There is also a number of cases where laser spectroscopy has just started. As an example, one should think of spectroscopy of molecules in the storage ring and a continuation of the laser Lamb shift excitations at medium mass ions under the improved conditions given at the storage rings. Laser cooling of stored ions is pursued as a subject in itself, aiming at beam crystallization. In applications it can provide the ultimate beam quality for precision experiments, but possibly also intense beams with ultra-high phase space density as needed for heavy-ion driven inertial fusion.

References

1. A. Bohr, V.F. Weisskopf: Phys. Rev. **77**, 94 (1950)
2. B. Franzke: Nucl. Instrum. Methods. B **24**, 18 (1987)
3. I. Klaft, S. Borneis, T. Engel, B. Fricke, R. Grieser, G. Huber, T. Kühl, D. Marx, R. Neumann, S. Schröder, P. Seelig, L. Völker: Phys. Rev. Lett. **73**, 2425 (1994)
4. S. Borneis, T. Engel, B. Fricke, R. Grieser, G. Huber, T. Kühl, I. Klaft, D. Marx, R. Neumann, S. Schröder, P. Seelig, L. Völker: Phys. Rev. Lett. **72**, 123 (1994)
5. M. Finkbeiner, B. Fricke, T. Kühl: Phys. Lett. A **176**, 113 (1993)
6. S.M. Schneider, W. Greiner, G. Soff: J. Phys. B **26**, L529 (1993)
7. M. Tomaselli, S.M. Schneider, E. Kankeleit, T. Kühl: Phys. Rev. C **51**, 2989 (1995)
8. M. Tomaselli: to be published (1998)
9. A. Rüetschi, L. Schellenberg, T.Q. Phan, G. Piller, L.A. Schaller, H. Schneuwly: Nucl. Phys. A **422**, 461 (1984)
10. J.R. Crespo Lopez-Urrutia, P. Beiersdorfer, D.W. Savin, K. Widmann: Phys. Rev. Lett. **77**, 826 (1996)
11. V.M. Shabaev, J. Phys. B **27**, 5825 (1994)
12. P. Seelig, S. Borneis, A. Dax, T. Engel, S. Faber, M. Gerlach, C. Holbrow, G. Huber, T. Kühl, D. Marx, K. Meier, P. Merz, W. Quindt, F. Schmitt, M. Tomaselli, L. Volker, H. Winter, M. Wurtz, K. Beckert, B. Franzke, F. Nolden, H. Reich, M. Steck, T. Winkler: to be published (1998)
13. P. Raghavan: At. Data Nucl. Data Tables **42**, 189 (1989)
14. H.M. Gibbs, C.M. White: Phys. Rev. **188**, 180 (1969)
15. H. Persson, S.M. Schneider, G. Soff, W. Greiner, I. Lindgren: Phys. Rev. Lett. **76**, 1433 (1996)
16. V.M. Shabaev, V.A. Yerokhin: Sov. Phys. - Pis'ma Zh. Eksp. Teor. Fiz. **63**, 309 (1996)
17. S.A. Blundell, K.T. Cheng, J. Sapirstein: Phys. Rev. A **55**, 1857 (1997)
18. V. M. Shabaev, M. Tomaselli, T. Kühl, A.N. Artemyev, V.A. Yerokhin: Phys. Rev. A **56**, 252 (1997)
19. H.E. Ives, G.R. Stillwell: J. Opt. Soc. Am. **28**, 215 (1938)
20. R. Grieser, R. Klein, G. Huber, S. Dickopf, I. Klaft, P. Knobloch, P. Merz, F. Albrecht, M. Grieser, D. Habs, D. Schwalm, T. Kühl: J. Appl. Phys. B **59**, 127 (1994); R. Grieser, T. Kühl, G. Huber: Am. J. Phys. **63**, 665 (1995)
21. J. Kowalski, R. Neumann, S. Noehte, H. Suhr, G. zu Putlitz, R. Herman: Z. Phys. A **313**, 147 (1983)
22. E. Riis, A.G. Sinclair, O. Poulsen, G.W.F. Drake, W.R.C. Rowley, A.P. Levick: Phys. Rev. A **49**, 207 (1994)
23. R. Grieser R. Grieser, P. Merz, G. Huber, M. Schmidt, V. Sebastian, V. Then, M. Grieser, D. Habs, D. Schwalm, T. Kühl: Hyp. Int. **99**, 145 (1996)
24. P. Merz, R. Grieser, G. Huber, P. Knobloch, M. Schmidt, V. Sebastian, F. Albrecht, M. Grieser, D. Habs, D. Schwalm, T. Kühl: GSI Scientific Report 1994 (GSI, Darmstadt), p. 142
25. H.J. Miesner: Dissertation (Heidelberg 1995) (unpublished)

IX Lifetimes of Excited States in Highly Charged Ions

Lorenzo J. Curtis and Indrek Martinson

Summary. The experimental measurement and theoretical formulation of atomic lifetimes are being extended to systems that have both many electrons stripped away and multiple electrons remaining. The dynamical structure of these systems can be significantly affected by interactions that are negligible in neutral and few-times ionized systems, and the lifetimes can sensitively probe interelectronic correlations, relativistic and quantum electrodynamic interactions, and effects of the finite size of the nucleus. These interactions often have characteristic dependences on the central charge, and studies of systematic trends can reveal both predictive regularities and theoretical insights. Besides providing new tests of atomic theory, these systems have important applications in astrophysics, in fusion plasma technology, and in XUV-laser development.

1. Introduction

The availability of fundamental atomic lifetime data has a significant impact on progress in other fields of science and in areas of modern technology, e.g., in fundamental physics and precise measurements, in the generation of coherent light, in solar physics and astrophysics, and in plasma diagnostics. The accurate specification of energy levels and transition probabilities for ionized atoms has important immediate applications. One application concerns the diagnosis and modeling of controlled fusion plasmas, both in the central region and in the region of the edge plasma. Transition probabilities are necessary both to determine concentrations of impurity elements and to predict the rates of dielectronic recombination, radiative cooling, etc. [1]. They are also needed in the development of XUV lasers and in the interpretation of atomic collision studies. These data are also essential to the determination of elemental abundances in the interpretation of the spectra of astrophysical objects. These objects are important both as working examples of energy production, and as light sources for spectroscopic investigations. This area has expanded rapidly since the deployment of the Goddard High Resolution Spectrograph on board the Hubble Space Telescope [2], and other facilities such as SUMER (the Solar Ultraviolet Measurement of Emitted Radiation) on SOHO (the Solar and Heliospheric Observatory) [3]. These instruments have yielded many interesting and sometimes unexpected results which require additional atomic data for their interpretation.

Progress in the spectroscopy and time resolved structure of multiply ionized atoms has been greatly enhanced by the development of new light sources and technical advances in the use of traditional light sources, motivated by urgent needs in specific applications. These light sources involve the use of fast ion beams, tokamak- and laser-produced plasmas, orbiting observation of astrophysical objects, recoil ions from ion–atom collisions, electron beam ion sources (EBIS) and ion traps (EBIT), electron cyclotron resonance (ECR) ion sources, ion storage rings and Fourier transform spectroscopy (FTS), as well as improved use of traditional spark, vacuum arc, and laser-produced plasma sources. Since the interpretation of time resolved measurements often requires a thorough understanding of the radiatively coupled energy level structure, they must proceed in concert with high resolution wavelength and energy level measurements.

Highly ionized atomic systems involve new theoretical regimes, where interactions that are negligible in neutral atoms can become dominant. Because they scale with different powers of the nuclear charge Z, relativistic and quantum electrodynamic (QED) interactions can become very large in these systems. Traditionally "forbidden" decay processes such as M1, E2, M2, and two-photon decay (2E1) can exceed "allowed" E1 processes because of their stronger Z scaling. Rather than being exotic, these highly ionized systems dominate high temperature environments such as the solar corona and stellar and laboratory plasmas, and thus almost all matter in the universe is in the ionized state. In such systems the electron binding energy is very large, and the strong Coulomb field has a long-range intensity that enhances interaction cross sections. The reduced electron screening causes inner shell effects to become more important. All of these properties are dependent on Z, which can be treated as a controllable experimental parameter if a sufficiently large and reliable data base can be followed along an isoelectronic sequence.

To prescribe these systems, calculational methods must sometimes undergo conceptual and computational refinements subject to experimental verification. Furthermore, the accurate specification of wavelength and energy level data does not insure correct predictions of transition probabilities and lifetimes. Measurements of the lifetimes are particularly important, since they provide absolute rate values necessary to normalize relative transition probabilities obtained by time-integrated techniques.

A key feature of the specification of transition probability rates in highly ionized atoms is the use of semiempirical parameterization as a predictive tool. The search for systematic behavior in existing data and the use of previously discovered systematics in the interpretation of new data has long been a primary methodology of atomic spectroscopy. These tendencies have been accelerated by the size and precision of the available data base. It is now possible to select subsets within large blocks of atomic data so as to probe a specific property, e.g. along an isoelectronic, isonuclear, homologous, or Rydberg sequence, or along an yrast chain. This can reveal unexpected regularities (or unexpected departures from known regularities) that would otherwise go unnoticed. This sensitivity can be enhanced using data-based semiempirical systematizations, which amplify subtle interactions by reexpressing them as an effective parameter in a simple independent particle mapping reduction.

This review will attempt to describe the current status of this field and to provide a few illustrative examples of recent results. For more details, the reader is referred to a number of recent reviews [4–13].

2. Experimental Methods

2.1 Excitation Methods

The most direct method for the experimental determination of level lifetimes is through the time resolved measurement of the free decay of the fluorescence radiation following a cutoff of the source of excitation. An important factor limiting the accuracy is the repopulation of the level of interest by cascade transitions from higher lying levels, which can produce multiexponential decay curves that are difficult to analyze with precision by standard curve fitting methods. To circumvent this limitation, two alternative methods are often employed. One involves selective excitation of the level of interest, thus eliminating cascading altogether; another uses correlations between cascade connected decays to account in detail for the effects of cascades.

While selective excitation methods totally eliminate the effects of cascade repopulation, they are generally limited to levels in neutral and singly ionized atoms that can be accessed from the ground state by strongly absorptive E1 transitions, and the selectivity itself is a limitation. Many very precise measurements have been made [5] by these techniques, frequently for Δn=0 resonance transitions in neutral alkali atoms and singly ionized alkali-like ions.

More general access can be obtained by nonselective excitation methods, such as pulsed electron beam bombardment of a gas cell or gas jet, or in-flight excitation of a fast ion beam by a thin foil. Pulsed electron excitation methods are well suited to measurements in neutral and near neutral ions (although for very long lifetimes in ionized species, the decay curves can be distorted if particles escape from the viewing volume through thermal migration or the Coulomb explosion effect [14]). Nonselective excitation techniques can also be applied to measurements such as the phase shift method and the Hanle effect, in which case cascade repopulation can also be a serious problem [15]. In the case of highly ionized atoms, the only generally applicable method of lifetime determination is by beam foil spectroscopy (BFS), the thin foil excitation of a fast ion beam. However, additional techniques, e.g., using recoil ions [16], ion traps and storage rings, have been employed in recent years for the determination of long lifetimes (cf. Sect. 6).

In beam-foil studies, excitation of the ions occurs in the dense environment of the foil, after which they emerge into a field free, collision free, high vacuum region. A time-resolved decay curve is obtained by translating the foil relative to the detection apparatus. The beam is a very tenuous plasma, which has both advantages and disadvantages. The low density avoids the effects of collisional de-excitation and radiation trapping, but produces relatively low light levels. This requires fast optical systems with reduced wavelength resolution, which can lead to blending of the Doppler broadened lines. Methods have been developed to refocus the monochromator to a moving light source, using the angular dependence of the Doppler shift to narrow and enhance the lines. Detailed descriptions of the various aspects of beam foil lifetime measurements can be found in Refs. [6, 8, 17].

2.2 Data Analysis

Due to the nonselective nature of beam-foil excitation, the level populations (and hence the decay curves) are affected by cascade repopulation. Thus, the decay curve involves a sum of many exponentials, one corresponding to the primary level, and one to each level that cascades (either directly or indirectly) into it. Decay exponentials do not comprise an orthogonal set of functions, and the representation of an infinite sum by a finite sum through curve fitting methods [18] can, in unfavorable cases, lead to significant errors, sometimes as large as 30–40%. Thus, while cascades which differ significantly in lifetime from that of the primary do not pose a serious problem, when the cascades and primary have similar lifetimes their contributions can seriously distort the decay curve.

Fortunately, alternative methods to exponential curve fitting exist, which permit the accurate extraction of lifetimes to be made from correlated sets of nonselectively populated decay curves. Situations in which cascading is dominated by a few strong decay channels are ideally suited to the arbitrarily normalized decay curve (ANDC) method [19, 17]. This method exploits dynamical correlations among the cascade-related decay curves, which arise from the rate equation that connects the population of a given level to those of the levels that cascade directly into it. The instantaneous population of each level is, to within constant factors involving the transition probabilities and detection efficiencies, proportional to the intensity of radiation emitted in any convenient decay branch. A joint correlated analysis of the decay curves of the primary level and those of the levels that directly repopulate it yields both the primary lifetime and the intensity normalizations of the cascades relative to that of the primary.

A schematic representation of the ANDC method, applied [20] to the resonance transition of Tl III in the Au isoelectronic sequence, is shown in Fig. 1. The study of lifetimes in ions of very heavy elements such as Tl has been motivated by their occurrence in the spectra of certain chemically peculiar stars [2]. The inset decay curves indicate that the arbitrarily normalized decay curves $I_{6p}(t)$ of the $6p\ ^2P_{3/2}$ level and its cascades from the $6d\ ^2D_{5/2}$ and

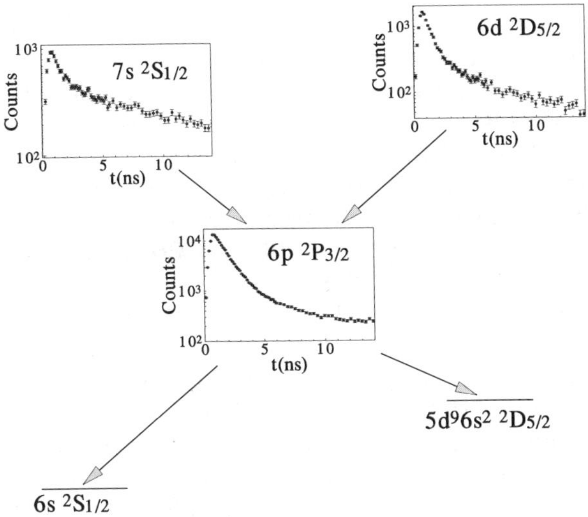

Fig. 1. Schematic plot of the ANDC analysis of the $6s$–$6p$ transition in Tl III

$7s\ ^2S_{1/2}$ are jointly analyzed using the population equation expressed in the form

$$\tau_{6p}\frac{dI_{6p}}{dt}(t_i) = \xi_{6d}I_{6d}(t_i) + \xi_{7s}I_{7s}(t_i) - I_{6p}(t_i)\ . \tag{1}$$

Here the primary lifetime τ_{6p} and the relative normalizations ξ_{6d} and ξ_{7s} for the cascade and primary decay curves are determined by simultaneous solution of the large set of relationships provided by (1) evaluated at each of the common points t_i on the decay curves. If all significant direct cascades have been included, the goodness-of-fit will be uniform for all time subregions, indicating reliability. If important cascades have been omitted or blends are present, the fit will vary over time subregions, indicating a failure of the analysis. Very rugged algorithms have been developed [21] that permit accurate lifetimes to be extracted even in cases where statistical fluctuations are substantial. The use of the ANDC method can provide a striking improvement in the reliability of the lifetime extracted from the decay curves. In a recent study of the lifetime of the $3s3p\ ^1P_1$ level in Mg-like Cl VI, *Engström* et al. [22] obtained a value 0.24±0.02 ns from curve fitting and a value of 0.165±0.009 ns from ANDC, a nearly 50% over-estimation by curve fitting.

2.3 Detection Methods

The use of position sensitive detection (PSD) to make simultaneous measurements of many different decay curves is an advance that has made possible many new types of experiments in fast ion beam spectroscopy. These multiplexed methods greatly enhance the data collection efficiency for this low density light source, and cause many possible systematic errors to cancel in differential measurements. Examples of new types of measurements that can be made using this technology include studies of: (a) the variation of lifetimes among fine structure components within a multiplet that occur due to, e.g., selective access to forbidden intercombination and autoionization channels; and (b) analysis of decay curves in a cascade chain using the ANDC method [19, 17]. If the intensities of a group of lines that are, e.g., within a multiplet, connected by a cascade scheme, overlapping in a blend, etc. are recorded simultaneously, effects such as fluctuations in the beam current, degradation of the foil, divergence of the beam, etc., will affect all of the decay curves in the same way, and vanish from the differential decay curves.

In these measurements the PSD is mounted at the exit focus of the analyzing monochromator. Since it records all lines within a wavelength interval simultaneously, it can greatly reduce the required accelerator time. The multichannel detection also provides the advantage of simultaneously measured reference lines (beam foil spectra exhibit sizable Doppler shifts, precluding the use of stationary calibration sources). Decay curves can be constructed by integrating over a line profile, and that profile can be examined for exponential content to eliminate blending. The time dependent backgrounds underlying the decay curves are directly available from the neighboring channels.

One application of this technique [23] concerned the precise measurement of the ratio of lifetimes of the $J = 3/2$ and $1/2$ resonance transitions in Li-like Si. This type of multiplexed decay curve measurement yielded an accuracy in the difference between the two lifetimes that was better than could be obtained for either lifetime separately. Another application [24] studied intercombination transitions (i.e., E1 transitions which violate the $\Delta S = 0$ selection rule) in Mg- and Al-like ions of Br. The time resolved wavelength spectrum provided a common exposition of decays with a wide range of lifetimes. The short-lived decays arising from transitions provided wavelength calibrations, and the long-lived lines arising from intercombination transitions could be used for wavelength and lifetime determinations.

3. Computational Methods

3.1 Definitions

The theoretical quantity that characterizes the spontaneous emission and absorption of radiation between levels i and k is the *line strength* S_{ik}, given in the electric dipole (E1) approximation by the expression [25]

$$S_{ik} = \sum_{m_i} \sum_{m_k} |\langle i|\boldsymbol{r}|k\rangle|^2 , \tag{2}$$

where m_i and m_k are magnetic quantum numbers.

Under the approximations and assumptions of the Weisskopf–Wigner formulation [26], the line strength can be related to the measured exponential lifetime τ_i, provided that the branching fraction B_{ik} for decay from the upper level i to a specific lower level k is known from other measurements. In such a case an experimental value for S_{ik} can be extracted from the measured data using the relationship

$$S_{ik} = [\lambda_{ik}(\text{Å})/1265.38]^3 g_i B_{ik}/\tau_i(\text{ns}), \quad (3)$$

where λ_{ik} is the transition wavelength and g_i is the upper level degeneracy. In terms of these quantities, the emission of radiation can be specified by the spontaneous transition probability rate A_{ik}

$$A_{ik} = B_{ik}/\tau_i \quad (4)$$

and the absorption of radiation can be specified by the absorption oscillator strength f_{ki}

$$g_k f_{ki} = [303.75/\lambda_{ik}(\text{Å})]S_{ik}, \quad (5)$$

where g_k is the lower level degeneracy.

3.2 Methodologies

In theoretical calculations of transition probabilities and lifetimes for highly charged ions, a number of effects must be considered. In addition to the electron correlation, which also plays an important role in neutral and a few times ionized ions, the effects of relativity, QED and sometimes even nuclear structure have to be incorporated in the calculations. The importance of these additional effects increases strongly with Z which means, for instance, that simplifications based on perturbative treatments may not be adequate in highly charged ions [27].

The theoretical techniques can be roughly divided into two categories, ab-initio calculations and semiempirical methods. In the latter category the wave functions are usually optimized by utilizing accurate experimentally determined data, e.g., excitation energies. This is done in the superposition of configuration (SOC) calculations with the *Cowan* code [25] by a suitable scaling of the Slater parameters. This method has been successfully applied to ions with complex electron structure where ab-initio calculations may encounter substantial computational difficulties. Another semiempirical technique, the Coulomb approximation with Hartree–Slater core (CAHS) method which incorporates a realistic model potential to account for polarizability, spin–orbit interaction and exchange effects [28] has yielded quite accurate f-value data, in particular for ions with a few active electrons. Extensive sets of data for multiply ionized atoms have also been obtained by means of the so-called MZ method, a Z-expansion technique based on perturbation theory [29], the results being quite accurate even for very high charge states.

Among the various ab-initio methods, those based on the multiconfiguration Hartree–Fock (MCHF) [30] or configuration interaction (CI) [31] treatments have in recent years provided very accurate data, e.g., for Be-like ions, where f-values with uncertainties as low as 2–3% have been predicted [32, 33].

The theoretical methods mentioned so far are basically nonrelativistic, but relativistic effects can usually be incorporated within the Breit–Pauli approximation [34], whereby the relativistic part of the Hamiltonian is added to the nonrelativistic one. However, fully relativistic calculations are now often performed for highly charged ions. The multiconfiguration Dirac-Fock (MCDF) method [35] is thereby frequently used. Another powerful program is the multiconfiguration relativistic random-phase approximation (MCRRPA), introduced in the early 1980s [36]. Recent results of calculations with this method concern Be-like ions [37]. Also, the many-body perturbation theory (MBPT), applied to atomic physics problems in the early 1960s, has in recent years been successfully extended to the relativistic regime [38, 39]. However, while the fully relativistic programs are very efficient when applied to highly charged ions, they are less powerful for lower ionization stages because they can usually treat electron correlation effects in a comparatively limited way.

4. Semiempirical Systematizations

Much progress in the evaluation, interpolation and extrapolation of atomic transition probability rate data has been achieved through systematic studies along isoelectronic sequences. For example, it has been observed [40] that for low to intermediate values of Z, a smooth isoelectronic variation is often obtained when the quantity f_{ik} is plotted for an isoelectronic sequence versus the quantity $1/Z$. However, as experimental measurements have been extended to very high stages of ionization, it has been found that a more extensive and more nearly linear exposition results when experimental values for line strengths are plotted versus an appropriately defined reciprocal effective screened charge. Thus large blocks of isoelectronic data can be described by the equation

$$Z^2 S \cong S_0 + b/(Z - C), \tag{6}$$

where S_0, b and C are empirical constants. In this manner a relatively small number of precision lifetime measurements can be used to interpolatively predict line strengths for entire isoelectronic sequences. A few of the types of sequences which have been systematized by these methods will be described below.

4.1 Alkali-Metal-Like Sequences

Isoelectronic regularities in line strength data for complex atoms were predicted theoretically [41, 42] and were first empirically applied [43] in the study of scaled line strengths for $\Delta n = 0$ resonance transitions in alkali-metal-like systems. For these systems, it has been found that the isoelectronic variation cannot only be linearized, but that the semiempirical fitting parameter S_0

effectively matches the corresponding hydrogenic $\Delta n = 0$ value S_{H}, which is given for an s–p transition by [34]

$$S_0 \to S_{\mathrm{H}} = 3n^2(n^2-1)g_i/4. \tag{7}$$

Expositions of the existing data bases for the Li, Na, Cu, and Ag sequences [44] are shown in Fig. 2. (Here, as well as in Figs. 3–6, all experimental data, except for some neutral and singly charged species, originate from beam-foil measurements.) The linearity is achieved by setting the screening parameters at $C = 2.17, 10.0, 25.9, 44.7$ for $n = 2, 3, 4, 5$, respectively. In these systems the ns–np resonance transitions are unbranched, so $B_{ik} = 1$ and the line strengths are determined directly from the measured lifetimes. If the linearity of (6) is valid, then accurate measurements for the lifetimes of a few members of the sequence at low stages of ionization can be interpolated to the hydrogenic limit at high Z, providing accurate estimates for all intervening ions.

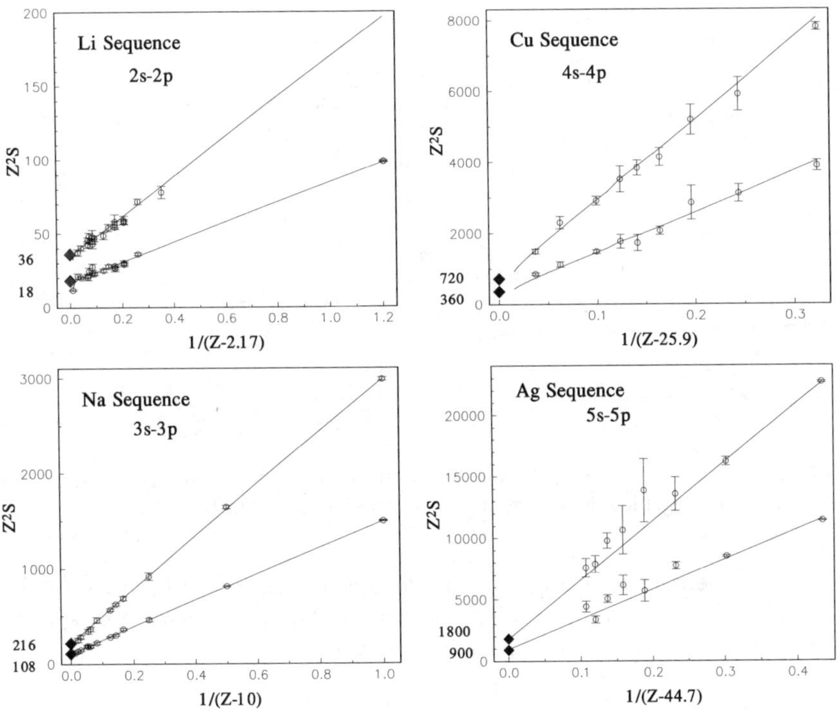

Fig. 2. Scaled line strength plots for the ns–np transitions in the alkali-like Li, Na, Cu and Ag sequences. Data sources are given in Ref. [44]

While these expositions exhibit empirical linearities for low to medium degrees of ionization, the hydrogen-like value given in (7) is nonrelativistically computed, and at very high Z this Schrödinger formulation would be expected to break down. Correspondingly, the solution to the Dirac equation for a pure $1/r$ potential is given by

$$Z^2S = S_{\rm H}[1 - \sum_i a_i(\alpha Z)^{2i}], \tag{8}$$

where $S_{\rm H}$ is as given by (7) and a_i is a set of coefficients given in Ref. [44]. An experimental test of this high Z behavior is available for the Li sequence, where the $2p\,^2P_{1/2}$ lifetime has been measured for lithium-like uranium U^{89+} [45]. As can be seen from the lithium-like case in Fig. 2 and the blowup of its high-Z region in Fig. 3a, the three-electron uranium measurement falls considerably below the linear trend. This discrepancy disappears, however, when the semiempirical exposition is modified to take into account the high-Z behavior of the corresponding hydrogen-like system, using the form

$$Z^2S/[1 - \sum_i a_i(\alpha\{Z - C\})^{2i}] \cong S_{\rm H} + b/(Z - C). \tag{9}$$

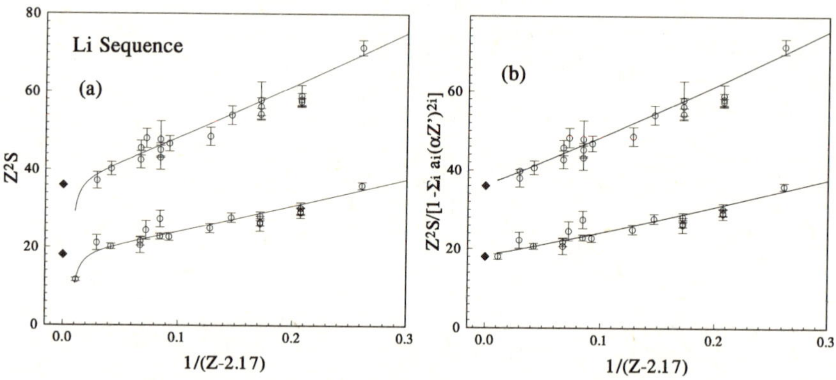

Fig. 3. Scaled line strengths for the Li $2s$–$2p$ sequence, without (a) and with (b) relativistic hydrogenic corrections. Data sources are given in Ref. [44]

As can be seen from Fig. 3b, this modification restores linearity to the exposition, and allows the existing data base to be used to accurately specify line strengths for the entire isoeletronic sequence.

4.2 Alkaline-Earth-Like Sequences

In the case of systems in which two valence electrons (or one electron and one hole) occur in an otherwise closed shell core, a simple extension of this formalism has been developed that combines lifetime data for the resonance and intercombination lines. In systems such as ns^2–$nsn'p$ and np^6–$np^5n's$, both the $^1S_0$–$^1P_1$ and $^1S_0$–$^3P_1$ transitions occur because intermediate coupling (IC) causes spin hybridization of the nominally singlet and triplet excited states. In the absence of strong configuration interaction (CI) effects, the IC amplitudes can be determined from spectroscopic energy level data and expressed in the form of a singlet-triplet mixing angle θ. The measured lifetime data for the unbranched resonance and intercombination lines are first converted to line strengths $S(\text{Res})$ and $S(\text{Int})$ using (3), and then reduced to the effective values

$$S_r(\text{Res}) \equiv S(\text{Res})/\cos^2\theta, \tag{10}$$

$$S_r(\text{Int}) \equiv S(\text{Int})/\sin^2\theta, \tag{11}$$

from which the effects of IC have been removed. Empirical plots of data that have been reduced in this way are shown in Fig. 4 for the ns^2–$nsnp$ transitions in the Mg [46], Zn [47], Cd [48], and Hg [49] sequences. Linear plots are achieved by use of the screening parameters C = 10.5, 28, 46, 75 for $n = 3, 4, 5, 6$. Here both $J = 1$ levels tend to converge to a value of S_H twice that predicted by (7) (consistent with the two equivalent ground term electrons).

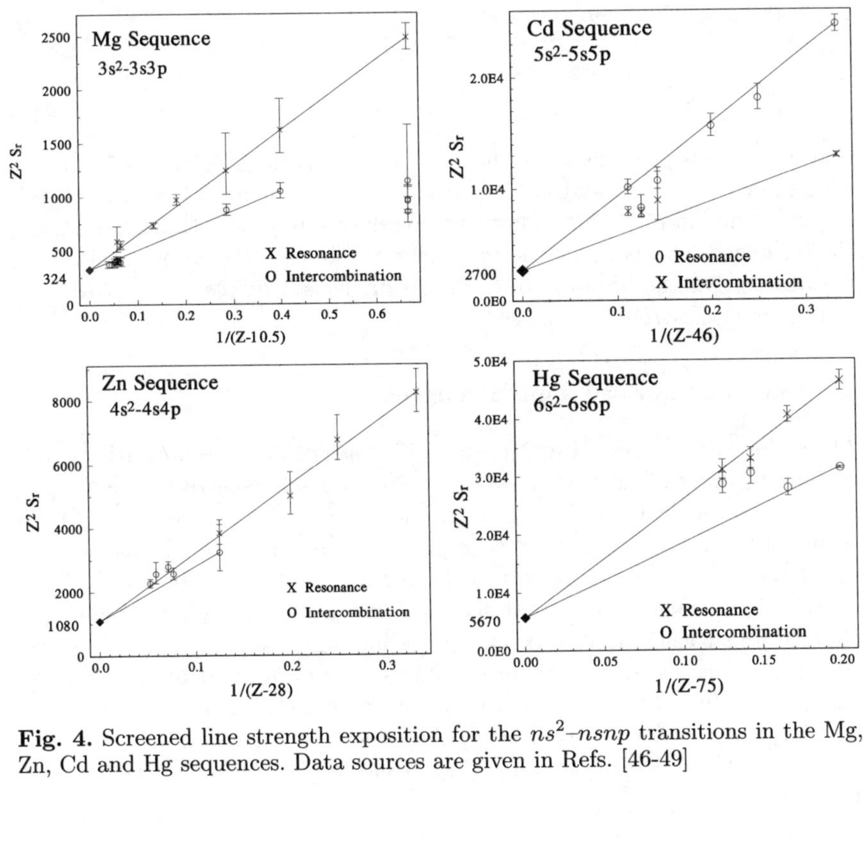

Fig. 4. Screened line strength exposition for the ns^2–$nsnp$ transitions in the Mg, Zn, Cd and Hg sequences. Data sources are given in Refs. [46-49]

In these two valence electron cases, the question arises as to whether the data should be extrapolated to high Z using the single-configuration hydrogenic value, or take into account the CI that can occur due to the hydrogenic degeneracy of levels within the same complex. *Curtis* and *Ellis* [50] have studied this for the Be sequence, which they have characterized by two mixing angles: ϕ (the high Z asymptotic limit of the CI mixing between $2s^2$ and $2p^2$) and θ (the IC mixing between the singlet and triplet J=1 levels in $2s2p$). They achieved a linear systematization of the data by theoretically calculating the nonrelativistic high Z limit of the $2s^2$–$2p^2$ mixing angle, which yielded ϕ=13° [50]. This predicts a high Z result $S_{\mathrm{H}} = 38.5$, which agrees very well with the trend of the measured data shown in Fig. 5a.

Studies have also been made [51] within this semiempirical formulation to characterize the differences between the radial dipole matrix elements for the resonance and intercombination transitions. *Curtis* et al. [51] have shown that for a Dirac relativistic formulation of the ns^2–$nsn'p$ transition, the line strengths can be specified in terms of the $J = 1/2$–$3/2$ and J=1/2 –1/2 radial matrix elements $R_{31}=\langle s_{1/2}|r|p_{3/2}\rangle$ and $R_{11}=\langle s_{1/2}|r|p_{1/2}\rangle$ by

$$S_{\mathrm{r}}(\mathrm{Res}) \equiv S(\mathrm{Res})/\cos^2(\theta - \xi), \tag{12}$$

$$S_{\mathrm{r}}(\mathrm{Int}) \equiv S(\mathrm{Int})/\sin^2(\theta - \xi), \tag{13}$$

where

$$\tan\xi \equiv \sqrt{2}(R_{31} - R_{11})/(2R_{31} + R_{11}). \tag{14}$$

Figure 5b shows the data for the $2s^2$–$2s2p$ transitions in the Be sequence with these additional relativistic corrections, which remove the gap between resonance and intercombination transition data. Thus, with these corrections, the combined data base for the resonance and intercombination lines can be used to predict the lifetimes of both the singlet and triplet $J = 1$ levels for the entire isoelectronic sequence.

4.3 Intershell $\Delta n = 1$ Line Strengths

All of the cases discussed above involve intrashell $\Delta n = 0$ transitions (ns–np for alkali-metal-like sequences, ns^2–$nsnp$ for alkaline-earth-like sequences) and the next logical step would be to extend the formalism to $\Delta n > 0$ transitions. Unfortunately, this is not generally possible because almost no measured transition probability data exist for $\Delta n > 0$ transitions in multiply charged ions. This results from the fact that decay rate data for multiply charged ions consist almost exclusively of lifetime measurements. These can be used to deduce transition probabilities for unbranched decays, but (as discussed in Sect. 5) virtually no branching fraction measurements exist for multiply charged ions. Two counterexamples exist, in which special conditions permit the determination of line strengths from lifetimes for $\Delta n = 0$ transitions.

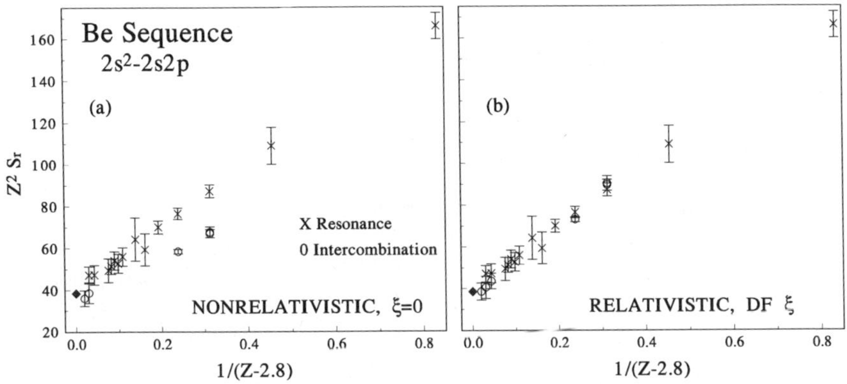

Fig. 5. Scaled line strengths for the Be $2s^2$–$2s2p$ sequence with corrections for (a) asymptotic CI between $2s^2$ and $2p^2$ and (b) with relativistic corrections for the J-dependence of the Dirac wave functions. Data sources are given in Ref. [50]

The first example involves the $2s^2\ {}^1S_0$–$2s3p\ {}^{1,3}P_1$ transitions in the Be isoelectronic sequence [52]. Here differential lifetime measurements in the $2s3s\ {}^3S_1$–$2s3p\ {}^3P_{2,1,0}$ decay channels determine the intercombination branch to ground as an extra channel available only to the ${}^3P_1$ level. This decay mode is made possible by the spin–orbit interaction which mixes the ${}^1P_1$ and ${}^3P_1$ levels. For the resonance transition, the ${}^1P_1$ level decays dominantly to the ground state, and other branching can be estimated theoretically without large error. The second example involves the $2p^6\ {}^1S_0$–$2p^5 3s\ {}^{1,3}P_1$ transitions in the Ne sequence [52], for which the first excited state is extrashell, giving rise to unbranched $\Delta n = 1$ resonance and intercombination transitions. Expositions of these data are shown in Fig. 6.

Here again the data appear to converge toward the hydrogenic limits at high Z ($S_{\rm H} = 18.79$ for $2s$–$3p$ and $S_{\rm H}$=1.76 for the $2p$–$3s$). Despite the obvious differences in their structures, striking similarities exist between these two sequences, and they are in some ways more closely related to each other than to other transitions within their own sequence. For example, the isoelectronic behavior of the singlet-triplet mixing angles for these two sequences are virtually identical. Unlike the case for the Be $2s2p$ sequence, both the Be $2s3p$ and the $2p^5 3s$ configurations pass through a value of 45° for the mixing angle, (where the singlet and triplet reverse their roles as the dominant amplitude in the wave function) just above $Z = 20$. These and other interesting properties motivate additional studies of $\Delta n \geq 1$ transitions.

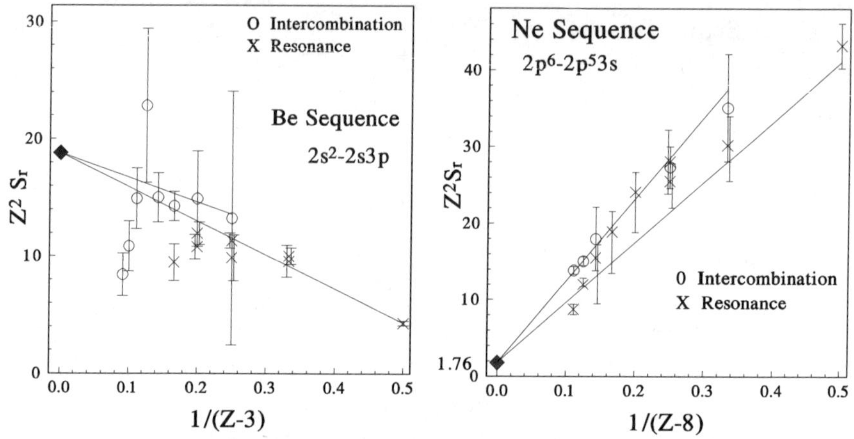

Fig. 6. Scaled line strengths for the $\Delta n = 1$ transitions in the (a) Be $2s^2$–$2s3p$ and (b) Ne $2p^6$–$2p^5 3s$ sequences. Data sources are given in Ref. [52]

5. Transition Probabilities and Branching Fractions

5.1 The Need for Branching Fractions

Branching fractions are needed to convert lifetime data to transition probabilities and absorption oscillator strengths. These quantities are needed to deduce elemental abundances from astrophysical spectra, to determine impurity concentrations from fusion plasmas spectra, and to make transition-by-transition comparisons between experiment and theory (summed comparisons can conceal the origin of discrepancies).

As discussed above, atomic lifetime measurements can be transformed into a slowly varying isoelectronic systematization by reducing each lifetime to a corresponding effective line strength factor. This is in contrast to the raw measurements, since the reciprocal lifetime of a branched decay consists of a sum of transition probabilities, each with a different wavelength dependence that possesses its own characteristic dependence on Z. Thus, while a branched line strength factor will usually have a simple and slowly varying isoelectronic behavior, a lifetime may have a rapid isoelectronic variation containing several different powers of Z. The line strength reduction requires knowledge of the branching fractions.

Much progress has been made in the determination of branching fractions in neutral and singly ionized atoms, through the use of absolute emission, absorption, or dispersion measurements, or through combined measurements of relative branching ratios and lifetimes [5]. In highly ionized atoms, many measurements of lifetimes in the 1–5% accuracy range have been made by ANDC analysis [19, 17] of beam-foil measurements. However, *branching ratio measurements in multiply charged ions are virtually nonexistent.* This dearth of $\Delta n \geq 1$ ionic transition probability data is not widely recognized, and deserves special emphasis.

5.2 Calibrated Intensity Measurements

The reason for this lack of branching fraction data in multiply charged ions is clear. Most measurement methods require a relative intensity calibration of the detection apparatus as a function of wavelength, and this is particularly difficult for multiply charged ions. Until recently, beam-foil excitation was the only general way of accessing highly ionized systems. With fast ion beam methods there are Doppler broadenings and shifts, polarizations due to anisotropic excitation, wavelengths not amenable to reflective and transmissive optical elements, differential downstream decays and repopulations of the levels, etc. These attributes are not well suited to standard techniques for calibrating grating spectrometers and detection systems through the use of standard lamps. Calibrations have been carried out using synchrotron radiation, or through in-beam lines with known branching fractions, but these techniques are highly specialized and not widely applicable. It is now possible to supplement time-resolved beam-foil studies with measurements of relative intensities from a common upper level using ECR sources, ion traps, and high effect lasers, coupled to both grating and FTS spectrometers, but many similar problems in calibration must also be overcome with these methods. While the calibration of the wavelength dependence of the detection efficiency of an optical system presents many difficulties, these are less severe in the study of neutral and singly charged ions. Here laboratory-fixed calibration sources can be used in some applications, as has been discussed by *O'Brian* and *Lawler* [10].

The use of Si(Li) detectors in the measurement of very short wavelength radiation in highly ionized atoms also offers possibilities for branching ratio measurements. Since these devices can specify the photon energy from pulse height information without the need for a spectrometer, they can be calibrated for detection efficiency as a function of energy. This type of detection was recently used to determine the branching ratio of the M1 channel to the two-photon decay channel in the $2s\ ^2S_{1/2}$ state in the one-electron krypton ion [53].

5.3 Specialized Methods for Determining Branching Fractions

When branching is within a multiplet between two terms that are each dominated by a single configuration, branching fractions can be empirically specified from IC amplitudes deduced from spectroscopic energy level data. Applications of this approach have recently been applied to the Pb sequence [51, 54].

Differential lifetime measurements can also sometimes be used to determine branching fractions. One application involves studies of the lifetimes of the individual fine structure components of a multiplet decay. If one fine structure level has a decay channel that is not available to the other levels (e.g., a spin-changing transition [52] or an autoionization mode [55] made possible by J-dependent IC) then the transition probability rate of the extra channel can be determined by differential lifetime measurements.

In addition, the correlated analysis of cascade-related decay curves by the ANDC method [17, 19] can yield both the lifetime of the primary level and the detection efficiencies of the cascade decay curves relative to that of the primary. The correlated time structure of the decay curves can therefore provide a relative calibration of the detection equipment as a by-product. This technique has been successfully used [56] to calibrate a detection system at discrete points over the wavelength region λ= 400–5000 Å by bootstrapping successive ANDC analyses in Na-like S VI.

6. Forbidden Transitions

6.1 Experimental Techniques

As indicated in the Introduction, the forbidden transitions (M1, M2, E2, 2E1, spin-forbidden E1, etc.) have very low transition probability rates in neutral or few-times ionized atoms where allowed (E1) transitions dominate. However, the decay rates of the latter scale approximately as Z (for $\Delta n = 0$) and as Z^4 (for $\Delta n = \pm 1$) to be compared with Z^6–Z^{10} for the various forbidden decay modes, which therefore become important in highly charged ions.

In beam-foil experiments the practical lifetime range is approximately 1 ps–200 ns. Measurements of longer lifetimes are cumbersome, partly because the 1/e path lengths become very large. This means that BFS can be used to determine f-values for allowed transitions from neutral atoms up to perhaps 70+ ions (in favorable cases) and for forbidden lines from approximately 15+ and higher charge states. Certain spin-forbidden E1 decays in lower state ions have also been studied by BFS, but such work has also been extended to highly charged systems [57]. However, long lifetimes in multiply charged ions can now be determined by using a number of techniques of more recent origin than BFS. These utilize: ion traps, for instance an EBIT facility, and storage/cooler rings for ions.

At the Livermore EBIT [58] the decay time of an M1 transition from the $1s2s\ ^3S_1$ level in He-like Ne and Mg were determined (cf. Fig. 7b) [59, 60]. This level was excited by a fast-switching electron beam and the fluorescent decay was recorded. The resulting lifetime values, 90.5±1.5 μs (Ne) and 13.61±0.49 μs (Mg), were found to be in good agreement with theoretical predictions. A more complicated system, Ti-like Xe^{32+}, was studied with the EBIT at NIST in Gaithersburg. Here a decay time of an M1 transition between two fine structure levels was determined [61]. The lifetime range that can be determined with EBIT facilities has been estimated to be 10^{-7}–10^{-2} s.

Storage/cooler rings for ions [62–64] are also well suited for the studies of long lifetimes. Using laser-induced fluorescence at the ESR heavy ion cooler ring at GSI, *Klaft* et al. [65] were able to observe the ground state hyperfine splitting of hydrogen-like $^{209}Bi^{82+}$ and measure the lifetime of the excited component. In studies of lifetimes of the $1s2s\ ^3S_1$ levels in He-like C^{4+} [66] as well as B^{3+} and N^{5+} [67] at the Heidelberg storage ring, fluorescence was not recorded. Instead, the lifetimes were studied by using dielectronic recombination (DR) resonances from collisions between the $1s2s\ ^3S_1$ level and electrons in the electron cooler. Very precise lifetimes, e.g., 20.59±0.05 ms for C^{4+}, were obtained which allowed detailed comparisons between theoretical values. Furthermore, this method has been extended to the Be-like ions C^{2+}, [68], N^{3+} and O^{4+} [69], for which the lifetime of the metastable level $2s2p\ ^3P_1$ has been accurately measured.

6.2 Forbidden Transitions in One- and Two-Electron Ions

Allowed and forbidden transitions between low-lying levels in H-like and He-like ions are depicted in Fig. 7. These levels have been extensively studied and lifetime data are available for a very large number of ions. The early work has been reviewed by *Marrus* and *Mohr* [70].

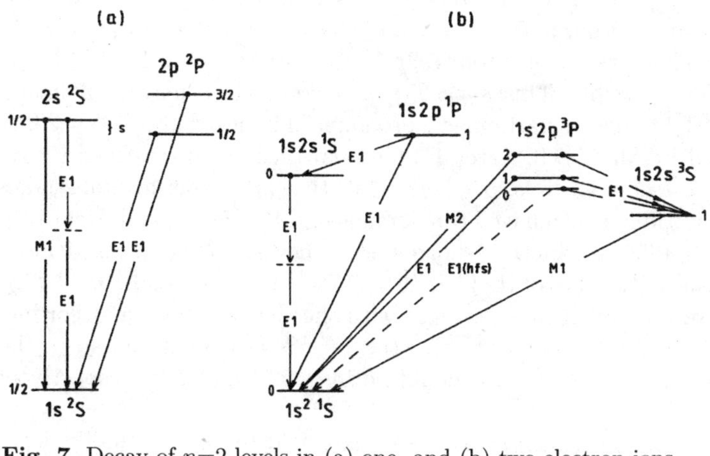

Fig. 7. Decay of n=2 levels in (a) one- and (b) two-electron ions

The $2s\ ^2S_{1/2}$ level in H-like ions decays by M1 or 2E1 emission to the ground state. These decay probabilities scale as Z^{10} and Z^6, respectively. For low values of Z, the 2E1 mode dominates while the probabilities are equal at about $Z = 41$. Experimental data, including the lifetime and branching ratio for Kr^{35}, as well as references to previous work, are given by *Cheng* et al. [53]. A 2E1 decay, similar to that for H-like ions, also takes place for the $1s2s\ ^1S_0$ level in He-like ions [71]. The photons from the 2E1 decays have a continuous frequency distribution but their sum equals the energy difference between the $n = 1$ and $n = 2$ levels.

The $1s2s\ ^3S_1$ level, already discussed above, exhibits a relativistically induced M1 decay to the $1s^2\ ^1S_0$ ground term, as first established from astrophysical observations [72]. For He-like Kr^{34+} the $^3S_1$ lifetime was determined by BFS [73]; the value, 171.0±2.0 ps, is in excellent agreement with theoretical predictions. As shown in Refs. [66, 67, 73] experimental data for this M1 transition are available for many species from He ($Z = 2$) to He-like Kr ($Z = 54$). In this interval the lifetimes range from about 8000 s (He) to 2.61 ps (Kr).

The levels $J = 0, 1, 2$ of the $1s2p\ ^3P$ term all make transitions to $1s2s\ ^3S_1$. For the $^3P_2$ level M2 transitions to the ground state are also possible. This rate scales as Z^8, and already for Ar^{16+} the M2 and E1 decay modes of $^3P_2$ are equally probable. Because of the spin–orbit interaction the wave functions of $1s2p\ ^3P_1$ and $1s2p\ ^1P_1$ are mixed, opening the spin-forbidden decay mode $1s^2\ ^1S$–$1s2p\ ^3P_1$, the transition probability of which scales as Z^{10}. Beam-foil data are here available up to He-like Ar [4].

6.3 Hyperfine Quenching

The $1s2p\ ^3P_0$ level in He-like ions has an allowed E1 decay mode to the triplet term $1s2s\ ^3S_1$. However, when the nuclear spin I differs from 0 the total angular momentum is $F = I + J$ and the $F = I$ states of the $1s2p\ ^3P_0$ and $^3P_1$ mix. Thus, transitions from $1s2p\ ^3P$ ($J = 0$, $F = 1$) to $1s^2\ ^1S$ ($J = 0$, $F = 1$) become possible. This *hyperfine quenching* which may significantly shorten the $^3P_0$ lifetime, is particularly pronounced in nuclei above $Z = 20$, as first predicted by Mohr [74]. Several additional theoretical treatments have appeared later, (see, e.g., Refs. [75, 76]), while the early experimental studies of this effect in He-like ions have been summarized by *Marrus* and *Mohr* [70].

In the early 1980s the lifetime shortening of the $1s2p\ ^3P_0$ level was demonstrated in beam-foil studies of He-like $^{19}F^{7+}$ [77], $^{27}Al^{11+}$ [78] and $^{31}P^{13+}$ [79]. The results were in excellent agreement with the theoretical data, according to which the ratio of the decay rate to the $1s^2\ ^1S_0$ ground state and to the $1s2s\ ^3S_1$ state in these three ions should be 14%, 53% and 25%, respectively [74].

In recent years experiments have been extended to heavier He-like ions, with $Z > 20$, where theories predict the hfs-induced decay branch to be much more probable than the decay to $1s2s\ ^3S_1$. As first shown by *Indelicato* et al. [75] measurements of the $1s2p\ ^3P_0$ lifetime will also allow fairly precise determinations of the $^3P_1$theoretical interest for high-Z ions where calculations are quite difficult. The first such determination, for He-like $^{107}Ag^{45+}$, was reported by *Marrus* et al. [80], resulting in a value of 0.74 eV, with an uncertainty of about 20%. This separation is, interestingly enough, smaller than the natural width of the $^3P_1$ level, 1.1 eV, the reason being that the two levels are nearly degenerate in this ion. Some years later, a more accurate beam-foil study was performed, using two isotopes, $^{107}Ag^{45+}$ and $^{109}Ag^{45+}$ [81]. The separation was now determined with a 5% uncertainty, the result being in excellent agreement with relativistic calculations. In a similar study of He-like Ni, *Dunford* et al. [82] used both $^{61}Ni^{26+}$ ($I = 3/2$) and $^{58}Ni^{26+}$ ($I = 0$) isotopes. Here the $1s2p\ ^3P_0$ lifetime was – as expected – quite different for the two isotopes, 0.47 ns and 2.5 ns, respectively, whereas the fine structure was determined with an uncertainty of 6%. Subsequently, a similar measurement was extended to the isotopes $^{155}Gd^{62+}$ and $^{157}Gd^{62+}$ [83]. While both nuclei have $I = \frac{3}{2}$, the nuclear magnetic moments are different, resulting in different $1s2p\ ^3P_0$ lifetimes. A very accurate value was now obtained for the fine structure splitting, the uncertainty being only 1%.

7. Conclusion

The examples of lifetime measurements for highly charged ions and the interplay of the experimental results with data from theoretical and semiempirical analyses discussed in this review should demonstrate that the field of lifetime studies for highly ionized atoms is still in a very active phase. The method of BFS which has been used for more than 30 years now produces data which are more reliable than ever before. At the same time new techniques (storage rings, ion traps) for determination of long lifetimes are yielding data with uncertainties below 1%, thereby providing accurate tests of advanced theoretical methods. This precision is thus comparable to that reached in laser-based lifetime experiments of neutral atoms such as Li, Na and Cs [84].

In view of the developments on the theoretical side, it now appears desirable to further develop experimental techniques in such a way that typical uncertainties in experimental data reach the 1–3% level, instead of 5–10% which appears to be common in most lifetime studies for highly ionized atoms. In this way it would be possible to obtain information about various subtle effects in such ions. While the cascading problems are largely understood and solved in the BFS case, there are additional difficulties when the ion energies are tens of MeV or even higher. Among these are line blends and high beam-dependent backgrounds. Thus, lifetime measurements must be preceded by or combined with detailed spectroscopic studies of the systems of interest whereas the background problems caused by electrons or x rays require careful designs of the experimental setup. Another practical limitation is caused by the fact that the available time for experiments at large ion accelerators is usually quite limited. However, here the introduction of position-sensitive detectors, in particular charge-coupled detectors (CCD), have increased the data-recording efficiency by orders of magnitude.

Acknowledgement. The authors are grateful to the US Department of Energy, Office of Basic Energy Sciences, Division of Chemical Sciences and to the Swedish Natural Sciences Research Council (NFR) for support during the preparation of this work.

References

1. R.K. Janev: Comm. At. Mol. Phys. **26**, 83 (1991)
2. D.S. Leckrone, S. Johansson, G.M. Wahlgren, S.J. Adelman: Phys. Scr. **T47**, 149 (1993)
3. U. Feldman, W.E. Behring, W. Curdt, U.Schüle, K. Wilhelm, P. Lemaire, T.M. Moran: Astrophys. J. Suppl. Ser. **113**, 195 (1997)
4. I. Martinson: Lithuanian J. Phys. **36**, 3 (1996)
5. L.J. Curtis: in *Atomic Molecular and Optical Physics Reference Book*, G.W.F. Drake ed. (AIP Press, New York 1996) pp. 206–212
6. E. Träbert: in *Accelerator-Based Atomic Physics: Techniques and Applications*, S.M. Shafroth and J.C. Austin, eds. (AIP Press, New York 1998) pp. 567–600
7. L.J. Curtis, I. Martinson: Comm. At. Mol. Phys. **24**, 213 (1990)
8. E.H. Pinnington: in *Atomic Molecular and Optical Physics Reference Book*, G.W.F. Drake ed. (AIP Press, New York 1996) pp. 213–219
9. R.K. Janev, L.P. Presnyakov, V.P. Shevelko: *Physics of Highly Charged Ions*, (Springer-Verlag, Berlin 1985)
10. T.R. O'Brian, J.E. Lawler: in *Atomic, Molecular and Optical Physics: Atoms and Molecules*, F.B. Dunning and R.G. Hulet, eds. (Experimental Methods in the Physical Sciences, R. Celotta and T. Lucatorto, eds., Vol. 29B, Academic Press, San Diego 1996) pp. 217–254
11. H.-P. Garnir, T. Bastin, P.-D. Dumont: Phys. Scr. **T65**, 36 (1996)
12. E. Träbert: Nucl. Instrum. Methods B **98**, 10 (1995)
13. L. Engström: Phys. Scr. **40**, 17 (1989)
14. L.J. Curtis, P. Erman: J. Opt. Soc. Am. **67**, 1218 (1977)
15. L.J. Curtis: J. Opt. Soc. Am. **64**, 495 (1974)
16. G. Hubricht, E. Träbert: Z. Phys. D **7**, 243 (1987)
17. L.J. Curtis: in *Beam Foil Spectroscopy*, S. Bashkin ed. (Berlin, Springer 1976) pp. 63–109
18. S.W. Provencher: J. Chem. Phys. **64**, 2772 (1976)
19. L.J. Curtis, H.G. Berry, J. Bromander: Phys. Lett. **34** A, 169 (1971)
20. M. Henderson, L.J. Curtis, R. Matulioniene, D.G. Ellis, C.E. Theodosiou: Phys. Rev. A **56**, 1872 (1997)
21. L. Engström: Nucl. Instrum. Methods **202**, 369 (1982)
22. L. Engström, P. Bengtsson, C. Jupén, A.E. Livingston, I. Martinson: Phys. Rev. A **51**, 179 (1995)
23. A.E. Livingston, F.G. Serpa, A.S. Zacarias, L.J. Curtis, H.G. Berry, S.A. Blundell: Phys. Rev. A **44**, 7820 (1991)
24. E. Träbert, J. Sulieman, S. Cheng, H.G. Berry, R.W. Dunford, E.W. Kanter, C. Kurtz, A.E. Livingston, K.W. Kukla, F.G. Serpa, L.J. Curtis: Phys. Rev. A **47**, 3805 (1993)
25. R.D. Cowan: *The Theory of Atomic Structure and Spectra*, (Univ. California Press, Berkeley 1981)
26. L.J. Curtis, R.T. Deck, D.G. Ellis: Phys. Lett. A **230**, 330 (1997)
27. I. Lindgren: Nucl. Instr. Meth. Phys. Res. B **31**, 102 (1988)
28. C.E. Theodosiou, L.J. Curtis: Phys. Rev. A **38**, 4435 (1988)

29. U.I. Safronova, V.S. Senashenko: Phys. Scr. **25**, 37 (1982)
30. C. Froese-Fischer: Comput. Phys. Commun. **64**, 369 (1991)
31. A. Hibbert: Comput. Phys. Commun. **9**, 151 (1975)
32. C. Froese-Fischer, M. Godefroid, J. Olsen: J. Phys. B **30**, 1163 (1997)
33. J. Fleming, A. Hibbert, R.P. Stafford: Phys. Scr. **49**, 316 (1994)
34. H.A. Bethe, E.E. Salpeter: *Quantum Mechanics of One- and Two-Electron Atoms* (Springer, Berlin 1957)
35. F. Parpia, C. Froese-Fischer, I.P. Grant: Comput. Phys. Commun. **94**, 249 (1996)
36. W.R. Johnson, K.-N. Huang: Phys. Rev. Lett. **48**, 315 (1982)
37. H.-S. Chou, H.-S. Chi, K.-N. Huang: Chin. J. Phys. B **32**, 261 (1994)
38. A.-M. Mårtensson-Pendrill: Phys. Scr. **T46**, 102 (1993)
39. J. Sapirstein: Phys. Scr. **T46**, 52 (1993)
40. M.W. Smith, W.L. Wiese: Astrophys. J. Suppl. Ser. **23**, 103 (1971)
41. M. Cohen, A. Dalgarno: Proc. Roy. Soc. A **280**, 258 (1964)
42. R.J.S. Crossley, A. Dalgarno: Proc. Roy. Soc. A **286**, 510 (1965)
43. B. Edlén: Phys. Scr. **17**, 565 (1978)
44. L.J. Curtis, D.G. Ellis, I. Martinson: Phys. Rev. A **51**, 251 (1995)
45. J. Schweppe, A. Belkacem, L. Blumenfeld, N. Claytor, B. Feinberg, H. Gould, V.E. Kostroun, L. Levy, S. Misawa, J.R. Mowat, M.H. Prior: Phys. Rev. Lett. **66**, 1434 (1991)
46. L.J. Curtis: Phys. Scri. **43**, 137 (1991)
47. E. Träbert, L.J. Curtis: Phys. Scr. **48**, 586 (1993)
48. L.J. Curtis: J. Phys. B **26**, L589 (1993)
49. M. Henderson, L.J. Curtis: J. Phys. B **29**, L629 (1996)
50. L.J. Curtis, D.G. Ellis: J. Phys. B **29**, 645 (1996)
51. L.J. Curtis, D.G. Ellis, R. Matulioniene, T. Brage: Phys. Scr. **56**, 240 (1997)
52. L.J. Curtis, S.T. Maniak, R.W. Ghrist, R.E. Irving, D.G. Ellis, M. Henderson, M.H. Kacher, E. Träbert, J. Granzow, P. Bengtsson, L. Engström: Phys. Rev. A **51**, 4575 (1995)
53. S. Cheng, H.G. Berry, R.W. Dunford, D.S. Gemmell, E.P. Kanter, B.J. Zabransky, A.E. Livingston, L.J. Curtis, J. Bailey, J.A. Nolen, Jr.: Phys. Rev. A **47**, 903 (1993)
54. M. Henderson, L.J. Curtis, D.G. Ellis, R.E. Irving, G.M. Wahlgren: Astrophys. J. **473**, 565 (1996)
55. A.E. Livingston, H.G. Berry: Phys. Rev. A **17**, 1966 (1977)
56. L. Engström: Phys. Scr. **28**, 68 (1983)
57. E. Träbert: Phys. Scr. **48**, 699 (1993)
58. D. Schneider: this volume
59. B.J. Wargelin, P. Beiersdorfer, S.M. Kahn: Phys. Rev. Lett. **71**, 2196 (1993)
60. G.S. Stefanelli, P. Beiersdorfer, V. Decaux, K. Widmann: Phys. Rev. A **52**, 3651 (1995)
61. F.G. Serpa, C.A. Morgan, E.S. Meyer, J.D. Gillaspy, E. Träbert, D.A. Church, E. Takács: Phys. Rev. A **55**, 4196 (1997)
62. M. Larsson: Rep. Prog. Phys. **58**, 1267 (1995)
63. P.H. Mokler, Th. Stöhlkler: Adv. At. Mol. Opt. Phys. **37**, 297 (1996)
64. A. Wolf: this volume
65. I. Klaft, S. Borneis, T. Engel, B. Fricke, R. Greiser, G. Huber, T. Kühl, D. Marx, R. Neumann, S. Schröder, P. Seelig, L. Völker: Phys. Rev. Lett. **73**, 2425 (1994)
66. H.T. Schmidt, P. Forck, M. Grieser, D. Habs, J. Kenntner, G. Miersch, R. Reprow, U. Schramm, T. Schüssler, D. Schwalm, A. Wolf: Phys. Rev Lett. **72**, 1616 (1994)
67. H. Schmidt: Ph.D. thesis, Aarhus University (1994)

68. J. Doerfert, E. Träbert, A. Wolf, D. Schwalm, O. Uwira: Phys. Rev. Lett. **78**, 4355 (1997)
69. J. Doerfert, E. Träbert: Nucl. Instrum. Methods B **99,** 124 (1995)
70. R. Marrus, P.J. Mohr: Adv. Atom. Molec. Phys **14**, 181 (1978)
71. R.W. Dunford, H.G. Berry, S. Cheng, E.P. Kanter, C. Kurtz, B.J. Zabransky, A.E. Livingston, L.J. Curtis: Phys. Rev. A **48**, 1929 (1993)
72. A.H. Gabriel, C. Jordan: Phys. Lett **32** A, 166 (1970)
73. S. Cheng, R.W. Dunford, C.J. Liu, B.J. Zabransky, A.E. Livingston, L.J. Curtis: Phys. Rev. A **49**, 2347 (1994)
74. P.J. Mohr: in *Beam-Foil Spectroscopy*, I.A. Sellin and D.J. Pegg, eds. (Plenum Press, New York 1976) p. 97
75. P. Indelicato, F. Parente, R. Marrus: Phys. Rev. A **40**, 3505 (1989)
76. A. Aboussaid, M.R. Godefroid, P. Jönsson, C. Froese-Fischer: Phys. Rev. A **51**, 2031 (1995)
77. L. Engström, C. Jupén, B. Denne, S. Huldt, Weng Tai Meng, P. Kaijser, U. Litzén, I. Martinson: J. Phys. B **13**, L143 (1980)
78. B. Denne, S. Huldt, J. Pihl, R. Hallin: Phys. Scr. **22**, 45 (1980)
79. A.E. Livingston, S. Hinterlong: Nucl. Instrum. Methods **202**, 103 (1982)
80. R. Marrus, A. Simionovici, P. Indelicato, D.D. Dietrich, P. Charles, J.-P. Briand, K. Finlayson, F. Bosch, D. Liesen, F. Parente: Phys. Rev. Lett. **63**, 502 (1989)
81. B.B. Birkett, J.-P. Briand, P. Charles, D.D. Dietrich, K. Finlayson, P. Indelicato, D. Liesen, R. Marrus, A. Simionovici: Phys. Rev. A **47**, R2454 (1993)
82. R.W. Dunford, C.J. Liu, J. Last, N. Berrah-Mansour, R. Vondrasek, D.A. Church, L.J. Curtis: Phys. Rev. A **44**, 764 (1991)
83. P. Indelicato, B.B. Birkett, J.-P. Briand, P. Charles, D.D. Dietrich, R. Marrus, A. Simionovici: Phys. Rev. Lett. **68**, 1307 (1992)
84. C.E. Tanner: in *Atomic Physics 14*, D.J. Wineland, C.E. Wieman and S.J. Smith, eds (AIP, New York 1995) pp. 130–145

X Beta Decay of Highly Charged Ions

Fritz Bosch

Summary. Nuclear beta-decay probabilities are influenced, sometimes dramatically, by the charge state of the corresponding atom, i.e., the number of bound electrons. In the ESR cooler ring at Darmstadt, beta-decay properties of highly charged ions could be measured for the first time. Therewith, it became possible to study beta decays under conditions similar to those prevailing in hot stellar plasmas during nucleosynthesis. An overview of these experiments is given, emphasizing the first observation of bound beta decay in the examples of bare ^{163}Dy and ^{187}Re ions . The latter nuclide, long-lived as a neutral atom (half-life 42×10^9 years), serves as a "clock" for the age of our galaxy. An experiment conducted at the ESR showed that for *bare* ^{187}Re ions the half-life is *shortened* by more than nine orders of magnitude. This result enables a "recalibration" of the rhenium aeon clock and thus provides a more reliable lower bound for the age of the galaxy and, a fortiori, for the age of the universe. An outlook is given on experiments which are related to beta decay of highly charged ions and could be conducted soon at ion storage rings.

1. Introduction

The decay probability of a radioactive nucleus appears to be almost independent of *external* conditions like temperature, pressure, electromagnetic field, velocity, and so on. This has been confirmed by a wealth of experiments which started as soon as radioactivity was detected in 1895 by Henri Becquerel. Radioactive samples were taken down into deep-seated mines or up to the highest mountains; they were compressed by pressures of thousands of atmospheres or distorted by the strongest electromagnetic fields available in the laboratories at those times; finally, they were cooled down to liquid-nitrogen temperatures or even hurled in high-speed centrifuges. Whatever one did with them, more than a fractional change of 10^{-3} in the nuclear lifetime was never observed [1]. Therefore, nuclear lifetimes are generally considered as being basic, invariable properties of the nuclei which can be tabulated once and for all in standard compilations like masses, energies, or angular momenta.

On the other hand, it was soon recognized that the half-life of a radioactive nucleus can be considerably altered by the atomic charge state (charge number q). This effect is smallest in α decay where the height of the Coulomb wall – and therewith the tunneling probability – is only marginally changed when removing some or even all of the bound electrons [2]. On the contrary, the half-life of an *excited* nuclear state may crucially depend on the number and configuration of bound electrons, as soon as the de-excitation is shared between the emission of a γ ray and of a previously bound electron (*internal conversion*).

Most important, however, becomes the atomic charge state in nuclear β decay. Already for the case of the well-known *continuum* $\beta^{\pm}$ decay both the phase-space and the maximum energy W_0 of the emitted charged lepton and, hence, the lifetime of the atomic nucleus noticeably depend on the number of bound electrons. This is mirrored by the *Fermi function* $f(Z, W_0)$ and, in particular, by the *screening factor* ϵ_f which denotes the modification of the wave function of the emitted (charged) lepton as a function of the atomic charge number q [3]. Orbital electron capture (EC), a special mode of β^+ decay, is simply proportional to the total probability density of all bound electrons at the nuclear surface and, therefore, determined by the number and configuration of them: The higher the charge number q of an atom, the smaller the EC decay probability. The bare nucleus, finally, becomes stable provided that EC is the only allowed decay channel. Also for the time-mirrored EC process, the *bound beta (β_b) decay*, where the created electron remains in a bound atomic orbit of the daughter atom, spectacular changes of the half-life may occur, depending on the atomic charge number q (Sect. 4).

Even though it is obvious after all that, in particular β half-lives are influenced considerably by the electron cloud, the question arises whether it is worthwhile to investigate painstakingly all this stuff, either by theory or by experiment. Are those efforts more than finger-exercises, because atoms are, as a rule, in the neutral charge state? Just the opposite is true: Nuclear β decay is *the basic process* in nucleosynthesis; it occurs in *hot stellar plasmas*, however, at temperatures

corresponding to a few keV (H-burning in the Sun), a few tens of keV (nucleosynthesis via the s process), or to some hundreds of keV (r process). At those temperatures, even the heaviest atoms are highly ionized. Hence, to be on firm ground in this field, it is mandatory to calculate and to measure β decay probabilities as a function of the atomic charge number q. Until quite recently, one had to be content with calculations, however, since highly charged ions were not yet available for experimental studies (all tabulated experimental β half-lives are taken for *neutral* atoms).

This situation has changed thoroughly by the advent of *storage-cooler rings* and ion traps a few years ago, where ions in a well-defined charge state can be accumulated, cooled, and stored for many hours. Therewith it became possible for the first time to study weak decays under conditions similar to those prevailing in hot stellar plasmas during nucleosynthesis. This chapter summarizes the first, yet hesitant, steps on this route.

Firstly, the *experimental method* to produce and to store radioactive ions of a well-defined charge state, and to observe their β decay daughters, will be outlined (Sect. 2). Because the experimental storage ring ESR at GSI is tailored for this purpose, some of its unrivalled features – like storing and cooling of *radioactive ions* or high-precision *Schottky spectrometry* – will be presented in this context. In Sect. 3 the first experiments related to β^+ half-lives of *fully ionized* atoms are described from which unambiguous information on β^+/EC branching ratios and decay energies (Q values) can be obtained.

As the main part of this chapter *bound beta* (β_b) *decay* is addressed in Sect. 4 by two examples: a) the very first observation of this special decay mode at the ESR (decay of bare $^{163}_{66}Dy^{66+}$ to hydrogen-like $^{163}_{67}Ho^{66+}$) and, b) the β_b decay of bare $^{187}_{75}Re^{75+}$. This nuclide with a long half-life of 42×10^9 years in the neutral charge state is one of the very few *aeon clocks* that can provide constraints for the duration of pre-solar nucleosynthesis in our galaxy and, a fortiori, for the age of the universe. An experiment conducted at the ESR showed, however, that the half life of *fully ionized* $^{187}Re^{75+}$ is more than *a billion times shorter*. This astonishing result requires us to calibrate anew the rhenium clock. How this can be done and how the recalibrated clock might provide a lower bound for the age of the universe will be outlined.

Finally, a brief outlook is given on forthcoming experiments at the ESR (or another storage device) that could shed new light on old questions in the domain of nuclear astrophysics.

2. Experimental Methods

So far, β decay of highly charged ions has been solely studied at the accelerator complex of GSI in Darmstadt. For that reason the following description is restricted to the specific methods applied at this facility to produce and to store highly charged radioactive ions, and to observe their β decay daughters.

2.1 Providing Highly Charged, Beta-Unstable Ions

In general, there are two kinds of β-unstable atoms which also require different production methods. The first group comprises those being – as neutral atoms or low-charged ions – stable or quasi-stable, but becoming unstable as soon as they are (highly) ionized. These can be extracted from standard ion sources and accelerated to energies of several hundred MeV/u by the linear accelerator UNILAC and the heavy ion synchrotron SIS. After a pulse-by-pulse extraction from SIS some or even all residual electrons of the ions can be optionally removed in a stripper foil.

Afterwards, a specific charge state can be selected for injection into the storage ring ESR (Fig. 1). The high energies attainable in the synchrotron ($\leq$ 1 GeV/u) ensure that even for very high nuclear charge numbers ($Z \approx 90$) still a fair number of ions in the highest charge states (bare, hydrogen-like, etc.) is leaving the stripper. Examples for this kind of β instability *induced* by stripping off the bound electrons are given in Sect. 4 (β_b decay of bare $^{163}Dy^{66+}$ and $^{187}Re^{75+}$). Because here the β decay of the *primary* beam is being analyzed, a large number of ions ($N \leq 10^8$) can be accumulated and stored in the ESR. On the other hand, this induced β decay is restricted to a few cases only.

The second, much larger group includes all *truly* β-unstable nuclides, i.e., atoms that are unstable irrespective of the condition of their electron cloud. Those are produced at GSI by in-flight fragmentation of primary ions from SIS in a target at the entrance of the *fragment separator* FRS (*projectile fragmentation*). If the FRS is operated as a pure magnetic-rigidity analyzer, ions with a similar mass-to-charge ratio m/q are filtered out and injected into the ESR. This allows one to measure *simultaneously* masses as well as half-lives of a series of nuclides which is the basis of an unambiguous calibration.

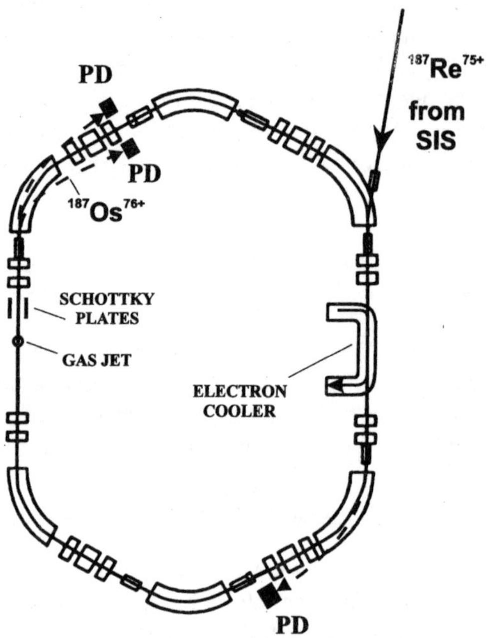

Fig. 1. The experimental storage ring ESR with its main installations (PD = particle detector). Also sketched is the experimental scheme for detection of β_b decay of bare ^{187}Re ions (see also Sect. 4.3)

2.2 Techniques of Beta Decay Measurements at the ESR

The ESR enables precise measurements of β half-lives of highly charged ions. This capability is based on some unique properties, most of which are provided by *electron cooling* (cf. *A. Wolf*, Chap. I): a long storage time of up to several hours (for a residual gas pressure of about 10^{-11} mbar) on the one hand, and an efficient, sensitive, accurate and time-resolved identification of the different species of circulating nuclides on the other hand.

Electron cooling in the ESR narrows the originally large relative momentum spread $\delta p/p$ of the beam in a short time to values $10^{-7} \leq \delta p/p \leq 10^{-4}$, depending on the number of stored ions. For low-intensity (up to a few thousand particles), "hot" secondary beams, for instance, the large initial momentum spread $\delta p/p \approx 10^{-1}$ can be reduced to less than 10^{-6} in about 20 s, at typical electron densities of 10^6 cm^{-3} in the electron-cooler section. A corresponding cooling is achieved in less than one second for already "cool" ($\delta p/p \approx 10^{-3}$) primary ions, as provided by the synchrotron. Electron cooling not only generates a brilliant monoenergetic beam of small size and small angular divergence, it also counteracts all heating processes caused by intra-beam scattering or by encounters with residual gas atoms. Because cooling not only brings about but also maintains a sharp *common velocity* of all circulating ions over the whole period of storage, cooler rings may serve as superb *mass spectrometers*, which has been convincingly demonstrated at the ESR within the last few years [4].

2.2.1 Schottky lifetime spectroscopy. The standard technique for a non-destructive beam diagnosis in storage rings is based on short pulses ("Schottky noise") that are induced by every circulating ion at each turn onto pickup plates mounted in the ring aperture ("Schottky plates" in Fig. 1). If many ions, randomly spread along the circumference, are in the ring, the accumulated time spectrum of all these signals is like *white noise*. When transformed to the frequency domain, however, distinct lines (*Schottky* lines) appear at the individual revolution frequency of every ion (i) $f_{\mathrm{i}}^0 = (\beta c)_{\mathrm{i}}/C_{\mathrm{i}}$ and at the corresponding harmonics $f_{\mathrm{i}}^n = n\, f_{\mathrm{i}}^0$ (n = 2,3, ...), where $(\beta c)_{\mathrm{i}}$ and C_{i} are the velocity and closed-orbit length, respectively, of the ion.

Because in an electron-cooled beam all the various ion species get, enforced by the electrons, the *same* velocity βc (with a very small spread), differences Δf in revolution frequencies can be caused only by differences in orbit lengths, ΔC. The latter depend in a simple way on the mass-to-charge ratio m/q of the ions, according to the basic equation:

$$B\rho = \frac{m}{q}\gamma\beta c, \tag{1}$$

where, for fast particles, the Lorentz factor $\gamma = (1-\beta^2)^{-1/2}$ has to be included. Equation (1) simply means: the higher is m/q, the larger is, for a given magnetic field strength B, the bending radius ρ and, thus, the circumference C of the closed orbit.

There is accordingly – for any electron-cooled circulating particle – a one-to-one correspondence of its revolution frequency f on the one hand, and of its mass-to-charge ratio m/q on the other hand; this relationship sets the starting point of *Schottky mass spectrometry* [4]. The connection between the difference $\Delta(m/q)_{12}$ of two mass-to-charge ratios m/q and the corresponding difference Δf_{12} of revolution frequencies reads as:

$$\frac{\Delta f_{12}}{f_1} = -\alpha_p \frac{\Delta(m/q)_{12}}{(m/q)_1}, \tag{2}$$

where α_{p}, the *momentum compaction*, denoting the ratio of a relative change of the circumference C to a relative change in the magnetic rigidity $B\rho$, is given by the structure and the ion-optical setting of the ring. For the ESR α_p has typically a value of about 0.14.

Whereas the frequency of a Schottky line renders m/q – supposing there are well-known calibration masses nearby – its width δf mirrors the momentum spread $\delta p/p$ according to

$$\frac{\delta f}{f} = \left(\frac{1}{\gamma^2} - \alpha_p\right)\frac{\delta p}{p}. \tag{3}$$

Most important, the *area of a Schottky line* is proportional to $Nq^2f_0^2$, where N is the particle number and q the atomic charge number of the corresponding ion species. This property brings about *Schottky lifetime spectroscopy,* since from the measured time evolution of the area of those Schottky lines, corrected for the common losses in the ring, nuclear lifetimes can be derived. The analysis will be the more precise the higher the charge number q, due to the dependence of the peak area on q^2. What has been achieved in the meantime with Schottky mass spectrometry at the ESR is shown in Fig. 2. Here, an isomeric nuclear state of bare ^{52}Mn at 378 keV excitation energy (^{52m}Mn) appears almost completely resolved from its ground state. The resolution corresponds to a mass resolving power $m/\Delta m$ of almost 300 000 [4] (in a recent experiment this value could be even sizably improved).

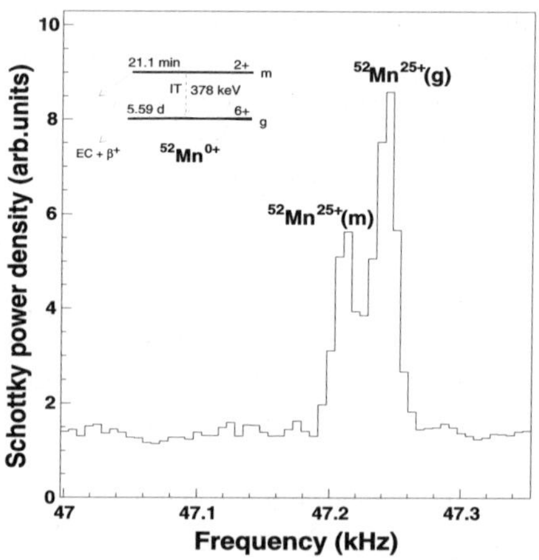

Fig. 2. Schottky lines vs. down-mixed revolution frequency of mass-resolved ground (^{52}Mn^{25+}(g)) and isomeric nuclear states at 378 keV of bare, cooled ^{52}Mn ions. A relative change $\delta f/f$ in the revolution frequency f corresponds, for cooled ions, to a change in the mass-to-charge ratio m/q. The partial decay scheme shown refers to neutral ^{52}Mn. From [5]

The future potential for both Schottky mass and lifetime spectroscopy, is indicated in Fig. 3. It shows what *P. Kienle* [6] calls *isobar spectroscopy,* for the example of ^{197}Au fragments provided by the fragment separator. The

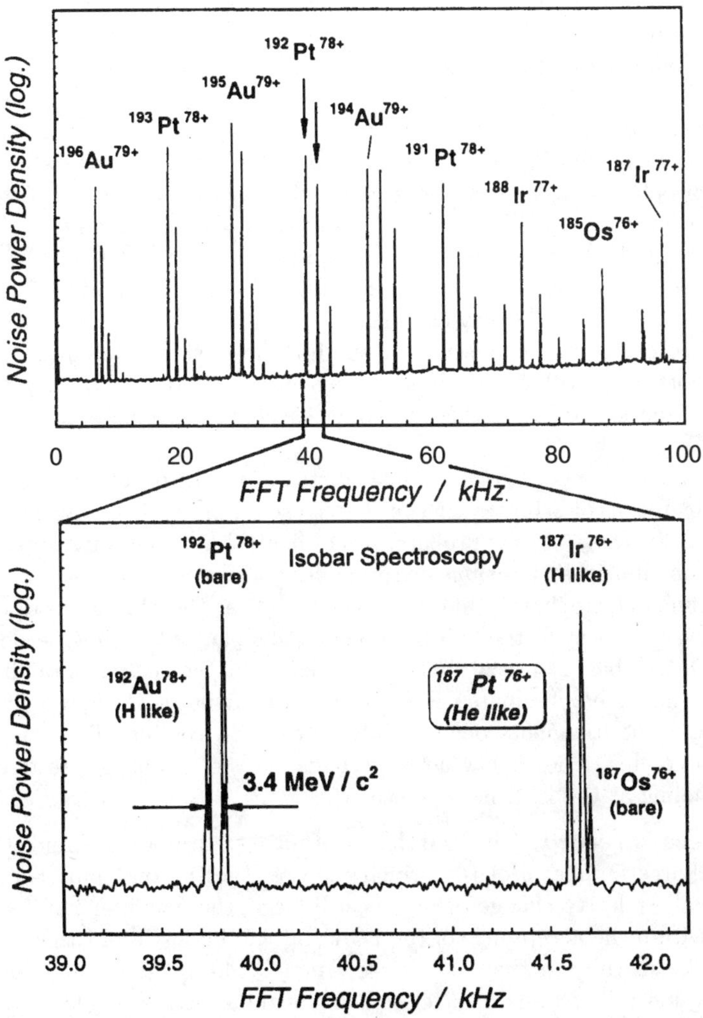

Fig. 3. Schottky lines vs. down-mixed revolution frequency of fragments, provided by the FRS, stored and cooled in the ESR and circulating simultaneously there (top). A 3 kHz detail of the upper spectrum shows well-resolved A = 192 and A = 187 multiplets containing (grand)mother and daughter nuclei from EC decay chains. From [4]

Schottky signals of all nuclei stored in the ESR were recorded at the 16th harmonic of their revolution frequency ($\sim$ 30 MHz), down-mixed to a 100 kHz band and frequency-analyzed by *fast fourier transform*. The data (top) comprise the well-resolved fingerprints of many dozens of simultaneously circulating nuclides, most of them dressed with 0, 1, or 2 electrons. The bottom part shows a 3 kHz detail of the full spectrum. It strikingly visualizes the isobar spectroscopy: there is a doublet at mass number $A = 192$ and charge $q = 78$ (left-hand side), and a triplet at $A = 187$ and $q = 76$, both of them well resolved.

The former shows a β-unstable mother atom, hydrogen-like ^{192}Au, and its EC daughter, bare ^{192}Pt. The separation of these lines *directly* reflects their mass difference, hence the energy release of the decay, Q_{EC}. The $A = 187$ isobar represents even a *triple* EC decay chain, consisting of the helium-like ^{187}Pt grandmother, the hydrogen-like ^{187}Ir mother, and the bare ^{187}Os daughter. By looking on the time development of the areas of the corresponding lines, one can single out a *pure* β decay channel, e.g., orbital EC for an atom with exactly one, respectively two, electrons bound in the K shell. Moreover, the number of both the mother and daughter nuclei can be observed *simultaneously* as a function of time, which helps decisively in reducing systematic errors.

A series of such investigations has just started at the ESR. The goal is to *disentangle* by means of Schottky lifetime spectroscopy of highly charged ions what in the inclusive β^+ decay probabilities of neutral atoms always appear mixed: the contributions of *various* decay modes and decay channels.

For the time being Schottky lifetime spectroscopy at the ESR has serious restrictions with respect to the half-life range accessible, as the time needed to cool *secondary* beams ($\sim$ 20 s) is rather long. In the near future this shortcoming should be eliminated in large part by virtue of *stochastic precooling* of the hot fragments before applying electron cooling. This could lead to total cooling times below a few seconds, as very promising tests of stochastic cooling at the ESR have shown [7].

2.2.2 $\beta^{\pm}$ decays recorded by particle detectors. In both, β^+ and β^- decays the charge number q of the primary ion changes by one unit. For $q \leq 40$, where the relative change of m/q is still large, the daughter ion does not remain within the acceptance of the ESR and will be lost by hitting the walls of the beam tube in the next dipole. For $q \geq 40$, on the other side, the daughter ion still travels on a closed, but significantly changed orbit. In both cases the β daughters can be easily recorded in situ by particle detectors moved appropriately into the ring aperture while not disturbing the primary orbit (marked in Fig. 1 as PD). This simple procedure is not applicable, neither for EC nor for β_{b} decay, because there only the nuclear mass is changing (slightly), but *not* the atomic charge number q. Therefore, the orbit remains virtually the same after the decay.

At the ESR presently five positions are equipped with particle detectors. Four of them monitor, at the outer side of the storage ring, β^+ decays whereas the one detector at the inner side records β^- decay products. Those positioned in the dipoles register β^+ decay daughters of *light* ions, with a restricted detection efficiency of about 16%, because only decays along the last straight section before the position of the detectors can be "seen". The other ones (see Fig. 1, PD), recording β decay of *heavy, highly charged* ions, provide an almost unrestricted detection of all decays wherever they might occur. As standard detectors position sensitive, multi-wire proportional chambers are mounted in movable pockets, separated by a thin window from the ultra-high vacuum.

This technique is comparatively simple and efficient. There is also, in principle, no obstacle to addressing half-lives even *below* 1 ms, far away from the capabilities of the Schottky technique. Those short-lived nuclei mark the way nature behaves in the r and rp branches of nucleosynthesis. To provide mono-isotopic beams of them for injection into the ESR is still a challenge.

3. Measurement of β^+ Half-Lives of Bare Nuclei

The measurement of the β^+ half-life of a *bare* nucleus is the cleanest disentangling of β^+ and EC decay branches which, for dressed atoms and for decay energies $Q_{\mathrm{EC}} > 2\, m_{\mathrm{e}}c^2$ ($m_{\mathrm{e}}c^2$ = electron rest energy of 511 keV), always appear unseparated. By comparing the result with the half-life of the corresponding neutral atom, one immediately obtains the β^+/EC branching ratio. If the latter is already known, the screening factor ϵ_{f} can be deduced which provides, on its part, also inclusive information on the Fermi function $f(Z, W_0)$ [3]. This is still a blind spot in the field of beta decay, since the Fermi function could never be "tested", but had to be taken all the time from theory.

For the example of $^{19}\mathrm{Ne}^{10+}$ for the first time the β^+ half-life of a *bare* nucleus was measured [8]. For this specific case one expects that the half-life (in the ion rest frame) should be almost equal to that of the neutral $^{19}\mathrm{Ne}$ atom as both the EC branch and the screening factor ϵ_{f} [3] should become negligible for the large Q_{EC} value of ~ 3.2 MeV and the comparatively small nuclear charge number $Z = 10$.

Fragments were produced in the target of the fragment separator FRS by a primary $^{20}\mathrm{Ne}$ beam of 350 MeV/u. The FRS was operated as a pure magnetic-rigidity analyzer. Ions with a mass-to-charge ratio A/q of about 1.9 (where A is the mass in atomic mass units (u)) were selected, and injected into the experimental storage ring ESR. Because electron cooling was not applied, an identification via Schottky spectroscopy of the various stored ion species was not possible. Figure 4 shows the counts of a detector placed in the aperture of the dipole next to the injection point, but outside the trajectory of the primary beam. The observed decay pattern is superimposed from 163 injection cycles with a period of 120 s.

A single-component exponential fit yielded a half-life of 18.5 s in the ion rest frame (observed half-life divided by the Lorentz factor $\gamma \approx 1.27$), exceeding by more than 7% (Fig. 4) the value of 17.3 s as measured for the neutral $^{19}\mathrm{Ne}$ atom [9]. This outcome is, most probably, due to a small

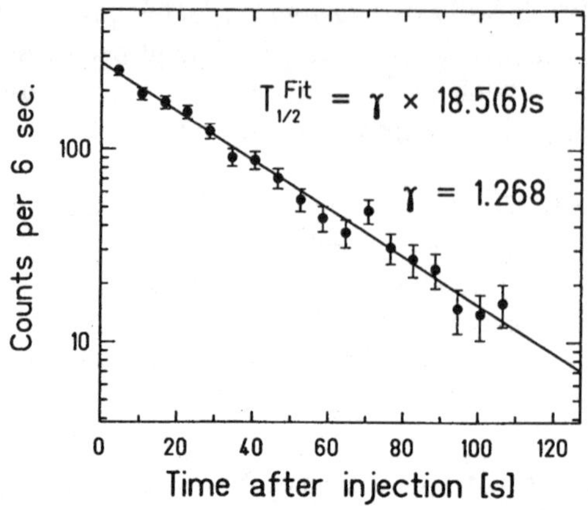

Fig. 4. Count rate, recorded by a particle detector in the ring aperture, of β^+ daughters $^{19}F^{9+}$ from the decay of fully stripped $^{19}Ne^{10+}$ ions. The neon ions were injected from the fragment separator into the ESR as short pulses every 120 seconds. From [8]

contamination with another stored (longer-lived) β^+ emitter, $^{15}O^{8+}$, whose A/q-value of 1.875 was still within the acceptance of the ring.

Hence, to measure reliably lifetimes in a time domain, where electron cooling and/or identification of the stored ions via resolved Schottky lines cannot be applied ($T_{1/2} \leq 30$ s), the experimental technique has to be improved. One possibility is to rely upon mono-isotopic secondary beams. These can be injected into the storage ring if the FRS is adjusted to a mode where, within a given m/q range, nuclei of *one* specific nuclear charge number Z will be singled out by means of a *degrader* mounted in the symmetry plane of the FRS. An even simpler way is to use Z-sensitive particle detectors by which specific β^+ decays may be picked out.

Longer β^+ half-lives were addressed with fragments originating from ^{58}Ni primary ions. Here, Schottky lifetime spectroscopy was tested first [5]. Figure 5 shows a part of the Schottky spectrum of stored and cooled bare fragments, all of which bear an identical number of protons and neutrons and, hence, a A/q value of about 2. None of the nuclides displayed in Fig. 5 has a known β^+ branch, except ^{52}Fe. EC decay, as allowed for the *neutral* ^{44}Ti, ^{48}Cr, and ^{56}Ni atoms, is ruled out for these bare nuclei, too. Therefore, they are all stable with regard to nuclear decay. Bare $^{52}Fe^{26+}$, on the other hand, should exhibit a significant change of its half-life, as compared to neutral ^{52}Fe, where the β^+ and EC decay-branches are almost equally strong [9].

The total decay constant λ_{Lab}, as observed in the laboratory, is given as

$$\lambda_{Lab} = \frac{1}{\gamma}\lambda_{\beta+} + \lambda^*, \tag{4}$$

where $\lambda_{\beta+}$ is the β^+ decay probability in the ion rest frame and λ^* denotes the common *beam loss* probability (measured in the laboratory). The latter is, typically, in the order of 0.2 h^{-1} for medium-heavy highly charged ions. Beam-loss is mainly due to recombination processes in the cooler section and

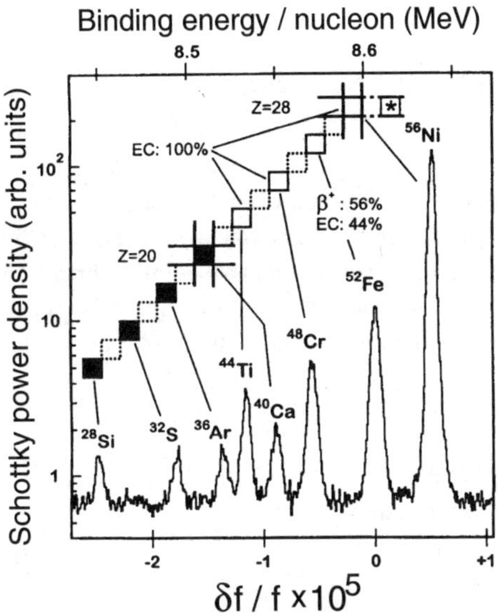

Fig. 5. Schottky lines vs. relative differences of revolution frequencies f of *bare* nuclei with a mass-to-charge ratio $A/q \approx 2$. Except $^{52}Fe^{26+}$, all nuclei are stable with respect to β decay and, thus, monitor common beam losses. The recording of the time dependence of the Schottky peak area of $^{52}Fe^{26+}$, yielding its β^+ half-life, was the first example of *Schottky lifetime spectroscopy*. From [10]

scales roughly with Z^2. Therefore, it can be reliably extrapolated for the β-unstable bare ^{52}Fe ions from the neighboring nuclides that are stable with respect to nuclear decay (Fig. 5).

The data, corrected for λ^*, yielded for the nuclear half-life of bare $^{52}Fe^{26+}$ a value of $T_{1/2}^{\beta+} = (12.5 \pm 1.3)$ h [5]. By comparing this result with the significantly shorter half-life of neutral ^{52}Fe ($T_{1/2} = (8.28 \pm 0.01)$ h) [9], a β^+/EC branching ratio of (2 ± 0.2) was derived, in *disagreement* with the (calculated) literature value of 1.3 [9]. The comparison shows that measurements of pure β^+ decay channels could pave the way for a *direct* determination of branching ratios and Q values, which is otherwise extremely cumbersome, if not impossible. Before that, however, the uncertainty, still too large in this very first experiment, has to be narrowed significantly.

4. Observation of Bound Beta Decay of Bare Atomic Nuclei

Since its detection, β decay has been addressed probably more often than any other topic of nuclear physics. At least two reasons might be quoted for that: nuclear β decay is, at once, the most frequent and most easily accessible manifestation of the still puzzling *weak interaction*. Moreover, it is of unrivalled importance in *nucleosynthesis*, where the abundance of the created nuclei is governed by a subtle interplay of β decay and neutron (proton) capture. It is no accident that most of the radionuclides, in use everywhere nowadays, are beta emitters.

Nuclear β decay can be condensed to the basic exchange

$$n + \nu_e \longleftrightarrow p + e^-_{(b)}, \tag{5}$$

where neutron n and proton p represent the hadronic part, and the electron neutrino ν_e and electron e^- the leptonic part. The electron occurs either as a *free* one (e^-) or as *bound* (e^-_b) in the atomic electron cloud. From (5) commonly five groups are derived by *crossing*, i.e., by reversing the arrow and/or by substituting particles on either side by their antiparticles on the other side:

$$\begin{array}{rll} \text{1a)} & n + \nu_e \to p + e^- & (\text{free } \beta^- \text{decay}, \beta_c), \\ \text{b)} & n + \nu_e \leftarrow p + e^- & (\text{free electron capture}), \\ \text{2a)} & n + \nu_e \to p + e^-_b & (\text{bound } \beta^- \text{decay}, \beta_b), \\ \text{b)} & n + \nu_e \leftarrow p + e^-_b & (\text{orbital electron capture, EC}), \\ \text{3)} & p \to n + e^+ + \nu_e & (\beta^+ \text{decay}). \end{array} \tag{6}$$

In a more symmetric form item 3 becomes a subset of 1b, with the *free* electron being captured from the *negative* continuum, however. The reversed arrows in the corresponding subsets a, b, indicate the inherent *time reversal symmetry*. Sets 1 and 3 comprise three-body decays, whereas from the two-body reactions of set 2 *monochromatic* (anti)neutrinos originate.

In contrast to β_c, β^+ and EC, free electron capture has not yet been observed, and β_b decay was not observed until 1992 [11]. The latter decay modes are marginal ones for neutral atoms; to be observed they demand either a high atomic charge state or a tremendous density of free electrons, or both of them. Vice versa, these decays are the more important the hotter and denser the stellar plasma.

4.1 General Survey on Bound Beta Decay

As already noticed, β_b decay and EC are time-mirrored processes (see set 2 of (6). Therefore, they are closely connected by *detailed balance*. In particular, they share identical nuclear transition matrix elements $|M_{fi}|^2$ for transitions $i \longleftrightarrow f$, connecting the *same* initial (i) and final (f) nuclear states, but in reversed time order. They differ, on the other hand, with respect to phase-space factors and Q values (decay energies).

Bound beta decay was first predicted by *R.Daudel, M.Jean*, and *M. Lecoin* [12] in 1947 and, 14 years later, thoroughly discussed by *J.N.Bahcall* [13]. He derived a useful expression for the ratio of β_b decay probability into an orbit with main quantum number n ($\lambda^n_{\beta_b}$) on the one hand, and of continuum β_c decay probability (λ_{β_c}) on the other:

$$\frac{\lambda_{\beta_b}^n}{\lambda_{\beta_c}} \approx \frac{n_f \cdot \left(Q_{\beta_b}^n\right)^2 |\psi^n(R)|^2}{f(Z, W_0)} \approx n_f (W_0 - 1)^2 \frac{(\alpha Z)^2}{n^3 f(Z, W_0)}, \tag{7}$$

where n_f is the number of free places in the shell, $Q_{\beta_b}^n$ the corresponding Q value, $\psi^n(R)$ the electron wave function at the nuclear surface R, W_0 the nuclear energy release in units of the electron rest energy, $f(Z, W_0)$ the Fermi function, and where, finally, α denotes the Sommerfeld fine-structure constant. The ratio sensitively depends on the nuclear charge number Z of the daughter nucleus ($\sim Z^2$) and on n ($\sim n^{-3}$). From the properties of the *Fermi function* it follows that the lower one is the energy release W_0, the higher one is the relative fraction of β_b decay.

In neutral atoms where the innermost atomic orbits are occupied (*Pauli-blocked*) β_b decay is almost negligible as compared to continuum decay. From equation (7) one easily derives, for instance, a relative fraction of $\approx 10^{-6}$ for β_b decay of a neutron into a *neutral* hydrogen atom. When going to highly ionized heavy atoms, where the strongly bound inner shells (K, L, ...) are completely or partially empty, the situation may change thoroughly: the total β^- decay probability will be enhanced considerably due to β_b transitions to deeply bound states with Q values enlarged by several tens of keV for high-Z atoms; nuclei that are stable, if embedded in their full electron cloud, might become *unstable* with respect to β_b decay if they get highly ionized (e.g., ${}^{163}_{66}\mathrm{Dy}$, ${}^{205}_{81}\mathrm{Tl}$, ${}^{160}_{64}\mathrm{Gd}$, ...). There might also be changes in β half-lives by many orders of magnitude, in cases where a β_b decay to an *excited* state in the daughter nucleus with favorable selection rules becomes possible, which is energetically *not* allowed in the corresponding neutral atom (e.g., ${}^{187}_{75}\mathrm{Re}^{75+}$, ${}^{205}_{81}\mathrm{Tl}^{81+}$, ...).

The energy release, $Q_{\beta_b}^{K,L,\ldots}$, for β_b decay to K, L, ... states strongly depends on the final electron state. For the case of a fully ionized mother atom with atomic mass number A and nuclear charge number Z, $(Z, \mathrm{A})^{q=Z}$, the relation to the Q value for continuum β_c decay of the corresponding neutral atom, Q_{β_c}, is especially simple [11]:

$$Q_{\beta_b}^{K,L} = Q_{\beta_c} - |\Delta B^\mathrm{e}| + \left|B_{Z+1}^{K,L}\right|. \tag{8}$$

Here, ΔB^e is the difference of the total electron binding energies of the neutral mother and daughter atoms, respectively, and $B_{Z+1}^{K,L}$ the corresponding K, L electron binding energy in the hydrogen-like daughter. $|\Delta B^\mathrm{e}|$ extends from 64 eV (${}^3\mathrm{H} \to {}^3\mathrm{He}$) to about 20 keV for high Z, whereas the K binding energy reaches, for instance, a value as large as 130 keV ($Z = 92$). The energy release Q_{β_c} in continuum β_c decay (or Q_EC for the time-reversed EC, with $Q_\mathrm{EC} = -\, Q_{\beta_c}$) is defined as the mass difference of the neutral atoms

$$Q_{\beta_c} = -Q_\mathrm{EC} = M(A, Z)^{q=0} - M(A, Z+1)^{q=0}. \tag{9}$$

By taking into account that the atomic mass is composed of the nuclear mass and of the rest mass of all bound electrons, diminished by their total binding energies, (8) becomes obvious. It also follows that β^--stable neutral atoms ($Q_{\beta_c} < 0$) can become *unstable* as bare ions via β_b decay, as soon as

$$\left| B_{Z+1}^{K,L,\ldots} \right| > |Q_{\beta_c}| + |\Delta B^e| \tag{10}$$

holds true. From similar conditions the minimum charge state q can be derived where partially stripped ions become unstable, or where EC-unstable neutral atoms become stable.

4.2 First Observation of Bound Beta Decay: The Case of $^{163}_{66}Dy^{66+}$

The very first observation of β_b decay took place, not by accident, at an ion storage ring [11]. An inspection of (8) suggests the following detection scenario: choosing first an EC-unstable, heavy neutral atom with a Q_{EC} value as small as possible; the daughter atom (stable in the neutral charge state) should become β_b-unstable, if most or all of its electrons are stripped off, provided only that the constraint of (10) is satisfied. These highly ionized atoms have been stored for a long time and, finally, their β_b decay products are counted as a function of the storage time t_s. If the latter is small compared to the β_b lifetime, the number of daughters at t_s, $N_d(t_s)$, should be in first order (for the exact analysis see [11]) merely proportional to the storage time t_s, to the β_b-decay probability λ_{β_b} observed in the laboratory, and to the number of mother ions still circulating, $N_m(t_s)$:

$$N_d(t_s) \approx \lambda_{\beta_b} \cdot t_s \cdot N_m(t_s). \tag{11}$$

The pilot experiment followed exactly the steps described above. As a most promising test case the couple $^{163}_{66}Dy/^{163}_{67}Ho$ was chosen. Neutral ^{163}Ho is EC-unstable, but with the smallest known Q_{EC} value of only 2.56 keV [9]. From (8) one calculates, taking $|\Delta B^e| = 12.5$ keV and $|B_{Ho}^K| = 65.1$ keV from the literature [14, 15], for the transition of bare $^{163}Dy^{66+}$ to the holmium K shell a $Q_{\beta_b}^K$ value of about 50 keV and of ~ 1.7 keV for a transition to the L shell. By a first guess the ratio of β_b decay probabilities to the K and L shell, $\lambda_{\beta_b}^K$ and $\lambda_{\beta_b}^L$, respectively, should be proportional to the ratio of the corresponding Q values squared, and to the respective densities of the electrons at the nuclear surface R of the daughter atom:

$$\frac{\lambda_{\beta_b}^K}{\lambda_{\beta_b}^L} \approx \frac{\left(Q_{\beta_b}^K\right)^2}{\left(Q_{\beta_b}^L\right)^2} \cdot \frac{|\psi_K^e(R)|^2}{|\psi_L^e(R)|^2}. \tag{12}$$

By inserting the numbers given above for $Q_{\beta_b}^{K,L}$, and with a rough estimate for the ratio of $K(\mathrm{n}_1{=}1)$ and $L(\mathrm{n}_2{=}2)$ electron densities as $(n_2/n_1)^3 = 8$, it follows that β_b decay to the K shell is, at least, three orders of magnitude stronger than the corresponding L-shell transition. Hence, one can regard the β_b decay of the couple $^{163}Dy^{66+}/^{163}Ho^{66+}$ as an almost *pure* K shell transition, i.e., $\lambda_{\beta_b} = \sum_n \lambda^n_{\beta_b} \approx \lambda^K_{\beta_b}$.

Similarly, an estimate for $\lambda^K_{\beta_b}$ can be given from the measured Q value and the partial half-life for EC from the M shell of neutral ^{163}Ho [16]. Since for both, EC- and β_b-transition the nuclear ground state of ^{163}Ho is in common, either as initial or as final state, the nuclear matrix element drops out in a comparison of the two decay probabilities. By taking the experimental values ($Q^M_{EC} \approx 0.6$ keV, $T^M_{1/2}(EC) \approx 18\,000$ yr [16]) one expects for the β_b half-life of fully ionized $^{163}Dy^{66+}$ a value of about *50 days* [17]. Since ^{163}Dy ions can be provided by a standard ion source, a realistic scenario is to store up to $\sim 10^8$ bare, cooled $^{163}Dy^{66+}$ ions for several hours in the ESR. Given the half-life estimate of 50 days, several tens of thousands of β_b decays should occur during the storage.

The only problem left is to "reveal" the decay products, hydrogen-like $^{163}_{67}Ho^{66+}$. As already pointed out in Sect. 2, neither β_b decay nor EC do change the atomic charge number q. After β_b decay the rest energy of the primary ^{163}Dy ion is changed by 50 keV ($Q^K_{\beta_b}$), which corresponds to a fractional change of only 3×10^{-7}. Resolving this tiny difference is not yet within the capabilities of Schottky mass spectrometry at the ESR, although there is well founded hope to achieve such a resolution in the not-too-distant future.

Until a *direct* observation of β_b decay can be realized for such *small* Q values, one has to be content with a more complicated detection technique: After the storage time t_s the bound electron, created by β_b decay, will be stripped off in the internal gas jet of the ESR which is turned on for a few minutes. The resulting change of the atomic charge number from $q = 66$ to $q = 67$ alters the trajectory and, hence, the revolution frequency of the daughter atom. That is easily detected by either Schottky spectroscopy or by a particle detector intercepting the new orbit. This procedure is simplest for *bare* mother nuclei where the β_b daughters become finally bare ions, too. But even in this case an auxiliary experiment is needed to determine the fraction of ionization in the gas jet ($\approx 75\%$ for $Z = 66$ and an energy of ~ 300 MeV/u) with respect to recombination. This number emerged from storing *hydrogen-like* ^{163}Dy ions and measuring, with the aid of the argon gas jet, the cross sections for recombination and ionization, respectively (Fig. 6, top). Figures 6 and 7 show typical spectra of the dysprosium experiment with both techniques applied.

All together, the ratio of holmium and dysprosium ions was taken for various storage times t_s, reaching from a few hundred seconds to about 100 minutes. The slope of this curve then directly rendered, according to (11), the β_b decay probability λ_{β_b} in the laboratory system, i.e., $\gamma \cdot \lambda_{\beta_b}$ in the ion rest frame ($\gamma \approx 1.3$). Equation (11) represents, however, only a first approximation. In a more accurate treatment [11] the ion numbers N_{Dy}, N_{Ho}, have to be corrected for the corresponding (slightly different) beam losses during storage and, in particular, for bare $^{163}Ho^{66+}$ ions produced by

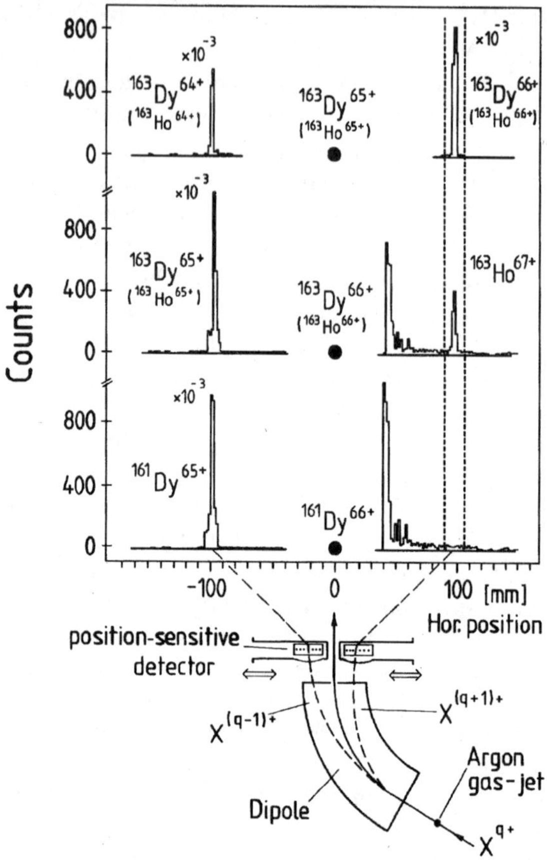

Fig. 6. Experimental scheme and spectra recorded by the particle detectors behind the gas jet for the three dysprosium experiments.
Top: Stored hydrogen-like-$^{163}Dy^{65+}$ ions either losing (r.h.s. detector) or capturing an electron in the jet. *Middle*: Main experiment: After long storage of bare $^{163}Dy^{66+}$ the electron of the β_b daughter $^{163}Ho^{66+}$ is stripped off in the jet and $^{163}Ho^{67+}$ appears at the r.h.s. *Bottom*: Control experiment with neighboring bare dysprosium ions, $^{161}Dy^{66+}$, where no corresponding peak at the r.h.s. appears. From [18]

nuclear (p, n) reactions in the gas jet. This kind of "background" that cannot be avoided in the chosen experimental scheme, sets a limit for a reliable detection of very weak β_b decays. The number of (p, n) reactions can be determined, on the other hand, from a measurement after a very short storage time where β_b daughters are still missing. In contrast to β_b decay the nuclear yield does *not* depend on the storage time t_s. Alternatively, the nuclear yield can be estimated from a neighboring β_b-stable isotope with comparable (p,n) cross section (Figs. 6, 7 bottom). It turned out, with both methods applied, that for the case of ^{163}Dy the nuclear background is almost negligible.

A final problem is left in the Schottky analysis. The Schottky area has to be calibrated in terms of the *absolute* particle number, when the storage time is small with respect to the β_b half-life, as anticipated in (11). This was done in the following way: Step by step the number of stored particles N_i was reduced from 2×10^5 to one single ion by turning on the argon gas jet for a short time. This procedure was controlled with all monitors available:

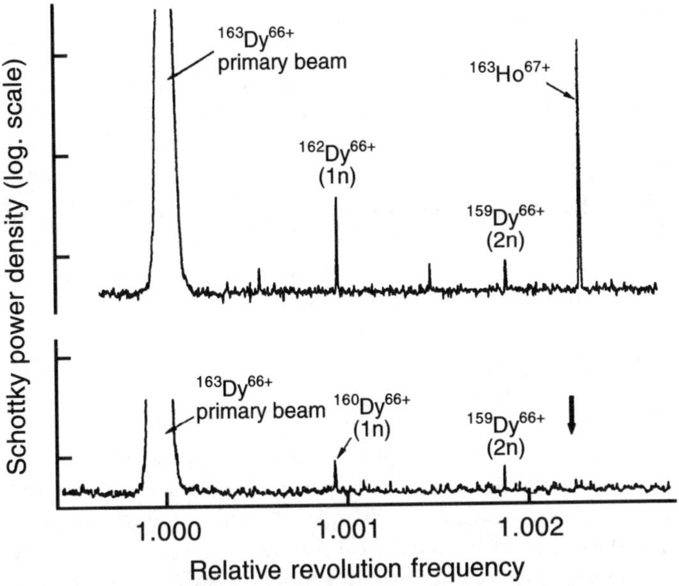

Fig. 7. Schottky spectrum of the primary $^{163}Dy^{66+}$ beam (*top*) and of some reaction products from its interaction with the gas jet. The approximately 2×10^4 $^{163}Ho^{67+}$ ions, signalling β_b decay, result from a 1-hour storage. *Bottom*: Control experiment for neighboring bare $^{161}Dy^{66+}$ ions: no corresponding $^{161}Ho^{67+}$ ions appear. From [10]

beam transformer, particle detectors and Schottky lines. For $N_i \leq 80$, discrete jumps in the Schottky peak area could be identified and assigned to specific changes ΔN_i of the particle number. A nearly perfect linear relation between Schottky area and particle number turned out – except for a small region around $N_i \approx 3000$, where, according to *M.Steck* [19] a phase transition to a *one-dimensional ordering* (linear chain) of the circulating ions may occur.

The two independent methods yielded, within their combined uncertainties, a consistent β_b half-life for bare $^{163}Dy^{66+}$ of $T_{1/2} = (48 \pm 3)$ days in the ion rest frame [18], also in accordance with the calculated value of 50 days [17].

There are some by-products of this first β_b experiment. The rather short half-life of 48 days for bare $^{163}Dy^{66+}$, now experimentally confirmed, may be responsible for the exceptionally high solar abundance (1.61%) of ^{164}Er [9]. Moreover, from the ratio of the β_b decay probability of bare ^{163}Dy on the one hand, and of the EC decay probabilities of the M_1 and M_2 sub-shells of neutral ^{163}Ho [16] on the other (the common nuclear matrix element drops out for that comparison), a constraint for the electron *neutrino* rest energy of $m_{\nu e} < 275$ eV could be derived [18]. The reason for this astonishing outcome – regarding the rather large uncertainty of about 6% of all corresponding experiments – is, that the EC decay probabilities of the M_1, M_2 subshells of holmium are extremely sensitive to the neutrino rest energy, due to the very small Q values involved ($Q_{\mathrm{EC}}^{M_1} \approx 0.6$ keV). Amazingly, the 275 eV still represent the smallest upper bound for the rest energy of the electron *neutrino* ever derived in a laboratory experiment. However, if *CPT-invariance* holds true, the rest energy of the electron *neutrino* should exactly equal the value of the corresponding *antineutrino*, whose upper limit lies below *10 eV* [20]. Nevertheless, a quite realistic improvement of the experimental accuracy to about 1% for both, the EC and the bound beta lifetimes could reduce the limit of 275 eV significantly and thus provide a valuable *independent* upper bound for the rest energy of the electron neutrino.

4.3 Setting the Cosmic Clock ^{187}Re

4.3.1 Cosmic clocks. Until well into this century most astronomers favored a stationary, that is – viewed on a large scale – an unchanging universe. In the late 1920s Edwin Hubble at the Mount Wilson Observatory made the epochal discovery of a non-stationary, expanding universe. He was the first to notice that the further spiral nebulae are away, the more rapidly they move away from us. From the foremost data of a universal recession velocity per distance – later called the *Hubble constant* H_0 – of 500 km per second and megaparsec (1 megaparsec (Mpc) = 3.26 million light years), Hubble deduced the origin of the universe to be two billion years ago. Nowadays, H_0 is believed to be much smaller and to lie somewhere between 50 and 100 km s^{-1} Mpc^{-1} which corresponds to an age T_{U} of the universe of between 10 and 20 billion years (constant expansion supposed).

Since Hubble's discovery astrophysicists searched for other clocks that could provide independent constraints for the age of the universe. Perhaps the most reliable of them is hidden in the famous *Hertzsprung–Russell diagram*, where the absolute luminosity of stars is depicted versus their color, i.e., temperature. The ages of the stars can be estimated from the point on this plot where the oldest stars, the *globular clusters*, leave the "main sequence": 15 billion years with a lower limit of 12 billion years, according to the most recent, comprehensive review of *D.A. VandenBerg* and co-workers [21].

Besides the globular clusters and other astronomical objects like "white dwarfs", there are also "true" chronometers, the *aeon clocks*. Nature has created them, radiochemists detected them and geologists and astrophysicists learned to understand and interpret their message. They are long-lived radioactive mother nuclei and their *direct* decay daughters. From the abundance ratio of them, measured at any arbitrary time, and from the known decay constant of the mother nuclei, the time elapsed can be deduced – provided only that no daughter atom already existed at the beginning of the decay.

Perhaps the most striking accomplishment of these aeon clocks is the remarkably precise setting of the *age of our solar system* amounting to $T_{\mathrm{S}} = (4.60 \pm 0.05) \times 10^9$ years. That was derived from today's relative abundance of ^{238}U and its (final) decay product ^{206}Pb, as found in meteorites that were formed just at the decoupling of the solar system from the galaxy and did *not* contain any lead at that time. This method cannot be extended straight away to the time before decoupling, when the matter which would become later the Sun, Earth and living beings was still scattered and mixed in the stars and the interstellar space of our galaxy. For an estimate of the duration T_{N} of *pre-solar* nucleosynthesis one had to rely upon the relative abundance of ^{232}Th and ^{238}U. Both of them are long-lived radionuclides but do *not* belong to a common decay chain. To get, from this couple, reliable information on the elapsed time, one has to know in addition the production probabilities during nucleosynthesis of both ^{238}U and ^{232}Th. They strongly depend, however, on the not yet understood details of the r process and are, for this reason, uncertain to a large extent. Accordingly, the literature values for the period T_{N} of pre-solar nucleosynthesis – inasmuch as they are based on the U/Th clock – are spread within a factor of *two*.

Already in 1964 *D.D. Clayton* [22] pointed out that this serious shortcoming could be removed by choosing a couple of radionuclides *not* loaded with the peculiarity of the U/Th clock. He proposed as an ideal clock the pair ^{187}Re/^{187}Os. For this *mother–daughter* relation any (unknown) details of the production of the mother nucleus ^{187}Re drop out. ^{187}Re undergoes β^- decay to the ground state of ^{187}Os with a half-life of 42×10^9 years. That is long enough, on the one hand, to cover the full scale of about 10×10^9 years, estimated for the duration of the pre-solar galaxy, but also short enough, on the other hand, to produce detectable amounts of ^{187}Os during that period of time.

Shortly after Clayton had proposed the Re/Os clock, *E.M.D. Symbalisty* and *D.N. Schramm* [23] developed a *mean-age* approach to determine the pre-solar galactic age. They showed that from the abundance ratio of a directly connected radioactive pair both an upper $\left(T_{\mathrm{N}}^{\mathrm{MAX}}\right)$ and a lower $\left(T_{\mathrm{N}}^{\mathrm{MIN}}\right)$ bound for the *pre-solar* galactic age T_{N} can be derived, including both the extreme cases, a sudden (*one-spike*) and a steady nucleosynthesis, respectively. These limits differ by a factor of two only and read for the couple Re/Os and for $\lambda T_{\mathrm{N}} \ll 1$ as:

$$T_{\mathrm{N}}^{\mathrm{MIN,MAX}} = \frac{1(2)}{\lambda} \left(\frac{^{187}\mathrm{Os}^{\beta}}{^{187}\mathrm{Re}} \right)_{\mathrm{dec.}} , \tag{13}$$

where $\lambda^{-1} = \tau$ is the mean lifetime of ^{187}Re, 61×10^9 years, and where the $^{187}\mathrm{Os}^{\beta}/^{187}\mathrm{Re}$ ratio has to be taken at *decoupling*, i.e. 4.6×10^9 years ago. This simple expression leaves "only" one problem to be solved: In (13) only the β-part, $^{187}\mathrm{Os}^{\beta}$, stemming from β decay of ^{187}Re has to be taken into account. There is, however, another contribution to the total ^{187}Os abundance, due to slow neutron capture of ^{186}Os during the s process.

Thus, in previous decades the effort has focused on determining this s fraction as reliably as possible. The main problem was to estimate the fraction of neutron capture into the first *excited* state of ^{187}Os at about 10 keV –

knowing that this level should be considerably populated at typical *s* process temperatures corresponding to 30 keV. Nowadays this battle seems to be over and it is commonly believed that the *s* process contribution (about 40% of the total ^{187}Os abundance) can be deduced with an accuracy of about 5% [24]. The half-life of ^{187}Re has meanwhile been precisely settled – after a long period of contradictory results – to $(42.3 \pm 1.3) \times 10^9$ years [25]. Only the relative abundances of both ^{187}Os and ^{187}Re still have a comparatively large uncertainty of about 6% [26]. From all these data one extracts, within the *mean-age* approach of (13), about 9 billion years as a lower limit $T_{\mathrm{N}}^{\mathrm{MIN}}$ for the *pre-solar* age of our galaxy. Therewith, and based on the well-known *solar* age T_{S}, a lower limit $T_{\mathrm{G}}^{\mathrm{MIN}}$ for the *whole* age of our galaxy can be deduced, where

$$T_{\mathrm{G}}^{\mathrm{MIN}} = T_{\mathrm{N}}^{\mathrm{MIN}} + T_{\mathrm{S}} = (9 + 4.6) \times 10^9 \,\text{years}. \tag{14}$$

This value for $T_{\mathrm{G}}^{\mathrm{MIN}}$ of about 14×10^9 years comprises a fortiori also a lower limit for the age T_{U} of the universe. The 14 billion years match fairly well the 12 billion years established from globular clusters [21] as a rigorous lower limit for T_{U}, keeping in mind that the dating methods are basically different.

4.3.2 Bound beta decay of bare ^{187}Re. The mean-age approach tacitly contains some simplifications that do *not*, most probably, correspond to the real scenario of nucleosynthesis. In particular, the ability of the rhenium clock to provide a lower limit for the duration of the pre-solar galactic age, relies on the assumption that there is *one* and only one decay constant λ (i.e., the one determined for *neutral* ^{187}Re with the full electron cloud).

As with all other atoms created by rapid neutron capture (*r* process), ^{187}Re is also formed within a matter of seconds, most likely during the explosive phase of a supernova and is immediately flung into interstellar space. At a later stage this galactic gas probably compresses again, with ^{187}Re – and the ^{187}Os produced in the meantime – as a small part of it, and builds up to a new star. Some of the atoms go through this *astration* several times. Depending on the location in the star and the temperatures prevailing there, the atoms become more or less strongly ionized.

The higher the atomic charge state, the more β_{b} decay into empty electron states competes with continuum β_{c} decay, keeping in mind the tiny Q value of the latter of only 2.66 keV (taken for neutral ^{187}Re). Starting from an atomic charge number q of about 40, continuum beta decay even becomes impossible and will be fully "substituted" by bound beta decay.

This possible acceleration of the decay in q-times ionized ^{187}Re^{q} has been known for a long time. *K. Takahashi* and co-workers [27] realized first, however, that – for bare and hydrogen-like ^{187}Re ions – also the first *excited* nuclear state of ^{187}Os at 9.75 keV can be fed by β_{b} decay. Because here only a change of the nuclear spin I of one unit ($\Delta I = 1$) is involved, this transition should be much stronger than the corresponding one to the ^{187}Os ground state, where the nuclear spin has to change by two units. For bare

$^{187}Re^{75+}$ mother ions and for the created electron being bound in the K shell of $^{187}Os^{75+}$, a $Q^K_{\beta_b}$ value of 63.22 keV for the transition to the first excited state follows from (8) (see also Fig. 8), where $|\Delta B^e| = 15.31$ keV and the K binding energy of hydrogen-like $^{187}Os^{75+}$, $|B^{Os}_K| = 85.62$ keV are taken from standard compilations [14, 15].

In contrast to the case of ^{163}Dy, the nuclear matrix element for this decay *cannot* be provided from the time-mirrored EC process, because the excited 9.75 keV state does not undergo any EC decay during its short lifetime. Only a rough guess might be deduced from known matrix elements of neighboring nuclear states of similar structure. From this comparison *K. Takahashi* and co-workers [17] estimated, for bare ^{187}Re, a half-life of 14 years for β_b transition to the 9.75 keV state (almost identical with the total half-life of bare ^{187}Re). This would be a spectacular change by more than 9 orders of magnitude with respect to the half-life of 42×10^9 years of neutral ^{187}Re.

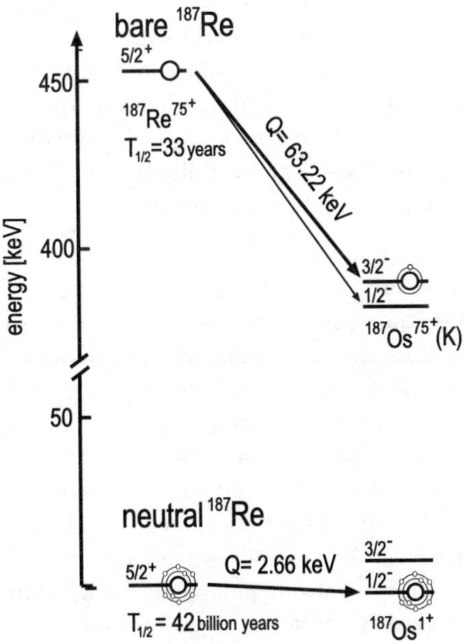

Fig. 8. Decay schemes for neutral ^{187}Re (bottom) and bare ^{187}Re (top). For the latter, β^K_b decay to the first excited state of ^{187}Os (9.75 keV, I=3/2$^-$) with the created electron being in the K shell of osmium, becomes energetically allowed (Q=+63.22 keV). The experiment [28] established a half-life of 33 years for that decay

To set this prediction on safe ground it was mandatory to measure [28] the half-life of bare ^{187}Re, all the more so since the ESR had already turned out to be tailored for this kind of experiment. The challenge was that a 100 to 1000 (!) times smaller decay rate had to be expected as compared to the β_b experiment on dysprosium. It was an open question whether such large β_b half-lives of several tens of years could be addressed at all with the same technique as applied for a 48-day half-life.

Like ^{163}Dy, the quasi-stable ^{187}Re (63% natural abundance) was introduced in an ion source, moderately ionized there, and subsequently accelerated to 370 MeV/u in the UNILAC and SIS facilities. After having left the latter, the remaining electrons were stripped off in a foil and the bare $^{187}Re^{75+}$ selected for injection into the storage ring ESR. The following steps like accumulation, cooling, storing, ionization of the hydrogen-like ^{187}Os daughters and, finally, detection by either Schottky spectroscopy or by a particle counter, were accomplished in the same manner as in the dysprosium experiment (Sect. 4.2). As the only difference – due to the much weaker β_b decay probability to be expected – storage times of up to 5 hours were taken and the particle detector was upgraded by a small ionization chamber for identifying the nuclear charge Z, too.

Against some fears, both detection devices showed clear-cut signals of β_b decay (Fig. 9), and within their respective experimental uncertainties both methods yielded the same half-life of (33 ± 2) years for the β_b decay of *bare* ^{187}Re [28]. This amazing result shows, indeed, that the half-life of ^{187}Re is more than a billion times shorter if all of the bound electrons are removed. From this half-life of the bare ion (the electron goes predominantly into the K shell, as in the case of dysprosium), the nuclear matrix element for the transition to the first *excited* state and, hence, the corresponding *log ft* value, can be deduced. Therewith, the half-life of $^{187}Re^{q}$ can be reliably calculated for *any* atomic charge number q, and, hence, at *any* temperature.

4.3.3 Probing the standard cosmological model. Does the dramatic difference in lifetime between neutral and fully ionized ^{187}Re render the rhenium clock useless? How can we possibly know all the details of the history of ^{187}Re, namely the location in the star after each astration, the temperature there (the distribution of atomic charge states) or the duration of stay? All that would be needed, however, to get a new *effective* decay constant λ_{eff}, is the proper mean of all the charge state dependent λ_q values. Even though the latter can *now* be reliably figured out, they still have to be weighted according to their respective distributions in time. This task is tremendous but less hopeless than it might seem at first sight: Astrophysicists have succeeded meanwhile in modeling the chemical evolution of our galaxy in a way that can explain to a large extent the observed abundance of all elements and isotopes.

Within this framework and based on the new knowledge of the half-life of any charge state whatever, *K. Takahashi* [29] has tried to develop the probable *life story* of the couple ^{187}Re/^{187}Os. Some facts might facilitate this mammoth work: As already mentioned, the decay to the first excited state and, hence, the most dramatic change in lifetime, is only allowed for *bare or hydrogen-like* ^{187}Re ions. This degree of ionization corresponds to temperatures of about 9×10^8 K that prevail only for short times and in the innermost zones of massive stars. In those extremely hot stellar plasmas the density of free, highly energetic electrons becomes, at the nuclear surface,

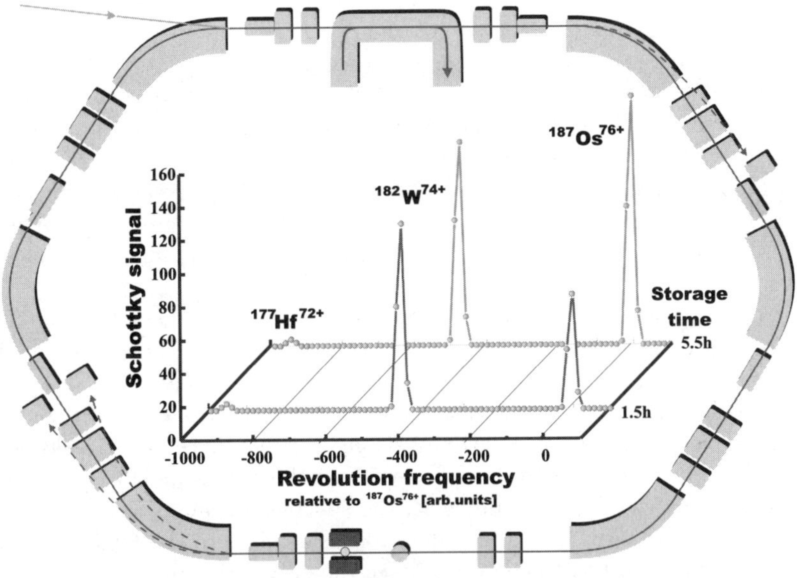

Fig. 9. Schottky spectra of the β_b decay daughter $^{187}Os^{76+}$ (after stripping off the electron) and of some nuclear products ($^{182}W^{74+}$, $^{177}Hf^{72+}$) from the interaction of the primary ^{187}Re beam with the argon gas jet. The spectra, taken for two storage times of 1.5 and 5.5 hours, respectively, were normalized onto the primary ^{187}Re current. Only the yield of the β_b daughter $^{187}Os^{76+}$ grows in proportion to the storage time

comparable to that of *bound* electrons. It happens therefore exactly in this exotic environment that the newly created ^{187}Os can *decay back* to ^{187}Re from its *excited* 9.75 keV state (strongly populated at temperatures of about 10^9 K, corresponding to 80 keV) by capturing a *free* electron. The probability of this back decay depends on the same matrix element as the β_b decay to the 9.75 keV state and can be easily derived, given only the density and temperature of the electron plasma.

As a very preliminary result *K. Takahashi* deduced [29] from the recalibrated rhenium clock and its stellar history an age for the galaxy of $(15^{+2}_{-3}) \times 10^9$ years. The lower limit of 12 billion years for the *galactic* age, being a fortiori a lower limit for the age of the *universe*, is significantly, albeit not dramatically, smaller than the 14 billion years as derived from the decay constant of neutral rhenium in the framework of the much simpler mean-age approach. On the other hand, Takahashi's preliminary result exactly matches the lower limit for the age of the universe as derived from globular clusters [21]. This outcome is perhaps surprising, insofar as one should expect a stronger decrease of the estimate for the pre-solar galactic age, due to the much enhanced β_b decays during that time. Apparently, the *back decay* by capturing *free* electrons (their energy has to be larger than about 64 keV) from the *excited* ^{187}Os state *counterbalances* to some extent the β_b decay, and this counterbalance is more complete the higher the temperature and charge state.

The value for the product of the Hubble constant, H_0, and of the age of the universe, T_U, depends on the energy-matter density, Ω, of the universe. Commonly, Ω is given as a fraction of the *critical density* $\Omega_c = 3H_0^2/(8\pi G)$, with G the gravitational constant. The critical density exactly defines the borderline between an infinitely expanding universe and a universe for which, somewhere in the future, the expansion will turn back to a contraction (closed universe).

Presently, the favorite cosmological model (*standard model*) is the *Einstein-de-Sitter* universe. For this cosmos the metrics is *Euclidean*, i.e. the curvature is zero, and the density is exactly the critical one ($\Omega = \Omega_c$). This choice is also most often required in models of the *inflationary universe* that just a few years ago received new and strong support by the COBE sky map of the 3-K microwave radiation, left over from the *Big Bang* [30].

In the Einstein–de Sitter universe, the so-called *cosmological constant* Λ is assumed to be zero. It was introduced by Einstein into the field equations of his *general relativity*, to counterbalance the gravitational attraction of the masses by a kind of repulsive energy density of the *vacuum*, because he sought for solutions providing a *stationary* universe. (Later on, Einstein called the postulate of Λ "the biggest blunder of my life".) A direct consequence of a critical density would be that about 90% of matter must be *dark*, because less than 10% of the critical density is visible.

If the cosmological constant Λ is taken to be zero, a simple relation between the Hubble constant H_0, the age of the universe T_U, and the density Ω holds true which, for practical purposes, reads:

$$\begin{aligned} H_0 \cdot T_U &\rightarrow 978 \quad (\Omega \rightarrow 0), \\ H_0 \cdot T_U &= 652 \quad (\Omega = \Omega_c), \end{aligned} \tag{15}$$

where H_0 is given in terms of the recession velocity (km/s) per megaparsec, and T_U in billions of years. If $H_0 \cdot T_U > 978$ then the cosmological constant Λ has, *necessarily*, a positive value and the universe would *expand* forever.

From (15) we may immediately learn what could be a serious approach probing the Einstein–de Sitter universe: To search for *independent* lower limits for both the Hubble constant H_0 and the age of the universe T_U, and to look thereafter at whether their product corresponds to the underlying assumption of a critical density and a vanishing cosmological constant. That is exactly why reliable lower limits for both H_0 and T_U are so important.

Very recently, the detection of *δ-Cepheids* in the *M100* galaxy of the *Virgo* cluster [31, 32] brought a breakthrough for determining the *true* distances of very far galaxies, the biggest obstacle so far in settling the Hubble constant. δ-Cepheids are variable stars but show a well-known correspondence between period and absolute luminosity and thus serve as calibration marks. The distance of the Virgo cluster could, for the first time, be reliably deduced to

15 Megaparsec and, thereafter, the Hubble constant H_0 to 80 km/(s Mpc) with a lower limit of 63 km/(s Mpc). These values were consistently derived with both the Hubble Space Telescope and with the Canada–France–Hawaii Telescope [31, 32].

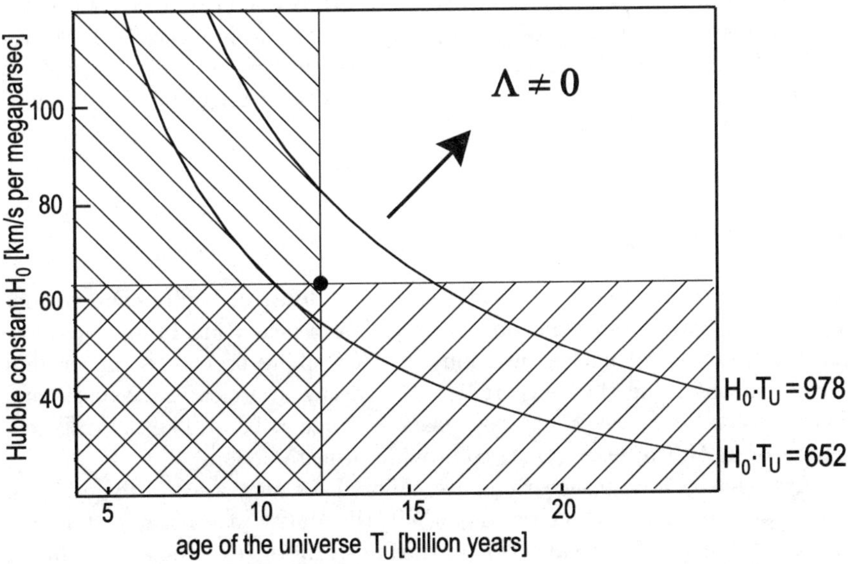

Fig. 10. Plot of the Hubble constant H_0 vs. the age of the universe T_U, for $\Lambda = 0$. For the standard model ($\Omega = \Omega_c$), the lower curve, for $\Omega \rightarrow 0$ the upper hyperbola is expected. Present lower limits of 63 km s^{-1} Mpc^{-1} for H_0, and of 12 billion years for T_U, respectively, exclude the dashed regions. They signal, if confirmed, a density *less* than the critical one and/or a *non*-vanishing cosmological constant Λ. Above the upper hyperbola Λ gets a positive value

For the time being almost all observations point to a lower limit of at least 12 billion years for the *age of the universe.* That is true – besides the globular clusters and the recalibrated rhenium clock – also for other clocks like the luminosity distribution of white dwarfs, or for the aeon clock $^{238}U/^{232}Th$ (with a large uncertainty, however). Combining this number of 12 billion years for the minimum age of the universe with the new lower limit of 63 km/(s Mpc) for the Hubble constant, one gets for their product a value of about 760, significantly *above* the value of 652 as predicted by the Einstein–de Sitter model (Fig. 10).

If both limits are eventually confirmed by future observations, the standard cosmological model would be ruled out in the sense that the density of the universe has to be *less* than the critical one and/or that there has to exist a *non*-vanishing cosmological constant. In that case the implications for cosmology would be, apparently, profound. Neither for the Hubble constant nor for the age of the universe can "watertight" limits be provided, however, keeping in mind all the intricate and peculiar methods needed to derive them. It is of the utmost importance to make further checks on the reliability and consistency of these clocks and, if possible, also to develop new ones. Maybe storage rings can be helpful for that purpose also in the future.

Each individual new result, either for the Hubble constant or for the age of the universe, is like a grain of sand in a thousand-piece puzzle. It seems, nonetheless, that the time is not too far off when we will get a first impression of the image hidden in this puzzle. No matter what it might be, it will reveal the future fate of the universe.

5. Outlook

Already the first experiments addressing beta decay of highly charged ions have shown that the storage-cooler ring ESR is a nearly ideal tool to explore this fresh ground of nuclear astrophysics. A striking proof was the prompt success in observing the long-predicted β_b decay in a clear-cut experiment. There has to be, however, for most cases a source at hand, like the fragment separator at GSI in Darmstadt, to provide β-unstable ions.

Asking why ion storage rings are tailored for that kind of experiment, one recognizes mainly three reasons: First, the ability to select a *well-defined charge state* either before injection or during storage. Next, the availability of *electron cooling* that produces brilliant, long-lived beams and enables high-resolution *Schottky lifetime spectroscopy.* The latter offers, for its part, the first opportunity to record *simultaneously* the time dependent number of both mother and daughter nuclei. Finally, the high efficiency of detecting β decay (between 16% to 100% of all decays) inherent in both the Schottky and the particle-detector technique.

One of the still present shortcomings of Schottky spectroscopy at the ESR, its restriction to half-lives ≥ 30 s, should be reduced considerably in the near future by applying *stochastic pre-cooling.* That would allow us, together with the intensity upgrading presently under way at GSI, to address a wealth of yet unexplored questions of nuclear astrophysics:

- β half-lives of *short-lived* ($T_{1/2} \leq 100$ ms) nuclei near the "waiting points" of both the r and rp processes (before that, however, one has to succeed in producing, injecting and storing *mono-isotopic* beams of both neutron- and proton-rich nuclei);
- *weak β^+ decay branches*, e.g., of ^{54}Mn, ^{56}Ni, … which serve as distance-ladders for supernovae;
- determination of EC decay probabilities of atoms with exactly *one, two … electrons*, to deduce from them the electron density at the origin together with the *screening factor*;

- measurement of β_b decay of *bare* $^{205}Tl^{81+}$ which alone provides the nuclear matrix element for *solar neutrino capture* to the first excited state of ^{205}Pb;
- search for *weak EC branches* in superallowed $0^+ \to 0^+$ transitions (e.g., ^{14}O).

There is, last but not least, a challenging proposal by *P.Kienle* [6] to use Schottky lifetime spectroscopy with *uncooled* beams at the *transition energy* of the ESR, where the revolution frequency depends only on m/q, but *not* on the energy. (The "transition energy" is defined by the Lorentz factor $\gamma_t^2 = \alpha_p^{-1}$ (see (3)).) If successful, the lifetime restrictions of Schottky spectroscopy could be circumvented for most practical cases.

Perhaps not all of these ideas can or will be realized soon. For the time being, however, only ion cooler rings can help to explore further this new and challenging field at the borderline of atomic physics and weak interaction physics.

References

1. M. Curie, M. Kammerling Onnes: Le Radium **10**, 181 (1913)
2. K.W.F. Kohlrausch: *Radioaktivität*, Handbuch der Experimentalphysik **15**, (Akad. Verlagsges., Leipzig, 1928)
3. H. Behrens, J. Jänecke: *Numerical Tables for Beta-Decay and Electron Capture*, Landolt-Börnstein I,4 (Springer-Verlag, Berlin, Heidelberg 1969)
4. B. Schlitt, K. Beckert, F. Bosch, H. Eickhoff, B. Franzke, Y. Fujita, H. Geissel, M. Hausmann, H. Irnich, O. Klepper, H.-J. Kluge, C. Kozhuharov, G. Kraus, G. Münzenberg, F. Nickel, F. Nolden, Z. Patyk, T. Radon, H. Reich, C. Scheidenberger, W. Schwab, M. Steck, K. Sümmerer, Th. Winkler, T. Beha, M. Falch, Th. Kerscher, K.E.G. Löbner, H.C. Jung, H. Wollnik, Yu. Novikov: Nucl. Phys. A **626**, 315c (1997)
5. H. Irnich, H. Geissel, F. Nolden, K. Beckert, F. Bosch, H. Eickhoff, B. Franzke, Y. Fujita, M. Hausmann, H.C. Jung, O. Klepper, C. Kozhuharov, G. Kraus, A. Magel, G. Münzenberg, F. Nickel, T. Radon, H. Reich, B. Schlitt, W. Schwab, M. Steck, K. Sümmerer, T. Suzuki, H. Wollnik: Phys. Rev. Lett. **75**, 4182 (1995)
6. P. Kienle: Nucl. Phys. A **626**, 591c (1997)
7. F. Nolden, D. Böhne, W. Bourgeois, B. Franzke, M. Steck, A. Schwinn: Nucl. Phys . A **626**, 491c (1997)
8. H. Geissel, K. Beckert, F. Bosch, H. Eickhoff, B. Franczak, B. Franzke, M. Jung, O. Klepper, R. Moshammer, G. Münzenberg, F. Nickel, F. Nolden, U. Schaaf, C. Scheidenberger, P. Spädtke, M. Steck, K. Sümmerer: Phys. Rev. Lett. **68**, 3412 (1992)
9. C.M. Lederer, J.M. Hollander, I. Perlman: *Table of Isotopes*, (Wiley, New York 1975)
10. O. Klepper: Nucl. Phys. A **626**, 199c (1997)
11. M. Jung, F. Bosch, K. Beckert, H. Eickhoff, H. Folger, B. Franzke, A. Gruber, P. Kienle, O. Klepper, W. Koenig, C. Kozhuharov, R. Mann, R. Moshammer, F. Nolden, U. Schaaf, G. Soff, P. Spädtke, M. Steck, Th. Stöhlker, K. Sümmerer: Phys. Rev. Lett. **69**, 2164 (1992)
12. R. Daudel, M. Jean, M. Lecoin: J. Phys. Radium **8**, 238 (1947)

13. J.N. Bahcall: Phys. Rev. **124**, 495 (1961)
14. J.P. Desclaux: At. Data Nucl. Data Tables **12**, 311 (1973)
15. W.R. Johnson, G. Soff: At. Data Nucl. Data Tables **33**, 405 (1985)
16. S. Yasumi, M. Ando, H. Maezawa, H. Kitamura, T. Ohta, A. Kikuni, K. Sera, T. Omori, G. Izawa, K. Masumoto, K. Shima, T. Mukoyama, I. Sugai, A. Masuda, O. Kawakami: Phys. Lett. B **181**, 169 (1986)
17. K. Takahashi, R.N. Boyd, G.J. Mathews, K. Yokoi: Phys. Rev. C **36**, 1522 (1987)
18. F. Bosch: Phys. Scr. **T59**, 221 (1995)
19. M. Steck: Nucl. Phys. A **626**, 473c (1997)
20. Particle Data Group: Phys. Lett. B **239**, 1 (1990), p. VI.2
21. D.A. VandenBerg, M. Bolte, P.B. Stetson: Ann. Rev. Astron. Astrophys. **34**, 461 (1996)
22. D.D. Clayton: Astrophys. J. **139**, 637 (1964)
23. E.M.D. Symbalisty, D.N. Schramm: Rep. Progr. Phys. **44**, 293 (1981)
24. R.R. Winters, R.F. Carlton, J.A. Harvey, N.W. Hill: Phys. Rev. C **34**, 840 (1986)
25. M. Lindner: Geochim. Cosmochim. Acta **53**, 1597 (1989)
26. E. Anders, N. Grevesse: Geochim. Cosmochim. Acta **53**, 197 (1989)
27. K. Yokoi, K. Takahashi, M. Arnould: Astron. Astrophys. J. **117**, 65 (1983)
28. F. Bosch, T. Faestermann, J. Friese, F. Heine, P. Kienle, E. Wefers, K. Zeitelhack, K. Beckert, B. Franzke, O. Klepper, C. Kozhuharov, G. Menzel, R. Moshammer, F. Nolden, H. Reich, B. Schlitt, M. Steck, T. Stöhlker, T. Winkler, K. Takahashi: Phys. Rev. Lett. **77**, 5190 (1996)
29. K. Takahashi: unpublished (1998)
30. D. Wilkinson: Science Spectra **4**, 52 (1996)
31. M.J. Pierce, L.W. Douglas, R.D. McClure, S. VandenBergh, R. Racine, P.B. Stetson: Nature **371**, 385 (1994)
32. W.L. Freedman, B.F. Madore, J.R. Mould, R. Hill, L. Ferrarese, R.C. Kennicutt Jr, A. Saka, P.B. Stetson, J.A. Graham, H. Ford, J.G. Hoessel, J. Huchra, S.M. Hughes, G.D. Illingworth: Nature **371**, 757 (1994)

Part C

Collisional and Radiative Processes

XI Relativistic Ion–Atom Collisions

Jörg Eichler and Thomas Stöhlker

Summary. With new developments in the design of accelerators and detectors, increasing attention is being focused on ion–atom collisions, in which the projectile velocity is comparable to the speed of light and in which inner shells of high-Z ions are involved, so that the electrons also move relativistically. We will regard collisions as relativistic, if the projectile energy exceeds 100 MeV/u. In this overview, we discuss recent improvements of experimental methods, the various processes that can occur, specifically direct reactions like ionization and excitation, nonradiative and radiative electron capture, and finally electron–positron pair production.

1. Introduction

In a relativistic ion–atom collision, we consider projectile ions with a velocity comparable to the speed of light, that is with a kinetic energy of 100 MeV/u or more. The most interesting situations occur for bare or almost bare high-Z ions as for example Au^{79+}, Pb^{82+}, or U^{92+}. In these cases, the electronic motion requires a relativistic quantum description by the Dirac equation. On the other hand, the nuclear motion can, to a very good approximation, be described by classical relativistic kinematics.

While physical interest is centered on the fate of the electrons during the collision, the nuclei merely serve as vehicles for the electrons and as sources of strong electromagnetic fields. They are surrounded by an electromagnetic field, which for slow collisions is produced by a spherically symmetric Coulomb potential. With increasing projectile velocity, however, the potential takes the shape of a Liénard–Wiechert potential, which shrinks in the longitudinal and extends in the transverse direction. In the limiting case of very high velocities, the projectile field can be represented by an ultrashort light pulse of (almost) real photons. This behavior of the potential, together with the sometimes unfamiliar transformation properties of the wave functions, forms the main difference to nonrelativistic collisions.

Most reactions that occur in relativistic collisions are basically similar to their nonrelativistic counterparts. We have to consider excitation and ionization, charge transfer and radiative electron capture. The only reaction type specific to relativistic encounters is the collisional production of electron–positron pairs.

In an experimental situation, the projectile may be bare or hydrogenic, but in all cases accessible so far (as long as colliders are not yet operational) the target is neutral it may be a gas or a foil. Therefore, the real object of observation is the emerging projectile. The basic analysis consists in measuring the charge. If the projectile charge is unchanged and an electron is detected in coincidence, one has identified target ionization. The decrease of the projectile charge by one unit can be associated with three different reactions: (a) charge transfer, if no or only characteristic projectile photons and no positrons are detected in coincidence, (b) radiative electron capture (REC), if a photon is found in coincidence with an energy beyond the characteristic projectile photon range, (c) bound–free pair production, if a positron is detected in coincidence, that is a reaction, in which an electron is produced in a bound state and a positron in a free state. A detailed account of the various reactions up to 1995 has been given by *Eichler* and *Meyerhof* [1].

In Sect. 2 of this brief review, we consider the experimental methods currently available, and in Sects. 3–6, we describe excitation and ionization, charge transfer, radiative electron capture and electron–positron pair production, respectively. Finally, in Sect. 7, we give a summary.

2. Experimental Methods

For a detailed discussion of the techniques most common in experiments at relativistic accelerators, such as beam handling, target preparation, photon or lepton detection, we refer to Ref. [1]. In the following, we concentrate on the most recent technical development in the field of relativistic heavy-ion physics, i.e. the relativistic heavy-ion storage ring ESR at GSI in Darmstadt. This electron-cooler ring provides the basis for atomic-collision experiments dealing with beams of unprecedented quality [2–5]. Here, the interaction of intense beams of cooled high-Z ions with low-density gaseous matter can be studied without any beam collimation, guaranteeing almost completely background-free experimental conditions. Typically, the beam diameter is close to 1 mm and the gas-jet target density amounts to about 10^{12} particle/cm^3 [6], compared to 10^{21} particle/cm^3 of solid-state targets. Nevertheless, due to the high revolution frequency of the ions of about 10^6 s^{-1}, even collision processes with tiny reaction cross sections (0.1 barn or even smaller) become accessible to experiments. Until now, most atomic-collision studies at the ESR were concentrating on the study of x-ray emission occurring in relativistic collisions [4, 7, 8].

Here, special emphasis is given to final-state resolved electron capture studies [7, 8], angular distribution [5, 9] and alignment investigations [10]. In order to demonstrate the clean conditions at the ESR jet target, we show in Fig. 1 an x-ray spectrum associated with electron capture measured for 358 MeV/u $U^{92+} \rightarrow N_2$ collisions at an observation angle of 132° [11]. Besides the Lyman ground-state transitions, the most prominent features observed in the spectrum are due to radiative electron capture into the ground- and excited projectile states. In Fig. 2, a scheme of the present experimental arrangement at the ESR gas-jet target is depicted. Various x-ray detectors view the gas-jet/beam interaction zone at different observation angles in the range from almost 0° to almost 180° [12]. In addition, the dipole magnet

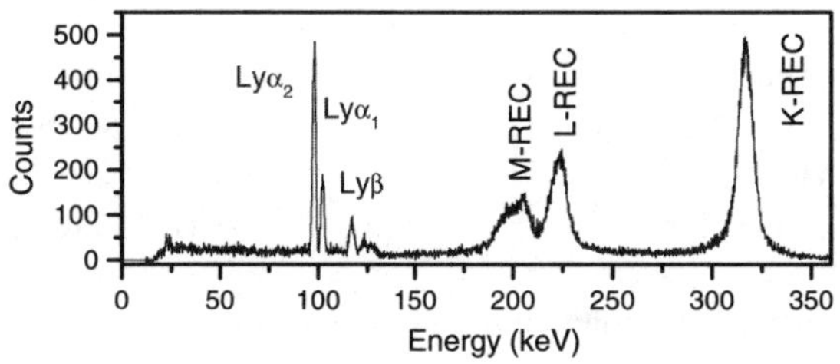

Fig. 1. An x-ray spectrum for 358 MeV/u U^{92+} on N_2 is shown following electron capture into the projectile. The data were taken at the ESR storage ring at an observation angle of 132°. The quoted energies refer to the emitter system

accomplishes the analysis of the charge states of the emerging projectiles and the x-ray emission can be measured in coincidence (electron capture) or in anti-coincidence (projectile excitation) with the charge changed projectiles. In the near future, similar arrangements can be anticipated which may allow the study of electron and electron–positron emission.

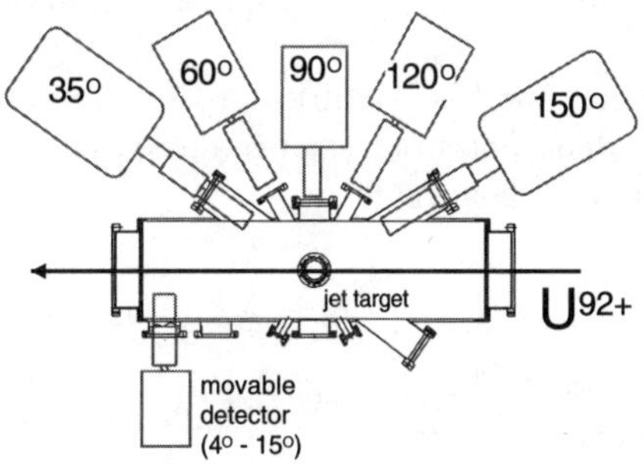

Fig. 2. Layout of the scattering chamber at the ESR gas-jet target. Various x-ray detectors view the target zone at different observation angles ranging from almost 0° to 150° [12]

3. Excitation and Ionization

In general, when considering relativistic collisions, we take the target atom as the system in which excitation, ionization or pair production occurs, while the projectile merely provides the time-dependent electromagnetic field that produces the transitions of interest. In an actual experimental situation, it may be the other way around. Suppose, the projectile carries a single electron. Then transitions in this hydrogenic system can be very well identified experimentally, so that it is convenient to regard the target as moving with respect to the projectile and providing the time-dependent electromagnetic field. In reality, of course, target and projectile should be treated on the same footing. This is certainly mandatory in the case of charge transfer, but is usually dispensable in direct (i.e. single-center) reactions.

Projectile and target are characterized by the nuclear charges eZ_{P} and eZ_{T}, respectively. The projectile is considered to move along a rectilinear classical trajectory $\boldsymbol{R}(t)$ in the z-direction with velocity $\boldsymbol{v}$ and an impact parameter $\boldsymbol{b}$ with respect to the target nucleus. Its motion is then described by $\boldsymbol{R}(t) = \boldsymbol{b} + \boldsymbol{v}t$. For relativistic collisions, one introduces the abbreviations $\beta = v/c$, where c is the speed of light, and $\gamma = 1/\sqrt{1-\beta^2}$ is the Lorentz factor.

With this notation, one may write down the Liénard-Wiechert potentials

$$\begin{aligned} \Phi(\boldsymbol{r},t) &= \frac{\gamma Z_{\mathrm{P}} e}{r'}, \\ \boldsymbol{A}(\boldsymbol{r},t) &= \frac{\boldsymbol{v}}{c}\Phi(\boldsymbol{r},t), \end{aligned} \tag{1}$$

produced by the projectile in the target frame [1] and satisfying the Lorentz gauge, where

$$r'(t) = \sqrt{(x-b)^2 + y^2 + \gamma^2(z-vt)^2} \tag{2}$$

is the distance of the electron (with coordinate (x,y,z,t)) from the origin of the moving frame.

A single electron bound to the target nucleus is subject to the target potential $Z_{\mathrm{T}}e/r$ and to the time-dependent potential (1) produced by the projectile. The corresponding Dirac equation has the form

$$\begin{aligned} &\mathrm{i}\hbar\frac{\partial}{\partial t}\Psi(\boldsymbol{r},t) \\ &= \left(-\mathrm{i}\hbar c\,\boldsymbol{\alpha}\cdot\nabla - \frac{Z_{\mathrm{T}}e^2}{r} - \gamma(1-\beta\alpha_z)\frac{Z_{\mathrm{P}}e^2}{r'(t)} + m_{\mathrm{e}}c^2\gamma^0\right)\Psi(\boldsymbol{r},t), \end{aligned} \tag{3}$$

where $\boldsymbol{\alpha}$ and γ^0 (sometimes denoted as β) are the Dirac matrices.

We now come to the processes of excitation and ionization [13–18]. Let φ_{i} and φ_{f} be the initial and final target eigenstates. Then the cross section for excitation is given by

$$\sigma_{\mathrm{fi}} = 2\pi \int_0^\infty |A^{\mathrm{e}}_{\mathrm{fi}}|^2\, b\, \mathrm{d}b, \tag{4}$$

while for ionization σ_{fi} is replaced by $\mathrm{d}\sigma/\mathrm{d}E_{\mathrm{f}}$. Within perturbation theory, the transition amplitude has the form

$$A^{\mathrm{e}}_{\mathrm{fi}}(b) = \mathrm{i}\frac{\gamma Z_{\mathrm{P}} e^2}{\hbar} \int \mathrm{d}t\, \mathrm{e}^{\mathrm{i}(E_{\mathrm{f}}-E_{\mathrm{i}})t/\hbar} \int \mathrm{d}^3 r\, \varphi^\dagger_{\mathrm{f}}(\boldsymbol{r})(1-\beta\alpha_z)\frac{1}{r'}\varphi_{\mathrm{i}}(\boldsymbol{r}). \tag{5}$$

Here, E_{i} and E_{f} are the energies of the initial and final states. The two contributions in the transition amplitude arise from the scalar and the vector part of the projectile potential and add up coherently.

Strictly speaking, the initial and final states φ_{i} and φ_{f} cannot be considered as unperturbed as is required by perturbation theory. The perturbing potential has a long range, so that even at infinite separations the atomic states are distorted. However, this defect can be remedied by imposing proper Coulomb boundary conditions or gauge transformations [19, 20], which amount to subtracting the leading long-range part and including a corresponding phase factor in the wave function. However, it has been shown [20] that for direct reactions like excitation and ionization, owing to orthogonality, the transition amplitudes (5) remain unchanged, and only the convergence in a numerical time integration is improved.

For high-Z projectiles and small impact parameters b, the electron–projectile interaction may be so strong that perturbation theory is no longer valid, and a nonperturbative approach is required. A widely used possibility is the coupled-channel method. If we still regard the projectile merely as a source of electromagnetic fields and not as a carrier of eigenstates, the single-center method consists in expanding the complete time-dependent wave function in terms of target eigenfunctions $\varphi_k(\boldsymbol{r})$ according to

$$\Psi(\boldsymbol{r},t) = \sum_k a_k(t)\varphi_k(\boldsymbol{r})\,\mathrm{e}^{-\mathrm{i}E_k t/\hbar} \tag{6}$$

and inserting this expansion into (3). By truncating the summation in (6), one obtains a set of coupled differential equations of the first order for the expansion coefficient $a_k(t)$. These equations can efficiently be solved numerically for a given initial condition, i.e. for a specification of the initially occupied state. Proper Coulomb boundary conditions or gauge transformations will speed up convergence (with respect to the size of the basis set), but do not change the results of converged calculations [21]. For a symmetric treatment of target and projectile, the sum over target states in (6) has to be supplemented by a sum over Lorentz transformed projectile states [22, 23].

Another nonperturbative quantum approach uses the finite-difference method for solving the time-dependent Dirac equation directly on a grid [24, 25]. Because of limitations in computing power, such calculations have mostly been performed for $b = 0$, thus saving one space dimension. Moreover, it is difficult to enclose the system in a box large enough to define asymptotically unperturbed initial and final states. The transition amplitudes and hence the cross sections are obtained by projecting on the final wave functions of interest. Alternatively, direct numerical solutions have been achieved in momentum space by solving the integro-differential equation corresponding to the Dirac equation (3) in position space [26]. For unbound states with a given asymptotic momentum, the advantage of the latter method consists in the localization of continuum wave functions in momentum space. However, also in this case large computing power is needed.

3.1 Comparison with Experimental Data

The experimental studies of projectile ionization and excitation in the relativistic velocity regime focus in general on collisions where high-Z one- and two-electron projectiles and low-Z targets are involved. Here, the perturbation caused by the nuclear charge of the target is small and, consequently, first-order perturbation theories should be appropriate with respect to the target. Also, the relativistic electron wave functions are exactly known for H-like ions and the assumption that the projectile moves on a straight line trajectory is valid. Therefore, relativistic collisions are a clean testing ground for first-order perturbation theories and the effects caused by the magnetic part of the Liénard–Wiechert potential are in particular addressed in experiments.

For ionization, total cross sections were studied systematically for a wide range of high-Z ions starting from xenon up to uranium [27]. As an example, in Fig. 3, the projectile-ionization cross sections measured for H-like Au^{78+} ions colliding with carbon atoms are plotted as a function of the beam energy. In the figure, the data are compared with PWBA calculations where corrections for the relativistic wave functions are considered in an approximate way [28]. The full line refers to the result obtained by taking also into account the so-called transverse contribution term. The latter was derived by adding incoherently the electric and magnetic amplitudes of the Liénard–Wiechert potential. For a detailed and critical discussion of the described approach see Ref. [1]. Within their uncertainties, the experimental data agree well with this theoretical approach applied.

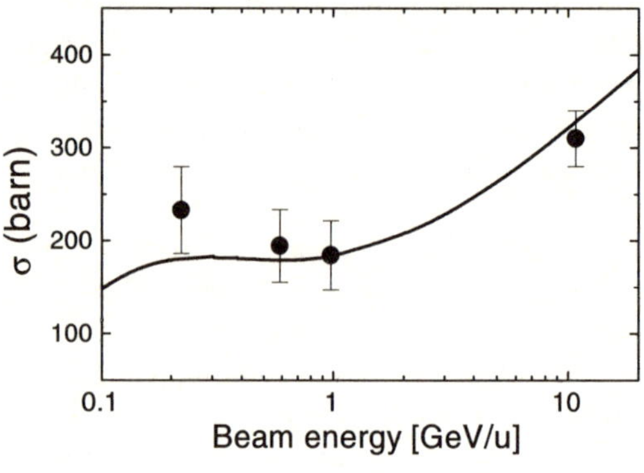

Fig. 3. Projectile ionization cross sections for H-like gold (Z=79) ions colliding with carbon targets

In contrast to ionization, almost no data are available for projectile excitation in relativistic collisions. Very recently, a first experimental study of projectile K-shell excitation for high-Z ions was reported [29, 30]. For the case of H-like Bi, the excitation process was unambiguously identified by observing the radiative decay of the excited levels to the vacant $1s$ shell in coincidence with ions that did not undergo any charge exchange in the reaction target. This allowed one to measure Coulomb excitation of the K-shell electrons in H-like bismuth ($Z = 83$) into specific L-shell sublevels. The results obtained provide a first indication of an interference between the electric and magnetic parts of the Liénard–Wiechert interaction which is most pronounced for the case of excitation to the $2p_{3/2}$ level. This is illustrated in Fig. 4. Here, the reduced cross-section values (σ/Z_T^2) for Ly-α_1 ($1s_{1/2} \rightarrow 2p_{3/2}$) and Ly-$\alpha_2$ ($1s_{1/2} \rightarrow 2s, p_{1/2}$) production are compared with a fully relativistic first-order perturbation approach [31] which considers the complete Liénard–Wiechert interaction without any further approximation (solid line). In addition results obtained by adding incoherently the electric and magnetic amplitudes (dashed line) are shown. All three measurements (for both Ly-α_1 and Ly-α_2) are in agreement with the solid line and are in variance with the dashed line. This finding is contrary to a commonly used assumption that the magnetic part of the interaction must always lead to a cross-section enhancement [28]. It points towards an interference between the amplitudes for the magnetic and electric parts of the interaction and to a break-down of a commonly used quasi-relativistic approach [13]. Until now, such interferences have not been discussed for high-Z ions, since K-shell ionization data agree with quasi-relativistic calculations at energies as high as 12 GeV/u [32].

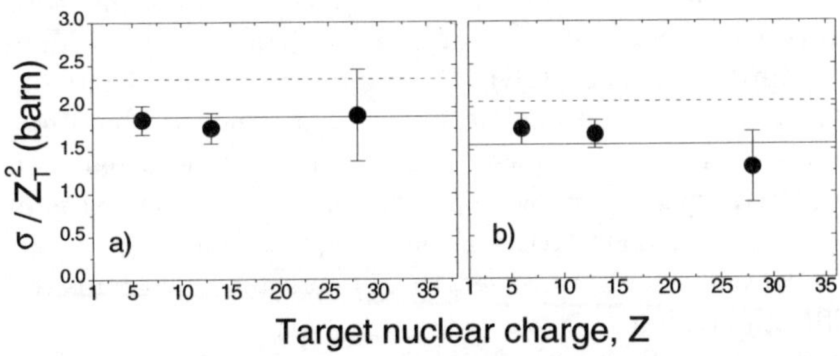

Fig. 4. Reduced Ly-α cross section (σ/Z_T^2) with statistical errors measured for Bi^{82+} ions at 119 MeV/u. The data are shown in comparison with theoretical calculations obtained with (solid line) and without (dashed line) the magnetic part of the Liénard–Wiechert interaction (a) Ly-α_1 and (b) Ly-α_2

4. Charge Transfer

In this section, we consider nonradiative charge transfer between a bound state of the target and a bound state of the projectile. Since there is no other particle to take up energy and momentum, the initial velocity of the electron in the target plus the velocity of relative target–projectile motion has to be accommodated in the final velocity distribution of the electron in the projectile. In other words, the transition amplitude for transfer is largely governed by the overlap of the electron momentum distributions in target and projectile displaced by the collisional momentum $m_e\boldsymbol{v}$. In nonrelativistic collisions, this leads to the well-known asymptotic energy dependence of the cross section as E^{-6} in first-order or $E^{-5/2}$ in second-order perturbation theory [19]. At relativistic projectile velocities, the spatial electronic distribution becomes Lorentz contracted, and hence the momentum distribution Lorentz extended in the direction of the projectile motion. As a result, the asymptotic energy dependence in first (and second) order behaves as E^{-1} [1].

The first-order Born transition amplitude, in analogy to (5), is given by

$$A_{\rm fi}^{\rm tr}(b) = {\rm i}\frac{e^2}{\hbar}\int {\rm d}t\, {\rm e}^{{\rm i}(E_{\rm f}t' - E_{\rm i}t)/\hbar}\int {\rm d}^3 r\, \varphi_{\rm f}'^{\dagger}(\boldsymbol{r}')\, S\, \frac{Z_{\rm P}}{r'}\, \varphi_{\rm i}(\boldsymbol{r})\,, \qquad (7)$$

where

$$S = \sqrt{(\gamma+1)/2}\,(1 - \delta\alpha_z) \qquad (8)$$

with $\delta = [(\gamma-1)/(\gamma+1)]^{1/2}$, is a 4×4 matrix that Lorentz transforms a four-spinor from the target to the projectile system. Although the first and second-order Born cross sections for charge transfer [33, 34] have the correct asymptotic energy dependence, the first-order Born cross sections are far too large, and it has been shown [35] that the exactly evaluated second-order cross sections are even worse. This can be ascribed to the failure of the Born approximation to take into account the asymptotic Coulomb distortion. In a distorted-wave Born approximation (B1B) taking into account the Coulomb boundary conditions [36], initial and final states are multiplied by phase factors, each containing a logarithmic phase, and the asymptotic separation dependence is subtracted from $Z_{\rm P}/r'$, so that the remaining potential is of short range. Similarly as in the nonrelativistic case [19], reasonable agreement with experimental data is obtained. However, the numerical evaluation is too complicated for wide application.

An accessible distorted-wave theory which yields good overall agreement with experimental data is provided by the relativistic eikonal approximation [37], which, in its asymmetric form, approximately includes the Coulomb distortion in one of the entrance or exit channels. A tabulation of cross sections for the most relevant target-projectile combinations is given in [38]. For the purpose of a simple first estimate, an approximate analytical formula has been given for $1s - 1s$ transitions [37].

A symmetric eikonal approximation [39] and a "continuum-distorted wave theory" [40] have also been proposed. Both approaches have desirable features in that they satisfy Coulomb boundary conditions, but pose additional problems in the fully relativistic limit [41]. Moreover, they cannot be evaluated without further approximations. Almost all relativistic theories of electron transfer predict an asymptotic charge and energy dependence as

$$\sigma_{\rm fi}^{\rm capt} \propto Z_{\rm P}^5 Z_{\rm T}^5\, \frac{1}{\gamma}, \qquad (9)$$

where the dominant Z^5 dependences already arise in nonrelativistic expressions from the Fourier transforms of the bound-state wave functions of the target and the projectile atom. They reflect the ability of high-Z atoms to accommodate high-momentum components of the electron translational motion.

Similarly as in Sect. 3 for excitation and ionization, one may obtain a nonperturbative solution of the time-dependent Dirac equation (3) with the coupled-channel method, in which, for transfer, both centers carry basis functions [23]. In this approach, the problem is treated rigorously within a limited space of basis functions. The degree of approximation then is essentially determined by the choice and the truncation of the basis set. While, in principle, the choice of the basis set is arbitrary, the convergence will be faster, if the long-range effect of the Coulomb field is already taken into account by imposing Coulomb boundary conditions [1, 20]. Numerical two-center coupled-channel calculations have been performed for $U^{92+} + U^{91+}$ collisions at 0.5 and 1.0 GeV/u [23]. It turns out that B1B, post and prior eikonal and coupled-channel calculations for total K-shell capture are all in fair agreement with one another.

Figure 5 shows experimental capture cross sections for La^{57+} projectiles impinging on Cu, Ag, and Au targets at 0.405, 0.96, and 1.3 GeV/u [42]. The theoretical cross sections represent the sum of nonradiative capture calculated in the prior form of the eikonal approximation according to [37, 38] and of radiative electron capture (Sect. 5) calculated [42] from tabulated photoelectric cross sections. An earlier comparison between experimental and theoretical capture data can be found in [1, 43].

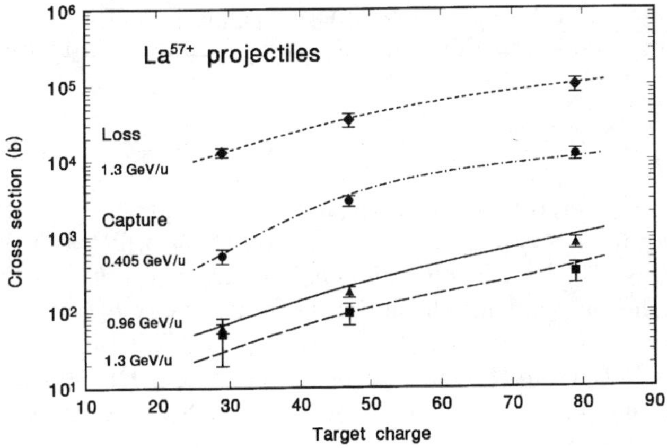

Fig. 5. Measured total electron-capture cross sections for 0.405, 0.96, and 1.3 GeV/u La^{57+} and electron loss cross section for 1.3 GeV/u La^{56+} as a function of target atomic number. The lines are theoretical values and, for capture, include contributions from nonradiative and radiative capture. From [42]

5. Radiative Electron Capture

For free electrons, nonradiative capture is forbidden by energy and momentum conservation. If, however, the electron transfer is accompanied by the emission of electromagnetic radiation, the emitted photon acts as a third body carrying away energy and momentum released by the formation of the final bound state. This means, qualitatively, that electrons loosely bound in low-Z target atoms or in outer shells are more likely to be captured *with* photon emission than *without*. Hence for low-Z target atoms at high projectile energies, the REC cross section exceeds the cross section for Coulomb capture. From this point of view, the REC mechanism deserves particular attention.

5.1 Reduction to Radiative Recombination

In the limit of a loosely bound target electron, REC is identical with radiative recombination (RR), in which an electron initially moving with the velocity $-\boldsymbol{v}$ towards the projectile is captured into a bound state with the simultaneous emission of a photon of energy $\hbar\omega'$ and wave number $\boldsymbol{k}'$.

Assuming that the cross section σ_{RR} for radiative recombination is known, it is natural to refer all momenta to the projectile frame. If the target electron has the momentum $\boldsymbol{q}$ with respect to the target nucleus, its momentum in the projectile frame $\boldsymbol{q}'$ is obtained by a Lorentz transformation. If the momentum $\gamma m_{\mathrm{e}} v$ of an electron traveling with the relative speed of the target towards the projectile is large compared to the electron momentum q in a low-Z target atom, one may use the impulse approximation [44] to write the double-differential REC cross section in the projectile frame (primed coordinates) as

$$\frac{\mathrm{d}^2\sigma_{\mathrm{REC}}}{\mathrm{d}\Omega'\,\mathrm{d}(\hbar\omega')} = \int \mathrm{d}^3q\,\frac{\mathrm{d}\sigma_{\mathrm{RR}}(q')}{\mathrm{d}\Omega'}\,|\tilde{\varphi}_{\mathrm{i}}(\boldsymbol{q})|^2\,\delta(\hbar\omega' - |\epsilon_{\mathrm{f}}'| + \gamma|\epsilon_{\mathrm{i}}| - T_{\mathrm{e}} + \gamma v q_z), \quad (10)$$

where $\tilde{\varphi}_{\mathrm{i}}(\boldsymbol{q})$ is the Fourier transform of the initial electronic target wave function. The delta function expresses the energy conservation between the final electronic energy $E_{\mathrm{f}}' = m_{\mathrm{e}}c^2 - |\epsilon_{\mathrm{f}}'|$ and the photon energy $\hbar\omega'$ in the projectile frame on the one hand and the initial electronic energy $E_{\mathrm{i}}' = \gamma(E_{\mathrm{i}} - vq_z)$ (also in the projectile frame) on the other. Here $|\epsilon_{\mathrm{i}}|$ and $|\epsilon_{\mathrm{f}}'|$ are the binding energies of initial target and final projectile states. Furthermore, $T_{\mathrm{e}} = m_{\mathrm{e}}c^2(\gamma - 1)$ is the kinetic energy of an electron traveling with the same speed as the projectile.

The initial momentum distribution of the target electron centered around $\boldsymbol{q} \approx 0$ gives rise to a peak in the double-differential cross section (10) around

$$\hbar\omega_0' = \gamma E_{\mathrm{i}} - E_{\mathrm{f}}' = |\epsilon_{\mathrm{f}}'| - \gamma|\epsilon_{\mathrm{i}}| + T_{\mathrm{e}}. \quad (11)$$

The Doppler broadening described by the integral over the electronic momentum $\boldsymbol{q}$ is approximately related to the Compton profile of the initial state.

The required momentum wave function can be obtained from a hydrogenic model or from a Hartree–Fock calculation [45, 46].

Here we have reduced the problem of calculating the cross section for REC to the corresponding task for radiative recombination. This process, in turn, is the inverse of the photoelectric effect. One may now put the four distinct steps together, needed to derive the differential cross section for radiative electron capture:

- first, we calculate the cross section of the photoelectric effect
- next, we calculate the cross section for radiative recombination, i.e. the capture of a free electron using the principle of detailed balance
- this cross section has to be convoluted with the momentum distribution of the active electron in the target atom according to (10)
- in order to compare with measured cross sections, one finally performs a Lorentz transformation from the projectile to the laboratory system.

5.2 The Atomic Photoelectric Effect

For the photoelectric effect, approximate formulae for the differential cross section are available [47], mostly for K-electron emission. When a photon with energy $\hbar\omega$ interacts with an atom, an electron with total energy E (including the rest mass) may be emitted. If $|\epsilon_{\mathrm{i}}|$ is the binding energy of the initially bound electron, conservation of energy requires that

$$E = \hbar\omega + m_{\mathrm{e}}c^2 - |\epsilon_{\mathrm{i}}|, \tag{12}$$

so that there exists a one-to-one relation between E and $\hbar\omega$.

Although we are dealing with relativistic atomic collisions, it turns out that a nonrelativistic dipole approximation for calculating the photoelectric cross section derived by Stobbe [48] is quite useful for some applications, far beyond the limits of validity of nonrelativistic criteria $\hbar\omega \ll m_{\mathrm{e}}c^2$ and $\alpha Z \ll 1$. Let $\nu = Ze^2/\hbar v$ be the Sommerfeld parameter with $\nu^2 = |\epsilon_b|/E_{\mathrm{kin}}$ for a nonrelativistic system with a hydrogenic K shell. Using exact nonrelativistic hydrogenic initial and final states, disregarding retardation effects, and adopting the dipole approximation, one obtains the photoelectric cross section per K-shell electron in the form [48, 49]

$$\sigma_{\mathrm{ph}}^{\mathrm{Stobbe}} = \frac{2^8\pi^2\alpha}{3}\,\frac{m_{\mathrm{e}}c^2}{\hbar\omega}\,\lambda\!\!\!^{-}{}_{\mathrm{c}}^2\left(\frac{\nu^2}{1+\nu^2}\right)^3 \frac{\mathrm{e}^{-4\nu\arctan(1/\nu)}}{1-\mathrm{e}^{-2\pi\nu}}, \tag{13}$$

where $\lambda\!\!\!^{-}{}_{\mathrm{c}} = \hbar/m_{\mathrm{e}}c$ is the Compton wavelength of the electron. The quantity α denotes the fine-structure constant. The corresponding differential cross section has a $\sin^2\theta$-dependence normalized to (13).

In a rigorous relativistic description one has to adopt the Dirac theory for an electron subject to a Coulomb potential in the initial bound and final continuum state. Assuming that the electron spin is not detected and the photons are unpolarized (the momenta of the photons define the quantization axis), one obtains the differential cross section for a single electron [47, 45]

$$\frac{d\sigma_{\rm ph}}{d\Omega} = \frac{\alpha\, m_e c^2}{8\,\hbar\omega}\, \frac{\lambda_c^2}{2j_i+1} \sum_{\mu_i}\sum_{m_s}\sum_{\lambda} \left|M_{\boldsymbol{p},i}(m_s,\lambda,\mu_i)\right|^2 , \tag{14}$$

where we have averaged over the $(2j_i + 1)$ angular momentum projections μ_i in the initial bound state and have summed over the spin components $m_s = \pm\, 1/2$ of the emitted electron. Furthermore, we have averaged over the circular polarizations $\lambda = \pm 1$ of the incoming photon.

The transition matrix element (adopting relativistic units $\hbar = c = m_e = 1$) is

$$M_{\boldsymbol{p},i}(m_s,\lambda,\mu_i) = \int \psi^{\dagger}_{\boldsymbol{p},m_s}(\boldsymbol{r})\, \boldsymbol{\alpha}\cdot\hat{\boldsymbol{u}}_{\lambda}\, e^{i\boldsymbol{k}\cdot\boldsymbol{r}}\, \psi_{j_i,\mu_i}(\boldsymbol{r})\, d^3r, \tag{15}$$

where $\psi_{j_i,\mu_i}(\boldsymbol{r})$ and $\psi_{\boldsymbol{p},m_s}(\boldsymbol{r})$ are exact bound and continuum Coulomb Dirac wave functions, the latter being given by a partial-wave expansion.

Equation (14) represents the rigorous relativistic cross section for photoionization of a hydrogenic electron. The effect of retardation and that of all multipole orders are included. The formulae have to be evaluated numerically [45, 50, 51].

5.3 From the Photoeffect to Radiative Recombination

The cross section $\sigma_{\rm RR}(E',\theta')$ for radiative recombination is related by the principle of detailed balance to the cross section $\sigma_{\rm ph}(\omega',\theta')$ of its inverse reaction, namely the photoelectric effect. By multiplying $\sigma_{\rm RR}(E',\theta')$ by the phase space ratio k'^2/p'^2, we replace the phase space factor of the outgoing electron by that of the emitted photon. For the transition of an electron into a specific substate of the final shell, we hence obtain

$$\frac{d^2\sigma_{\rm RR}(E',\theta')}{dE'\,d\Omega'_{\rm ph}} = \frac{(\gamma - 1 + |\epsilon_b|/m_e c^2)^2}{\gamma^2 - 1}\, \frac{d^2\sigma_{\rm ph}(E',\theta')}{dE'\,d\Omega'_{\rm el}} . \tag{16}$$

In applications to relativistic ion–atom collisions, the z-direction is usually defined as the direction of the projectile (or photon) momentum. This is opposite to the direction of the electron (or photon) momentum as seen from the projectile. Hence for REC, the angle θ' of the photoelectric effect or of the radiative recombination has to be replaced by $\pi - \theta'$, or $\cos\theta'$ is replaced by $-\cos\theta'$. Quite generally, the asymptotic behavior of the cross section for radiative electron capture from a neutral target into the projectile for large values of γ is given by

$$\sigma_{\rm fi}^{\rm REC} \propto Z_{\rm P}^5 Z_{\rm T}\, \frac{1}{\gamma}, \tag{17}$$

where the factor Z_P^5/γ arises from the capture process, compare (9), and the factor Z_T simply counts the number of electrons available in the neutral target atom. As a final step, one has to transform all relevant quantities from the projectile into the laboratory system [1, 45]. After having performed all four steps outlined above, one has succeeded to compute a REC cross section in the laboratory system from an originally derived photoelectric cross section (14).

5.4 Systematics and Comparison with Experimental Results

The importance of the REC process for relativistic collisions involving high-Z projectiles and low-Z targets can already be deduced from the x-ray spectrum shown in Fig. 1, i.e. under such collision conditions REC is the only relevant charge exchange process. For the high-energy domain, total electron capture cross sections as well as total K-REC cross sections were studied systematically over a wide range of beam energies and projectile charges. This is depicted in Fig. 6a,b, where all available total charge-exchange cross sections related to REC, normalized to the number of quasi-free electrons, are plotted versus the adiabaticity η-parameter, defined by $\eta = 1/\nu^2$ [8].

The accuracy of the experimental data is typically on the 20% level. In addition, the lower x-axis shows the energy scale for the particular case of Z=79. The predictions of the Stobbe theory are given by the dotted line whereas the solid line refers to the rigorous relativistic calculations performed for Z=79. For energies below 400 MeV/u, a good accordance between both theories and the experimental data can be stated. In the high-energy regime, the application of the exact relativistic theory is certainly more appropriate and at the very high energy of 12 GeV/u, the dipole approximation significantly underestimates the experimental result. Note that by applying a correct velocity description, the dipole approximation would already fail completely at intermediate beam energies. Obviously, by using the nonrelativistic η-scaling, several relativistic factors cancel accidentally. For completeness, the predictions of the Sauter theory [1] are given in the figure by the dashed line. As one may expect from a plane-wave-Born approach, the latter fails in the low-energy regime, but is in reasonable agreement with the experimental data for not too high collision energies. One of the relativistic corrections naturally not considered in a nonrelativistic treatment of the photoeffect are spin-flip transitions. At high energies, the latter contribute significantly to the total charge exchange cross sections. This is shown in the lower part of Fig. 6a,b, where the contributions of spin-flip and non-spin-flip transitions to the total-capture cross sections are given separately. The calculations were done in a relativistic exact way for K-RR into bare Au^{79+} ions [51].

At much lower energies, spin-flip transitions are already of substantial importance when angular distributions are considered. For capture into projectile s-states, e.g., the K-shell, only spin-flip transitions can produce photon emission at 0° and 180°. The latter follows directly from angular conservation

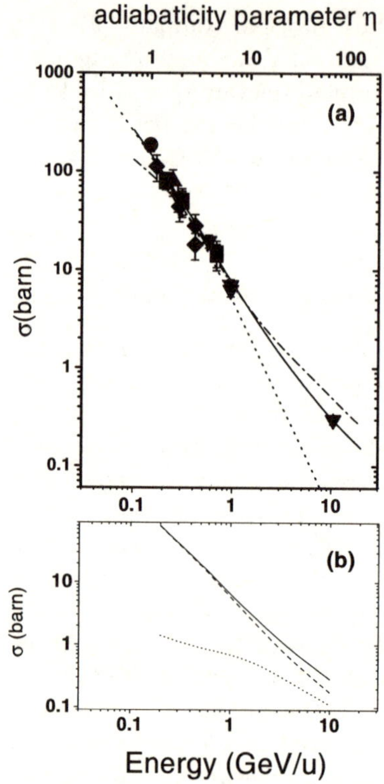

Fig. 6a,b. (a) K-shell REC cross sections vs. adiabaticity in comparison with experimental data. The predictions of the dipole approximation are given by the dotted curve. The solid curve shows the results of rigorous relativistic calculations for Z_p=79. The dash-dotted curve refers to the relativistic Sauter prediction. In (b) the result of rigorous relativistic calculations for Z_p=79 is depicted along with the spin-flip (dotted line) and the non-spin-flip (dashed line) contribution

laws for electric multipole transitions [50]. Note that there is no other process known at present where magnetic effects show up so unambiguously in relativistic collisions [5]. Currently, experiments are aimed at an identification of this. In Fig. 7 a preliminary x-ray spectrum is displayed, measured very recently at almost 0° observation angle at the ESR storage ring. Close to an x-ray energy of 660 keV, the K-REC is clearly visible. Already the presence of a measurable photon intensity at this energy along with an observation close to 0° is a clear evidence for magnetic transitions from continuum states into a bound projectile state.

Angular differential REC cross sections are extremely sensitive to details of the atomic wave functions and therefore provide detailed insight into the atomic structure of one- and few-electron systems. For such studies high-Z ions are of particular interest. Here, the large fine-structure splitting provides direct experimental access even for the investigation of the angular distributions associated with REC into the j-sublevels of the L-shell. A very first experimental study of the angular distributions for REC into the L-shell sublevels was conducted at the GSI Fragment Separator for 89 MeV/u $U^{90+}\rightarrow C$ collisions [9]. The results obtained within this experiment are depicted in Fig. 8a,b. In the figure the measured angular distributions for REC into

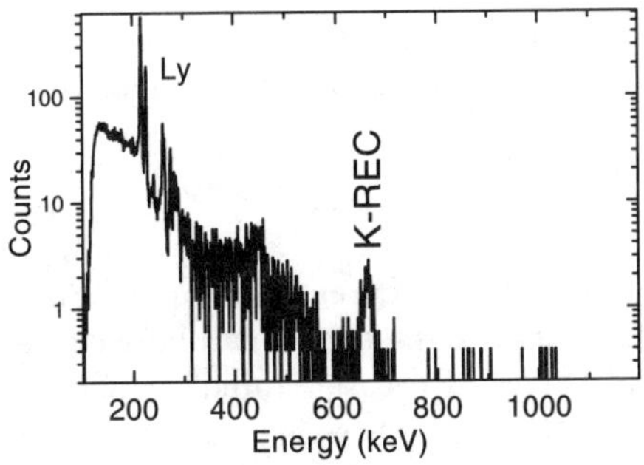

Fig. 7. X-ray spectrum obtained at almost 0° observation angle for U^{92+} on N_2 collisions. The spectrum was measured at the ESR jet target

the j=1/2 and j=3/2 levels of the L shell of He-like uranium are plotted in comparison with the predictions of the exact relativistic theory. The data for capture into the j=1/2 levels show a considerable bending of the angular distribution into the forward direction whereas the distribution for capture into the pure p-state (j=3/2) exhibit a slight enhancement at backward angles.

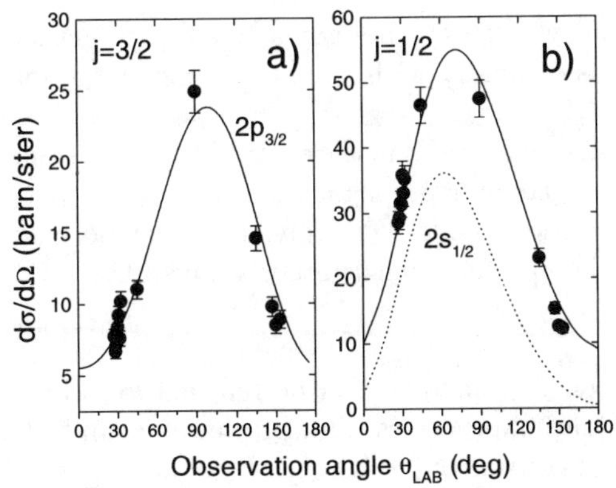

Fig. 8a,b. Experimental angular distribution of L-REC radiation for (a) REC into the $2p_{3/2}$ level and (b) for the $2s_{1/2}$ and $2p_{1/2}$ states. The solid line shows the result derived from relativistic exact calculations. The theoretical result for capture into the $2s_{1/2}$ state (dashed line in (b)) is shown separately [9]

6. Electron–Positron Pair Production

The production of electron–positron pairs in collisions of charged cosmic ray particles with nuclei has been a subject of experimental and theoretical interest since soon after the discovery of the positron in the early 1930s. In an intuitively appealing picture, the equivalent-photon or Weizsäcker–Williams method, see e.g., [1, 52, 53], the rapidly changing electromagnetic field of the moving particle with respect to a particle at rest is Fourier-decomposed into ultrashort pulses (subattosecond, i.e. $< 10^{-18}$ s) of polarized radiation. The projectile is hence replaced by a swarm of high-energy, almost real, photons corresponding to an extremely high energy density during the passage time.

If b is the impact parameter, the approximate duration of the passage is given by $\Delta t \approx b/\gamma v$, so that the photon spectrum extends up to a maximum frequency

$$\omega_{\mathrm{max}} = \frac{\gamma c}{b}, \tag{18}$$

which increases dramatically for high values of the Lorentz factor γ. While the collision time is shortest and the frequency highest for small impact parameters, one has to realize that small values of b entail rapid changes of the fields with b, so that a representation by plane waves is no longer valid. Moreover, photons will be approximately real ($k = \omega/c$) only for the highest frequencies [1]. This limits the validity of the equivalent-photon method to very high projectile energies. In general, a more detailed theory is needed for describing the rapid passage of charges.

A simple estimate based on the equivalent-photon method [52] for the total pair production cross section at large values of γ is provided by

$$\sigma_{\mathrm{e^+e^-}} = \frac{28}{27\pi}(\alpha Z_{\mathrm{T}})^2(\alpha Z_{\mathrm{P}})^2 \lambda\!\!\!^{-}_{\mathrm{c}}{}^2 \left(\ln\frac{\gamma}{4}\right)^3, \tag{19}$$

where $\lambda\!\!\!^{-}_{\mathrm{c}}$ is the electron Compton wavelength and α is the fine-structure constant. The cross section increases rapidly with γ and becomes large for high-Z collision partners.

In the near future, very high collision energies in the center-of-mass system will be reached in colliders, in which counter-propagating beams of like particles with equal and opposite velocities collide. For a given collider Lorentz factor γ_{coll}, the equivalent projectile γ_{FT} for a fixed target is given by

$$\gamma_{\mathrm{FT}} = 2\gamma_{\mathrm{coll}}^2 - 1. \tag{20}$$

In this way, extreme relativistic collision energies can be reached for which pair production is one of the dominating processes. Pair production can best be visualized by considering the spectrum of the Coulomb Dirac equation (Fig. 9). The creation of a pair corresponds to the excitation of an electron from the negative-energy sea to a bound or a continuum state of positive energy. The former case is known as *bound–free pair production* (sometimes called *capture from pair production*, however, since this term suggests a two-step process, we do not use it); the latter is known as *free pair production.*

6.1 Production of Free Electron–Positron Pairs

Energy-momentum conservation forbids the creation of an electron–positron pair by a single moving charge A. Therefore, a second particle B is needed to take up energy and momentum, and the lowest-order QED process can be described as follows: Nucleus A emits a virtual photon, which disintegrates into a $\mathrm{e^+e^-}$ pair. One of the leptons leaves the collision region freely, while the other lepton interacts with nucleus B by exchanging a virtual photon

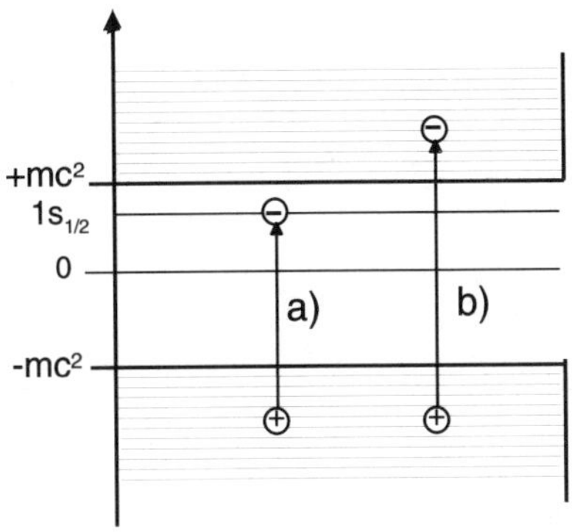

Fig. 9. Schematic representation of (a) bound-electron–free-positron pair production and (b) free pair production

and subsequently leaves the interaction region freely. This description [54] is adequate for high collision and hence lepton energies.

For lower relativistic energies, say from 1 GeV/u up, and high-Z nuclei, one has to take into account the Coulomb distortion of the lepton waves by a distorted-wave Born approximation (DWBA) [55–57]. In line with Fig. 9, the transition amplitude is analogous to (5) for excitation and reads

$$A_{p_+p_-}(\boldsymbol{b}) = \mathrm{i}\frac{\gamma Z_\mathrm{P} e^2}{\hbar}\int \mathrm{d}t\, \mathrm{e}^{\mathrm{i}(E_{p_+}+E_{p_-})t/\hbar}\int \mathrm{d}^3r\, \varphi^\dagger_{\boldsymbol{p}_+}(1-\beta\alpha_z)\frac{1}{r'}\varphi_{\boldsymbol{p}_-}, \qquad (21)$$

where E_{p_+}, $\boldsymbol{p}_+$ and E_{p_-}, $\boldsymbol{p}_-$ are the electron and positron energies and momenta. The sixfold-differential pair-production cross section can be written in the form

$$\frac{\mathrm{d}^6\sigma}{\mathrm{d}E_{p_+}\,\mathrm{d}E_{p_-}\,\mathrm{d}\Omega_{p_+}\mathrm{d}\Omega_{p_-}} = \frac{p_+p_-E_{p_+}E_{p_-}}{c^4}\sum_{s_+s_-}\int \mathrm{d}^2b\, |A_{p_+p_-}(\boldsymbol{b})|^2, \qquad (22)$$

where the summation extends over the unmeasured spin projections of electron and positron.

Compared to the case of excitation or ionization, the evaluation of the amplitudes (21) is complicated by the presence of two continuum Coulomb–Dirac wave functions. By integrating over the directions of emission of positron and electron, one obtains the double-differential cross section $\mathrm{d}^2\sigma/\mathrm{d}E_{p_+}\,\mathrm{d}E_{p_-}$ depending only on the electron and positron energies. Inserting exact continuum states, one obtains an *incoherent* summation over partial waves [55]. In practice, owing to computer limitations, the summation over partial waves has to be truncated at a certain maximum angular momentum for electron and positron, thus limiting the maximum projectile energy to which this method can be applied.

The partial-wave expansion can be avoided by using approximate *Furry–Sommerfeld–Maue* wave functions [56, 57], valid for lepton energies large compared to the rest energy $m_\mathrm{e}c^2$. In Fig. 10, we show the behavior of the single-differential cross section obtained by integrating over electron energies [57]. It is seen that the positron spectra extend to increasingly high energies as the projectile energy increases. The total cross section is obtained by integrating also over the positron spectrum. Typical theoretical results for U^{92+} + U^{92+} collisions are presented in Fig. 11.

Higher-order and nonperturbative treatments have also been performed [58–60].

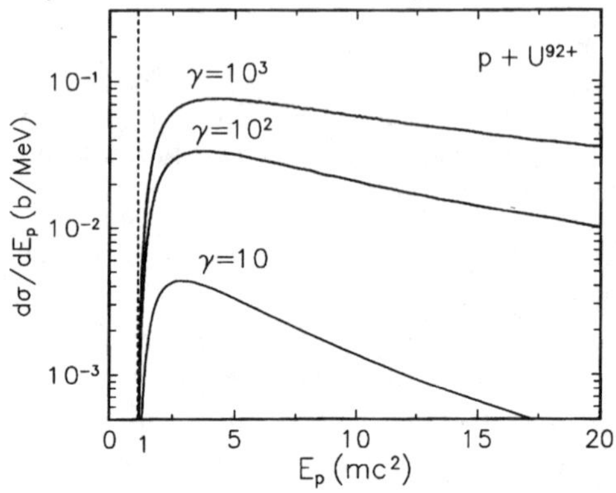

Fig. 10. Single-differential cross section in p + U^{92+} collisions as a function of the positron total energy E_{p} for several values of the Lorentz factor γ. For other projectiles, the cross section scales with Z_{P}^2. From [57]

6.2 Bound-Electron–Positron Pair Production

In a pair-production process, the electron may be created in a *bound state* of the target or the projectile, rather than in a continuum state. Production in the K-shell is dominant since it has the broadest momentum distribution to accommodate the boosted wave functions. Experimentally, this new process has been detected by the decrease of the projectile charge state by one unit [62–64] and has therefore been interpreted as a capture process, see [65].

In contrast to free pair production, for which the asymptotic dependence on the projectile energy and charge states involved is given by (18), the asymptotic dependence of the cross section for pair production with the electron emerging as bound to the projectile is given by [28]

$$\sigma_{\mathrm{fi}}^{\mathrm{b-f,\ pair}} \propto Z_{\mathrm{T}}^2 Z_{\mathrm{P}}^5 \ln\left(\frac{\gamma}{\gamma_0}\right), \tag{23}$$

where γ_0 varies between 6.81 and 8.23 as the projectile charge increases from $Z_{\mathrm{P}} = 1$ to $Z_{\mathrm{P}} = 92$. The factor Z_{P}^5 reflects the availability of high-momentum components in the projectile bound-state wave function and similarly occurs for charge transfer, see (9) and (17).

Bound–free pair production is of particular relevance for the design and operation of colliders, because it changes the charge state of the colliding ions, which then get lost from the beam in the storage ring, thus limiting its luminosity and lifetime. The process has been extensively treated with perturbative [66–68] and nonperturbative methods. *Baltz* et al. [67] predict the total cross section for bound-electron–free-positron pair production to be of the form

$$\sigma_{\mathrm{fi}}^{\mathrm{b-f\ pair}} = A \ln\gamma + B, \tag{24}$$

where $\gamma = \gamma_{\mathrm{FT}}$ is the equivalent fixed-target Lorentz parameter (19) and A as well as B are independent of γ (to within higher orders of $1/\gamma^2$). The quantity A has to be estimated from perturbation theory, while B receives contributions both from the perturbative and the nonperturbative range of impact parameters.

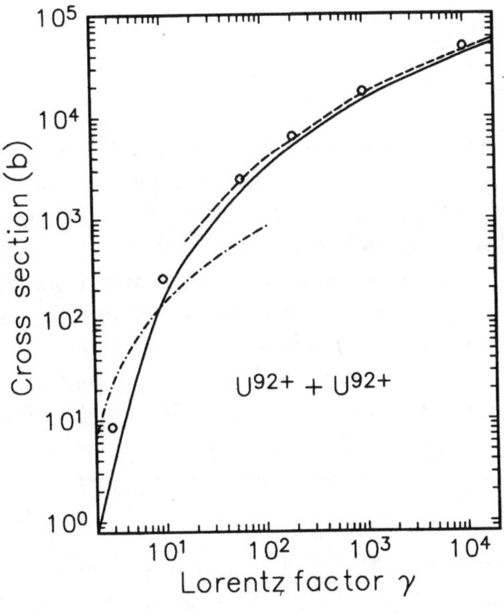

Fig. 11. Total cross section for pair production as a function of the Lorentz factor γ in a U^{92+} + U^{92+} collision. Dashed-dotted curve, first-order perturbation theory with partial-wave expansion [55]. Dashed curve, QED calculation [54]. Circles, combination of perturbative QED calculation with the equivalent-photon method [61]. Solid curve, first-order perturbation theory with Sommerfeld–Maue wave functions. From [57]

Since for bound–free pair production only one lepton is in the continuum, this case is more easily amenable to nonperturbative treatments. Single-center coupled-channel calculations have been performed in [69, 18], a finite-difference calculation for a direct numerical solution of the time-dependent Dirac equation in [25] and lattice calculations in [70]. Similarly, the Dirac equation in *momentum space* has been solved numerically in [26]. Another treatment has been based on the Magnus approximation to the time-evolution operator [71].

A great deal of attention has been given to modifying the original Dirac equation in such a way that the effective interaction becomes short-range and one hence achieves better convergence in coupled-channel calculations or in direct numerical solutions of the time-dependent Dirac equation. The method applied in various treatments is essentially the same but has been given different names. It has been denoted as an approach satisfying *Coulomb boundary conditions* by introducing a suitable phase transformation (for a review see e.g. [19]) and was first analyzed in coupled-channel calculations for relativistic collisions in [20]. The method can also be interpreted as a *gauge transformation* [67, 72–74] formulated to render the problem more tractable. In [73], an exact numerical solution of the time-dependent Dirac equation was obtained, and in [74] a new light-fronts approach for treating nonperturbative effects was derived. It has been pointed out in connection with coupled-channel calculations [21] that a gauge transformation cannot change the physical content or parametric dependences [72], but may improve the convergence properties of basis expansions. A detailed analysis of phase transformations of this kind is given in [75].

6.3 Multiple Pair Production

For highly charged collision systems as for example $U^{92+} + U^{92+}$ and extremely high collision energies, one expects that the creation of multiple pairs in a single collision will occur. There are several theoretical treatments of this process [76–80]. While these approaches are quite different in detail, they all adopt the basic assumption, without which a theory would become exceedingly complicated, namely that the pairs are created completely independently of each other. This means that electrons and positrons do not interact with one another (this is a $1/Z$ correction) and are not subject to the Pauli principle. The resulting multipair production probability is described by a Poisson distribution

$$P(N, b) = \frac{[\mathcal{P}(b)]^N}{N!} \, \mathrm{e}^{-\mathcal{P}(b)}, \tag{25}$$

where N denotes N-pair production and $\mathcal{P}(b)$ is the impact-parameter dependent lowest-order perturbation result for the total single-pair production probability. This quantity can also be interpreted as the average number of electrons produced out of the vacuum. The experimental detection of this process will be very difficult, and so far no evidence has been found.

7. Conclusion

The possibility of accelerating high-Z ions to relativistic velocities and to strip them of all or almost all of their electrons, has opened up a new field of atomic physics. First, one has gained access to the spectroscopy of heavy one-electron and few-electron ions up to uranium and hence to new tests of quantum electrodynamics in strong fields. Second, one may study new atomic collision processes that are strongly influenced by the Lorentz contraction of the Coulomb fields and by the interaction between the relativistically induced magnetic field of the projectile and the Dirac magnetic moment of the electron. In contrast to nonrelativistic collisions, the effects of spin-flip processes may be identified in a unique fashion. Third, at very high projectile energies, ion–atom collisions are increasingly well represented by photon collisions resulting from ultrashort (subattosecond) light pulses with the highest energy density. In the present review, we have mainly concentrated on the second aspect, namely on the various collision mechanisms, which form the basis of all other studies.

References

1. J. Eichler, W.E. Meyerhof: *Relativistic Atomic Collisions* (Academic Press, San Diego 1995)
2. F. Bosch: Nucl. Instrum. Methods B **23**, 190 (1987)
3. F. Bosch: Nucl. Instrum. Methods A **314**, 1992 (1992)
4. P.H. Mokler, Th. Stöhlker: Adv. At. Mol. Opt. Phys. **37**, 297 (1996)
5. Th. Stöhlker, P.H. Mokler, C. Kozhuharov, A. Warczak: Comments At. Mol. Phys. **33**, 271 (1997)
6. H. Reich, W. Bourgeois, B. Franzke, A. Kritzer, V. Varentsov: Nucl. Phys. A **626**, 417c (1997)
7. Th. Stöhlker, P.H. Mokler, K. Beckert, F. Bosch, H. Eickhoff, B. Franzke, H, Geissel, M. Jung, T. Kandler, O. Klepper, C. Kozhuharov, R. Moshammer, F. Nickel, F. Nolde, H. Reich, P. Rymuza, C. Scheidenberger, P. Spädtke, Z. Stachura, M. Steck, A. Warczak: Nucl. Instrum. Methods B **87**, 64 (1994)
8. Th. Stöhlker, C. Kozhuharov, P.H. Mokler, A. Warczak, F. Bosch, H. Geissel, C. Scheidenberger, R. Moshammer, C. Scheidenberger, J. Eichler, A, Ichihara, T. Shirai, Z. Stachura, P. Rymuza: Phys. Rev. A **51**, 2098 (1995)
9. Th. Stöhlker, H. Geissel, H. Irnich, C. Kozhuharov, P.H. Mokler, G. Münzenberg, F. Nickel, C. Scheidenberger, T. Suzuki, M. Kucharski, A. Warczak, P. Rymuza, Z. Stachura, A. Kriessbach, D. Dauvergne, R.W. Dunford, J. Eichler, A. Ichihara, T. Shirai: Phys. Rev. Lett. **73**, 352 (1994)
10. Th. Stöhlker, F. Bosch, A. Gallus, C. Kozhuharov, G. Menzel, P.H. Mokler, H.T. Prinz, J. Eichler, A. Ichihara, T. Shirai, R.W. Dunford, T. Ludziejewski, P. Rymuza, Z. Stachura, P. Swiat, A. Warczak: Phys. Rev. Lett. **79**, 3270 (1997)
11. Th. Stöhlker: Phys. Scr. **T37**, 29 (1997)
12. Th. Stöhlker: to be published (1998)
13. R. Anholt: Phys. Rev. A **19**, 1004 (1979)
14. P.A. Amundsen, K. Aashamar: J. Phys. B **14**, 4047 (1981)
15. S.R. Valluri, U. Becker, N. Grün, W. Scheid: J. Phys. B **17**, 4359 (1984)
16. U. Becker, N. Grün, W. Scheid: J. Phys. B **18**, 4589 (1985)
17. G. Mehler, G. Soff, K. Rumrich, W. Greiner: Z. Phys. D **13**, 193 (1989)
18. K. Rumrich, G. Soff, W. Greiner: Phys. Rev. A **47**, 215 (1993)
19. D.P. Dewangan, J. Eichler: Phys. Rep. **247**, 59 (1994)
20. N. Toshima, J. Eichler: Phys. Rev. A **42**, 3896 (1990)
21. J. Eichler, A. Belkacem: Phys. Rev. A , **54**, 4954 (1996)
22. N. Toshima, J. Eichler: Phys. Rev. Lett. **60**, 573 (1988)
23. N. Toshima, J. Eichler: Phys. Rev. A **38**, 2305 (1988)
24. U. Becker, N. Grün, W. Scheid: J. Phys. B **16**, 1967 (1983)

25. J. Thiel, A. Bunker, K. Momberger, N. Grün, W. Scheid: Phys. Rev. A **46**, 2607 (1992)
26. K. Momberger, A. Belkacem, A.H. Sørensen: Phys. Rev. A **53**, 1605 (1996)
27. Th. Stöhlker, D.C. Ionescu, P. Rymuza, T. Ludziejewski, P.H. Mokler, H. Geissel, C. Scheidenberger, F. Bosch, B. Franzke, O.Klepper, C. Kozhuharov, R. Moshammer, F. Nickel, H. Reich, Z. Stachura, A. Warczak: Nucl. Instrum. Methods B **124**, 160 (1997)
28. R. Anholt, U. Becker: Phys. Rev. A **36**, 4628 (1987)
29. Th. Stöhlker, D. C. Ionescu, P. Rymuza, F. Bosch, H. Geissel, C. Kozhuharov, T. Ludziejewski, P.H. Mokler, C. Scheidenberger, Z. Stachura, A. Warczak, R.W. Dunford: Phys. Lett. A **238**, 43 (1998)
30. Th. Stöhlker, D.C. Ionescu, P. Rymuza, F. Bosch, H. Geissel, C. Kozhuharov, T. Ludziejewski, P.H. Mokler, C. Scheidenberger, Z. Stachura, A. Warczak, R.W. Dunford: Phys. Rev. A **57**, 845 (1998)
31. D.C. Ionescu: unpublished (1997)
32. N. Claytor, A. Belkacem, T. Dinneen, B. Feinberg, H. Gould: Phys. Rev. A **55**, R842 (1997)
33. W.J. Humphries, B.L. Moiseiwitsch: J. Phys. B **17**, 2655 (1984); *ibid*, **18**, 2295 (1985)
34. B.L. Moiseiwitsch: Phys. Rev. A **39**, 5609 (1989)
35. F. Decker, J. Eichler: Phys. Rev. A **44**, 377 (1991)
36. J. Eichler: Phys. Rev. A **35**, 3248 (1987); Erratum: **37**, 287 (1988)
37. J. Eichler: Phys. Rev. A **32**, 112 (1985)
38. A. Ichihara, T. Shirai, J. Eichler: At. Data Nucl. Data Tables **55**, 63 (1993)
39. B.L. Moiseiwitsch: J. Phys. B **19**, 3733 (1986); *ibid* **20**, L171 (1987)
40. G.R. Deco, R.D. Rivarola: J. Phys. B **20**, 317 (1987)
41. J. Eichler, N. Toshima: in *The Physics of Electronic and Atomic Collisions*, L.J. Dubé, J.B.A. Mitchell, J.W. McConkey, Ch.E. Brion, eds., p. 847 (1995); N. Toshima, J. Eichler: Comments At. Mol. Phys. **31**, 109 (1995)
42. A. Belkacem, H. Gould, B. Feinberg, R. Bossingham, W.E. Meyerhof: Phys. Rev. A **56**, 2806 (1997)
43. W.E. Meyerhof, R. Anholt, J. Eichler, H. Gould, Ch. Munger, J. Alonso, P. Thieberger, H.E. Wegner: Phys. Rev. A **32**, 3291 (1985)
44. M. Kleber, D.H. Jakubassa: Nucl. Phys. A **252**, 152 (1975)
45. A. Ichihara, T. Shirai, J. Eichler: Phys. Rev. A **49**, 1875 (1994)
46. E. Clementi, C. Roetti: At. Data Nucl. Data Tables **14**, 177 (1974)
47. R.H. Pratt, A. Ron, H.K. Tseng: Rev. Mod. Phys. **45**, 273 (1973)
48. M. Stobbe: Ann. Phys. (Leipzig) **7**, 661 (1930)
49. H.A. Bethe, E.E. Salpeter: *Quantum Mechanics of One- and Two-Electron Atoms* (Plenum, New York 1977)
50. J. Eichler, A. Ichihara, T. Shirai: Phys. Rev. A **51**, 3027 (1995)
51. A. Ichihara, T. Shirai, J. Eichler: Phys. Rev. A **54**, 4954 (1996)
52. W. Heitler: *The Quantum Theory of Radiation* (Oxford University Press, Oxford 1954).
53. C.A. Bertulani, G. Baur: Phys. Rep. **163**, 299 (1988)
54. C. Bottcher, M.R. Strayer: Phys. Rev. D **39**, 1330 (1989)
55. U. Becker, N. Grün, W. Scheid: J. Phys. B **19**, 1347 (1986)
56. F. Decker: Phys. Rev. A **44**, 2883 (1991)
57. D.C. Ionescu, J. Eichler: Phys. Rev. A **48**, 1176 (1993)
58. K. Hencken, D. Trautmann, G. Baur: Phys. Rev. A **49**, 1584 (1994)
59. K. Hencken, D. Trautmann, G. Baur: Phys. Rev. A **51**, 998 (1995); *ibid* **51**, 1874 (1995)

60. J.C. Wells, V.E. Oberacker, A.S. Umar, C. Bottcher, M.R. Strayer, J.-S. Wu, G. Plunien: Phys. Rev. A **45**, 6296 (1992)
61. P.B. Eby: Phys. Rev. A **43**, 2258 (1991)
62. A. Belkacem, H. Gould, B. Feinberg, R. Bossingham, W.E. Meyerhof: Phys. Rev. Lett. **71**, 1514 (1993)
63. A. Belkacem, H. Gould, B. Feinberg, R. Bossigham, W.E. Meyerhof: Phys. Rev. Lett. **73**, 2432 (1994)
64. C.R. Vane, S. Datz, E.F. Deveney, P.F. Dittner, H.F. Krause, R. Schuch, H. Gao, R. Hutton: Phys. Rev. A **56**, 3682 (1997)
65. H. Gould: in *Physics of Electronic and Atomic Collisions*, Proceedings of the XX ICPEAC, Vienna 1997
66. U. Becker, N. Grün, W. Scheid: J. Phys. B **20**, 2075 (1987)
67. A.J. Baltz, M.J. Rhoades-Brown, J. Weneser: Phys. Rev. A **44**, 5569 (1991)
68. J. Eichler: Phys. Rev. Lett. **65**, 3653 (1995)
69. K. Rumrich, K. Momberger, G. Soff, W. Greiner, N. Grün, W. Scheid: Phys. Rev. Lett. **66**, 2613 (1991)
70. J.C. Wells, V.E. Oberacker, M.R. Strayer, A.S. Umar: Phys. Rev. A **53**, 1498 (1996)
71. D.C. Ionescu: Phys. Rev. A **49**, 3188 (1994)
72. A.J. Baltz, M.J. Rhoades-Brown, J. Weneser: Phys. Rev. A **47**, 3444 (1993); *ibid* **50**, 4842 (1994); Phys. Rev. E **54**, 2433 (1996); A.J. Baltz: Phys. Rev. A **52**, 4970 (1995)
73. A.J. Baltz: Phys. Rev. Lett. **78**, 1231 (1997)
74. B. Segev, J.C. Wells: Phys. Rev. A **57**, 1849 (1998)
75. J.C. Wells, B. Segev, J. Eichler: Phys. Rev. A, to be published (1998)
76. G. Baur: Phys. Rev. A **42**, 5736 (1990)
77. M.J. Rhoades-Brown, J. Weneser: Phys. Rev. A **44**, 330 (1991)
78. Ch. Best, W. Greiner, G. Soff: Phys. Rev. A **46**, 261 (1992)
79. M.C. Güclü, J.C. Wells, A.S. Umar, M.R. Strayer, D.J. Ernst: Phys. Rev. A **51**, 1836 (1995)
80. A. Alscher, K. Hencken, D. Trautmann, G. Baur: Phys. Rev. A **55**, 396 (1997)

XII Electron-Impact Ionization and Recombination of Highly Charged Ions

Alfred Müller

Summary. Due to new technical developments, much progress has been achieved recently in the field of electron–ion collision physics. In particular, the use of heavy-ion storage rings and traps has provided access to very detailed measurements on electron-impact ionization and recombination of ions from H^- to U^{91+}. In addition to mere cross sections and rate coefficients, the measurements provide insight into fundamental issues of atomic structure and atomic collisions. A new spectroscopy of very highly charged ions by the observation of resonance phenomena is being developed, and the influence of external electromagnetic fields, of interference phenomena and of plasma effects on electron–ion recombination presently receive much attention.

With their fundamental character and unique properties, electron–ion collisions constitute serious challenges to the theory of atomic structure and atomic interactions. On the basis of advanced detailed experimental results calculations can now be tested on an increasingly critical level. The frontiers of this field have thus been pushed far out during the last decade.

1. Introduction

Charge-changing electron–ion collisions are important fundamental processes in all kinds of plasmas. They determine the charge-state balance of the ions and, hence, also the spectrum of electromagnetic radiation emitted by the plasma. Understanding and diagnosing the state of a plasma, whether of astrophysical origin or man made, relies on information about cross sections and rate coefficients for electron–ion interactions.

Electron–ion collisions comprise four classes of processes distinguished by the final state of the ion: recombination, excitation, ionization and elastic scattering. All of these can occur either directly in one step by a single interaction or indirectly in two or more steps including several successive processes. Detailed experiments can distinguish between different contributions to the total cross sections and thus provide an ideal testing ground for basic atomic collision and atomic structure theory. Rate coefficients, cross sections, lifetimes of states, transition probabilities, and level energies are accessible to electron–ion collision experiments and they are needed for the modeling and understanding of ionized matter. The applied interest in plasmas has therefore spurred the theoretical and experimental investigation of electron–ion interactions over decades [1]. More recently, storage ring and accelerator physics [2] as well as the possible production of antihydrogen by recombination [3] of antiprotons and positrons provided additional impetus to the field.

Colliding beams of electrons and ions [4, 5] have been used since 1961 [6] for studies of electron–ion collision processes. By far most of the available data were obtained with small-scale equipment, i.e., with an ion source on an electrostatic potential of typically several kV, providing beams of slow ions, and with an (intersecting) electron beam of eV to keV energies in combination with the necessary equipment to characterize the beams and their overlap and to accomplish signal recovery. With these techniques a considerable body of data has been accumulated particularly for electron-impact ionization of ions in low and intermediate charge states [7, 8]. Data have also become available for electron-impact excitation [9, 10], for electron–ion scattering [11, 12] and for electron–ion recombination in experiments with low-energy ion beams [13].

By the combination of the well-known merged-beam approach [14] with ion accelerator and ion storage ring technology a new era of electron–ion collision studies began little over a decade ago [15]. Single-pass merged-beam experiments and heavy ion storage rings equipped with electron-cooling devices have provided excellent opportunities to study interactions of electrons with atomic or molecular ions.

A number of alternative approaches to electron–ion cross section and rate measurements exist and have yielded important data. These approaches differ by the way in which a target of electrons is provided in the experiments. Electrons are present in atoms, molecules and solids (crystals). When ions with high velocities interact with these neutral targets, the bound electrons can be treated as quasi-free particles and hence, electron–ion collision phenomena can be studied [7, 16, 17]. Examples for such studies are the so-called REC (radiative electron capture) and RTE (resonant transfer and excitation) experiments. REC is treated in Chap. XI.

Another approach deals with interactions of ions with free electrons in a plasma environment (see, e.g., [18]). Typically, radiation from the plasma is observed and the plasma properties, i.e., temperatures and densities, are characterized. From collisional models of the plasma and the related rate equations the expected radiation intensities are calculated under simplifying assumptions and fitted to the observations using the unknown electron–ion collision rate coefficients as free parameters. This technique has provided the earliest experimental electron–ion collision data on highly charged ions. The results, however, cannot compete in both quality and detail with data that have become available by the application of mono-energetic electron beams incident on ions in beams or in traps.

A particular plasma environment is present in dense mono-energetic electron beams in which slow ions are trapped [19]. Here the electron beam has several functions at a time. It serves as an ionizing agent producing ions (from neutrals) in a certain range of charge states, it provides the negative space charge for radial trapping of the slow ions and it serves as a projectile beam for electron–ion collision experiments. By charge-state analysis of extracted ions or by the observation of radiation from the trap in connection with electron-beam energy manipulations, detailed information on electron–ion interactions can be extracted. Much progress has been accomplished by the use of advanced electron beam ion sources and traps (EBIS and EBIT). The most prolific of these devices in terms of electron–ion collision data, the Livermore EBIT, is reviewed in Chap. II.

The emphasis of the present article is on direct observation of electron–ion collision processes in colliding beams experiments. The discussion concentrates on recombination and ionization mechanisms.

2. Basic Considerations and Techniques

2.1 Mechanisms

In all the experiments with projectiles A^{q+} reported here, the ions which have changed their charge state either by recombination or electron removal are detected behind a charge separator downbeam from the interaction region. The normalized count rate of these ions is measured as a function of the electron laboratory energy. With the known ion and electron energies the electron–ion center-of-mass energy is determined. Absolute total cross sections and rates are measured with high energy resolution. Since photons or electrons are not detected, the observation of recombination and ionization comprises a number of different processes which all lead to the same net production of ions in a new charge state, $A^{(q-1)+}$ or $A^{(q+n)+}$ (with $n = 1, 2, ...$), respectively. While these charge states are identified in the measurements, the different mechanisms producing these charge states are not experimentally distinguished from each other and thus they all contribute to the measured cross sections and rates.

Mechanisms of recombination, i.e., net production of $\mathrm{A}^{(q-1)+}$ ions from A^{q+} ions, are:

– radiative recombination (RR)

$$e + A^{q+} \rightarrow A^{(q-1)+} + h\nu \ , \tag{1}$$

where the excess energy released by the binding of an initially unbound electron is carried away by a photon in a direct process. After radiative recombination the captured electron can be in a highly excited state and hence further radiation will be emitted until the electron is in its ground level;

– dielectronic recombination (DR) where, in a first step,

$$e + A^{q+} \rightarrow \left[A^{(q-1)+}\right]^{**} \ , \tag{2}$$

the excess energy released by the capture of the electron is absorbed within the ion by the excitation of a core electron; and where, in a second step, the intermediate multiply excited state decays by the emission of two or more photons

$$\left[A^{(q-1)+}\right]^{**} \rightarrow \left[A^{(q-1)+}\right] + h\nu_1 + h\nu_2 \ . \tag{3}$$

The first step of the DR process, the dielectronic capture (2), can only occur if the kinetic energy of the projectile electron matches the difference $E_\mathrm{i} - E_\mathrm{f}$ of total binding energies of all electrons in the initial and final states of the ion;

– three-body (or ternary) recombination (TR)

$$e + e + A^{q+} \rightarrow A^{(q-1)+} + e, \tag{4}$$

where one of the two electrons can carry away the excess energy released by the recombination. TR becomes important at high electron densities and low energies.

None of the present experiments provides the electron densities and the electron temperatures for which TR to low lying bound levels of the recombined ions would be expected with measurable rates. TR into high Rydberg states is conceivable at low electron energies, however, the $\mathbf{v}\times\mathbf{B}$ fields ($\mathbf{v}$: ion velocity, $\mathbf{B}$ magnetic flux density) present in the experiments (e.g., in charge/mass analyzing and storage-ring dipole magnets) strip off the Rydberg electrons of such states. And yet, some of the observations indicate a strong involvement of TR in the recombination rates measured at zero center-of-mass energy $\mathrm{E_{cm}} = 0$ eV.

Although not experimentally distinguished, the different recombination mechanisms lead to very characteristic energy dependences in the measured cross sections and rates. DR produces narrow resonances and can thus easily be identified in a measured recombination spectrum. The RR rate coefficient has its maximum at $\mathrm{E_{cm}} = 0$ eV and drops off as a smooth function of energy. Only in specific cases where RR and DR have identical initial and final states the two recombination mechanisms cannot be isolated. In that case interference effects may be observed. Additional recombination, possibly involving ternary processes with capture into high Rydberg states, is observed at relative energies below about 10 meV.

Similarly, different ionization mechanisms leading to net production of $\mathrm{A}^{(q+1)+}$ from A^{q+} ions produce distinct features in the total cross section allowing the experimentalist to quantify individual contributions [7]. The most important of all the possible single-ionization processes are

– direct ionization (DI)

$$e + A^{q+} \rightarrow A^{(q+1)+} + 2e \tag{5}$$

proceeding in a single "knock-on" event;

– excitation–autoionization (EA)

$$e + A^{q+} \to [A^{q+}]^{**} + e \tag{6}$$

$$[A^{q+}]^{**} \to A^{(q+1)+} + e \tag{7}$$

involving direct excitation of an inner-shell electron to a bound configuration of the A^{q+} ion and subsequent autoionization of the intermediate multiply excited $[A^{q+}]^{**}$ ion;

– resonant-excitation-double-autoionization (REDA)

$$e + A^{q+} \to [A^{(q-1)+}]^{**}, \tag{8}$$

$$[A^{(q-1)+}]^{**} \to A^{(q+1)+} + 2e, \tag{9}$$

involving radiationless (dielectronic) capture of the incident electron by the A^{q+} ion and subsequent sequential emission of two electrons. REDA proceeds via intermediate compound states of the projectile electron and the parent ion and thus is closely related to the DR process (in fact, (2) and (8) are identical). Multiple ionization can also proceed via direct and indirect channels similar to the ones listed above for single ionization. In particular, inner-shell ionization with subsequent electron emission often contributes considerably to net multiple ionization of ions.

2.2 Measurement of Cross Sections and Rate Coefficients

When electrons with a spatial density $n_\mathrm{e}(\boldsymbol{r})$ and ions with a spatial density $n_\mathrm{i}(\boldsymbol{r})$ interact in a volume τ via a cross section $\sigma(v)$, where $\boldsymbol{v}$ is the vector of relative velocity between the collision partners and $v = |\boldsymbol{v}|$ its absolute value, then the rate R of interactions (number of interactions per unit time) is given by

$$R = \int_\tau \int_v n_\mathrm{e}(\boldsymbol{r})\, n_\mathrm{i}(\boldsymbol{r})\, \sigma(v)\, v\, f(\boldsymbol{v})\, \mathrm{d}^3v\, \mathrm{d}^3r. \tag{10}$$

For the deduction of cross sections from measured counting rates R the spatial density distributions of the particles and the distribution function $f(\boldsymbol{v})$ of the relative velocities have to be determined. In the equation it has been assumed that f is independent of the location $\boldsymbol{r}$. In electron–ion colliding-beam experiments the integral over spatial density distributions is usually taken into account by introducing a form factor [4, 5].

The integral over the velocity distribution $f(\boldsymbol{v})$ can be carried out separately by introducing a rate coefficient α. In plasma experiments, $f(\boldsymbol{v})$ is determined by the temperature of the plasma electrons and is often described by a Maxwellian distribution. In colliding-beam experiments with electrons and ions, f is usually a distribution centered at an average relative velocity v_rel of the colliding particles. Thus, α is determined by

$$\alpha(v_{\mathrm{rel}}) = <\sigma\, v_{\mathrm{rel}}> = \int \sigma(v)\, v\, f(v_{\mathrm{rel}}, \boldsymbol{v})\, \mathrm{d}^3 v\ . \tag{11}$$

The quantity α is called a "rate coefficient" if $f(\boldsymbol{v})$ is an isotropic Maxwellian distribution characterized by a temperature T. In a colliding-beam experiment α has a slightly different meaning. Its definition is given by (11). In such experiments, v_{rel} and f have to be specified in order to provide an unambiguous physical quantity α.

For the particular case of merged electron and ion beams, e.g., in storage cooler rings, two velocity coordinates are commonly used to describe the electron-velocity distribution in the rest frame of the ions: $v_{\|}$, the velocity component in the electron-beam direction, and $v_{\perp}$, the velocity component perpendicular to the electron-beam direction. The energy (or velocity) spreads are therefore characterized by two corresponding temperatures $T_{\|}$ for the longitudinal and $T_{\perp}$ for the transverse direction. In the accelerated electron beam, these temperatures are quite different with $T_{\|} \ll T_{\perp}$, so that $f(\boldsymbol{v})$ is highly anisotropic. Its mathematical form is given by

$$f(v_{\mathrm{rel}}, \boldsymbol{v}) = \frac{m_{\mathrm{e}}}{2\pi k T_{\perp}} \exp\left(-\frac{m_{\mathrm{e}} v_{\perp}^2}{2kT_{\perp}}\right) \sqrt{\frac{m_{\mathrm{e}}}{2\pi k T_{\|}}} \exp\left(-\frac{m_{\mathrm{e}}(v_{\|} - v_{\mathrm{rel}})^2}{2kT_{\|}}\right), \tag{12}$$

where m_{e} denotes the electron rest mass. The quantity v_{rel} in this formula is the average longitudinal center-of-mass velocity

$$v_{\mathrm{rel}} = \frac{|\, v_{e,\|} - v_{i,\|}\,|}{1 + v_{i,\|} \cdot v_{e,\|} c_{\mathrm{o}}^{-2}}, \tag{13}$$

where c_{o} is the vacuum speed of light, and $v_{e,\|}$ and $v_{i,\|}$ are the longitudinal velocity components of the electron and ion beams, respectively. They are determined from

$$v_{e,\|} = c_{\mathrm{o}} \sqrt{1 - \left[1 + \frac{E_{\mathrm{e}}}{m_{\mathrm{e}} c_{\mathrm{o}}^2}\right]^{-2}} \tag{14}$$

and

$$v_{i,\|} = c_{\mathrm{o}} \sqrt{1 - \left[1 + \frac{E_{\mathrm{i}}}{m_{\mathrm{i}} c_{\mathrm{o}}^2}\right]^{-2}}\ . \tag{15}$$

The ion rest mass is represented by m_{i}. The energies E_{e} and E_{i} are determined by electron and ion acceleration voltages, respectively. The relative velocity v_{rel}, as defined by (13), can be different from the velocity v_{cm} in the electron–ion center-of-mass frame. This is especially true for low energies, where E_{cm} comes close to $kT_{\perp}$ and $kT_{\|}$. Since only E_{rel} is directly accessible to a measurement, experimental data are usually displayed as a function of the relative energy E_{rel}.

When $E_{\mathrm{rel}} \gg kT_{\perp}, kT_{\|}$ the velocity v_{rel} approaches v_{cm} and hence (11) can be rewritten as

$$\alpha(v_{\rm rel}) \approx v_{\rm rel} \int \sigma(v)\, f(v_{\rm rel}, \boldsymbol{v})\, d^3v = v_{\rm rel}\, \sigma_{\rm app}. \qquad (16)$$

The apparent cross section $\sigma_{\rm app}$ results from the convolution of the real cross section σ with the experimental velocity (or energy) distribution. This quantity makes physical sense when the width of the distribution function is much smaller than $v_{\rm rel}$ (or $E_{\rm rel}$, respectively). In this case apparent cross sections

$$\sigma_{\rm app} = \alpha(v_{\rm rel}) \,/\, v_{\rm rel} \qquad (17)$$

are used instead of rate coefficients α.

3. Dielectronic Recombination

Studies of DR have greatly profited from the development of very cold electron beams with temperatures $T_\perp$ and $T_\parallel$ reaching down as far as 1 meV and less than 0.1 meV, respectively. As a result, recombination resonances can be measured with superb energy resolution, particularly at low energies which are not accessible with techniques other than merged beams. The observation of radiation from EBIT devices, using high-resolution spectrometers, complements colliding-beam experiments on DR with respect to detailed assessment of intermediate multiply excited states and their decay processes. DR mea-

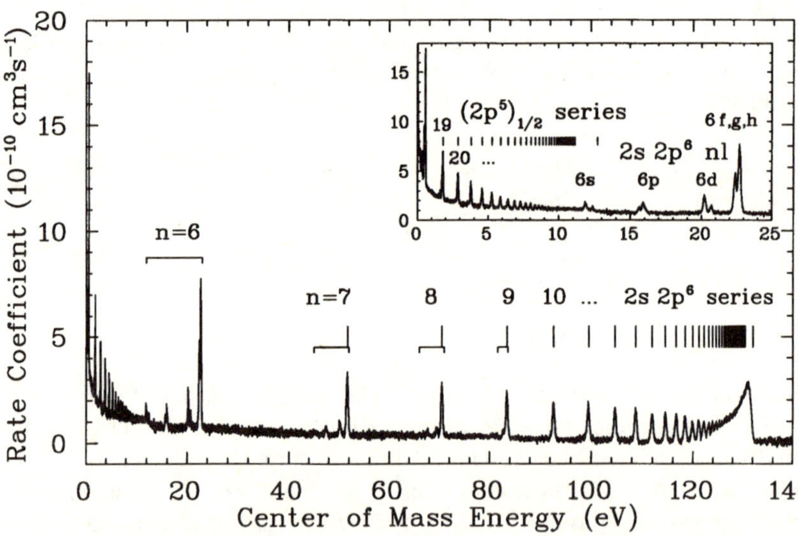

Fig. 1. Recombination rates of Fe^{17+} ions measured [20] at the *Test Storage Ring* (TSR) of the Max-Planck-Institute for Nuclear Physics in Heidelberg

surements have been concentrated mostly on ions with few active electrons. In particular, Li-like ions have been studied in a wide range of charge states from B^{2+} to U^{89+}. Due to their relevance in plasma and accelerator physics, detailed measurements are now also performed for more complex electron configurations. As an example, Fig. 1 shows measured recombination rates for Fe^{17+} ions from an experiment carried out at the heavy ion storage ring TSR (*Test Storage Ring*) of the Max-Planck-Institute for Nuclear Physics in Heidelberg. Different from previous expectations, the experiment showed the importance of M1 excitations of the $^2P_{3/2}$ core in DR. Hence, in the cold plasma of x-ray driven interstellar nebulae, the recombination rate can be much higher than previously calculated [20].

3.1 Influences of External Fields

An important issue in research on DR is the effect of external electric and magnetic fields. Previous measurements on Mg^+ [21] demonstrated a clear dependence of DR cross sections on known ambient electric fields in the collision region. Discrepancies by factors of 2–10 between theory and experiment [13] were partially removed [22, 23] by considering electric field mixing of angular momentum states.

Quantitative investigations of field effects on DR require controlled fields in the electron–ion collision region. Since an electrostatic external electric field would corrupt the energy definition of the electrons, motional electric

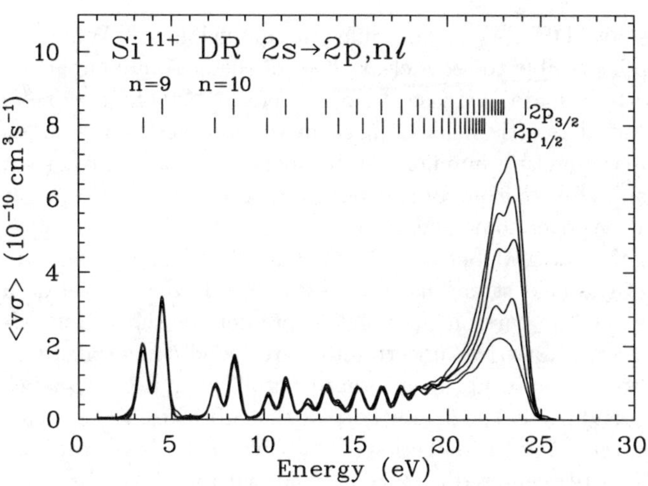

Fig. 2. Rates for DR of Si^{11+} ions measured [24] for five different external electric fields E_c = 0.0, 46.0, 91.5, 137.5, and 183.1 V/cm. For this comparison, the resonances were artificially broadened in order to simulate equal energy resolution of all spectra. The experiment clearly shows an increase of the rate coefficient with increasing E_c

fields $\boldsymbol{E}_c = \boldsymbol{v} \times \boldsymbol{B}$ are the best choice for those studies. External motional electric fields can be introduced in the cooler of a storage ring by tilting the axis of the longitudinal magnetic field that guides the electron beam. An accurately known angle between the ion beam and the magnetic field lines facilitates the precise knowledge of the size of the external electric field experienced by the ions in the collision region.

In an experiment at the CRYRING facility in Stockholm, absolute rates for DR of Li-like Si^{11+} ions were measured for different transverse magnetic fields [24]. The experiments covered the range of resonances characterized by the dielectronic capture processes $e+(1s^2\,2s) \rightarrow (1s^2\,2pn\ell)$ with $n = 9, 10, \ldots$ up to about 50.

The maximum Rydberg quantum number of contributing resonant states is limited by field stripping. As the recombined $Si^{10+}(1s^2 2sn\ell)$ ions travel toward the detector, they encounter the field of one of the bending magnets. The resulting Lorentz field ionizes Rydberg states with quantum numbers n greater than some cutoff value which can be expressed [21] as: $n_c = ((7.3 \times 10^8)q^3/F)^{1/4}$, where F is the field strength in V/cm. Thus, only recombined ions with $n < n_c$ would reach the detector. The simple formula gives $n_c = 39$, but, of course, there is radiative decay of Rydberg states in the (2.15 ± 0.51) m between the cooler (1 m interaction length) and the detector position, so that the formula for n_c is only indicative. Using hydrogenic Rydberg lifetimes to estimate the decay of Rydbergs to all lower states in the flight between the center of the cooler and the front edge of the dipole magnet, Bartsch et al. [24] obtained $n_c = 47 \pm 4$ (depending on where in the cooler the ions have recombined).

Figure 2 shows five DR spectra for different applied fields. While for $E_{cm} < 20$ eV it is not possible to see a clear effect of the field, a comparison of the spectra shows a dramatic increase of DR rates for $n \geq 20$ ($E_{cm} > 20$ eV) with increasing external electric field strength in the interaction region. In this range of principal quantum number, the Rydberg states are unresolved, and it is most meaningful for comparison purposes to consider the sum or integral of contributions over some energy range.

Figure 3 shows the integral between 20 eV and 25 eV of the rate coefficients for each of the field strengths E_c investigated. Experimental data are the solid points and the bars on the points represent the relative uncertainty of normalization. Statistical uncertainties are inconsequential. Since field ionization does not set in abruptly ("hard cutoff") for a given n but starts gradually, depending on the symmetries of the substates, a "soft cutoff" was adopted representing the detection probability of Rydberg states by a Fermi-function $f(n, \Delta n)$ centered at $n = n_c = 50$, with a width $\Delta n \approx 10$. The DR calculations are presented in Fig. 3 with a "soft cutoff" and a "hard cutoff" at $n_c = 50$ and, considering the uncertainty of the cutoff, also at $n_c = 45$ with a hard cutoff. The results for a hard cutoff at $n_c = 39$ (i.e., Rydberg states up to $n = 38$ are included) and $n_c = 60$ are also displayed.

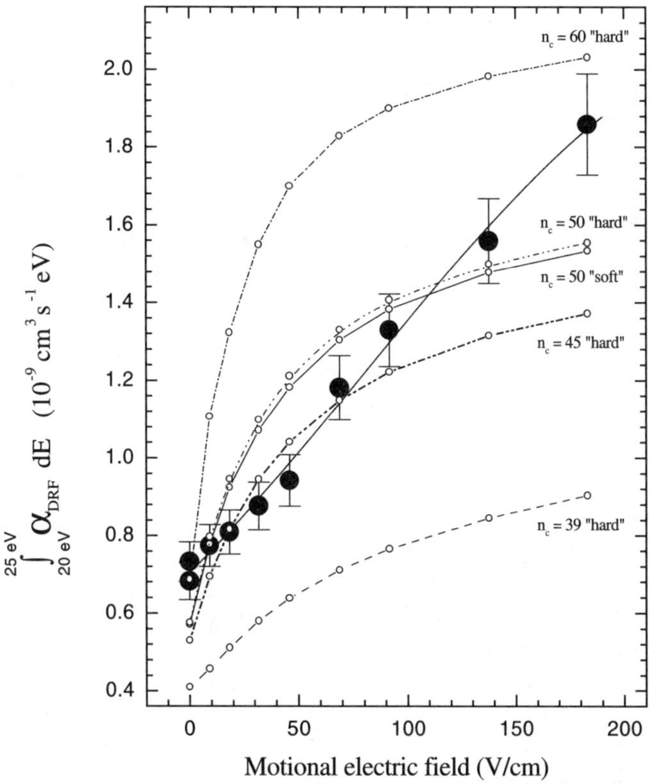

Fig. 3. Comparison between theory and experiment concerning field effects on DR of Li-like Si^{11+} ions. The rate coefficient of DR is integrated from 20 eV to 25 eV, and the values obtained are displayed vs corresponding electric field strengths E_c. Experimental data are black dots with error bars showing the relative uncertainty of normalization. Theory is given by open circles with connecting lines depending on the choice of cutoff (see text). From [24]

A striking feature of the data compared with theory is that the experimental integrated DR collision strength continues to climb at a roughly constant slope for all fields investigated, in conspicuous contrast to all the theoretical curves which are relatively flat beyond 50 V/cm, independent of whichever cutoff is supposed. A speculation was made to rationalize the puzzling disagreement between theory and experiment: the calculations are done for a static, purely electric field. In the experiment, the guiding magnetic field of 0.030 T is at right angles to the motional electric field found from Lorentz transforming the transverse magnetic field. This arrangement of crossed electric and magnetic fields in the ion frame may provide additional magnetic sublevel mixing and further enhancement of the cross section [25].

3.2 Spectroscopy and Lifetime Measurements

The measurement of DR with good energy resolution provides a unique new tool for the spectroscopy of intermediate doubly excited states of very highly charged ions. Experiments have been initiated aiming at precise information on the structure of high-Z ions in order to provide a sensitive test of QED calculations. For this purpose DR measurements have been recently performed with Au^{76+} ions. Within the center-of-mass energy range from 0 to 225 eV, two groups of resonances are found. One corresponds to a core excitation $2s_{1/2} \rightarrow 2p_{3/2}$ and capture of the incident electron into the $n = 6$ shell, leading to $1s^2 2p_{3/2} 6\ell_j$ doubly excited states. The second one manifests itself in a $2s_{1/2} \rightarrow 2p_{1/2}$ excitation and capture into Rydberg states with $n \geq 20$. The latter process is of special interest, since it renders possible extraction of the $2s_{1/2} - 2p_{1/2}$ splitting (Lamb shift) for Li-like heavy ions.

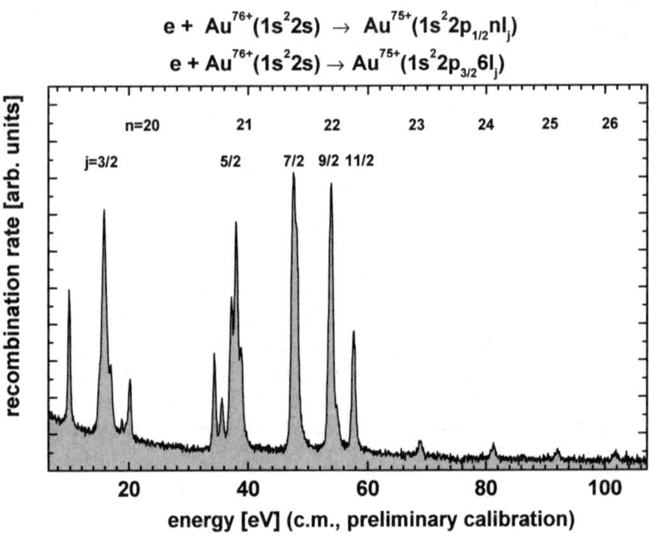

Fig. 4. Experimental rate coefficients for DR processes with Au^{76+} ions. The assignment of center-of-mass energies is preliminary [26]

The resonance condition for DR requires that the kinetic energy of the incoming electron plus its gain of binding energy must be equal to the excitation energy of the core electron. In particular, in the case of Li-like Au^{76+} ions, the resonance condition is fulfilled if the excitation $2s_{1/2} \rightarrow 2p_{1/2}$ is accompanied by the capture of the electron into Rydberg states with $n \geq 20$. With increasing energy one measures a whole series of Rydberg resonances, and the series limit ($n = \infty$) marks the point where the kinetic energy is equal to the core excitation energy, i.e., the $2s_{1/2} - 2p_{1/2}$ energy splitting. The main idea for the measurement is to determine precisely the position of all $1s^22p_{1/2}n\ell$ resonances and to extrapolate the related sequence of energies to the series limit. The potential of this method has been demonstrated in an experiment [27] where, even with moderate counting statistics and energy resolution, resonances up to $n = 26$ have been detected, and where the "Lamb shift" could be deduced with a statistical accuracy of 0.6 eV plus an additional systematic uncertainty of 0.5 eV. The new results were measured with much better counting statistics and energy resolution. Resonances up to n=41 and beyond have been observed [26]. The data have not been fully analyzed yet but promise to yield a significant improvement in the precision of measured level energies. Since one deals with a series of resonances, the redundancy in the measured sequence of energies minimizes the uncertainty of the measured $2s_{1/2} - 2p_{1/2}$ splitting further. Figure 4 shows the measured DR spectrum at energies between 7 and 107 eV (preliminary energy scale). It contains resonances of the $1s^22p_{1/2}n\ell$ sequence with $n = 20, 21, ..., 26$ and of the $1s^22p_{3/2}6\ell_j$ manifold with $j = 3/2, 5/2, 7/2, 9/2$, and $11/2$.

Storage rings have been extensively used also for lifetime measurements of excited atomic, molecular and nuclear states [2]. The time dependent observation of DR starting from excited long-lived parent states has provided high precision data for transition probabilities in few-electron atomic ions [28]. An example is the lifetime of (20.589 ± 0.045) ms of $C^{4+}(1s\,2s\ {}^3S)$. The technique used in such experiments is widely applicable. Experiments were recently also carried out at the TSR to determine the lifetime of metastable 2^3S He-like Li^+ [29]. The preliminary result of the analysis is (47.3 ± 2.9) s.

4. Non-Resonant Recombination

Unexpected high recombination rates at $E_{\rm rel} = 0$ eV were observed in merged-beam experiments with U^{28+} ions employing a cold dense electron target at the linear accelerator UNILAC of the GSI in Darmstadt [30]. Within an energy range between 0 and only about 1 meV the observed recombination rate dropped by a factor of 2 from a maximum of about 2×10^{-7} cm^{3}s^{-1}, while a realistic estimate of the RR rate yields approximately 1×10^{-9} cm^{3}s^{-1} [3]. It was immediately recognized that the lifetime of a stored cooled beam of U^{28+} ions would only be seconds considering the normal density of electron coolers, and yet this had not been widely noticed until after a series of experiments at the Low Energy Accumulator Ring LEAR at CERN in which Pb^{53+} ions were stored and cooled for preparation of further acceleration [31]. For $n_{\rm e} = 4.4\times10^7$ cm^{-3} the beam lifetime was only 2 s which is intolerable for efficient handling and acceleration of the ion beam.

While recombination rates for Pb^{53+} and its neighboring charge states were determined only indirectly from lifetime measurements at LEAR, recently a special effort was made at the TSR [33] to investigate recombination of Au^{50+} ions which are isoelectronic with Pb^{53+}. An extremely sharp peak was found at $E_{\rm rel} = 0$. Within approximately 4×10^{-4} eV the recombination

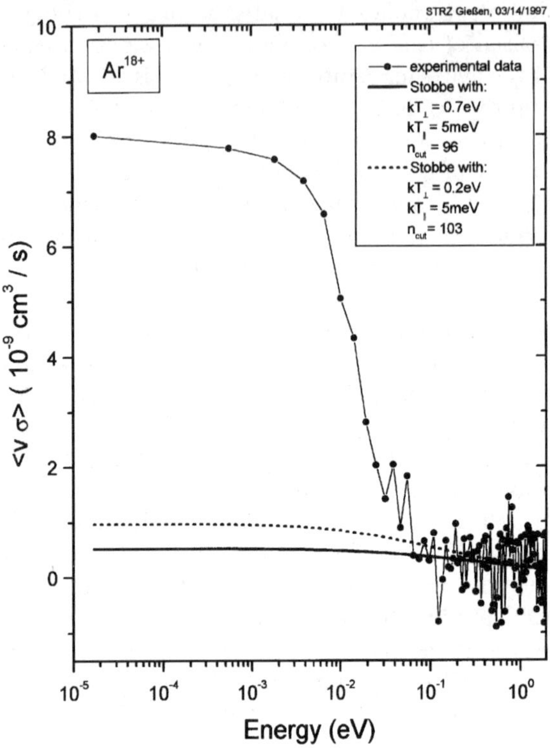

Fig. 5. Experimental and theoretical (Stobbe calculations) rate coefficients for recombination of Ar^{18+} ions. The curves are calculated RR rates obtained with different electron beam temperatures and Rydberg cutoffs [32]

rate drops from its maximum which is as high as $1.8 \times 10^{-6}\ \mathrm{cm^3\,s^{-1}}$ to half this value. This is the highest recombination rate coefficient ever observed in merged-beam experiments. The enhancement factor beyond the expectation of RR theory is about 60. Apart from the recombination maximum at $E_{\mathrm{rel}} = 0$ huge DR peaks were detected. Already at $E_{\mathrm{rel}} = 30$ meV a first resonance could be distinguished whose full width at half maximum is only about 15 meV. This is the smallest width ever observed in electron collision experiments with multiply charged ions. The DR resonances are so densely spaced in energy that one can easily imagine strong resonance contributions also to the peak at zero energy. Apparently, excessive recombination enhancement beyond the expectations of RR theory is most likely due to DR resonances close to zero energy. However, the phenomenon of enhanced recombination at low energies is also found with completely stripped ions which do not support DR (see, e.g., Ref. [3]). An example is shown in Fig. 5 where the recombination rate of completely stripped Ar^{18+} ions measured [32] at the

UNILAC is shown as a function of E_{rel}. The statistical uncertainties of the data above 1 meV are rather high. Below 1 meV, however, the measured rate exceeds the RR calculations by one order of magnitude. This phenomenon is not understood yet and needs further investigation.

5. Ionization of Ions by Electron Impact

Removal of electrons from atomic ions by incident free electrons has been a subject of intense research for almost 40 years. Access to increasingly higher ionic charge states was made possible by the development of advanced ion sources, ion traps and storage rings. By the observation of x rays from an EBIT and modeling of the charge state balance of trapped ions, cross sections for ionization of ions as highly charged as U^{91+} became available [34]. The use of energy scanning techniques was the key to very detailed measurements of structure in the ionization cross sections due to indirect processes [7]. These measurements were started in crossed-beam experiments and later transferred to storage ring measurements [35] with more highly charged ions such as Cl^{6+}, Fe^{15+} and Se^{23+}.

5.1 Direct Single Ionization

Although direct single ionization of atoms has been an issue of theoretical and experimental research since the beginning of this century, the case is not settled yet. The reason is the three body nature of the problem already in the simplest of these processes, the ionization of a hydrogenic ion. Clearly, studies on H-like ions provide the most straightforward access to DI. Crossed-beam experiments with electrons and ions started with the investigation of DI of H-like He^{+} [6]. Measurements of a similar nature were carried out with ions up to O^{7+} [36, 37]. The measured cross sections are in very good agreement with theory. Trap measurements with much more highly charged ions up to U^{91+} [34, 38] stimulated the development of the theoretical techniques towards inclusion of relativistic effects. In particular, the Breit interaction has been shown to be very important in high-Z collisions besides the Coulomb interaction and good agreement with experiments on H-like ions up to U^{91+} is obtained [39, 40]. However, serious discrepancies are found between the most advanced theoretical calculations and experimental DI cross sections of Li- and Na-like ions [41–43]. In the light of the perfect agreement in the case of H-like systems, the origin of that disagreement is presently not clear.

5.2 Indirect Ionization Processes

The possible importance of indirect ionization mechanisms was already being debated more than 30 years ago [44, 45]. For the astrophysically relevant Na-like Fe^{15+} ion strong contributions of EA were theoretically suggested. First measurements with Mg^+ ions [46], also Na-like, did not show evidence for the expected EA contributions. More elaborate calculations predicted smaller EA cross sections but suggested the increasingly stronger relative importance for ions in higher charge states [47, 48]. In particular, for Fe^{15+} ions EA was calculated to exceed DI. An even more striking prediction was made by *LaGattuta* and *Hahn* [49] 16 years ago. Their analysis of the ionization of Fe^{15+} suggested that REDA should be very important and could even dominate the total ionization cross section at certain energies.

It was quite a step in the development of experimental techniques that led to the first measurement of electron-impact ionization of Fe^{15+} ions by Gregory et al. [50], six years after the first theoretical predictions of resonant contributions to electron-impact ionization of ions. Intensity and background problems, however, did not permit the gathering of unambiguous evidence for the REDA mechanism. The origin of the excessive background was traced back to ion-beam contaminations by metastable autoionizing Fe^{15+} ions with $2p^5 3s3p$ configurations which can survive the μs flight to the collision region.

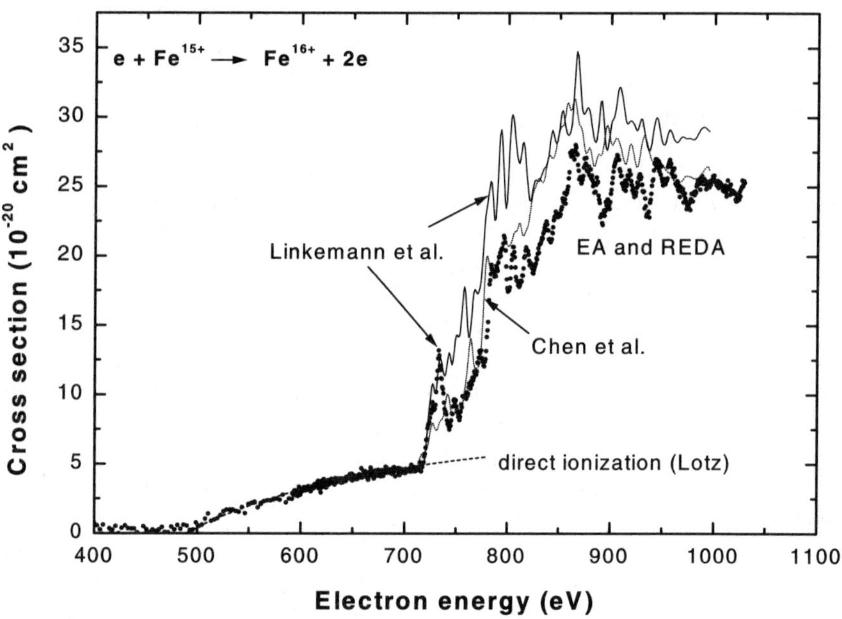

Fig. 6. Cross sections for electron-impact ionization of Fe^{15+} ions. The experimental data [35] are indicated by solid dots. The solid and dotted lines are the results of extensive theoretical calculations [35, 51]. DI was estimated from the Lotz formula [52]

With the advent of heavy ion storage rings, an entirely new approach to the problem has become feasible. The two main drawbacks of previous measurements with multiply charged Na-like ions – the low signal rate and the high background due to decaying metastable ions – can be circumvented by using a storage ring equipped with an electron cooler. Thus, even within short storage times of the order of only 1 s before the measurement, practically all metastable ions in the injected beam decay, which avoids the background problems inherent in the previous small-scale colliding-beam experiments. Furthermore, signal rates are substantially enhanced as compared to conventional experiments. Therefore, the Heidelberg storage ring TSR was used to measure electron-impact ionization of Fe^{15+} ions stored at a beam energy of 300 MeV [35].

The experimental results are shown in Fig. 6 by the solid dots. Relative uncertainties of these measurements are about a factor 40 lower than those of the previous work [50] and the density of the data points clearly reveals the rich structure caused by EA steps and REDA peaks in the cross section function. Between the threshold of direct outer shell ionization at 489 eV and the onset of EA processes at about 710 eV, the ionization cross section rises to about 5×10^{-20} cm^2. The Lotz formula [52], which can only describe DI, is in excellent agreement with the measured cross section in this energy range. Above 710 eV, the cross section rises by about a factor 5 within a narrow energy span of little more than 100 eV. Obviously, the indirect ionization mechanisms dominate the ionization cross section of Fe^{15+} ions at energies above 800 eV.

6. Summary and Outlook

Collisions of ions with free electrons lead to interesting fundamental interactions. They play a significant role in all kinds of plasmas and therefore have vast applications in plasma diagnostics and plasma modeling. Due to their fundamental character they provide an ideal testing ground for our understanding of atomic structures, transition probabilities and collision dynamics. In particular, DR can be employed for collisional spectroscopy of multiply excited states. Extrapolation of Rydberg resonance energies to their series limit can even provide precise excitation thresholds for singly excited states and thus open new pathways towards high precision tests of QED calculations of level energies in very highly charged few-electron systems. The physics of electron–ion recombination also involves unique interesting phenomena, such as interference of different recombination channels or different but overlapping resonances of equal symmetry. Effects of external fields on DR provide an intriguing field of research. Surprising results are obtained at the very low relative energies which can now be accessed by the merged-beam technique in storage rings using the cold electron beams of the cooling devices. At least for energies up to about 1 keV the storage cooler rings provide the method of choice for studies of electron–ion recombination. These studies will be further extended in the near future.

Electron impact ionization studies at storage rings are more difficult. Only very selected ion species will be studied using rings. The extension of accessible electron–ion center-of-mass energies is required for that. A big step in the potential and the quality of measurements will be possible when additional, dedicated electron targets in the rings become available.

Acknowledgement. The author acknowledges very fruitful interaction with numerous colleagues and co-workers who have made possible the work to which this overview is related. Support by the Gesellschaft für Schwerionenforschung (GSI), Darmstadt, by the Max-Planck-Institut für Kernphysik, Heidelberg, by the Bundesministerium für Erziehung, Wissenschaft, Forschung und Technologie (BMBF), Bonn, and by the Human Capital and Mobility Programme of the European Community is gratefully acknowledged.

References

1. M. Inokuti (ed.): Adv. At. Mol. Opt. Phys. **33,** (1994)
2. A. Müller, A. Wolf: in *Accelerator-Based Atomic Physics Techniques and Applications*, S. Shafroth, J. Austin, eds. (AIP, New York 1997), pp. 147–182
3. A. Müller, A. Wolf: in Proceedings of the International Workshop on *Antimatter Gravity and Antihydrogen Spectroscopy*, Sepino, Italy, May 1996, M. H. Holzscheiter, ed., Hyp. Int. **109,** 233 (1997)
4. F. Brouillard and J. W. McGowan (eds.): *Physics of Ion–Ion and Electron–Ion Collisions* (Plenum, New York 1983)
5. F. Brouillard (ed.): *Atomic Processes in Electron–Ion and Ion–Ion Collisions*, (Plenum, New York 1986)
6. K.T. Dolder, M.F.A. Harrison, P.C. Thonemann: Proc. Roy. Soc. **A 264**, 367 (1961)
7. A. Müller: in *Physics of Ion Impact Phenomena*, D. Mathur, ed. (Springer, Berlin, Heidelberg 1991), pp. 13–90
8. D.L. Moores, K.J. Reed: Adv. At. Mol. Opt. Phys. **34,** 301 (1994)
9. R.A. Phaneuf: in *The Physics of Electronic and Atomic Collisions*, T. Andersen, B. Fastrup, F. Folkmann, H. Knudsen, N. Andersen, eds., AIP Conference Proceedings **295** (AIP, New York 1993), pp. 405–414
10. M.E. Bannister, Y.-S. Chung, N. Djurić, B. Wallbank, O. Woitke, S. Zhou, A.C.H. Smith: in *Photonic, Electronic and Atomic Collisions*, Proc. 20th Int'l. Conf. on The Physics of Electronic and Atomic Collisions, Vienna, Austria, 23–29 July 1997, F. Aumayr, H. Winter, eds. (World Scientific, Singapore 1998), pp. 335–338
11. C. Bélenger, P. Defrance, R. Friedlein, C. Guet, D. Jalabert, M. Maurel, C. Ristori, J. C. Rocco, B.A. Huber: J. Phys. B **29,** 4443 (1996)
12. I.D. Williams: in *Photonic, Electronic and Atomic Collisions*, Proc. 20th Int'l Conf. on The Physics of Electronic and Atomic Collisions, Vienna, Austria, 23–29 July 1997, F. Aumayr, H. Winter, eds. (World Scientific, Singapore 1998), pp. 313–322
13. G.H. Dunn: in *Recombination of Atomic Ions*, W. G. Graham, W. Fritsch, Y. Hahn, J.H. Tanis, eds., NATO ASI Series B: Physics Vol. 296, (Plenum, New York 1992), pp. 115–132

14. D. Auerbach, R. Cacak, R. Caudano, T.D. Gaily, C.J. Keyser, J.W. McGowan, J.B.A. Mitchell, S.F.J. Wilk: J. Phys. B **10**, 3797 (1977)
15. W.G. Graham, W. Fritsch, Y. Hahn, J.H. Tanis (eds.): *Recombination of Atomic Ions*, NATO ASI Series B: Physics Vol. 296, (Plenum, New York, 1992)
16. T.J.M. Zouros, D.-H. Lee: in *Accelerator-Based Atomic Physics Techniques and Applications*, S. Shafroth and J. Austin, eds. (AIP, New York 1997), pp. 427–479
17. A. Müller: Nucl. Instrum. Methods. B **124**, 281 (1997)
18. J.S. Wang, H.R. Griem, R. Hess, W.L. Rowan: Phys. Rev. A **38**, 4761 (1988)
19. M.P. Stöckli: in *Accelerator-Based Atomic Physics Techniques and Applications*, S. Shafroth and J. Austin, eds. (AIP, New York 1997), pp. 67–116
20. D.W. Savin, T. Bartsch, M.H. Chen, S.M. Kahn, D.A. Liedahl, J. Linkemann, A. Müller, M. Schmitt, D. Schwalm, A. Wolf: ApJ. Letters **489**, L115 (1997)
21. A. Müller, D.S. Belić, B.D. DePaola, N. Djurić, G.H. Dunn, D. Mueller, C. Timmer: Phys. Rev. Lett. **56** 127 (1986); Phys. Rev. A **36**, 599 (1987)
22. K. LaGattuta, I. Nasser, Y. Hahn: Phys. Rev. A **33**, 2782 (1986)
23. C. Bottcher, D.C. Griffin, M.S. Pindzola: Phys. Rev. A **34**, 860 (1986)
24. T. Bartsch, A. Müller, W. Spies, J. Linkemann, H. Danared, D.R. De Witt, H. Gao, W. Zong, R. Schuch, A. Wolf, G.H. Dunn, M.S. Pindzola, D.C. Griffin: Phys. Rev. Lett. **79**, 2233 (1997)
25. F. Robicheaux, M.S. Pindzola: Phys. Rev. Lett. **79**, 2237 (1997)
26. C. Brandau, F. Bosch, G. Dunn, B. Franzke, A. Hoffknecht, C. Kozhuharov, P.H. Mokler, A. Müller, F. Nolden, S. Schippers, Z. Stachura, M. Steck, T. Stöhlker, T. Winkler, A. Wolf: Hyp. Int. **114**, 45 (1998)
27. W. Spies, O. Uwira, A. Müller, J. Linkemann, L. Empacher, A. Frank, C. Kozhuharov, P. H. Mokler, F. Bosch, O. Klepper, B. Franzke, M. Steck: Nucl. Instrum. Methods. B **98,** 158 (1995)
28. H.T. Schmidt, P. Forck, M. Grieser, D. Habs, J. Kenntner, G. Miersch, R. Repnow, U. Schramm, T. Schüßler, D. Schwalm, A. Wolf: Phys. Rev. Lett. **72**, 1616 (1994)
29. A. Saghiri, J. Linkemann, M. Schmitt, D. Schwalm, A. Wolf, T. Bartsch, A. Hoffknecht, A. Müller, W.G. Graham, J. Tanis, N. Badnell, A. Price: in *Proceedings of the XXth International Conference on the Physics of Electronic and Atomic Collisions*, Vienna, Austria, 23.07. – 29.07.1997, Abstracts of Contributed Papers, TU 143
30. A. Müller, S. Schennach, M. Wagner, J. Haselbauer, O.Uwira, W. Spies, E. Jennewein, R. Becker, M. Kleinod, U. Pröbstel, N. Angert, J. Klabunde, P.H. Mokler, P. Spädtke, B. Wolf: Phys. Scr. **T37**, 62 (1991)
31. S. Baird, J. Bosser, C. Carli, M. Chanel, P. Lefèvre, R. Ley, R. Maccaferri, S. Maury, I. Meshkov, D. Möhl, G. Molinari, F. Motsch, H. Mulder, G. Tranquille, F. Varenne: Phys. Lett. B **361**, 184 (1995)
32. O. Uwira, A. Müller, W. Spies, A. Frank, J. Linkemann, C. Brandau, T. Cramer, C. Kozhuharov, J. Klabunde, N. Angert, P.H. Mokler, R. Becker, M. Kleinod, N.R. Badnell: Hyp. Int. **108**, 167 (1997)
33. O. Uwira, A. Müller, J. Linkemann, T. Bartsch, C. Brandau, M. Schmitt, A. Wolf, D. Schwalm, R. Schuch, W. Zong, H. Lebius, W.G. Graham, J. Doerfert, D.W. Savin: Hyp. Int. **108**, 149 (1997)
34. R.E. Marrs, S.R. Elliott, D.A. Knapp: Phys. Rev. Lett. **72**, 4082 (1994)
35. J. Linkemann, A. Müller, J. Kenntner, D. Habs, D. Schwalm, A. Wolf, N.R. Badnell, M.S. Pindzola: Phys. Rev. Lett. **74**, 4173 (1995)
36. K. Tinschert, A. Müller, G. Hofmann, K. Huber, E. Salzborn, R. Becker, D.C. Gregory: J. Phys. B **22**, 531 (1989)

37. K. Aichele, U. Hartenfeller, D. Hathiramani, G. Hofmann, F. Melchert, V. Schäfer, M. Steidl, M. Stenke, R. Völpel, E. Salzborn: Phys. Scr. **T73**, 125 (1997)
38. R.E. Marrs, S.R. Elliott, J.H. Scofield: Phys. Rev. A **56**, 1338 (1997)
39. C.J. Fontes, D.H. Sampson, H.L. Zhang: Phys. Rev. A **51**, R12 (1995)
40. D.L. Moores, K. J. Reed: Phys. Rev. A **51**, R9 (1995)
41. M.S. Pindzola, F. Robicheaux, N.R. Badnell, T.W. Gorczyca: Phys. Rev. A **56** 1994 (1997)
42. K. Bartschat, I. Bray: J. Phys. B **30**, L109 (1997)
43. N.R. Badnell, M.S. Pindzola, I. Bray, D.C. Griffin: J. Phys. **31**, 911 (1998)
44. L. Goldberg, A.K. Dupree, J.W. Allen: Ann. D'Astrophysique **28**, 589 (1965)
45. O. Bely: J. Phys. B **1**, 23 (1968)
46. S.O. Martin, B. Peart, K.T. Dolder: J. Phys. B **1**, 537 (1968)
47. D.L. Moores, H. Nussbaumer: J. Phys. B **3**, 161 (1970)
48. D.C. Griffin, C. Bottcher, M.S. Pindzola: Phys. Rev. A **25**, 154 (1981)
49. K.J. LaGattuta, Y. Hahn: Phys. Rev. A **24**, 2273 (1981)
50. D.C. Gregory, L.-J. Wang, F.W. Meyer, K. Rinn: Phys. Rev. A **35**, 3256 (1987)
51. M.H. Chen, K.J. Reed, D.L. Moores: Phys. Rev. Lett. **64**, 1350 (1990)
52. W. Lotz: Z. Phys. **216**, 241 (1968)

XIII Excitation of Atoms by Multiply Charged Ions

Ratko K. Janev

Summary. A review of recent studies of excitation processes in collisions of atoms with multiply charged heavy ions is presented. Special attention is given to the excitation mechanisms in various energy regions and to the scaling properties of the cross sections with respect to the ionic charge and electron transition energy. Problems related to the further understanding of excitation mechanisms in the low-energy region are discussed.

1. Introduction

The interest in excitation processes in collisions of atoms with multiply charged ions A^{q+} ($q \geq 2$) has significantly increased during the last 10–15 years due to the needs of fusion energy research [1–3]. Energetic neutral hydrogen, helium or lithium beams are currently injected into fusion plasmas either for heating or diagnostic purposes and the attenuation dynamics of these beams is strongly affected by the heavy-particle excitation processes. For neutral beam energies of interest to fusion applications (ten to several hundred keV/u), the collision times of inelastic heavy-particle processes are of the same order of magnitude as the radiative decay times of excited beam atoms, or smaller. Therefore, multistep collisional processes (e.g., upward excitation cascade followed by an electron removal process) may significantly contribute to the beam stopping cross section [4, 5]. The importance of excitation (as well as electron removal) processes of beam atoms with multicharged plasma impurity ions derives from the strong (linear or higher) dependence of the excitation cross section with the ionic charge.

The scientific interest for the study of excitation processes in atom–multicharged ion collisions lies in understanding the excitation mechanisms, particularly in the low and intermediate energy regions where many states of the colliding system are strongly coupled [6–8]. When the target atom contains many electrons, the complexity of the collision dynamics increases dramatically both because of the multitude of competing processes (including multielectron transition processes) and because of the effects of electronic correlations.

Most of the experimental studies of atom–multicharged ion excitation collisions have so far been performed with hydrogen [9], helium [10–12] and lithium [13] target atoms. The theoretical investigations of atom–multicharged ion excitation processes have been much more numerous and are partially reviewed in [6, 7, 14]. In the present chapter, we shall give an overview of the recent work on atom excitation by multiply charged ions with emphasis on the general results related to the excitation mechanisms and the cross section scaling properties. The presentation will be limited to one-electron excitation and in theoretical discussions the target will be treated in the one-electron approximation, while the ion will be considered structureless.

Atomic units ($m_e = \hbar = e = 1$) will be used throughout in the text, unless explicitly stated otherwise.

2. Theoretical Models for Excitation Processes

As is well known, there is no a unique theoretical method which has the potential to describe adequately the heavy-particle collision dynamics in the entire space of dynamical and interaction parameters of the system. For the inelastic heavy-particle collision processes involving multiply charged ions, such as the excitation of atoms

$$A^{q+} + B(\alpha) \rightarrow A^{q+} + B^*(\beta), \;\; q \geq 2, \tag{1}$$

where α and β are the sets of quantum numbers describing the state of the active electron before and after the collision, respectively, the collision dynamics is, roughly speaking, strongly influenced by the values (and their mutual ratios) of the collision energy, transition energy and the strength of interparticle interactions in the reaction zone. For the excitation process (1), a suitable combination of dynamical and interaction parameters is the *reduced* velocity parameter [15]

$$\xi = v/(\omega q)^{1/2}, \tag{2}$$

where v is the relative collision velocity, ω is the transition energy and q is the ionic charge. If $\xi \gg 1$, a perturbational approach can be applied, while for $\xi \ll 1$ the process can be considered in the framework of an adiabatic approach to heavy-particle collision dynamics. In the region $\xi \sim 1$, where the electron transition takes place most effectively, neither of these approaches is applicable and the use of the close-coupling method is the only appropriate quantum-mechanical alternative for description of the excitation process. Because of the large number of states involved in the process when $\xi \sim 1$, the classical dynamics description also proved to be fairly successful in describing the excitation process.

We shall now outline the basic ideas of the approximate methods used in the studies of the process (1) and review the results of some of the recent theoretical investigations.

2.1 Adiabatic Region

In the parametric region defined by the inequality $\xi \ll 1$, the most appropriate description of the collisional dynamics is the close coupling of molecular states of the collision system. The close-coupling method based on a molecular orbital (MO) expansion of the total electron wave function [16, 17] has, however, not been so far extensively applied for studies of the excitation processes involving multiply charged ions. One of the reasons is the small population of the excitation channels on the background of very intense electron capture, particularly when q is large. The associated numerical difficulties of close coupling calculations are also enormous. For instance, the calculation of $1s \to 2l$ excitation in the $Li^{3+} + H(1s)$ system needs inclusion of all states having the principal quantum numbers in the united atom limit $N \leq 7$, as well as some of the states with $N = 8$. This gives a basis of about 150 molecular orbitals, which imposes formidable difficulties in the solution of the corresponding system of coupled equations for the transition amplitudes. Large-scale MO close-coupling cross-section calculations have been performed only for the $1s \to 2s, 2p$ excitation of hydrogen by He^{2+} impact [18].

A more practical framework for treating the excitation processes (1) in the adiabatic region is provided by the advanced adiabatic method [19] which describes the evolution of the system in the complex plane of internuclear distance R and includes all the radial and rotational transitions within the group of molecular states with $N \leq N_0$, where N_0 is the principal quantum number of the highest molecular state included in the basis. The radial transitions between the molecular states take place in the regions $\mathrm{Re}R \simeq \mathrm{Re}R_c$, where R_c is a point in the complex plane at which the electron wave function has a singularity and the corresponding adiabatic energy surface has a square-root branch point (*hidden crossing* of the potential energy surfaces of the coupled states). The advanced adiabatic method (or the Hidden Crossing Theory – HCT) has been used for excitation of ground-state hydrogen atoms to 2ℓ and 3ℓ states in slow collisions with He^{2+} [20], Li^{3+} [21] and Be^{4+} and B^{5+} [22] ions.

The *hidden-crossing* approach provides a transparent picture of the excitation mechanism in the adiabatic region. During the outgoing part of the trajectory (in a semiclassical picture) the system, initially being in the separated atomic state $n_0\ell_0m_0$, follows "superseries" of Q-type transition points [19] until it reaches the molecular states which at $R \to \infty$ correlates to the Stark states associated with the final state $n\ell m\,(m = m_0)$. Assuming that only one Q-superseries of transitions gives a dominant contribution to the $n_0\ell_0 m \to n\ell m$ transition, the excitation probability is proportional to the product of elementary transition probabilities at each of the traversed Q-type transition points, $R^{(Q)}_{c,k}$

$$P_{n_0 n} \sim \prod_{k=n_0}^{n} \exp\left(-\frac{2}{v}\Delta^{(Q)}_{\mathrm{k}}(b)\right) = \exp\left(-\frac{2}{v}\sum_{k=n_0}^{n}\Delta^{(Q)}_{\mathrm{k}}(b)\right), \tag{3}$$

where, for simplicity, we have dropped the lm quantum numbers, $\Delta^{(Q)}_{\mathrm{k}}(b)$ is the generalized Massey parameter and b is the impact parameter. For $q \gg 1$ and $n \ll n_0$ (i.e., the number of Q-type transition points involved in the

$n_0 \to n$ transition is large), the sum in (3) can be replaced by an integral and with the available approximate expressions for $\Delta_{\mathrm{k}}^{(Q)}(b)$ [23], one obtains for the leading term of $P_{n_0 n}$ (to within an accuracy $O(n_0^2/n^2)$) [24]

$$P_{\mathrm{n_0 n}} \sim \exp\left[-a\frac{(q\omega_{\mathrm{n_0 n}})^{1/2}}{v}\left(1+\frac{b^2}{2|R_{\mathrm{n,c}}^{(Q)}|^2}\right)\right], \quad a = \text{constant}, \tag{4}$$

where $R_{\mathrm{n,c}}^{(Q)}$ is the outermost point of the Q-superseries. After integration over the impact parameters (up to $b = \mathrm{Re}R_{\mathrm{n,c}}^{(Q)} \sim q^{1/2}n^2$) the excitation cross section in the adiabatic region is obtained in the form [24]

$$\sigma_{\mathrm{n_0 n}}^{(ad)} \sim qn^4 \frac{v}{(q\omega_{\mathrm{n_0 n}})^{1/2}}\chi(v,q,n_0,n)\ \exp\left(-a(q\omega_{\mathrm{n_0 n}})^{1/2}/v\right), \tag{5}$$

where $\chi(v,q,n_0,n)$ is a slowly varying function which originates from the unspecified pre-exponential factor in (3). The expression (5) provides the correct dominant dependence of the excitation cross section on the reduced velocity $\xi = v/(q\omega_{\mathrm{n_0 n}})^{1/2}$.

If during the incoming part of the trajectory, another Q-type superseries of transitions is excited (e.g., via a rotational transition in the united atom region) that leads to population of another Stark state of the $n\ell$-manifold, then the interference of these two reaction pathways may produce an oscillatory structure in the excitation cross section. Such a structure has been indeed seen in the MO close-coupling cross section calculations for the $1s \to 2p$ excitation of hydrogen by He^{2+} impact [18] and parts of it even observed in experiment [25, 26].

2.2 The Region of High-ξ Parameters

In the region $\xi \gg 1$, a perturbational approach to the excitation problem should be applicable. In this region, the electron excitation process can be considered as being caused by the ion–atom interaction

$$V(\mathbf{r},\mathbf{R}) = -\mathbf{q}/|\mathbf{R(t)} - \mathbf{r}|, \tag{6}$$

where $\mathbf{r}$ is the radius vector of the optical electron with respect to the center of the atom and $\mathbf{R}$ is the internuclear distance. The standard perturbational approach to excitation assumes that the interaction (6) is small with respect to the kinetic energy of the relative motion. For collisions involving highly charged ions, this condition is satisfied only for very high collision energies (where relativistic effects may also become important) and the first term of the perturbational (Born) series is insufficient to provide an adequate description of the process in the high-energy region attainable in most of the experiments. Inclusion of higher-order terms in the Born series is, therefore, required to describe the atomic excitation by highly charged ions. Among the methods that include higher-order effects in the description of the excitation process, those based on the distorted-wave approach have proved to be the most effective. From those that introduce distortion in the initial and final channel, the symmetric-eikonal (SE) model [27] is the simplest one which still provides a very good description of the process. The SE model has been applied to the excitation of the hydrogen atom ($n_0 \to n$ transitions, $n_0, n =$ 1–4, $n > n_0$) by fully stripped ions with charges in the range $q =$ 2–26 [28]. Similar cross section calculations have been recently performed also for the He target atoms [29].

It is important to note that when using the perturbational approach to the excitation problem in atom–multicharged ion collisions, special care has to be devoted to the choice of the expansion parameter. For instance, the use of v_{e}/v as a small parameter in the theory (v_{e} being the bound-electron velocity in the initial state) could lead to incorrect results in the higher orders of the perturbational series for high values of the ionic charge or the transition energy $\omega_{n_0 n}$. Thus, the second-order (in v_{e}/v) perturbational theory for excitation (derived from the Schwinger variational principle, [30]) predicts in the high energy region a saturation of the excitation cross section when q increases to very high values, in contrast to the q^2-dependence given in the first Born approximation. This is inconsistent with the fact that the perturbational series for direct processes (such as excitation) is always convergent, provided the expansion parameter is properly chosen. As shown in [24], an appropriate expansion parameter for the excitation problem, at least within the dipole approximation, is the inverse reduced velocity ξ^{-1}. For $(q\omega_{n_0 n})^{1/2} \sim v/v_{\mathrm{e}}, \xi^{-1}$ is of the order of unity, and the perturbational treatment becomes invalid independent of the value of v/v_{e}. A perturbation theory approach to the excitation problem in terms of the parameter ξ^{-1} with the exact interaction potential (6) has still not been formulated. This, however, can be done in the dipole approximation for $V(\mathbf{r}, \mathbf{R})$, which for an $n_0 \to n$ dipole allowed transition gives the following form of the excitation cross section in the first-order (Born) approximation [6, 15]

$$\sigma_{n_0 n}^{(B)} \sim \frac{q f_{n_0 n}}{\omega_{n_0 n}^2 \xi^2} \ln\left(\frac{c\xi^2}{\lambda}\right), \; \lambda = \left(\frac{f_{n_0 n}}{2\omega_{n_0 n}}\right)^{1/2}, \; c = \text{constant}, \tag{7}$$

where $f_{n_0 n}$ is the oscillator strength for the $n_0 \to n$ transition.

2.3 The Region of Intermediate ξ Parameters

The excitation process in the $\xi \sim 1$ region involves strong coupling of many electronic states and the quantum-mechanical description of the process should be based on a coupled-channel formalism. In this region, the expansion of the electron wave function over traveling atomic orbitals (AO), centered on both the target and projectile nuclei (to account for the effects of electron capture channels), provides a good basis for the description of the process. The expansion can further be augmented by suitably chosen pseudostates to describe the effects of the continuum (AO^+ basis) or the molecular effects (triple-centre expansion). Theoretically, if the expansion basis is complete, then the target AO are sufficient for description of the process (single-center expansion).

Extensive single-center AO close-coupling calculations have been performed for the $1s \to n$ ($n = 2$–4) and $n_0 \to n$ ($n_0 = 2,3$, $n > n_0$) transitions in the hydrogen atom in collisions with fully stripped ions with charge in the range $q = 2$–26 [28]. More elaborate two-center expansion AO close-coupling calculations have been performed for the $1s \to 2\ell, 3\ell$ transitions in He^{2+} + $H(1s)$ [31] Li^{3+} + $H(1s)$ [32] and Be^{4+}, C^{6+} + H $(1s)$ [14] collisions. Similar types of calculations were recently performed for the $1s \to 2\ell, 3\ell$ transitions in He^{2+} + He collisions [33] and for the $2s \to 2p$ transition in He^{2+}, Be^{4+}, C^{6+} + $Li(1s^2 2s)$ collision systems [34]. For a limited number of collision velocities, the $1s \to 2p$ excitation of H and He by highly charged ions ($q > 3$) has also been treated within the multistate AO close-coupling method [35–39].

It is interesting to note that for the $n_0 s \to np$ transitions, the coupled equations for the transition amplitudes can be solved analytically in the dipole approximation for the potential (6) [15]. This solution shows that within this approximation (termed DCC), the $n_0 s \to np$ excitation cross section exhibits distinct scaling properties with respect to the ionic charge, transition energy and the oscillator strength. The reduced velocity parameter ξ naturally appears in the theory and the cross section (in certain reduced form) is expressed as a function of ξ only. In the region of high ξ, the DCC cross-section behavior is given by (7), while in the adiabatic region ($\xi \ll 1$) it behaves as

$$\sigma_{\mathrm{DCC}}^{(ad)} \sim \frac{q f_{\mathrm{n_0 n}}}{\omega_{\mathrm{n_0 n}}^2 \xi^2} \exp\left(-a\frac{\lambda^{1/2}}{\xi}\right), \quad a = constant. \tag{8}$$

While the pre-exponential ξ-dependence of $\sigma_{\mathrm{DCC}}^{(ad)}$ appears to be too strong (cf. (5)), its exponential dependence on ξ is correct. The question of the scaling of the excitation cross section will be discussed in more detail in the next section.

The large number of coupled states involved in the excitation process in the $\xi \sim 1$ region justifies the use of the classical dynamics for description of the process. In the quasi-classical limit, the motion of a classical particle can be represented by a wave packet (in terms of the *Ehrenfest theorem*) formed by superposition of a large number of discrete levels within a narrow energy range of the spectrum. The averaging over the quantum numbers in this energy range leads to the disappearance of the logarithmic cross section energy dependence associated with the dipole matrix elements. The relation between the classical and quantum descriptions of the excitation process has been extensively discussed elsewhere [40]. In this reference the earlier applications of classical dynamics to inelastic atomic processes are also described. Among the most frequently used classical methods for the calculation of inelastic atomic collision cross sections are the classical impulse (CI) approximation and the numerical integration of the equations of motion (the so-called Classical Trajectory Monte Carlo (CTMC) method [41]).

The CTMC method has been applied to calculate the $1s \to n\,(n = 2\text{–}4)$ and $n_0 \to n\,(n_0 = 2, 3, n > n_0)$ excitation cross sections in collisions of hydrogen atoms $H(n_0)$ with multiply charged ions with charge in the range $q =$ 2–26 [28]. These calculations have shown good agreement of CTMC cross sections with those of the single-center AO close-coupling method in the energy region of the cross section maximum and with the results of the symmetrized eikonal approximation at higher energies. In the region of high energies, the CTMC results become identical with those of the classical impulse (or binary encounter) approximation. For an $n_0 \to n$ transition, the CI approximation gives the following expression of the excitation cross section [42, 43]

$$\sigma_{\mathrm{CI}} = \gamma \frac{q^2 n_0^4 n(7n^2 - 3n_0^2)}{v^2(n^2 - n_0^2)^3}, \quad \gamma = \frac{8\pi}{3}. \tag{9}$$

In terms of the variable ξ ($2\omega_{n_0 n} = n_0^{-2} - n^{-2}$) and for $(n_0/n)^2 \ll 1$, (9) becomes

$$\sigma_{\mathrm{CI}} \simeq \gamma' \frac{q}{n_0^2 n^3 \omega_{n_0 n}^4} \frac{1}{\xi^2}, \quad \gamma' = \frac{7\pi}{3}. \tag{10}$$

We note that the classical impulse approximation allows us to obtain an approximate analytic expression also for the cross section of $n_0\ell_0 \to n\ell$ transitions in one-electron systems [43]. The σ_{CI} $(n_0\ell_0 \to n\ell)$ becomes a maximum for $n = n_0$, when ℓ_0 and ℓ are fixed, and for $\ell = \ell_0$, when n_0 and n are fixed. σ_{CI} $(n_0\ell_0 \to n\ell)$ decreases from these maximum values as $|n_0 - n|^{-4}$ and $|\ell_0 - \ell|^{-4}$, respectively.

3. Scaling Properties of the Excitation Cross Section

The extensive systematic experimental studies of the $1^1S \to n^1P(n = 2\text{–}5)$ transitions in H induced by collisions with He^{2+}, Si^{q+} $(q = 2\text{–}9)$ and Cu^{q+} $(q = 3\text{–}11)$ [9] and for the transitions $1^1S \to n^1L$ $(n = 2 - 4, L = S, P, D)$ in He induced by a large variety of multicharged ions of various species with $2 \le q \le 45$ [10–12] have demonstrated that in the investigated energy range (from $\sim$ 15q keV/u up to $\sim$ 200q keV/u), the measured excitation cross sections obey (within their experimental uncertainty of 20–30%) the scaling relationship

$$\frac{\sigma}{q} = f(E/q), \tag{11}$$

predicted earlier by the DCC theory [15] for the $S \to P$ transitions. The multistate close-coupling calculations performed more recently for the 2^1P excitation of $H(1^1S)$ [14, 35–37, 39] and $3^1P, 4^1S, 4^1P$ and 4^1D excitation of He (1^1S) [38] have also confirmed the cross section scaling (11) in the energy region above 15–20q keV/u. The scaling relationship (11) is a direct consequence of the dipole approximation for the potential (6) [36] but follows also from classical dynamics considerations [35]. The physical basis of the scaling relationship (11) lies in the fact that for multiply charged ions, the distant collisions give a dominant contribution to the excitation process, which justifies the use of the dipole potential approximation.

The DCC model indicates that, at least for the dipole allowed transition, the excitation cross section has scaling properties also with respect to the transition energy, ω_{n_0n}, and the oscillator strength. A detailed analysis of the scaling properties of excitation cross sections for both dipole-allowed and dipole-forbidden transitions has been performed recently [24]. This analysis shows that to an accuracy consistent with the uncertainties of existing experimental results, the excitation cross section in the energy region above $\sim 15q$keV/u can be represented in the scaled form

$$\tilde{\sigma} = F(\tilde{E}), \ \tilde{\sigma} = g(q, \omega_{n_0n}, \eta)\sigma, \ \tilde{E} = h(\xi, \zeta), \tag{12}$$

where g and h are some functions of the indicated variables, and η and ζ collectively designate other parameters of the system (such as the oscillator strength, effective charge of the target ionic core, etc.). The theoretical models discussed in the preceeding section all show that, at least in the asymptotic regions, the function $F(\tilde{E})$ can be determined uniquely. The DCC theory shows that the scaling function $F(\tilde{E})$ exists also in the intermediate range of ξ, which is confirmed also by the experimental data [24]. The determination of the functions $g(q, \omega_{n_0n}, \eta)$ and $h(\xi, \zeta)$ is discussed below for the dipole-allowed and dipole-forbidden transitions separately.

3.1 Dipole-Allowed Transitions

The expressions (7) and (8) for the asymptotic behavior of the excitation cross section obtained by the DCC theory give the following expressions for the functions g and h:

$$g_{\mathrm{DCC}} = \frac{\omega_{n_0n}^2 \lambda}{q f_{n_0n}}, \ h_{\mathrm{DCC}} = \xi^2/\lambda. \tag{13}$$

The DCC approximation is not expected to be valid in the adiabatic region $\xi \ll 1$ where the process involves many coupled molecular states.

From the more accurate result (5) of the hidden crossing theory it follows that $h_{\mathrm{HCT}} = \xi^2$, i.e. that one should set λ=1 in the expression h_{DCC}. This redefinition of h_{DCC} leads automatically to the disappearance of λ from the expression for g_{DCC} (cf. (7) and (8)). Therefore, at least for low and high values of ξ, the scaling cross section and energy parameters for the dipole-allowed transitions are

$$g_0 = \frac{\omega_{n_0n}^2}{q f_{n_0n}}, \ h_0 = \xi^2 = \frac{v^2}{q\omega_{n_0n}}. \tag{14}$$

The scaled exprimental cross sections $\tilde{\sigma} = g_0\sigma$ vs scaled energy $\tilde{E} = 25h_0$ keV/u for the $1^1S \to n^1P$ transitions in H and He are shown in Figs. 1 and 2, respectively [24].

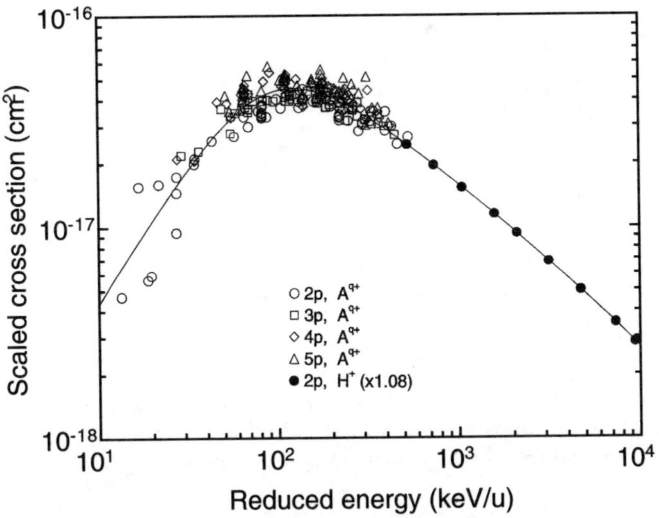

Fig. 1. Scaled excitation cross sections for $1s \to np$ ($n = 2$–5) transitions in atomic hydrogen induced by multicharged ion impact (open symbols, Ref. [9]) and proton impact (closed symbols, Ref. [28]). The solid curve is the best fit to the data

The open symbols in Fig. 1 are the data for multiply charged ions [9], while the filled circles are the SE data for proton impact [28]. The experimental data for the He target shown in Fig. 2 are taken from Refs. [10–12] (for the multicharged ions) and Ref. [44] (for protons). The solid lines in these figures represent best fits of the data, which for small and large values of $\tilde{E}$ behave in accordance with the theoretical predictions, (5) and (7), respectively [24].

3.2 Dipole-Forbidden Transitions

The hidden crossing theory result (5) for the adiabatic excitation cross section is valid for both dipole-allowed and dipole-forbidden transitions and, therefore, suggests the form of the function $h(\xi, \zeta)$ in the form given by (14). Because of the uncertainties involved in the pre-exponential factor in (5), one can determine the form of the scaling function $g(q, \omega_{n_0n}, \eta)$ from the cross section expression (10) of the classical impulse approximation and assume that its validity can be extended in the adiabatic region. This assumption has been verified a posteriori [24]. The scaling function g and h for the dipole-forbidden transitions are, thus, taken in the form

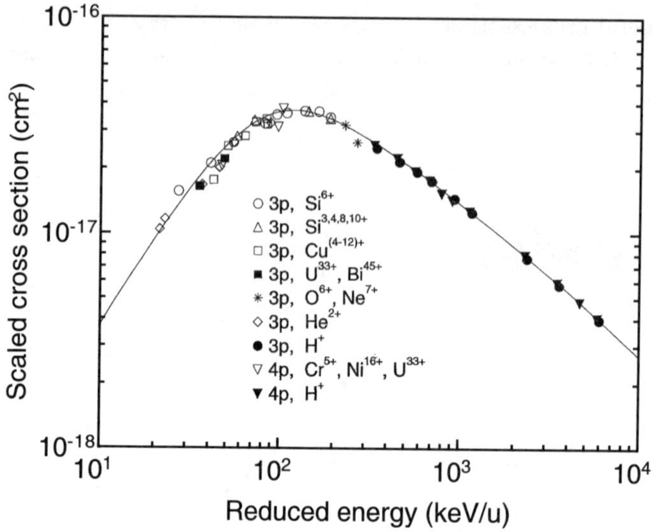

Fig. 2. Scaled excitation cross sections for $1^1S \to n^1P$ ($n = 3, 4$) transitions in He. The data for multicharged ion impact are from Refs. [10–12] and those for proton impact are from Ref. [44]. The solid curve is the best fit to the data

$$g_{\rm CI} = \frac{n_0^2 n^3 \omega_{n_0 n}^4}{q}, \quad h_{\rm CI} = h_0 = \xi^2 = \frac{v^2}{q\omega_{n_0 n}}. \tag{15}$$

It should be noted that the expression (10) for $\sigma_{\rm CI}$ is obtained by summing over the orbital momentum quantum numbers of the initial and final states. Therefore, the scaling function g_{CI} should be valid only for the $n_0 S \to nS$ transitions. However, one can expect that it remains approximately valid also for the $n_0 S \to nL$ ($L \neq 0, 1$) transitions, provided L is not too large.

In Figs. 3 and 4 the experimental cross sections for the $1^1S \to n^1S$ and $1^1S \to n^1D$ transitions in He, scaled in accordance with (15) ($n_0 = 1, \tilde{E} = 25\xi^2$ keV/u), are shown. The open symbols represent the data for impact by multiply charged ions [10–12], while the closed symbols are the proton impact data [44]. The lines represent the best fits of the data and have asymptotic behavior in agreement with (5) and (10). It is apparent from Fig. 3 that the data for excitation of the 4^1S state departs from the general curve for reduced collision energies above $\tilde{E} \simeq 80$ keV/u. In the adiabatic energy region ($\tilde{E} \ll \tilde{E}_{\rm m}$ ~50-60 keV/u), the reduced cross section $\tilde{\sigma}(^1D)$ shows a weak n-dependence which originates from the undefined pre-exponential factor in (5).

3.3 The q Dependence of Excitation Cross Sections for Fixed Energy

The scaling relationships (14) and (15), particularly the energy scaling function $h_0 = v^2/q\omega_{n_0 n}$, provide a basis for resolving the so-called *saturation* prob-

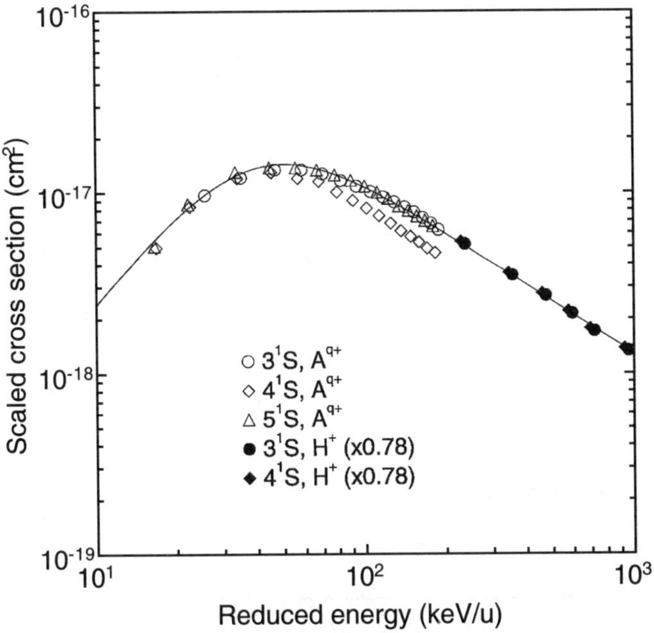

Fig. 3. Scaled excitation cross sections for $1^1S \to n^1S$ ($n = 3$–5) transitions in He induced by multicharged ions (open symbols, Refs. [10–12]) and protons (closed symbols, Ref. [44])

lem of the excitation cross section with increasing q for a fixed collision energy [30, 37, 38]. The problem has originally been formulated within the second-order perturbational approach (using the Schwinger variational principle) with v_e/v taken as a small parameter. As discussed in Sects. 2.2, 3.1 and 3.2, and demonstrated by Figs. 1–4, only the reduced velocity $\xi = v/q(\omega_{\mathrm{n_0n}})^{1/2}$ appears to be an appropriate dynamical variable for the excitation problem. When varying either of the parameters v, q and $\omega_{\mathrm{n_0n}}$ and keeping the other two fixed, the value of ξ can be varied from $\xi \ll 1$ to $\xi \gg 1$, passing through the region $\xi \simeq \xi_{\mathrm{m}} \sim 1$ where the reduced cross section $\tilde{\sigma}$ takes its maximum. For the dipole-allowed $1^1S \to n^1P$ transitions in H and He, for instance, the value of ξ_{m} is about $\xi_m \simeq 2.3$ (see Figs. 1 and 2). When the collision and transition energies are fixed, the scaled cross section $\tilde{\sigma}$ (as well as σ) as a function of q passes through a maximum at certain $q = q_{\mathrm{m}}$. Therefore, with increasing q, the excitation cross section for a given transition and at a given energy first increases, passes through a maximum and then decreases. An example is given in Fig. 5 for the H($1s \to 2p$) excitation cross section shown as a function of q for different collision energies. The open and solid symbols in this figure are calculated results taken from Ref. [24] (with normalization to the experiment) and from Ref. [37], respectively.

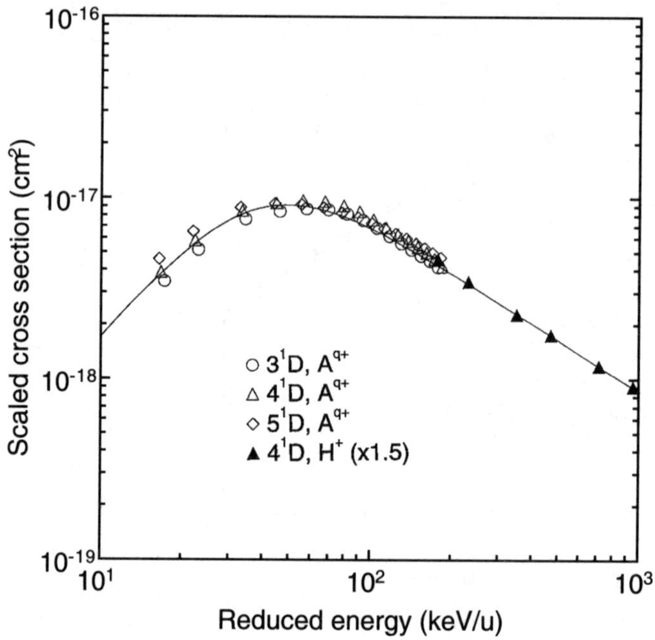

Fig. 4. Scaled excitation cross sections for $1^1S \rightarrow n^1D$ (n = 3–5) transitions in He induced by multicharged ions (open symbols, Refs. [10–12]) and protons (closed symbols, Ref. [44])

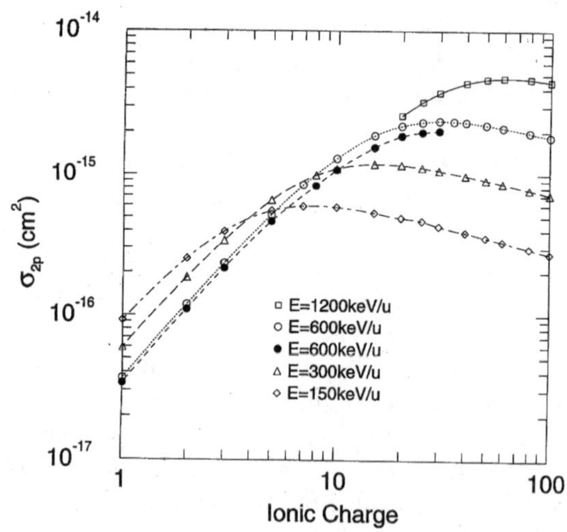

Fig. 5. Ion charge dependence of $1s \rightarrow 2p$ excitation cross section of H. Open and closed symbols are from Ref. [24] and Ref. [37], respectively

4. Conclusions

We have presented the results of recent studies of excitation processes in collisions of atoms with multiply charged ions. Emphasis was given to the excitation mechanisms in various collision energy regions and the associated theoretical methods for description of the processes. The experimental cross-section measurements for this class of processes have been performed mainly with the H and He targets and the corresponding results are shown in Figs. 1–4 in scaled form. Despite the fact that the excitation mechanisms differ considerably in various regions of the collision energy and other interaction parameters of the collision system, the excitation cross section shows distinct scaling properties for reduced energies $\tilde{E} = 25v^2/(q\omega_{\mathrm{n_0n}})$ keV/u greater than about 15–20 keV/u. In the adiabatic region ($\tilde{E} \leq 15$ keV/u) the excitation process may involve strong coupling of many molecular states (particularly when q is large), most of which are associated with the electron capture channels. The interference of various reaction paths, as well as the coupling of excitation and electron capture channels, make the collision dynamics in this region very complex. For better understanding of the physics of excitation processes in the adiabatic region, sophisticated MO-based close-coupling calculations and new, dedicated experimental efforts are highly desirable.

References

1. A. Boileau, M. von Hellermann, L.D. Horton, H.P. Summers: Plasma Phys. Control Fusion **31**, 779 (1989)
2. R.K. Janev: Comments At. Mol. Phys. **26**, 83 (1991)
3. J. Schweinzer, E. Wolfrum, F. Aumayr, M. Pöckl, H.P. Winter, R.P. Schorn, E. Hintz, A. Unterreiter: Plasma Phys. Control. Fusion **34**, 1173 (1192)
4. R.K. Janev, C.D. Boley, D.E. Post: Nucl. Fusion **29**, 2125 (1989)
5. A.A. Korotkov, R.K. Janev: Phys. Plasma **3**, 1512 (1996)
6. R.K. Janev, L.P. Presnyakov, V.P. Shevelko: *Physics of Highly Charged Ions* (Springer, Berlin 1985)
7. W. Fritsch: in *Review of Fundamental Processes and Applications of Atoms and Ions*, ed. C.D. Lin (World Scientific, Singapore 1993)
8. B.H. Bransden, M.R.C. McDowell: *Charge Exchange and the Theory of Ion–Atom Collisions* (Oxford Univ. Press, Oxford 1992)
9. D. Detleffsen, M. Anton, A. Werner, K.-H. Schartner: J. Phys. B **27**, 4195 (1994)
10. K. Reyman, K.-H.Schartner, B. Sommer: Phys. Rev. A **38**, 2290 (1988)
11. M. Anton, D. Detleffsen, K.-H. Schartner: Nucl. Fusion. Suppl. **3**, 51 (1992)
12. M. Anton, D. Detleffsen, K.-H. Schartner, A. Werner: J. Phys. B **26**, 2005 (1993)
13. R. Brandenburg, F. Aumayr, HP. Winter: unpublished (1997)
14. W. Fritsch: in *Proceedings of the VIth International Conference on the Physics of Highly Charged Ions*, P. Richard and C.D. Lin, eds. (AIP, New York 1993)
15. R.K. Janev, L.P. Presnyakov: J. Phys. B **13**, 4233 (1980)
16. M. Kimura, N.F. Lane: Adv. At. Mol. Phys. **26**, 79 (1990)

17. W. Fritsch, C.D. Lin: Phys. Rep. **202**, 1 (1991)
18. L.F. Errea, C. Harel, H. Jouin, L. Mendez, B. Pons, A. Riera: J. Phys. B **27**, 3603 (1994)
19. E.A. Solov'ev: Sov. Phys. Usp. **32**, 228 (1989)
20. P.S. Krstic, R.K. Janev: Phys. Rev. A **47**, 3894 (1993)
21. R.K. Janev, E.A. Solov'ev, Y. Wang: J. Phys. B **29**, 2497 (1996)
22. P.S. Krstic, M. Radmilovic, R.K. Janev: Nucl. Fusion Suppl. **3**, 113 (1992)
23. M. Pieksma, S.Yu. Ovchinnikov: J. Phys. B **24**, 2699 (1991)
24. R.K. Janev: Phys. Rev. A **53**, 219 (1996)
25. M.P. Hughes, J. Geddes, H.B. Gilbody: J. Phys. B **27**, 1143 (1994)
26. D.P. Higgins, J. Geddes, R.W. McCullough, H.B. Gilbody: Nucl. Instrum. Methods B **103**, 120 (1995)
27. G.R. Deco, P.D. Fainstein, R.D. Rivarola: J. Phys. B **19**, 213 (1987)
28. C.O. Reinhold, R.E. Olson, W. Fritsch: Phys. Rev. A **41**, 4837 (1990)
29. L. Gulyás, P.D. Fainstein: unpublished (1997)
30. B. Brendlé, R. Gayet, J.P. Rozet, K. Wohrer: Phys. Rev. Lett. **54**, 2007 (1985)
31. W. Fritsch, R. Shingal, C.D. Lin: Phys. Rev. A **44**, 5686 (1991)
32. B.H. Bransden, C.J. Noble: J. Phys. B **15**, 451 (1982)
33. W. Fritsch: Nucl. Fusion Suppl. **3**, 41 (1992)
34. J. Schweinzer, D. Wutte, HP. Winter: J. Phys. B **27**, 137 (1994)
35. W. Fritsch, K.-H. Schartner: Phys. Lett. **126**A, 17 (1987)
36. V.D. Rodriguez, J.E. Miraglia: J. Phys. B **25**, 2037 (1992)
37. V.D. Rodriguez, A. Salin: J. Phys. B **25**, L467 (1992)
38. F. Martin, A. Salin: J. Phys. B **28**, 671 (1995)
39. A. Macias, F. Martin, A. Riera, J.L. Sanz: Phys. Rev. A **53**, 2869 (1996)
40. I.C. Percival, D. Richards: Adv. At. Mol. Phys. **11**, 2 (1975)
41. R.E. Olson, A. Salop: Phys. Rev. A **16**, 531 (1977)
42. G.I. Ivanovski, E.A. Solov'ev: Sov. Phys.-JETP **77**, 887 (1993)
43. G.I. Ivanovski, R.K. Janev, E.A. Solov'ev: J. Phys. B **28**, 4799 (1995)
44. F.J. de Heer, R. Hoekstra, H.-P. Summers: Nucl. Fusion Suppl. **3**, 47 (1992)

XIV Projectile Ionization and Capture Reactions in Ion–Atom Collisions

Viatcheslav P. Shevelko, D. Böhne, B. Franzke and Th. Stöhlker

Summary. Projectile charge-changing processes – electron capture and ionization – are reviewed for collisions of fast, many-electron positive ions with neutral atoms and molecules at relative energies of 1–500 MeV/u. Experimental data and theoretical methods for estimation of the corresponding charge-changing cross sections in this energy range are discussed. Recently measured beam lifetimes of stored heavy ions are compared with theoretical predictions.

1. Introduction

Charge-changing processes arising in collisions of fast positive ions with atoms, ions and molecules play a key role in many problems related with physics of atomic collisions [1–4], plasma fusion [5], physics of stored ions [6–8], physics of the Earth's magnetosphere [9] and others.

Investigations of charge-changing processes in ion–atom collisions provide not only a test for the theory of atomic collisions dynamics but are also of general importance for the operation of heavy ion accelerators and storage rings. For example, the intensity of the low-charged ion beams in a storage ring or a synchrotron depends critically on the projectile charge-changing cross sections which define the lifetimes of stored ions.

In this chapter, the charge-changing processes of fast, many-electron positive ions colliding with neutral atoms are considered from theoretical and experimental viewpoints. Special attention is paid to the estimation of the beam lifetimes of stored ions in accelerator devices such as a storage ring or a synchrotron, in terms of the projectile charge-changing cross sections.

2. Basic Processes: Electron Capture and Ionization

Electron capture and ionization processes of targets and projectiles arising in ion–atom collisions have been considered in many papers (see, e.g., review articles [1–12]). The bibliography on electron transfer processes updated 1997 is given in [13].

A fast positive ion colliding with an atom or a molecule can lose or gain its charge via two main processes: by *ionization* (stripping) of its own electrons

$$X^{m+} + A \to X^{(m+a)+} + \dots \qquad a \geq 1\,, \tag{1}$$

or by *capture* of the target electrons

$$X^{m+} + A \to X^{(m-b)+} + \qquad \dots \qquad b \geq 1. \tag{2}$$

Here X^{m+} is the projectile ion and A is the target atom or molecule. The total cross section σ_{tot} for changing the projectile charge is defined by a sum of the cross sections for the corresponding single and multielectron ionization and electron-capture processes

$$\sigma_{\text{tot}} = \sum_{a \geq 1} \sigma_{\text{ion}}^{(a)} + \sum_{b \geq 1} \sigma_{\text{ec}}^{(b)}. \tag{3}$$

As a rule, at high projectile energies one-electron processes ($a = b = 1$) give the main contribution to the sum (3). However, under certain collision conditions, the multielectron processes can contribute up to 30% to the total cross sections (see, e.g., [14, 15]).

In general, projectile ionization and capture cross sections have different dependences on the atomic parameters such as the relative velocity v, the projectile charge Z_{P} and the target nuclear charge Z_{T}. At high relative velocities $v/Z_{\text{P}} > 1$ a.u. (1 a.u. of velocity $= 2.2 \times 10^8$ cm/s), cross sections of both processes have the following scaling:

$$\sigma_{\text{ion}} \propto Z_{\text{T}}^2/(Z_{\text{P}}^4 v^2), \qquad \sigma_{\text{ec}} \propto Z_{\text{T}}^5 Z_{\text{P}}^5 / v^{11}. \tag{4}$$

A typical behavior of the cross sections for different charge-changing processes is shown in Fig. 1 for O^{7+} + He collisions.

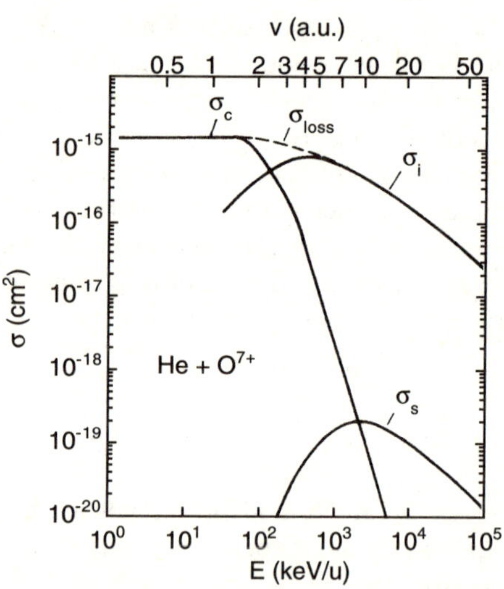

Fig. 1. Comparison of the electron capture (σ_{c}), target ionization (σ_{i}) and projectile stripping (σ_{s}) cross sections for O^{7+} + He collisions. From [16]

3. Theory

In this section, a brief description of the theoretical approximations used for practical estimation of the experimental cross sections in the high-energy (but nonrelativistic) region is given.

3.1 Projectile Ionization

For calculation of the projectile-ionization cross sections, the nonrelativistic Born approximation (Plane-Wave Born Approximation, PWBA) is used [17–19]

$$\sigma_{\mathrm{ion}} = \frac{8\pi a_0^2}{v^2} \sum_{\mathrm{P}} \int_0^\infty \mathrm{d}\epsilon \sum_{\mathrm{i(T)}} \int_{q_{\min}}^\infty \frac{\mathrm{d}q}{q^3} |F_{\mathrm{P}}(q)|^2 |F_{\mathrm{T}}(q)|^2, \tag{5}$$

where q is the momentum transfer, $F_{\mathrm{P,T}}(q)$ are the projectile and target form factors and ϵ is the energy of the ejected electron. The sum over P runs over all electrons of the projectile and the sum over $i(T)$ represents all final states i of the target atoms including the continuum (target ionization). The minimum momentum transfer is given by

$$q_{\min} = (I_{\mathrm{P}} + \epsilon + \Delta E_{\mathrm{T}})/v, \tag{6}$$

where I_{P} is the binding energy of the projectile electron and ΔE_{T} is the excitation energy of the target electron.

The form factors $F_{\mathrm{P,T}}(q)$ are given by

$$|F_{\mathrm{P}}(q)|^2 = |\langle\epsilon| \exp(\mathrm{i}\boldsymbol{qr})|0\rangle_{\mathrm{P}}|^2, \tag{7}$$

$$|F_{\mathrm{T}}(q)|^2 = |Z_{\mathrm{T}}\delta_{io} - \sum_{j=1}^{N_{\mathrm{T}}} \langle i| \exp(\mathrm{i}\boldsymbol{qr}_j)|0\rangle_j|^2, \tag{8}$$

where $|0\rangle_{\mathrm{P}}$ and $|0\rangle_{\mathrm{j}}$ denote the ground-state wave functions of the projectile and of the j-th target electron, and N_{T} is the total number of target electrons. For a neutral target, $N_{\mathrm{T}} = Z_{\mathrm{T}}$. The form factor $|F_{\mathrm{T}}(q)|^2 = Z_{\mathrm{T}}^2$ for a bare nucleus as a target, and for targets with $1s$-electrons, one has [19]

$$F_{1s,1s}(q) = Z_{\mathrm{T}} - N_{\mathrm{T}} \frac{1}{[1 + (q/2d)^2]^2}, \tag{9}$$

where d is the effective atomic number reflecting the screening of the target nucleus by the $1s$ electron. For a hydrogen target $Z_{\mathrm{T}} = N_{\mathrm{T}} = 1$, $d = 1$ and for helium $Z_{\mathrm{T}} = N_{\mathrm{T}} = 2$, $d = 1.69$. The form factors for arbitrary atoms are tabulated in [20].

Calculations of the cross sections using (5) are quite complicated even for H-like targets because the lower limit $q_{\min}$ depends on the energy of the final state i of the target and the energy of the ejected electron. If the target excitation energy is much smaller than the projectile binding energy (electrons are at rest), i.e.,

$$\Delta E_{\rm T} \ll I_{\rm P} + \epsilon, \tag{10}$$

then one can simplify the calculation and use the so-called *closure* approximation introduced in [21, 22]. Under this condition, the summation over the target ground and excited states i with the momentum transfer integration in (5) can be interchanged, and one has

$$\begin{aligned}
\sigma_{\rm ion}(v) &= \frac{8\pi a_0^2}{v^2}\sum_{\rm P}\int_0^\infty {\rm d}\epsilon \sum_{\rm T}\int_{q_{\rm min}}^{\infty}\frac{{\rm d}q}{q^3}|F_{\rm P}(q)|^2 S_{\rm T}(q), && (11)\\
S_{\rm T}(q) &= S_{\rm s}(q) + S_{\rm a}(q)\\
&= [Z_{\rm T} - \sum_j^{N_{\rm T}} F_{j{\rm T}}(q)]^2 + [N_{\rm T} - \sum_j^{N_{\rm T}} |F_{j{\rm T}}(q)|^2], && (12)\\
F_{j{\rm T}}(q) &= \langle i|\exp({\rm i}\boldsymbol{qr})|0\rangle_{\rm j}, \qquad q_{\rm min} = (I_{\rm P}+\epsilon)/v. && (13)
\end{aligned}$$

Physically, the two terms in (12) are interpreted as ionization by a screened target nucleus (first term, $S_{\rm s}$, screening) and the ionization by the N_T target electrons (second term, $S_{\rm a}$, antiscreening). The interaction between target and projectile electrons can increase or decrease the total ionization cross section. At small momentum transfers $q \to 0$ (large impact parameters R), one has $F_{j{\rm T}} \to 1$, and at high-momentum transfers $q \to \infty$ (small impact parameters), $F_{j{\rm T}} \to 0$. Respectively, the effective perturbing charge vanishes, $S_{\rm T}(q) \to 0$ at $q \to 0$, and has its maximum value at large momentum transfers

$$S_{\rm max}(q) \to Z_{\rm T}^2 + Z_{\rm T}, \qquad q \to \infty. \tag{14}$$

Therefore, the ionization cross section of the projectile by a neutral target atom at high energies is roughly proportional to $Z_{\rm T}^2 + Z_{\rm T}$, where $Z_{\rm T}^2$ is the contribution due to the target nucleus and $Z_{\rm T}$ comes from the number of target electrons.

The main contribution to the stripping cross sections is given by close collisions. It follows from the fact that the ionization probability decreases as $P(R) \propto R^{-2}$ [23] but (14) also requires a condition $P(R) \ll 1$ when the Born approximation is valid. The quantum mechanical estimation of the impact parameters at which collisions occur with a large probability is given by [24]

$$R < \hbar c_{\rm o}/(I^2 + 2Im_{\rm e}c_{\rm o}^2)^{1/2}, \tag{15}$$

where I is the binding energy of the projectile electron, $c_{\rm o}$ is the vacuum speed of light and $m_{\rm e}$ is the electron rest mass. For light projectiles ($Z_{\rm P} \ll Z_{\rm T}$) and intermediate projectile velocities, the screening effects are quite large and the stripping cross sections are roughly proportional to $Z_{\rm T}^{2/3}$ (see [11]).

Taking into account (11)–(14), the projectile-ionization cross section can be presented in the form

$$\begin{aligned}
\sigma_{\rm ion}(v) &= Z_{\rm T}^2\sigma_{\rm p}(v) + Z_{\rm T}\sigma_{\rm e}(v), && (16)\\
&\quad \sigma_{\rm p}(v) \approx \sigma_{\rm e}(v), \quad v > I^{1/2}, \quad Z_{\rm P} \gg Z_{\rm T},
\end{aligned}$$

where $\sigma_{\rm p}(v)$ and $\sigma_{\rm e}(v)$ are proton- and electron-impact ionization cross sections of the projectile. Equation (16) represents the *Bohr approximation* [25] valid for relative velocities v much higher than the orbital electron velocities when the target nucleus and electrons are treated as being at rest (eq. (10)), and when the interference effects between the target electrons and the nucleus may be neglected. Equation (16) is used for estimation of the ionization cross sections at high relative velocities provided that the projectile size is much smaller than the size of the K shell of the target atom (see [24]).

Another (ad hoc) approximation was suggested in [22] which saves the q-dependence of the effective perturbing charge and the threshold behavior of the electron-ionization cross section $\sigma_{\rm e}(v)$ in the form

$$S_{\rm T}(q) = [Z_{\rm T} - \sum_{j}^{N_{\rm T}} F_{j\rm T}(q)]^2 + [N_{j\rm T} - \sum_{j}^{N_{\rm T}} |F_{j\rm T}(q)|^2][\sigma_{\rm e}(v)/\sigma_{\rm p}(v)]. \tag{17}$$

3.2 One-Electron Capture

In the high-velocity range, the one-electron capture cross section has the following structure (the nonrelativistic Brinkman–Kramers approximation, see [2]):

$$\sigma_{\rm ec}(v) = \frac{8\pi a_0^2}{v^2} \sum_{\rm P}\sum_{\rm T} \int_{q_{\rm min}}^{\infty} q{\rm d}q\, |F_{\rm P}(q)|^2 |F_{\rm T}(q)|^2, \tag{18}$$

with

$$F_{\rm P}(q) = \int \Psi_{\rm P}(\boldsymbol{r}){\rm e}^{{\rm i}\boldsymbol{qr}}{\rm d}\boldsymbol{r}, \quad F_{\rm T}(q) = \int \Psi_{\rm T}(\boldsymbol{r})V{\rm e}^{-{\rm i}\boldsymbol{q}'\boldsymbol{r}}{\rm d}\boldsymbol{r}, \tag{19}$$

$$q_{\rm min} = \left|\frac{\Delta E}{v} + \frac{v}{2}\right|, \quad q^2 - q'^2 = 2\Delta E, \tag{20}$$

where $\Delta E = E_{\rm T} - E_{\rm P}$ is the energy difference or *energy defect*, and $\Psi(\boldsymbol{r})$ are the wave functions of the electron which makes a transition. The sum over P runs over projectile excited states and the sum over T means summation over all target electrons.

This representation corresponds to nonradiative electron capture (NRC) and is valid at fast, but not too high collision velocities. Although this simple first-order approximation is known to be in disagreement with experimental data by up to an order of magnitude, it already provides the correct cross-section scaling dependence, i.e.

$$\sigma_{\rm NRC} \sim Z_{\rm P}^5 Z_{\rm T}^5 / v^{11}. \tag{21}$$

A more refined theoretical treatment of the NRC process can be obtained within the Eikonal approximation which is described in detail in Refs. [26, 27]. The strong scaling of the NRC cross section with the nuclear charge of the target and the projectile implies that at not too high collision velocities this capture process is by far the dominant one in collisions of high-Z projectiles with medium- or high-Z atoms.

For practical purposes, the one-electron capture cross sections $\sigma_{\rm ec}$ are usually estimated by applying the semiempirical formula of *Schlachter* et al. [28] derived from experimental data for fast positive ions on H_2, He, N_2, Ne, Ar, Kr and Xe gases. It describes about 70% of experimental data within a factor of 2 and has the form:

$$\begin{aligned} \sigma_{\rm ec}(E) &= 1.1\times 10^{-8}\,{\rm cm}^2\, Z_{\rm P}^{0.5}/Z_{\rm T}^{1.8}\, u^{-4.8} \\ &\quad \times[1-\exp(-0.037u^{2.2})]\,[1-\exp(-2.44\times 10^{-5}u^{2.6})], \qquad (22)\\ u &= E[{\rm keV/amu}]/(Z_{\rm P}^{0.7}Z_{\rm T}^{1.25}), \qquad u\ge 10,\ Z_{\rm P}\ge 3, \qquad (23) \end{aligned}$$

where $Z_{\rm P}$ is the projectile charge and $Z_{\rm T}$ is the nuclear charge of the target atom.

At high reduced energies $u \gg 10$, (22) reads:

$$\sigma_{\rm ec}(E) \approx 1.1\times 10^{-8}\,{\rm cm}^2\, Z_{\rm P}^{3.9} Z_{\rm T}^{4.2}\, E^{-4.8}[{\rm keV/u}]. \qquad (24)$$

In the opposite case (high energies and light targets but highly charged projectiles), one has to include a contribution from radiative electron capture (REC) [29–31]. This capture process is accompanied by the emission of a photon and can be treated as the time-reversed photoionization process. For bare projectiles, an estimate for REC occurring in fast ion–atom collisions can already be obtained within the nonrelativistic dipole approximation. Here, the general result for radiative recombination into an arbitrary final n,ℓ state of a bare ion has already been given by *Stobbe* [29] in the form

$$\sigma_{nl}^{\rm Stobbe} = \frac{\pi^2}{3}\alpha^3 a_0^2 \left(\frac{\nu^3}{1+\nu^2}\right)^2 \left((l+1)\,[C_{nl}^{l+1}(\nu)]^2 + l\,[C_{nl}^{l-1}]^2\right), \qquad (25)$$

where α is the fine-structure constant, $\nu = \alpha Z_{\rm T}/(n\beta)$ is the Sommerfeld parameter, $\beta = v/c_{\rm o}$, and a_0 is the Bohr radius. The quantities $C_{nl}^{l\pm 1}$ are the dipole matrix elements for transitions between the bound nl states and continuum states with angular momenta $\ell\pm 1$. This expression has been evaluated by various authors (see, e.g., [29]). An estimate for REC cross sections for very light target atoms ($\alpha Z_{\rm T} \ll 1$) can be obtained from the Stobbe theory by multiplying $\sigma^{\rm Stobbe}$ with the number $Z_{\rm T}$ of quasi-free electrons in the target [31]. If only capture into the $1s$-state is considered, the equation simplifies to the well-known expression [32]

$$\begin{aligned} \sigma_{1s}^{\rm Stobbe} &= 9.2\times 10^{-21}{\rm cm}^2 \left(\frac{\nu^3}{1+\nu^2}\right)^2 \frac{{\rm e}^{-4\nu\arctan(1/\nu)}}{1-{\rm e}^{-2\pi\nu}}, \qquad (26)\\ \nu &= \alpha Z_{\rm T}/(n\beta). \end{aligned}$$

Within the dipole approximation, one obtains for the overall REC cross-scaling the following dependences (cf. (21) for NRC process):

$$\sigma_{\rm REC} \propto Z_{\rm T} Z_{\rm P}^5/v^{5/2}. \qquad (27)$$

4. Comparison with Experiment

4.1 Electron Capture Cross Sections

Radiative electron capture has been investigated intensively during recent years whereby a special emphasis was given to fast collisions of high-Z projectiles with low-Z targets (see, e.g., [12, 31, 33–35] and references therein). Here, REC is the only relevant charge-exchange process, which is illustrated in Fig. 2 for the particular case of 295 MeV/u bare uranium ions colliding with various neutral gas (squares) and solid atoms (circles) [31].

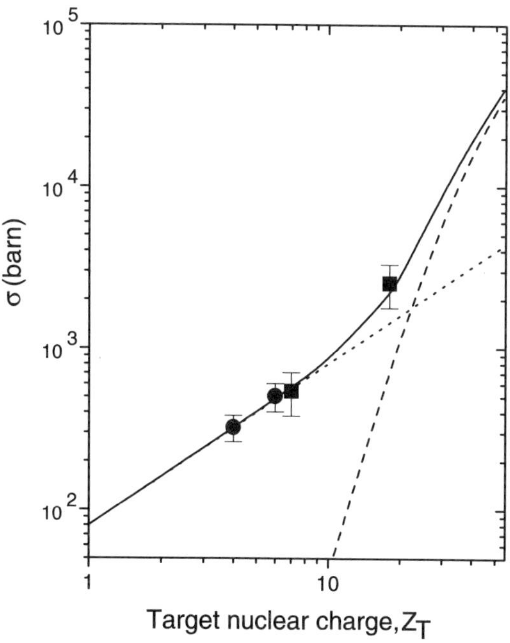

Fig. 2. Total electron-capture cross sections for bare U^{92+} ions colliding with gaseous (squares) and solid (circles) targets at an energy of 295 MeV/u. The results are compared with the theoretical cross section predictions for the NRC and the REC process (dashed and dotted lines, respectively). The sum of both cross sections is depicted by the solid line. From [31]

In addition, the theoretical results for NRC and REC are depicted. For REC, the simple dipole approximation was applied (dotted line) [29, 31] whereas NRC cross sections were derived on the basis of the relativistic Eikonal approach (dashed line) [27]. The solid line refers to the sum of both predictions where capture into projectile states up to the principal quantum numbers $n = 20$ were considered. As observed in the figure, the experimental data are in excellent agreement with the theoretical approaches used and the results confirm that for target charges $Z_T < 13$, NRC can be neglected. For REC the agreement between experiment and the nonrelativistic dipole approach must be regarded as fortuitous and attributed to the cancellation of various relativistic effects as discussed in detail in Ref. [36]. Therefore, the simple dipole approach provides a useful tool for reliable cross-section predictions at relativistic collision conditions and beam energies up to about 500 MeV/u [12]. The Eikonal approach is in general also in good overall agreement with experimental data and usually the deviation between experimental data and theory does not exceed a factor of two. This also holds true for collisions where targets with high nuclear charges are involved.

For predictions of the total single-electron capture cross sections in the low and medium collision velocity regime, i.e. at collision velocities where the velocity of the projectile K-shell electron is much larger than the collision velocity, n-body Classical Trajectory Monte Carlo calculations (nCTMC) appear to be most appropriate [37]. Here, for practical reasons, also the empirical Schlachter formula provides valuable cross-section estimates.

In Fig. 3, cross sections for single electron capture observed in collisions of $Ge^{31+} \rightarrow Ne$ are given (solid circles) and are compared with the nCTMC calculations of *Olson* et al. [37] (open circles), the Eikonal approach (dotted line) and the result of the empirical scaling formula of Schlachter et al. Both the nCTMC results and the empirical scaling law describe well the experimental data. Whereas the Eikonal approximation is still in moderate agreement with the data at energies around 8 MeV/u, it starts to deviate more markedly at the lower energies. Since the Eikonal approach is a high-energy approximation, the application of this theory at relatively slow collisions should be considered with caution.

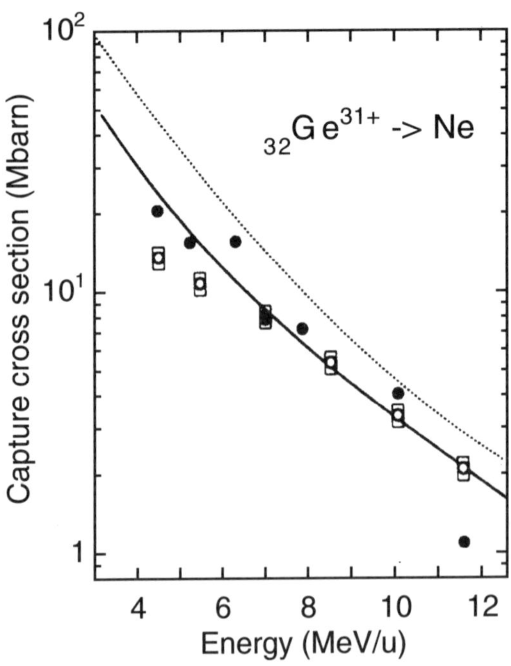

Fig. 3. Cross sections for single-electron capture in Ge^{31+} + Ne collisions plotted against projectile energy. Solid circles show the experimental data. The open circles refer to the nCTMC calculations. The full line gives the result of the Schlachter formula and the dotted line represents predictions based on the Eikonal theory. From [38]

4.2 Projectile-Ionization Cross Sections

Experiments on the projectile-ionization cross sections for collisions of fast positive ions with atomic and molecular targets are reported in many papers (see, e.g., [39–51]). Numerical calculations of the loss cross sections for light and heavy projectiles have been performed in [46, 48] and for fast many-electron ions in [52, 53] for estimation of the beam lifetimes in a synchrotron ion source (see Sect. 4.3).

The authors of [46, 48] have measured and calculated ionization cross sections for Li^{2+}, C^{5+}, O^{7+} projectiles on H_2 and He targets, Au^{52+} on H_2, He, C and N_2, Au^{75+} on H_2 and N_2, U^{86+} on H_2, He, and U^{90+} on H_2 using the PWBA in the form (11)–(13) and (17). They showed that with the projectile charge increasing, the screening cross sections approach the nuclear Born cross sections because with Z_P increasing the minimum momentum transfer increases, i.e., the maximum impact parameter decreases. If this minimum impact parameter, above which no projectile ionization can take place, falls significantly below the target K-shell radius, the form factor which characterizes the target electron's screening action approaches zero.

In [48], the effect of the target electrons on the projectile-ionization cross sections has been investigated. It was shown that the target electrons can act not only coherently by screening the target nucleus but may also exhibit incoherent behavior by directly ionizing a projectile and that the screening-antiscreening effects are important for targets with a low nuclear charge. The effect of the interaction between target and projectile electrons over a wide range of projectiles and collision energies was demonstrated and it was shown that this interaction can lead to a significant increase in the projectile-ionization cross section. The largest increases relative to the nuclear PWBA cross section which were observed are 89–90% for 1–3 MeV/u Li^{2+} and 70–80% for 380 MeV/u Au^{52+} on H_2. Some results are reproduced in Figs. 4 and 5 where the projectile-ionization cross sections are shown for light and heavy incident ions on atomic and molecular targets.

In [52, 53], the Bohr approximation given by (16) was used to calculate the charge-changing cross sections and the beam lifetimes for stored Mg^{1+}, Ar^{2+}, Pb^{29+}, U^{4+}, U^{10+} and U^{28+} ions in the energy range of 1–500 MeV/u. Electron- and proton-impact ionization cross sections $\sigma_p(v)$ and $\sigma_e(v)$ have been calculated in the Born and Coulomb–Born–Exchange approximations, respectively, using the ATOM code described in [54]. In calculations, ionization and excitation-autoionization processes of the inner-shell projectile electrons have been taken into account. The projectile radial wave functions have been found by solving the radial Schrödinger equation with the given binding electron energies which have been taken from Refs. [55, 56]; for uranium ions the data of [57] have been used. Calculated contributions from different atomic processes to the total charge-changing cross sections at different energies are given in Table 1. It is seen that in the energy range of 1–500 MeV/u, electron capture processes are important for ions with a large

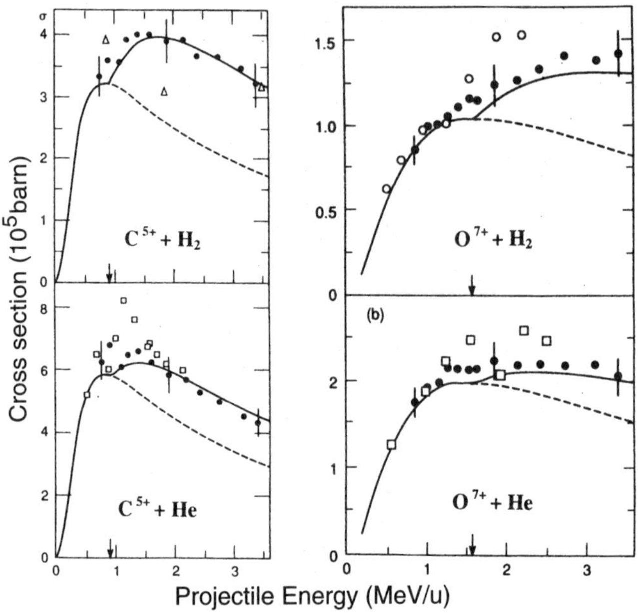

Fig. 4. Projectile ionization cross sections for C^{5+} and O^{7+} incident on H_2 and He. Symbols – experiment: solid circles – [48], triangles – [44], open circles and squares – [45]. Theory – PWBA: solid curves – (11)–(13), dashed curve – (11)–(13) with $S_a(q) = 0$. Arrows correspond to the energy threshold for electron-impact ionization. From [48]

ionic charge (Pb^{29+}, U^{28+}) and at rather low projectile energies $E < 10$ MeV/u.

Semiempirical formula for ionization cross sections. In the energy range of 10–500 MeV/u the charge-changing cross sections of many-electron projectiles are mainly attributed to ionization processes (electron capture is small, see Table 1). For this energy range, a semiempirical formula for the total ionization cross sections $\sigma_p(E)$ of many-electron projectiles by proton impact (direct ionization plus excitation-autoionization) was derived on the basis of numerical calculations [52]:

$$\sigma_p(E) \approx 3.3 \times 10^{-19}\mathrm{cm}^2\, N^{0.63}\, I[\mathrm{keV}]^{-0.86} E[\mathrm{MeV/u}]^{-1}, \tag{28}$$

$$N > 5, \quad E/I > 4 \times 10^3, \tag{29}$$

where N and I are the total number of electrons and the ionization potential of the projectile. This formula has an accuracy of a factor of 2 in the energy range $E/I > 4 \times 10^3$ (see Table 2). It is seen that (28) gives quite a good estimation for projectile-ionization cross sections until conditions (29) are violated.

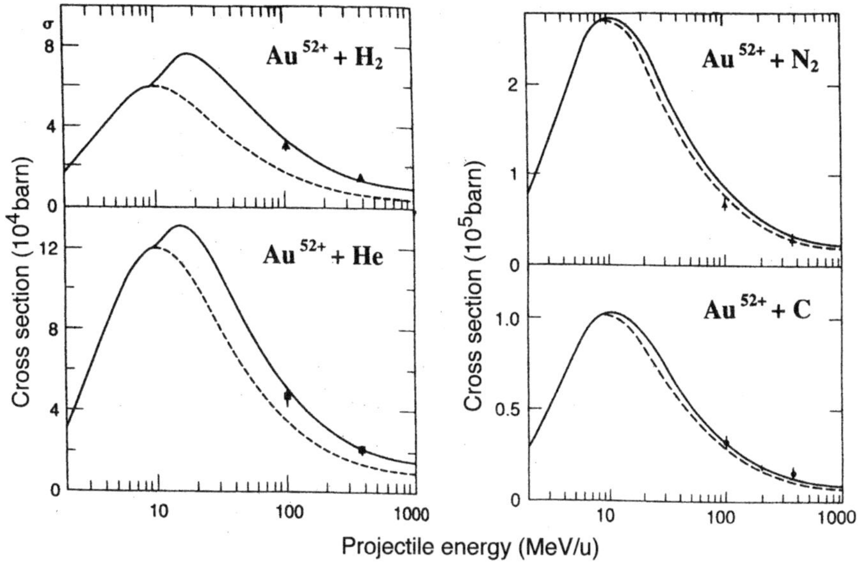

Fig. 5. Projectile ionization cross sections for Au^{52+} incident on N_2 and C. Symbols – experiment [48]. Theory – PWBA: solid curves – (11)–(13) and (17), dashed curve – (11)–(13) and (17) with $S_a = 0$. From [48]

4.3 Beam Lifetimes of Fast Many-Electron Ions

Recently, the beam lifetimes of stored many-electron Mg^{1+}, Ar^{2+}, Pb^{29+}, U^{10+} and U^{28+} ions at relatively high energies of E = 1–250 MeV/u have been measured at SIS/GSI [58, 59]. Corresponding calculations of the total charge-changing cross sections and the beam lifetimes have been performed in [52, 53]. The lifetime τ of an ion beam in an accelerator device is directly related with the total charge-changing cross section σ_{tot} by

$$\tau(v) = [\rho v \sigma_{\text{tot}}(v)]^{-1}, \tag{30}$$

where σ_{tot} is defined by (3), v is the projectile velocity and ρ is the total density of the residual gas. Therefore, knowing the total charge-changing cross sections, it is possible to make a comparison with the beam lifetimes. However, a very important question arises associated with the residual-gas composition since the projectile-ionization cross section is roughly proportional to the value $S_{\max} = Z_T^2 + Z_T$, where Z_T is the atomic number of the residual-gas atom. The heavier the rest-gas atom, the larger contribution it gives to the charge-changing processes. For example, for H_2 molecules, $S_{\max}(H_2) = 4$ and for Ar atoms $S_{\max}(\text{Ar}) = 342$. Therefore, to make an accurate comparison of the experimental beam lifetimes with theoretical calculations, it is necessary to know the vacuum conditions in the accelerator machine – the total residual-gas pressure and the fractions of the gas components.

Table 1. Calculated contributions of the direct ionization, $\sigma_{\text{ion}}^{\text{dir}}$, excitation-autoionization, $\sigma_{\text{ion}}^{\text{auto}}$, and the electron capture, σ_{ec}, processes to the total projectile charge-changing cross sections. From [52]

Projectile	$\sigma_{\text{ion}}^{\text{dir}}$, %	$\sigma_{\text{ion}}^{\text{auto}}$, %	σ_{ec}, %	Energy, MeV/u
Mg^{+}	100	0	0	> 0.2
Ar^{2+}	70	30	0	> 0.2
Pb^{29+}	15	5	80	1.6
Pb^{29+}	45	15	40	3.2
Pb^{29+}	60	20	20	6.4
Pb^{29+}	75	25	0	> 12.0
U^{4+}	98	1	1	0.25
U^{4+}	75	25	0	1.0
U^{4+}	70	30	0	4.0
U^{4+}	60	40	0	> 10.0
U^{10+}	65	15	20	0.48
U^{10+}	75	20	5	1.93
U^{10+}	80	20	0	> 5.0
U^{28+}	20	5	75	1.5
U^{28+}	45	20	35	3.0
U^{28+}	60	30	10	6.0
U^{28+}	70	30	0	> 12.0

In Fig. 6, the charge-changing cross sections for stored Mg^{1+} ions are displayed deduced from the experimental beam-lifetime data at SIS/GSI using the known average total pressure of $(6.4\pm1.3) \times 10^{-11}$ mbar. Since the composition of the residual gas was not measured precisely, it was estimated by comparison of the calculated charge-changing cross section for Mg^{1+} ion (as a relatively simple projectile) with that evaluated from experimental beam lifetimes.

As seen from the figure, calculated cross sections with an Ar concentration in the range between 1% and 3% give the best fit to the experimental data. In the calculations of the beam lifetimes for other ions, a 77%H_2+20%N_2+3%Ar mixture was adopted which seemed to be close to experimental conditions (see the discussion in [53]); for a more optimistic mixture of 97.7%H_2+2%N_2+0.3%Ar, a charge-changing cross section would be about 5 times smaller, i.e. the beam lifetime would be 5 times larger.

Beam lifetimes for different ions as a function of the beam energy are displayed in Figs. 7 and 8. Although the energy dependence is well reproduced by the theory, the calculated values are by a factor of 2–3 larger than the experimental data. This is the typical accuracy that can be achieved by the Bohr–Born approximation. It was found [53] that because of the $Z_T^2 + Z_T$ scaling for ionization cross sections, the heavy components of the residual gas (N_2 and Ar) although having small fractions, play a key role in charge-changing reactions and are mainly responsible for the beam lifetime values.

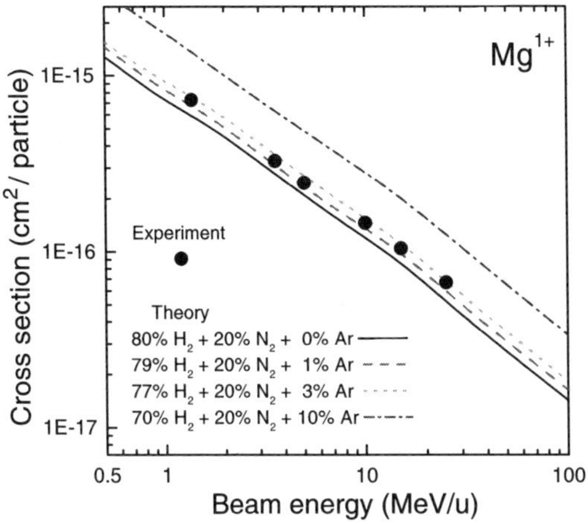

Fig. 6. Effective charge-changing cross sections for Mg^{1+} ions as a function of the beam energy: circles – experimental data [59], curves – calculations with different gas compositions [52]

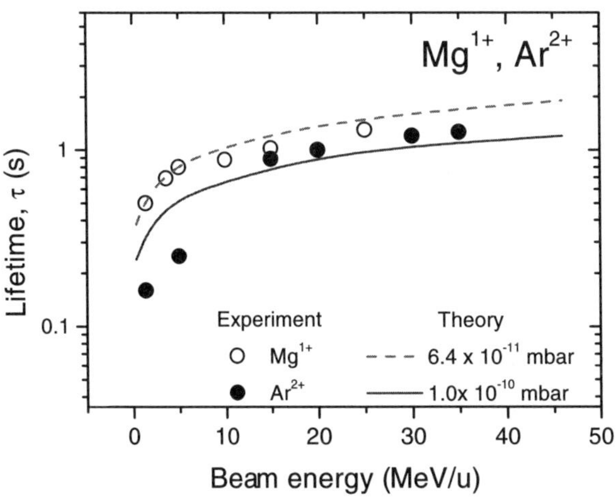

Fig. 7. Lifetime of Mg^{1+} and Ar^{2+} ion beams calculated [52] with the residual gas composition at two different gas pressures (curves), circles – experiment [58]

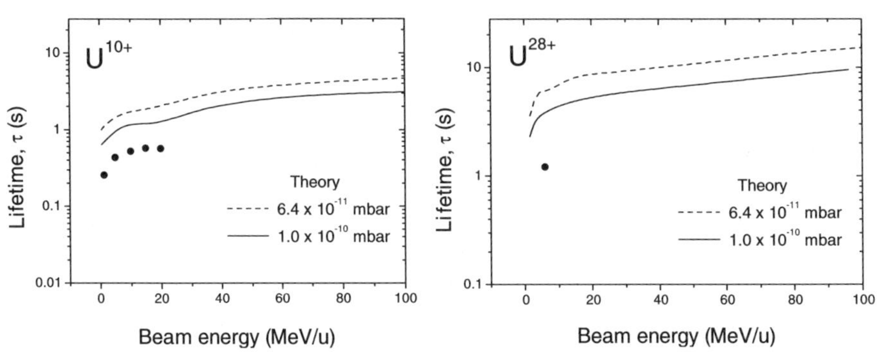

Fig. 8. Lifetime of U^{10+} and U^{28+} ions as a function of the beam energy: circles – experiment [58], curves – calculated with a 77%H_2+20%N_2+3%Ar gas composition at two different gas pressures [53]

Table 2. Experimental projectile-ionization cross sections (cm^2) in collisions of fast positive ions with atoms and molecules in comparison with the results of the ATOM code (present work) and the semiempirical formula (28). For molecular targets, the cross sections are multiplied by 2

Process	E, MeV/u	I, keV	σ, exp.	Ref.	ATOM	Eq.(28)
$Pb^{40+}+N_2$	5.9	1.95	1.6(−17)	[41]	1.2(−17)	3.7(−17)
$Pb^{55+}+N_2$	5.9	5.16	8.6(−18)	[41]	1.3(−17)	1.2(−17)
$U^{40+}+H_2$	7.8	1.83	1.0(−19)	[41]	6.4(−19)	1.1(−18)
$U^{63+}+H_2$	10	4.69	1.6(−19)	[41]	4.6(−20)	2.9(−19)
$U^{63+}+N_2$	10	4.69	8.0(−18)	[41]	1.3(−19)	8.2(−18)
$U^{86+}+H_2$	405	29.8	3.7(−22)	[24]	2.9(−22)	5.4(−22)
$U^{86+}+He$	405	29.8	5.5(−22)	[24]	4.4(−22)	8.2(−22)
$U^{86+}+N_2$	405	29.8	–	[24]	8.1(−19)	1.5(−20)
$Au^{52+}+H_2$	100	4.67	3.1(−20)	[48]	2.1(−20)	3.0(−20)
$Au^{52+}+He$	100	4.67	4.8(−20)	[48]	3.2(−20)	4.5(−20)
$Au^{52+}+C$	100	4.67	2.9(−19)	[48]	2.2(−19)	2.9(−19)
$Au^{52+}+N_2$	100	4.67	6.9(−19)	[48]	5.9(−19)	8.4(−19)
$Au^{52+}+H_2$	380	4.67	1.5(−20)	[48]	5.3(−21)	7.4(−21)
$Au^{52+}+He$	380	4.67	2.1(−20)	[48]	8.0(−21)	1.1(−20)
$Au^{52+}+C$	380	4.67	1.3(−19)	[48]	5.6(−20)	7.7(−20)
$Au^{52+}+N_2$	380	4.67	3.1(−19)	[24]	1.5(−19)	2.1(−19)
$Au^{75+}+H_2$	380	21.9	1.6(−22)	[24]	2.2(−22)	5.9(−22)
$Au^{75+}+N_2$	380	21.9	6.1(−21)	[24]	6.2(−21)	1.6(−20)
$V^{18+}+He$	8.6	1.34	3.0(−19)	[42]	2.8(−19)	4.9(−19)
$V^{19+}+He$	8.6	1.48	1.6(−19)	[42]	1.8(−19)	3.6(−19)
$V^{20+}+He$	8.6	1.57	8.1(−20)	[42]	8.6(−20)	7.5(−19)
$V^{21+}+He$	8.6	6.59	1.2(−20)	[42]	1.1(−20)	7.1(−20)
$V^{22+}+He$	8.6	7.00	5.4(−21)	[42]	4.6(−21)	4.3(−20)

The calculated charge-changing cross sections for fast many-electron ions are presented in Fig. 9 in comparison with the SIS data. The largest total charge-changing cross section corresponds to incident U^{4+} ions and the smallest to the Pb^{29+} ions.

5. Conclusion

A brief review of charge-changing reactions of fast heavy ions colliding with neutral atoms and molecules in the nonrelativistic projectile energy range of 1–500 MeV/u is presented. Experimental data on one-electron capture cross sections are well described by a combination of contributions from nonradiative and radiative electron-capture processes. Projectile-ionization cross sections can be calculated with an accuracy of a factor of 3 within the Born approximation or its modifications which, however, do not account for exci-

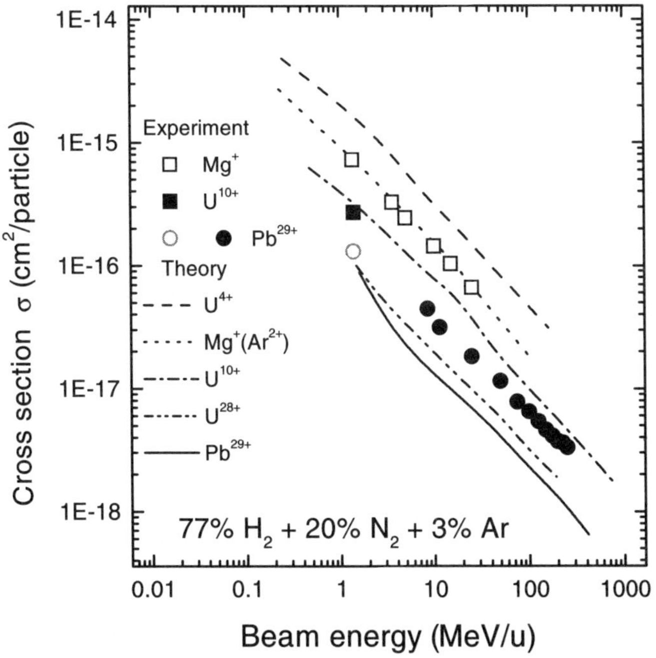

Fig. 9. The total charge-changing cross sections of many-electron ions. Experiment: open squares, Mg^{1+} and solid circles, Pb^{29+}, from SIS/GSI data [58]; solid square – U^{10+}, estimated from [41], open circle – Pb^{29+}, estimated from [40]. Theory: curves – calculations using a 77%H_2+20%N_2+3%Ar rest-gas mixture in (16) [52]

tation and ionization of the target, i.e., treat the target electrons and nucleus as being at rest.

Special attention is paid to estimation of the beam lifetimes τ of stored ions. First numerical calculations of τ values are presented for many-electron heavy ions including low-charged uranium ions. It was pointed out that to make a detailed comparison of experimental beam lifetimes with theoretical predictions, it is necessary to measure values of pressure and relative fractions of the rest gas together with the beam lifetimes.

For investigation of the beam lifetimes of highly charged ions, one has also to take into account the change of the projectile charge due to dielectronic recombination and radiative recombination processes occurring in collisions with *free* electrons which can be presented in the incoming beam.

Acknowledgement. We very much appreciate helpful discussions with W. Jacoby and M. Rau concerning vacuum conditions in the ESR and SIS at GSI. We also thank U.I. Safronova for providing us with calculations of the binding energies of highly charged ions prior to publication. V.P. Shevelko was supported in part by Russian Grant RFBR 96-02-16090.

References

1. H.-D. Betz: Rev. Mod. Phys. **44**, 465 (1972)
2. R.K. Janev, L.P. Presnyakov, V.P. Shevelko: *Physics of Highly Charged Ions* (Springer, Berlin, Heidelberg 1985)
3. C.L. Cocke, R.E. Olson: Phys. Rep. **205**, 153–219 (1991)
4. N. Stolterfoht, R.D. DuBois, R.D. Rivarola: *Electron Emission in Heavy Ion–Atom Collisions*, Spr. Ser. Atom. Plasmas Vol. 20 (Springer, Berlin, Heidelberg 1997)
5. R.K. Janev: *Atomic and Molecular Processes in Fusion Edge Plasmas* (Plenum, New York 1995)
6. D. Habs: *Proc. Topical Course on Frontiers of Particle Beams*, Lecture Notes in Physics 296 (Springer, Berlin 1988)
7. D. Habs, W. Baumann, J. Berger, P. Blatt, A. Faulstich, P. Krause, G. Kilgus, R. Neumann, W. Petrich, R. Stokstad, D. Schwalm, E. Szmola, K. Welti, A. Wolf, S.Zwickler, E. Jaeschke, D. Krämer, G. Bisoffi, M. Blum, A. friedrich, C. Geyer, M. Griese, H.W. Heyng, B. Holzer, R. Ihde, M. Jung, K. Matl, W. Ott, B. Povh, R. Repnow, M. Steck, E. Stefens, D. Dutta, T. Kühl, D. Marx, S. Schröder, M. Gerhard, R. Grieser, G. Huber, R. Klein, M. Krieg, N. Schmidt, R. Schuch, J.F. Babb, L. Spruch, W. Arnold, A. Noda: Nucl. Instrum. Methods B **43**, 390 (1989)
8. A. Wolf, V. Balykin, W. Baumann, J. Berger, G. Bisoffi, P. Blatt, M. Blum, A. Faulstich, A. Friedrich, M. Gerhard, C. Geyer, M. Griese, R. Grieser, D. Habs, H.W. Heyng, B. Hochadel, B. Holzer, G. Huber, E. Jaeschke, M. Jung, A. Karafillidis, G. Kilgus, R. Klein, D. Krämer, P. Krause, M. Krieg, T. Kühl, K. Matl, A. Müller, M. Music, R. Neumann, G. Neureither, W. Ott, W. Petrich, B. Povh, R. Repnow, S. Schröder, R. Schuch, D. Schwalm, P. Sigray, M. Steck, R. Stokstad, E. Szmola, M. Wagner, B. Wanner, K. Welti, S. Zwickler: In: *The Physics of Electronic and Atomic Collisions*, AIP Proc. (AIP, New York 1990), p. 378
9. W.N. Spjeldvik: Space Science Reviews **23**, 499 (1979)
10. R.K. Janev, L.P. Presnyakov: Phys. Rep. **70**, 1 (1981)
11. V.S. Nikolaev: Sov. Phys. - Uspekhi **8**, 269 (1965)
12. P.H. Mokler, Th. Stöhlker: Adv. At. Mol. Opt. Phys. **37**, 297 (1996)
13. H. Tawara: Report NIFS-DATA-42 (Nagoya 1997)
14. J.H. McGuire: Adv. At. Mol. Opt. Phys. **29**, 217 (1992)
15. V.P. Shevelko, H. Tawara: *Atomic Multielectron Processes*, Spr. Ser. Atoms and Plasmas (Springer, Berlin, Heidelberg 1998) (in preparation)
16. H. Knudsen, P. Hvelplund, L.H. Andersen, S. Bjørnelund, M. Frost, H.K. Haugen, E. Samsø: Phys. Scr. **T3**, 101 (1983)
17. D.R. Bates, G. Griffing: Proc. Phys. Soc. A **66**, 9611 (1953); **67**, 663 (1953); **68**, 90 (1955)
18. D.R. Bates (ed): *Atomic and Molecular Processes* (Academic, New York 1962)
19. J.S. Briggs, K. Taulbjerg: *Structure and Collisions of Ions and Atoms* in: Topics Curr. Phys., Vol. 5, I. Sellin, ed. (Springer, Berlin, Heidelberg 1978)
20. J.H. Hubbell, W.J. Veigele, E.A. Briggs, R.T. Brown, D.T. Comer, R.J. Howerton: J. Phys. Chem. Ref. Data **4**, 471 (1975)
21. J.H. McGuire, N. Stolterfoht, P.R. Simony: Phys. Rev. A **24**, 97 (1981)
22. R. Anholt: Phys. Lett. A **114**, 126 (1986)
23. J.D. Jackson: *Classical Electrodynamics*, 2nd ed. (Wiley, New York 1975), Ch. 13.1

24. B. Feinberg, H. Gould, W.E. Meyerhof, A. Belkacem, H.-P. Hülskötter, J.R. Alonso, L. Blumenfeld, E. Dillaed, N. Guardala, G.F. Krebs, M.A. McMahan, M.J. Rhoades-Brown, B.S. Rude, J. Schweppe, D.W. Spooner, K. Street, P. Thieberger, H. Wegner: Phys. Rev. A **47**, 2370 (1993)
25. N. Bohr: K. Dan. Vidensk. Selsk., Mat.-Fys. Medd. **18**, 8 (1948), Sects. 3.3 and 4.2
26. J. Eichler, F.T. Chen: Phys. Rev. A **20**, 104 (1979)
27. J. Eichler: Phys. Rev. A **32**, 112 (1985)
28. A.S. Schlachter, J.W. Stears, W.G. Graham, K.H. Berkner, R.V. Pyle, J.A. Tanis: Phys. Rev. A **27**, 3372 (1983)
29. M. Stobbe: Ann. Phys. **7**, 661 (1930)
30. A. Ichihara, T. Shirai, J. Eichler: Phys. Rev. A **49**, 1875 (1994)
31. Th. Stöhlker, C. Kozhuharov, P.H. Mokler, A. Warczak, F. Bosch, H. Geissel, R. Moshammer, C. Scheidenberger, J. Eichler, A. Ichihara, T. Shirai, Z. Stachura, P. Rymuza: Phys. Rev. A **51**, 2098 (1995)
32. H.A. Bethe, E.E. Salpeter: *Quantum Mechanics of One- and Two-Electron Atoms* (Plenum, New York 1977)
33. R. Anholt, H. Gould: Adv. At. Mol. Phys. **22**, 315 (1986)
34. J. Eichler, W.E. Meyerhof: *Relativistic Atomic Collisions* (Academic, San Diego 1995)
35. H.F. Beyer, H.-J. Kluge, V.P. Shevelko: *X-Ray Radiation of Highly Charged Ions* Spr. Ser. Atoms and Plasmas, Vol. 19 (Springer, Berlin, Heidelberg 1997)
36. J. Eichler, Th. Stöhlker: *Ion–Atom Collisions at Relativistic Velocities*, this volume
37. R. Olson: *Electronic and Atomic Collisions* (Elsevier, Amsterdam 1988) p. 271
38. Th. Stöhlker, C. Kozhuharov, P.H. Mokler, R.E. Olson, Z. Stachura, A. Warczak: J. Phys. B **25**, 4527 (1992)
39. J.R. Macdonald, F.W. Martin: Phys. Rev. A **4**, 1965 (1971); *ibid*, **4**, 1974 (1971)
40. W. Erb: Report GSI-P-7-78 (Darmstadt 1978)
41. B. Franzke: *IEEE*, Vol. NS-28, 2116 (1981)
42. W.G. Graham, K.H. Berkner, E.M. Bernstein, M. Clark, R.H. McFarland, T.J. Morgan, A.S. Schlachter, J.W. Stearns, M.P. Stockli, J.A. Tanis: J. Phys. B **18**, 2503 (1985)
43. R. Hippler, S. Datz, P.D. Miller, P.L. Pepmiller, P.F. Dittner: Phys. Rev. A **35**, 585 (1987)
44. R. Anholt, X.-Y. Xu, Ch. Stoller, J.D. Molotoris, B.S. Rude, R.J. McDonald: Phys. Rev. A **37**, 1105 (1988)
45. T.N. Tipping, J.M. Sanders, J. Hall, J.L. Shinpaugh, D.H. Lee, J.H. McGuire, P. Richard: Phys. Rev. A **37**, 2906 (1988)
46. H.-P. Hülskötter, W.E. Meyerhof, E. Dillard, N. Guardala: Phys. Rev. Lett. **63**, 1938 (1989)
47. S.A. Bowman, E.M. Bernstein, J.A. Tanis: Phys. Rev. A **39**, 4423 (1989)
48. H.-P. Hülskötter, B. Feinberg, W.E. Meyerhof, A. Belkacem, J.R. Alonso, L. Blumenfeld, E.A. Dillard, H. Gould, N. Guardala, G.F. Krebs, M.A. McMahan, M.E. Rhoades-Brown, B.S. Rude, J. Schweppe, D.W. Spooner, K. Street, P. Thieberger, H.E. Wegner: Phys. Rev. A **44**, 1712 (1991)
49. P. Rymuza, Th. Stöhlker, C.L. Cocke, H. Geissesl, C. Kozhuharov, P.H. Mokler, R. Moshammer, F. Nickel, C. Schneiderberger, Z. Stachura, J. Ullrich, A. Warczak: J. Phys. B **26**, L169 (1993)
50. O. Heber, G. Sampoll, B.B. Bandong, R.J. Maurer, R.L. Watson, I. Ben-Itzhak, J.M. Sanders, J.L. Shinpaugh, P. Richard: Phys. Rev. A **52**, 4578 (1995)

51. W. Wu, K.L. Wong, E.C. Montenegro, R. Ali, C.Y. Chen, C.L. Cocke, R. Dörner, V. Frohne, J.P. Giese, V. Mergel, W.E. Meyerhof, M. Raphaelian, H. Schmidt-Böcking: Phys. Rev. A **55**, 2771 (1997)
52. V.P. Shevelko, D. Böhne, Th. Stöhlker: *Charge-Changing Processes Involving Fast Low-Charged Ions* in: Nucl. Instrum. Methods A (1998) (in press)
53. V.P. Shevelko, O. Brînzanescu, W. Jacoby, M. Rau, Th. Stöhlker: *Beam Lifetimes of Low-Charged Ions in Storage Rings* in: Hyp. Int. **114**, 289 (1998)
54. V.P. Shevelko, L.A. Vainshtein: *Atomic Physics for Hot Plasmas* (IOP, Bristol 1993)
55. W. Lotz: J. Opt. Soc. Am. **60**, 206 (1970)
56. T.A. Carlson, C.W. Nestor, Jr., N. Wasserman, J.D. McDowell: At. Data Nucl. Data Tables, **2**, 63 (1970)
57. K. Rashid, M.Z. Saadi, M. Yasin: At. Data Nucl. Data Tables, **40**, 365 (1988)
58. K. Blasche, H. Eickhoff, P. Forck, B. Langenbeck, G. Moritz, P. Moritz, U. Oeftiger, A. Peters, P. Spiller, P. Strehlet: *GSI Scientific Report 1996* (Darmstadt 1997), p. 159
59. P. Strehl, H. Reeg, D. Wilms: unpublished (SIS/GSI, Darmstadt, April 1997)

XV Charge Changing Processes in Ion–Ion Collisions

Frank Melchert

Summary. Collisions between ions belong to the elementary processes occurring in all types of plasmas in astrophysical objects as well as in laboratory discharges. In recent years, the impetus for accurate cross section data for such collisions has become greater due to research in thermonuclear fusion using either magnetic or inertial confinement.

Compared with ion–atom collisions, experimental investigations of ion–ion collisions are quite scarce since the ion beams are very tenuous due to the mutual repulsion within both the projectile and the target ion beams. Nevertheless, modern crossed-beam experiments overcome these difficulties and give a detailed insight into a variety of ion–ion collisions at keV collision energies.

In this chapter, experimental techniques relevant to modern *crossed-beam* experiments will be discussed. A survey will be given of a variety of collisions. For light ions, where only one or two electrons are present besides the two nuclei, theoretical and experimental results can be compared. In the case of heavy and multiply charged ions, however, data are rare but recent progress will be reviewed.

1. Introduction

The investigation of charge-changing ion–ion collisions at keV energies is of considerable interest for several reasons. First, modeling of non-equilibrium plasmas requires details about all relevant atomic processes of which ion–ion collisions are one fundamental group. While dense gases or plasmas obey the laws of equilibrium thermodynamics and their properties can be discussed without recourse to atomic physics, most laboratory plasmas as well as stellar and planetary atmospheres [1] are far from thermodynamic equilibrium.

Second, thermonuclear fusion research requires data for charge-changing ion–ion collisions. In magnetically confined fusion plasmas, the ion of one element may be particularly effective in neutralizing fuel ions or inducing other undesirable effects. Early identification of these elements can help to exclude these elements. Furthermore, the fusion plasma needs auxiliary heating by injection of intense neutral hydrogen beams. Negative hydrogen ions H^- accelerated to kinetic energies of about 10^5 eV can be neutralized most effectively in so-called plasma neutralizers [2] where the interaction between H^- and multiply charged plasma ions X^{q+} dominates the neutralization process. If inertial confinement fusion is driven by heavy-ion impact, large storage rings have to hold intense ion beams up to GeV energies [3]. Intra-beam ion–ion collisions at keV energies cause intensity losses [4] which can be minimized by a proper choice of the driver-ion species.

Third, more fundamental reasons motivate the study of ion–ion collisions which are different from those between ions and atoms. When ions and atoms collide at low velocities $v < 1$a.u. (1 a.u. = 2.2×10^8 cm/s), the charge clouds of the target atom and projectile ion merge and a transient molecule forms. Electron capture and ionization processes can be explained in terms of transitions between states of these quasi-molecules. When two ions collide, the process is effected by long-range Coulomb forces absent in ion–atom collisions. At very low velocities it is clear that the positive ions' charge clouds do not merge and no molecular ions form. Therefore trajectory and polarization effects of the two ions may be important.

While ion–atom, electron–atom and, more recently, electron–ion collisions have been widely investigated, the study of ion–ion collisions is still in its infancy (see, e.g., [5]). Collisions between ions occur in all kinds of plasmas at high rates, but the preparation of a single collision event requires isolation of ions in the given charge states. It is a straightforward idea to investigate ion–ion collisions in the so-called *crossed-beam* technique where two ion beams are made to intersect. In the next section, essential aspects of *crossed-beam* experiments will be elucidated before the different collision processes are discussed.

2. Experimental Methods and Techniques

2.1 Basic Considerations

Ion–ion collisions can be studied in the well-defined setup of a *crossed-beam* experiment. Although this technique, in principle, seems to be straightforward, inherent difficulties arise from the small target thickness provided by typical ion beams.

Mutual Coulomb repulsion between the ions within their beams causes them to expand along their beam path. Under certain circumstances, a mutual deflection of both ion beams can also be important. To ensure experimental resolution, the beam diameter and divergence have to be limited which, as a consequence, limits the ion beam currents. For example, a beam of singly charged Ar^+ ions (energy 5 keV, diameter 4 mm, maximum divergence 1 deg) cannot transport more than a few μA of ion-beam current. This beam contains less than 10^7 ions per cm^3 which is comparable to the particle density within ultrahigh vacuum (UHV). While two ion beams penetrate each other, ion–ion collision rates can be expected that are equal to the rates observed when an ion beam travels through UHV, and that is almost nothing. Depending on the particular cross section, only 10^{-10}–10^{-14} of the incoming flux undergoes an ion–ion collision. Furthermore, the ion beams travel through UHV along their complete trajectory, while they overlap only in the interaction region which has typically mm dimensions for a *crossed-beam* arrangement. That is why the ion–ion reaction signal is masked by 10^2–10^4

times more intense background contributions which arise from ionic collisions with the residual gas.

The experimental challenge of *crossed-beam* experiments is to single out ion–ion events, which occur at absolute rates of 10^{-2}–10^{1} per second, from background rates of ions in the same charge state in the order of 10^{2}–10^{4} per second, and from the incoming flux in the order of 10^{10}–10^{13} per second. For this purpose, coincidence [6] and beam modulation [7] techniques have been developed. Further techniques will be discussed in Sect. 2.3.

In spite of these difficulties and the need for complex apparatus, the *crossed-beam* technique is very attractive. The reactants and the type of reaction can be clearly defined and the range of accessible energies spans seven or more orders of magnitude. If two ion beams cross at an interaction angle α, the collision energy E_{cm} in the center-of-mass frame (cm) is given by

$$E_{\mathrm{cm}} = \mu \left(\frac{E_1}{m_1} + \frac{E_2}{m_2} - 2\sqrt{\frac{E_1 E_2}{m_1 m_2}} \cos\alpha \right) , \tag{1}$$

where E_{i} and m_{i} are lab-frame energies and masses of beams 1, 2 and μ is the reduced mass $\mu = (m_1 \cdot m_2)/(m_1 + m_2)$. The interaction angle α determines the cm-energy range that is accessible for given lab frame energies. When both ion beams merge (α =0 deg), very low cm energies are attainable and special kinematic effects occur [5]. In Fig. 1 energies E_{cm} are plotted for a *merged-beam* setup ($\alpha = 0$ deg) and for *crossed-beam* setups using acute ($\alpha = 17.5$ deg) or obtuse ($\alpha = 137$ deg) angles. All energies are normalized by the kinetic energy E_1 in beam line 1. The energy resolution ΔE_{cm} stems from the lab frame energy spreads ΔE_1 and ΔE_2 in both ion beams.

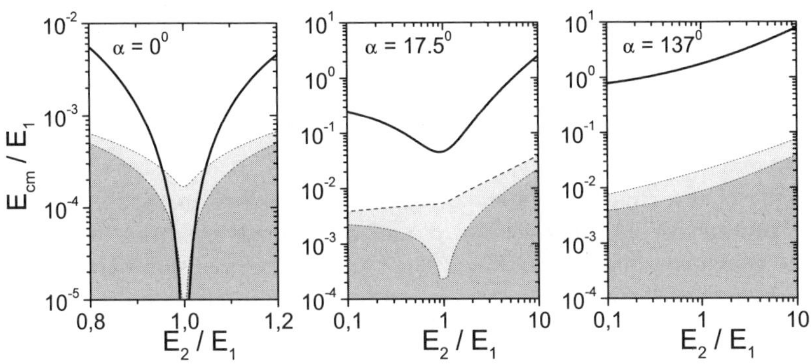

Fig. 1. Center-of-mass energy E_{cm} (solid line) as a function of the lab frame kinetic energy E_2 of ion beam 2, measured in units of the lab frame kinetic energy E_1, for three different interaction angles α between the two ion beams: $\alpha = 0$ deg (left), $\alpha = 17.5$ deg (center) and $\alpha = 137$ deg (right). The shaded areas represent contributions to the energy resolution ΔE originating from the lab frame energy spreads $\Delta E_{\mathrm{i}}, i = 1, 2$ (dark gray) and from the collision angle uncertainty $\Delta\alpha$ (light gray). $\Delta E_{\mathrm{i}} = 0.005 \cdot E_{\mathrm{i}}$, $\Delta\alpha = 1$ deg and $m_1 = m_2$ were assumed

Furthermore, the collision angle uncertainty $\Delta\alpha$ which is a consequence of individual ion trajectories spread around the beam symmetry axis contributes to ΔE_{cm}. To access low cm energies in *merged-beam* experiments, a careful beam collimation is required to reduce the angular spread $\Delta\alpha$ within the ion beams. Before experimental techniques are surveyed in Sect. 2.3, kinematic aspects of *crossed-beam* experiments are discussed which differ from the usual situation where the target is at rest in the laboratory frame.

2.2 Crossed-Beam Kinematics

The asymptotic solutions of an ion–ion collision process, which is a quantum mechanical event, have to satisfy classical laws of energy- and momentum conservation. We consider ion 1 (mass m_1, lab-frame velocity $\boldsymbol{v}_1$, cm velocity $\boldsymbol{u}_1$) which happens to collide at time $t = 0$ with ion 2, outlined in Fig. 2. The vector $\boldsymbol{u}_1$ is plotted at three different times: $t = -t_1$, $t = 0$ and $t = t_1$. During the collision, $\boldsymbol{u}_1$ transforms into $\boldsymbol{u}_1'$ as a consequence of the scattering process, described by angles ϑ_{cm} and φ_{cm} in the cm frame. Furthermore, $\boldsymbol{u}_1'$ is shorter than $\boldsymbol{u}_1$ if the reaction has a negative energy defect Q, i.e., if kinetic energy is transferred into potential energy. At a later time ($t = t_1$), the vector $\boldsymbol{u}_1'$ determines the velocity $\boldsymbol{v}_1'$ of mass 1 after the collision. For a given ϑ_{cm}, the endpoints of $\boldsymbol{v}_1'$ form a circle, parameterized by φ_{cm}. Viewed in the lab frame along the beam symmetry axis $\boldsymbol{v}_1$, this circle is projected to an elliptical structure on the detector with its center shifted from the symmetry axis. The shift $g(Q)$ is a function of the reaction Q-value. It should be noted that the *crossed-beam* kinematics produce elliptical symmetry of lab frame scattering distributions which are shifted from the symmetry axis, given by the beam initial velocity, as a function of the reaction Q-value. The magnitude of the eccentricity and shift depends on the mass and velocity ratios of the colliding ions.

2.3 Measuring Techniques

If two ion beams cross at an angle α, the event rate R of an ion–ion reaction is the product of its cross section σ and an integral over the interaction volume which contains the relative velocity v_{rel} of the collision partners and the particle densities n_{i} within the two ion beams. For well collimated *crossed beams* the integral can be solved and the cross section σ can be calculated from the equation

$$\sigma = \frac{v_1 v_2 \sin\alpha}{v_{\mathrm{rel}}} \cdot \frac{R\, q_1 q_2}{I_1 I_2} \cdot F, \tag{2}$$

where v_{i}, q_{i} and I_{i} are the velocities, charges and ion currents of both ion beams $i = 1, 2$. The so-called form factor F describes the spatial overlap of the ion beams and is determined by sliding a slotted shutter along the z-axis, which is perpendicular to the interaction plane, through the ion beams. Measurement of the transmitted ion beam currents as a function of the shutter position yields one-dimensional current densities $j_i(z)$ which can be recorded in both beams simultaneously. Then F is obtained from $j_i(z)$ by numerical integration which is described in full detail in Ref. [6].

Among all quantities in (2), the event rate R of ion–ion reactions is the most difficult to determine. Far more particles originating from interactions other than ion–ion collisions reach the detectors. If both collision partners alter their charge state, e.g., in an electron capture reaction (3), both reaction products can be detected in coincidence. Since they are generated simultaneously and since their flight times from the beam intersection to the detectors are fixed, the corresponding output pulses of single particle detectors show a fixed time delay in the case of true ion–ion signals, whereas there is no time

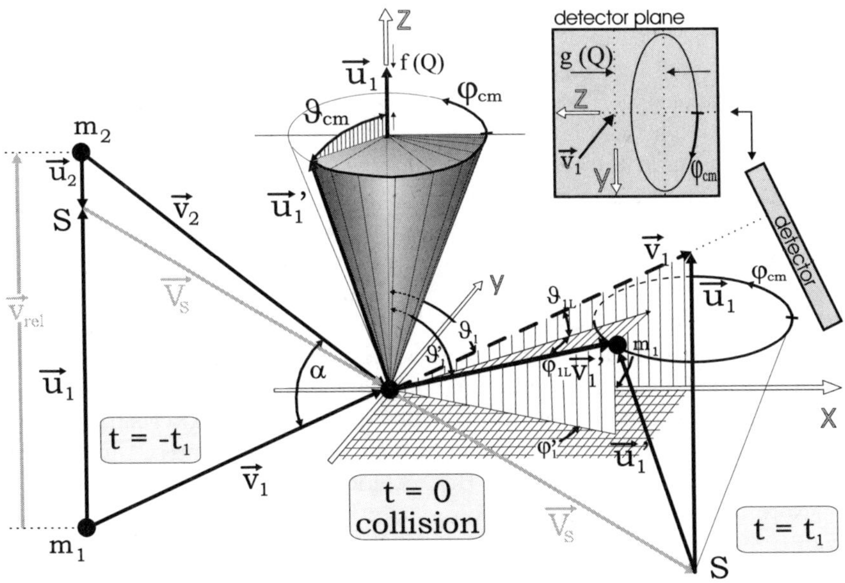

Fig. 2. Kinematics of an ion–ion collision in the x–z plane with z aligned along the relative ion velocity $\boldsymbol{v}_{\rm rel}$. The y-axis points into the paper plane. The insert (top right) views the detector surface along the symmetry axis $\boldsymbol{v}_1$ of beam 1 (b1). For a given $\vartheta_{\rm cm}$, the endpoints of $\boldsymbol{v}_1'$ form a circle which is projected to an elliptical structure on the detector surface. Q is the energy defect of the reaction

m_1, m_2 :	ion masses	$f(Q)$:	velocity change, cm system
S :	center of mass, cm	$g(Q)$:	horizontal shift on detector
$\boldsymbol{V}_{\rm S}$:	velocity of cm	α :	beam crossing angle
$\boldsymbol{v}_{\rm rel}$:	relative ion velocity	$\vartheta_{\rm cm}, \varphi_{\rm cm}$:	cm scattering angles
$\boldsymbol{v}_1, \boldsymbol{v}_2$:	lab vel. before collision	$\vartheta_{1L}, \varphi_{1L}$:	lab scattering angles (b1)
$\boldsymbol{v}_1'$:	lab vel. (b1) after collision	ϑ_1 :	polar angle of $\boldsymbol{v}_1$
$\boldsymbol{u}_1$:	cm vel. (b1) before collision	ϑ_1', φ_1' :	polar and azimuthal angles of $\boldsymbol{v}_1'$
$\boldsymbol{u}_1'$:	cm vel. (b1) after collision		

correlation between background events. Hence, time-to-amplitude-converter (TAC) spectra display a sharp peak produced by ion–ion collisions on top of a flat background due to random coincidences. A typical TAC spectrum is shown in the upper right part of Fig. 3; details of this technique are discussed in Ref. [6].

If only one collision partner alters its charge state, e.g., in an ionization reaction (4), a coincidence technique does not apply. The ion–ion signal can be separated from background events by means of a beam pulsing technique. Both beams are chopped, e.g., by electrostatic deflectors upstream of the interaction region, and the time dependence of the detector count rate is recorded. When both beams are switched on, ion–ion reactions are recorded on top of background contributions produced by either of the beams, and by electronic noise. In additional time periods, only beam 1, beam 2 or neither of the two is switched on in order to record the different background contributions. The true ion–ion signal is obtained by suitable subtraction. This method is described in detail in Ref. [7].

Recently, the first angular differential cross sections were measured for ion–ion collisions [8]. These measurements require position sensitive product ion detection. The idea is outlined in Fig. 3 for the resonant electron capture reaction (6). As in the coincidence technique described above, a TAC spectrum is measured which is started by He^+ reaction products, separated from their parent He^{2+} ion beam, and stopped by He^{2+} products, which stem from their parent He^+ beam. Detector 2 consists of a channel-plate array and a position sensitive resistive anode which allows one to record the event time and position simultaneously. The TAC coincidence time window (dark gray) contains signal and background events which were accumulated in the left position matrix 1, while noncoincident windows (light gray) contain background events only, accumulated in the right position matrix 2. After suitable normalization procedures, matrix 2 can be subtracted from matrix 1 in order to obtain the spatial distribution of He^+ reaction products.

Note that a spatial distribution matrix is accumulated at rates ranging from 10^{-2} to 10^1 events per second. For a single distribution, data acquisition times in the order of a few days are usually required.

The extraction of differential cross sections $d\sigma/d\Omega$ from a measured spatial distribution matrix requires, on the one hand, the transformation between lab- and cm system, discussed in Sect. 2.2. On the other hand, the primary ion beam is not a mathematical line which is infinitely narrow, but has a finite beam profile $\mathcal{P}$ (beam radius less than 1 mm, divergence half angle $\vartheta_d \approx$ 0.1 deg) which broadens the observed scattering distribution and smears out structures. The experimental data can be deconvoluted with respect to $\mathcal{P}$, if sufficient angular scattering occurs, i.e. if the main part of the distribution is scattered into solid angles $d\Omega(\vartheta)$ with ϑ larger or at least comparable to ϑ_d. Bayesian deconvolution [9] or Fourier methods were introduced by *Krüdener* [10] for ion–ion angular differential cross section measurements.

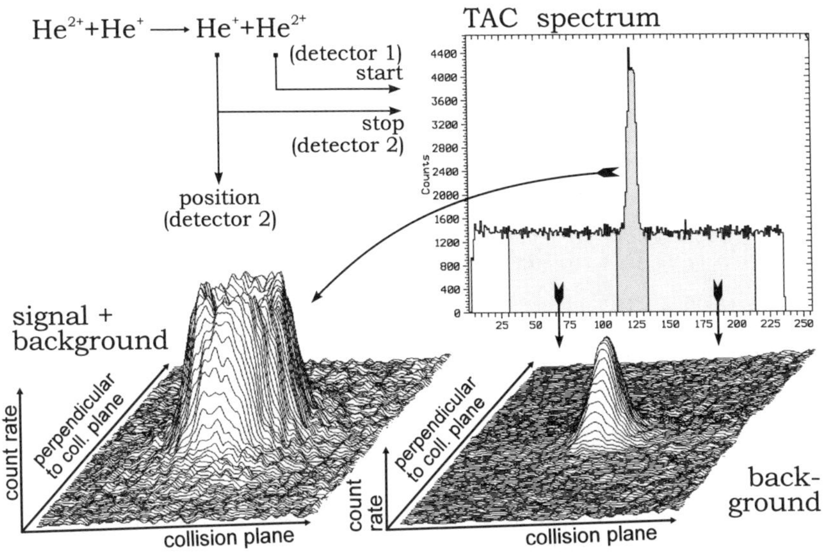

Fig. 3. Position-sensitive detection of reaction products, illustrated for the electron capture reaction $He^{2+}+He^{+} \longrightarrow He^{+}+He^{2+}$. He^{2+} product ions, recorded by single particle detector 1, start a time-to-amplitude-converter (TAC) which is stopped by He^{+} product ions recorded by single particle detector 2. Simultaneously, detector 2 records the event position. All events in the TAC coincidence window (dark gray) are accumulated in one position matrix (signal + background), the other events (light gray) in a second position matrix (background). The spatial distribution of ion–ion events is obtained by suitable matrix subtraction. The matrix area is 40 mm × 40 mm

A typical example of a modern *crossed-beam* experiment is the Giessen ion–ion collision facility which has been described in full detail by *Meuser* et al. [11]. Ions are produced in ECR sources and accelerated to energies up to $q_1 \cdot 300$ kV and $q_2 \cdot 20$ kV, respectively. Both beams are momentum-analyzed by magnets and pass through differentially pumped vacuum tanks providing increasingly lower residual gas pressure before they intersect in an UHV vacuum chamber which is operated at pressures in the low 10^{-11} mbar region. The interaction angle α can be adjusted at 17.5 deg or 137.5 deg which gives access to the cm energy range plotted in Fig. 1. The ion optical system of the apparatus must allow one to distinguish a true ion–ion event from the incoming flux which can be up to 10^{14} times more intense. The complete system is ion-optically optimized in extensive computer simulations and test measurements. Detailed information can be found in Ref. [11].

3. Collisions between Light Ions

3.1 One-Electron Systems

The investigation of collisions between protons and singly charged helium ions is of considerable interest in plasma physics, especially in thermonuclear fusion research. Not only do these encounters occur inevitably in fusion plasmas but they also represent the simplest ion–ion collision system involving only one electron. Description of electron capture σ_C

$$H^+ + He^+ \longrightarrow H^0 + He^{2+} \tag{3}$$

and ionization σ_I

$$H^+ + He^+ \longrightarrow H^+ + He^{2+} + e^- \tag{4}$$

is therefore an ideal testing ground for different theoretical models. Since the wave functions are known exactly, the accuracy of the calculated cross sections directly reflects the quality of the theoretical approach. Of special interest is the intermediate energy range where the relative velocity of the colliding particles is comparable with that of the orbiting electron.

Three experimental groups [6, 13, 12] have studied these collisions independently. While capture cross sections σ_C were measured directly by coincident detection of H^0 atoms and He^{2+} ions, ionization cross sections σ_I were measured indirectly by subtraction $\sigma_I = \sigma_{2+} - \sigma_C$ from σ_{2+}

$$H^+ + He^+ \longrightarrow He^{2+} + \ldots \tag{5}$$

which is the cross section for the total production of He^{2+} ions which is obviously the sum of reactions (3) and (4). It can be measured by means of a beam pulsing technique [7]. For medium collision energies the cross sections σ_{2+} and σ_C are almost equal: subtraction yields considerably larger errors for σ_C.

The experimental cross sections are compared with theory in Fig. 4. In the high energy limit, the Born approximation is valid while for lower energies the continuum distorted wave calculation CDW-EIS with eikonal initial states by *Martinez* et al. [23] agrees as perfectly with the ionization data as the atomic orbital AO^+ calculation by *Fritsch* and *Lin* [19] with the capture results. Since various authors confirm these findings, total cross sections for the reactions (3) and (4) are well described by theory.

For decades, theory was tested by comparing calculated and measured total cross sections. The first angular differential cross section was measured by *Krüdener* et al. [8] for the electron capture of α-particles from He^+ ions:

$$He^{2+} + He^+ \longrightarrow He^+ + He^{2+} \,. \tag{6}$$

This one-electron, charge-symmetric collision is resonant for $1s \rightarrow 1s$ electron capture and provides a unique testing ground for the study of the long-range Coulomb interaction in the quantum three-body problem. Krüdener

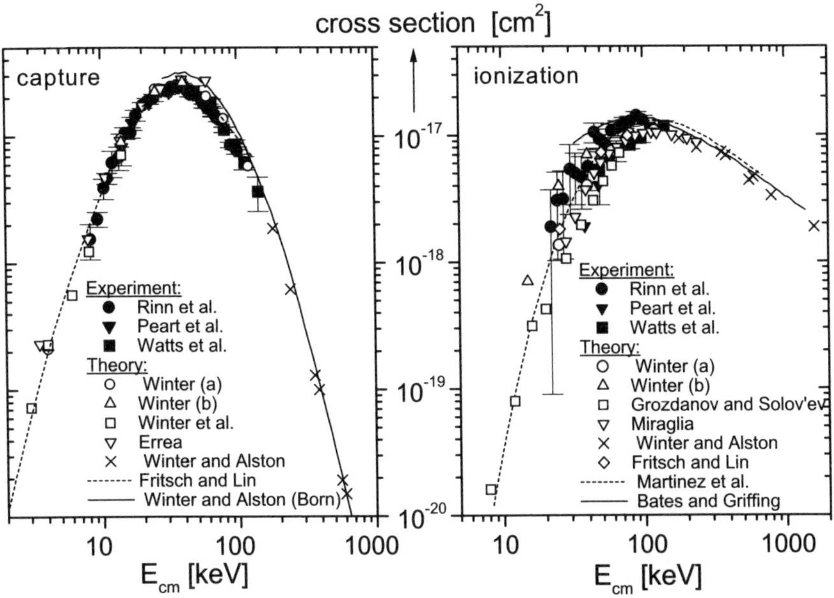

Fig. 4. Cross sections for electron capture (3) and ionization (4) in $H^+ + He^+$ collisions, as a function of the cm collision energy E_{cm}. Experimental *crossed-beam* results by *Rinn* et al. [6, 7], *Peart* et al. [12] and *Watts* et al. [13]. Most theoretical results are obtained by close-coupling (CC) calculations: *Winter* (a) [14]: Sturmian basis and pseudostates; *Winter* (b) [15]: three-center atomic orbital (AO) basis; *Winter* et al. [16]: MO basis; *Errea* [17]: molecular orbital (MO) basis; *Winter* and *Alston* [18]: high energy CC; *Fritsch* and *Lin* [19, 20]: AO^+ basis; *Grozdanov* and *Solov'ev* [21]: MO basis; *Miraglia* [22]: CDW; *Martinez* et al. [23]: CDW-EIS; *Winter* and *Alston* [18] and *Bates* and *Griffing* [24]: Born approximation

et al. found an oscillatory structure in the scattering distribution of He^+ ions produced in reaction (6) which is displayed in Fig. 5. The distribution is deconvoluted using procedures discussed in Sect. 2.3. The minimum in the forward direction ($\boldsymbol{v}_1$ in Fig. 2) is surrounded by symmetric structures with rapidly decreasing amplitudes (note the log-z scale). For larger scattering angles numerical fluctuations due to very low signal rates are introduced by the deconvolution algorithm. Integration over the azimuthal angle φ averages these fluctuations more or less and yields the differential cross section $d\sigma/d\Omega$, plotted in Fig. 6 as a function of the cm scattering angle ϑ_{cm}.

In addition to the minimum in the forward direction, another four minima of $d\sigma/d\Omega$ are resolved experimentally. The oscillation period increases with scattering angles. *Forster* et al. [25] used a semiclassical eikonal calculation which describes the experimental data for low scattering angles ϑ_{cm}, but "runs out of phase" for larger ϑ_{cm}. The remarkable oscillatory structures in the differential electron capture cross section can be interpreted as interference between *gerade* and *ungerade* electronic states of the ionic molecule.

Uskov and *Presnyakov* [8] presented a quantum approach based on a partial wave analysis using a molecular basis. They also included the nonadiabatic rotational $2p\sigma_u - 2p\pi_u$ coupling describing capture into the excited $2p$ state and related scattering matrix element corrections and finally delivered a good description of the experimental data. We note that theory and experiment are compared on an absolute scale. It is mainly counting statistics and related deconvolution effects which limit the experimental resolution.

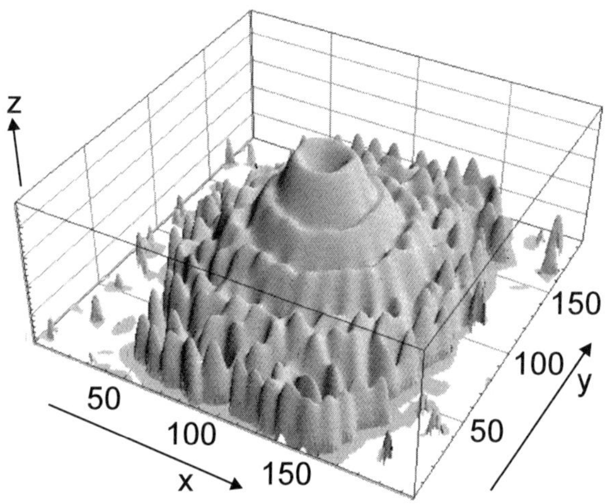

Fig. 5. Three-dimensional view of the scattering distribution of He^+ ions produced in $He^{2+} + He^+$ collisions at $E_{cm} = 2.5$ keV, drawn on a natural logarithmic z scale. Channel numbers on x and y axes equal 0.156 mm/channel, i.e. 0.00815°/channel. Data are deconvoluted with respect to the primary He^{2+} ion beam width. From Ref. [8]

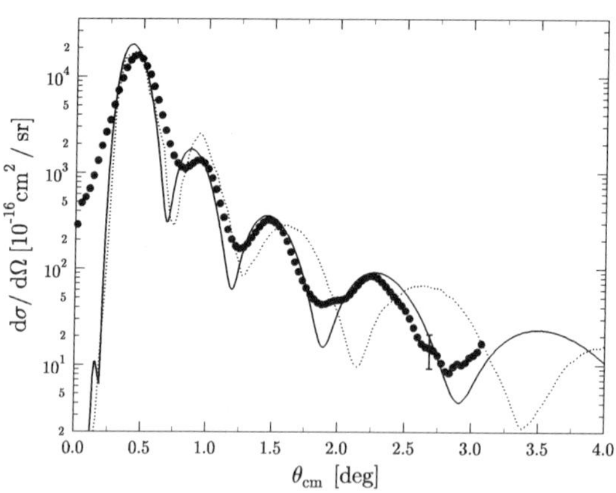

Fig. 6. Angular differential cross sections for the electron capture (6) in $He^{2+} + He^+$ collisions at $E_{cm} = 2.5$ keV. •: experiment, *Krüdener* et al. [8]; dotted line: semiclassical calculation, *Forster* et al. [25]; solid line: quantal calculation, *Uskov* and *Presnyakov* [8]. From Ref. [8]

3.2 Two-Electron Systems

The electron–electron interaction is introduced by the second electron into ion–ion collisions which enhances the difficulties of various theoretical approaches. The subject of fundamental studies is either the neutral (H – H) complex consisting of two electrons and two protons, or the charged (He – He)$^{2+}$ complex where two α-particles attract two electrons. We will first concentrate on the charged complex.

The (He – He)$^{2+}$ complex can be investigated either by colliding α-particles with atomic helium or, more sophisticated, by ion–ion collisions. The principle of detailed balance suggests that the cross section for electron capture σ_{C}

$$\mathrm{He}^{+} + \mathrm{He}^{+} \longrightarrow \mathrm{He}^{2+} + \mathrm{He}^{0} \tag{7}$$

in $\mathrm{He}^{+} + \mathrm{He}^{+}$ collisions equals the cross section for the reverse reaction

$$\mathrm{He}^{2+} + \mathrm{He}^{0} \longrightarrow \mathrm{He}^{+} + \mathrm{He}^{+} \tag{8}$$

in $\mathrm{He}^{2+} + \mathrm{He}^{0}$ collisions. This argument holds for state-specified transitions only. If two He^{+} ions collide, the electrons can orientate their spins parallel or antiparallel to each other, forming spin singlet or triplet states in the exit channels with statistical weights 1:3. Only 25% of the incoming flux will therefore populate singlet states. As helium gas is usually completely in the singlet ground state, the ion–atom cross section (8) is expected to be four times larger than the ion–ion cross section (7). Furthermore, it is clear that at very low collision energies mutual Coulomb repulsion between the ions suppresses electron transfer (7) whereas the Coulomb wall is absent in ion–atom collisions (8).

Modern close-coupling calculations are able to predict electron capture in the intermediate energy range, displayed in Fig. 7. *Gramlich* et al. [26] used Gaussian type orbitals, whereas *Kimura* [27] employed molecular, and *Fritsch* and *Lin* [28] atomic orbitals. All calculations correspond within the experimental uncertainty with *crossed-beam* results by *Melchert* et al. [29, 30], *Peart* et al. [31] and *Murphy* et al. [32].

Gramlich et al. and *Fritsch* and *Lin* also calculated partial cross sections for electron capture into singlet states. They showed that at low energies almost 50% of the incoming flux populates singlet states, and this fraction increases with collision energy. Afrosimov's measured cross sections for electron capture (8) slightly overestimate the singlet results when divided by the statistical weight of four.

Electron capture (7) into the singlet ground state He^{0} $(1s^2)$ has a Q-value of -30 eV while capture into triplet states $\mathrm{He}^{0}(1s, n = 2)$ requires $Q \approx -50$ eV. Translational energy spectroscopy is an extremely sophisticated attempt in ion–ion collisions and has not yet been done, but the nature of *crossed-beam* kinematics allows an estimate of Q. *Pfeiffer* et al. [33] observed a scattering distribution $\mathcal{D}$ (Fig. 8) of He^{0} atoms which were produced in reaction (7).

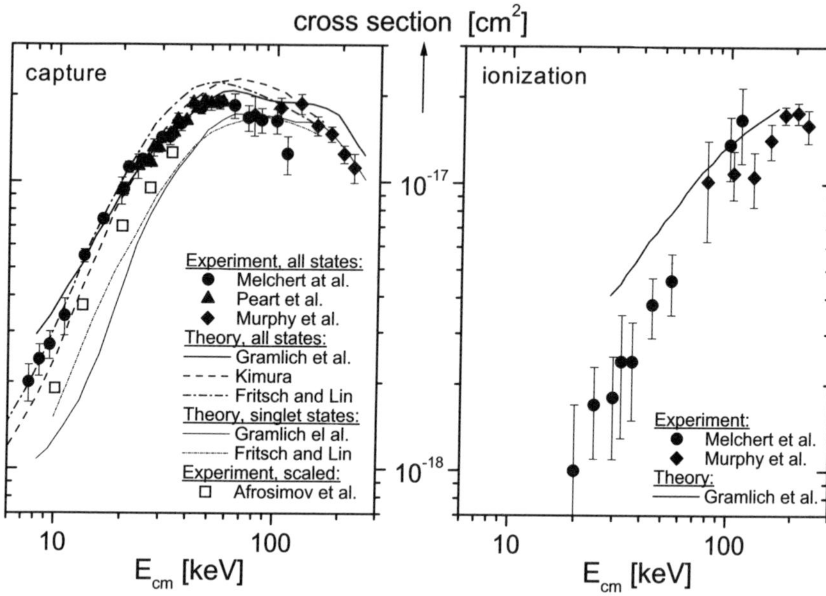

Fig. 7. Cross sections for electron capture (7) and ionization (9) in $He^+ + He^+$ collisions, as a function of the cm collision energy E_{cm}. Experimental *crossed-beam* data by *Melchert* et al. [29, 30], *Peart* et al. [31] and *Murphy* et al. [32]. Theoretical results by *Fritsch* and *Lin* [28] (atomic orbitals), *Kimura* [27] (molecular orbitals) and *Gramlich* et al. [26] (Gaussian orbitals). Partial cross sections denote electron capture into singlet states only. Cross sections (8) for $He^{2+} + He^0$ collisions, measured by *Afrosimov* et al. [34], are plotted divided by a factor of four

$\mathcal{D}$ is shifted with respect to the primary He^+ ion beam axis ($\boldsymbol{v}_1$ in Fig. 2), defined experimentally by He^0 atoms produced in collisions with the residual gas which form the distribution $\mathcal{P}$ (Fig. 8). Cuts through $\mathcal{D}$ and $\mathcal{P}$ exhibit a shift between them from which $Q = -(36 \pm 9)$ eV can be calculated. As a consequence, the intense inner part of $\mathcal{D}$ can be associated with charge transfer into the singlet ground state. The structure reflects the primary beam profile. $\mathcal{P}$-deflection by the small scattering angles can be neglected in comparison with the primary beam dimensions.

$\mathcal{D}$ shows a second feature: an elliptical structure perpendicular to the horizontal collision plane is observed (light gray in Fig. 8). *Crossed-beam* kinematics (Fig. 2) determines its symmetry since the relevant scattering angles are large in comparison with the beam profile for which $\mathcal{P}$ is a measure. This feature probably originates from electron capture into the triplet states, because these states contribute to the total cross section as discussed above. The proof would be a shift of seven channels ($Q \approx -50$ eV) for this feature, but poor counting statistics of $\mathcal{D}$ hinders the exact verification.

Ionization in $He^+ + He^+$ collisions

$$He^+ + He^+ \longrightarrow He^{2+} + He^+ + e^- \qquad (9)$$

is treated by *Gramlich* et al. [26] within the same framework as electron capture (7). They generate two-electron eigenstates with positive energies which they interpret as pseudostates describing the ionized continuum of He^+. Figure 7 shows good agreement of their results with experimental work of *Melchert* et al. [30] and *Murphy* et al. [32].

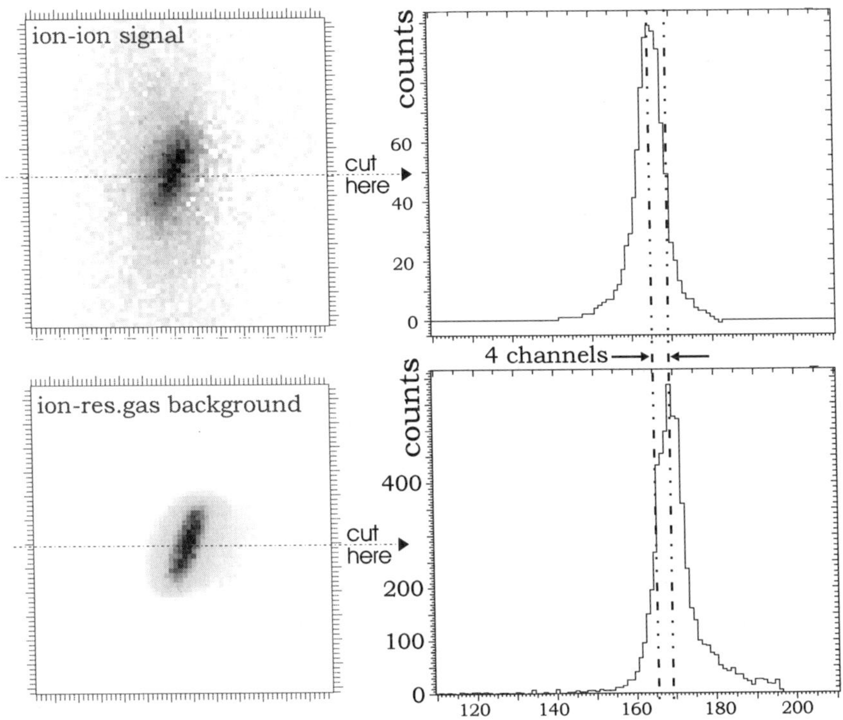

Fig. 8. Scattering distribution $\mathcal{D}$ of He^0 atoms produced in $He^+ + He^+$ collisions (7) (top left) and distribution $\mathcal{P}$, produced in collisions of He^+ with the residual gas (bottom left). Both distributions were measured simultaneously by *Pfeiffer* et al. [33]. Horizontal cuts through the same vertical channel are displayed on the right side. The intense parts of both distributions are shifted by four channels (0.62 mm) which corresponds to $\Delta\vartheta = (0.084 \pm 0.021)$ deg and $Q = (-36.5 \pm 9.1)$ eV. All scales are linear

The second basic two-electron collision complex (H – H) has been widely studied: inherent experimental and theoretical effort has been concentrated on this system. Detailed discussion would be beyond the scope of this chapter; therefore, in Fig. 9, we show only experimental data for ion pair production

$$H^0 + H^0 \longrightarrow H^+ + H^- \qquad (10)$$

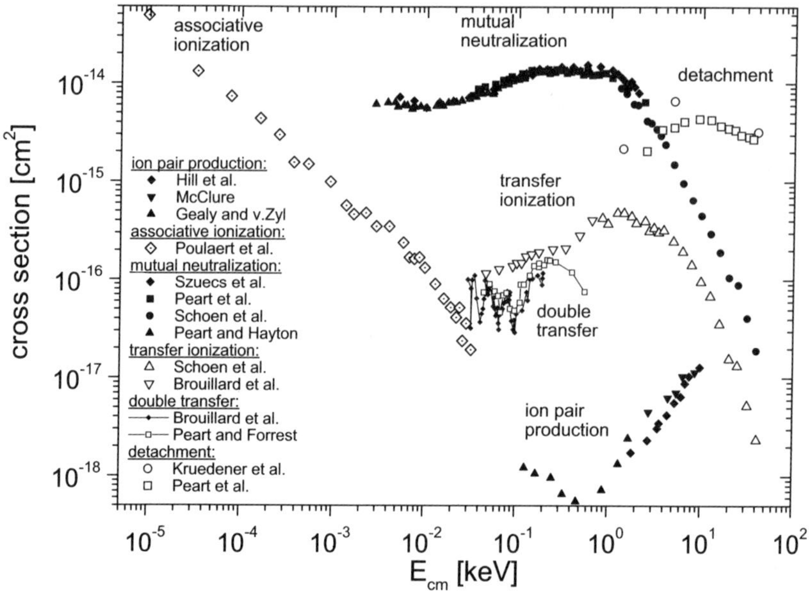

Fig. 9. Measured cross sections for various processes within the H-H collision complex as a function of the cm collision energy E_{cm}. Ion pair production (10): *Hill* et al. [35], *McClure* [36], *Gealy* and *v.Zyl* [37]; associative ionization (11): *Poulaert* et al. [38]; mutual neutralization (12): *Szücs* et al. [39], *Peart* et al. [40], *Schön* et al. [41], *Peart* and *Hayton* [42]; transfer ionization (13): *Schön* et al. [43], *Brouillard* et al. [43]; double transfer (14): *Brouillard* et al. [44], *Peart* and *Forrest* [45]; detachment (15): *Krüdener* et al. [46], *Peart* et al. [47]

in $H^0 + H^0$ collisions and for various processes in $H^+ + H^-$ collisions:

$$\text{associative ionization:} \quad H_a^+ + H_b^- \longrightarrow H_2^+ + e^- \tag{11}$$

$$\text{mutual neutralization:} \quad H_a^+ + H_b^- \longrightarrow H_a^0 + H_b^0 \tag{12}$$

$$\text{transfer ionization:} \quad H_a^+ + H_b^- \longrightarrow H_a^0 + H_b^+ + e^- \tag{13}$$

$$\text{double transfer:} \quad H_a^+ + H_b^- \longrightarrow H_a^- + H_b^+ \tag{14}$$

$$\text{detachment:} \quad H_a^+ + H_b^- \longrightarrow H_a^+ + H_b^0 + e^- . \tag{15}$$

Indices a and b indicate the fact that in *crossed-beam* experiments an ion of beam b can be distinguished by its momentum vector from an ion of beam a.

At low collision energies, associative ionization (11) occurs with cross sections almost as large as 5×10^{-14} cm^2, but it decreases rapidly with E_{cm}. In the medium energy range, mutual neutralization (12) is the dominating process with cross sections in the order of 10^{-14} cm^2. Data obtained in *crossed-beam* and *merged-beam* experiments by different experimental groups correspond almost perfectly. At collision energies above 2.5 keV, the H^- electron is not primarily captured by the proton, but ionized instead; electron detachment (15) surpasses mutual neutralization (12). Measurements by *Krüdener* et al. [46] show deviations from the data by *Peart* et al. [47] for the detachment (15). An actual calculation by *Belkic* [48] seems to confirm Krüdener's results.

A variety of calculations has been performed for the (H – H) collision complex; for clarity, they have been omitted from Fig. 9.

4. Heavy-Ion Collisions

The following paragraph is subdivided into three sections. First, collisions between ions with identical nuclei which are called homonuclear will be discussed. The main experimental work deals with collisions between singly charged ions. The second part contains experimental investigations of heteronuclear collisions between protons and singly charged heavy ions. Finally, the latest results of collisions between multiply charged ions will be reviewed.

4.1 Homonuclear Heavy-Ion Collisions

An understanding of electron capture and ionization in collisions between identical heavy ions is relevant to several aspects of controlled thermonuclear fusion research. Pellets of deuterium and tritium are proposed to be bombarded with intense GeV ion beams which are accumulated in large storage rings. Within these rings, electron capture σ_C

$$X^+ + X^+ \longrightarrow X^{2+} + X^0 \tag{16}$$

and ionization σ_I

$$X^+ + X^+ \longrightarrow X^{2+} + X^+ + e^- \tag{17}$$

occurs in collisions between the stored particles X^+ at keV relative energies. Two particles are lost from the stored ion beam X^+ per capture event (16), one particle per ionization event (17). For an accurate assessment of the expected particle loss rates, the cross sections σ_C and σ_I are needed. Cross sections σ_{2+} for the sum of capture (16) and ionization (17)

$$X^+ + X^+ \longrightarrow X^{2+} + \ldots \tag{18}$$

which is called the total X^{2+} production were measured for a variety of colliding ions X^+, plotted in Fig. 10. Contributions to σ_{2+} from reactions in which more than one electron is ionized are mostly negligible in the investigated energy range. With the exception of a few Li^+ data points, all cross sections σ_{2+} were measured at velocities less than an atomic unit and rise with the cm collision velocity v_{rel}. For v_{rel} fixed, σ_{2+} increases with the atomic number Z. The production of doubly charged alkali ions requires the break-up of a closed electronic shell, which is e.g., 75 eV for a Li^+ ion, but only 25 eV for a Cs^+ ion; the latter therefore possesses a larger cross section σ_{2+}.

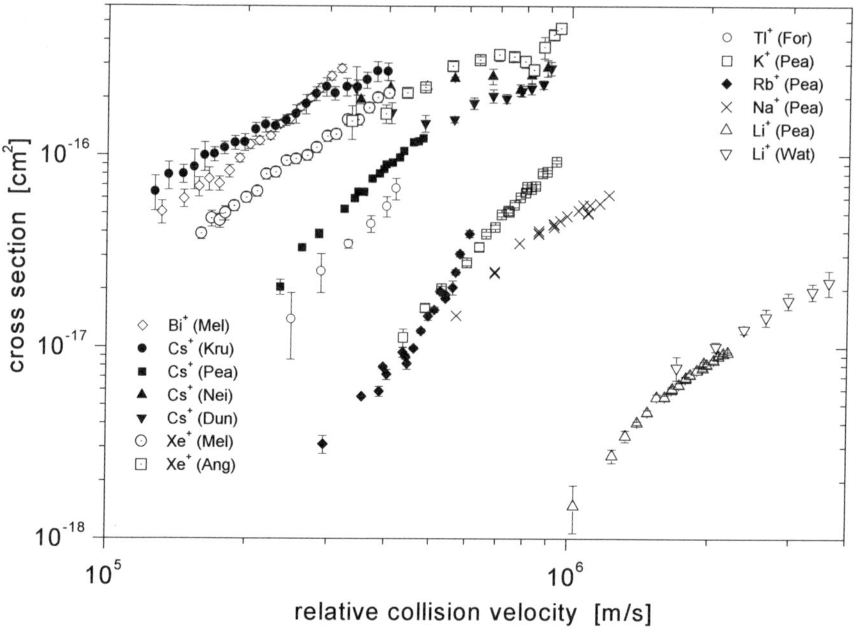

Fig. 10. Measured cross sections for the total X^{2+} production (18) in $X^+ + X^+$ collisions. $X^+ = \mathrm{Li}^+$: *Watts* et al. ([49]), *Peart* et al. [50] ; $X^+ = \mathrm{Na}^+, \mathrm{K}^+, \mathrm{Rb}^+$: *Peart* et al. [51]; $X^+ = \mathrm{Xe}^+$: *Angel* et al. [52], *Melchert* et al., unpublished ; $X^+ = \mathrm{Cs}^+$: *Krüdener* et al. [53], *Peart* et al. [54], *Dunn* et al. [55], *Neill* et al. [56]; $X^+ = \mathrm{Tl}^+$: *Forrest* et al. [57]; $X^+ = \mathrm{Bi}^+$: *Melchert* et al. [58]

Peart et al. [54] measured cross sections σ_{2+} for collisions between Cs^+ ions which are sufficiently smaller than the datasets by *Neill* et al. [56] and *Dunn* et al. [55] from the Belfast group, which themselves underestimate the data by Krüdener et al. [53] slightly. The deviations arise from incomplete detection of all Cs^{2+} ions due to large angular scattering in slow heavy ion collisions: *Krüdener* et al. measured angular differential cross sections for $\mathrm{Cs}^+ + \mathrm{Cs}^+$ collisions and found remarkable angular scattering of more than 10 deg in the cm frame for ionization (17) in $\mathrm{Cs}^+ + \mathrm{Cs}^+$ collisions at $E_{\mathrm{cm}} = 7.5$ keV.

Estimation of particle loss within storage rings as well as better physical understanding require knowledge of both the capture (16) and ionization (17) channels individually. These channels are only measured for a few ions X^+, plotted in Fig. 11. The closed Xe-like shell of Cs^+ is responsible for the relatively small cross section for capture, while the ionization cross section exceeds the Xe^+ and Bi^+ data. An atomic orbital (AO) calculation by *Ermolaev* et al. [59] indicates that, at intermediate energies, the main contribution to σ_{C} is capture of the $\mathrm{Cs}^+(5p)$ electron to the $6p$ state of Cs^0. For low energies, capture to $\mathrm{Cs}^0(6s)$ is the dominant channel. There exist no experimental data which can be compared with the molecular orbital (MO)

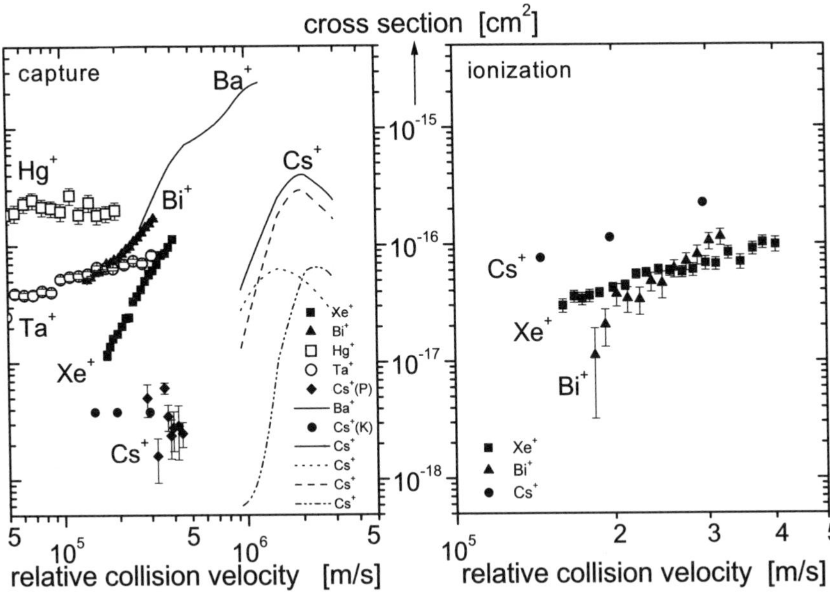

Fig. 11. Measured cross sections for electron capture (16) and ionization (17) in $X^+ + X^+$ collisions. Experimental data: Xe^+: *Melchert* et al., unpublished; Cs^+: *Krüdener* et al. [53]; *Peart* et al. [60]; Ta^+, Hg^+: *Melchert* et al., unpublished; Bi^+: *Melchert* et al. [58]. MO-calculation for $X^+ = Ba^+$: *Sramek* et al. [61]. *Ermolaev* et al. [59] calculated the total cross section for $X^+ = Cs^+$ within an AO basis, plotted as the solid line; the dashed line is the partial cross section for capture of the $Cs^+(5p)$ electron to the $Cs^0(6p)$ state. Analogously, the dotted line represents $5p \to 6s$ capture, the dash-dotted line $5p \to 5d$ capture

calculation by Sramek [61] for the Ba^+ ion. An experimental uncertainty is the presence of electronically excited states in the ion beams which were used for *crossed-beam* experiments. Plasma ion sources usually generate a whole spectrum of excited states. Metastable states may survive the flight time from the ion source to the collision region which usually is in the μs regime. The measured cross sections in Fig. 11 are averaged between the ground state and eventually excited states, weighted by the individual populations. The large cross section for electron capture (16) of the Hg^+ ion might be due to a large cross section of metastable fractions within the incoming ion beams. So far, no technique has been established in ion–ion *crossed-beam* experiments to state-select the incoming ion beams. On the other hand, ions used or encountered in technical applications may have similar excitation to those generated in the laboratory.

4.2 Heteronuclear Heavy-Ion Collisions

The main emphasis in heteronuclear ion–ion collisions is put on collisions of protons with singly charged heavy ions. In thermonuclear fusion plasmas, fuel ions H^+ (or D^+) can capture an electron from heavy impurity ions, leading to the loss of the neutralized fuel and, furthermore, to an enhanced charge of the impurity ion which increases the radiative energy loss of the plasma. Cross sections for electron capture σ_C

$$H^+ + X^+ \longrightarrow H^0 + X^{2+} \tag{19}$$

and ionization σ_I

$$H^+ + X^+ \longrightarrow H^+ + X^{2+} + e^- \tag{20}$$

are plotted in Fig. 12. Small cross sections were observed for light ions $X^+ = He^+$ or Li^+, while heavy ions $X^+ = Tl^+$ or Ba^+ exhibit large cross sections. For collision energies 50 keV $\leq E_{cm} \leq$ 200 keV, the collision velocity v_{rel} is neither low nor high in comparison with the orbiting velocity v_e of the active electron. In heavy ion collisions, however, v_e is not a well defined quantity as inner electrons can be involved: they can be captured directly or, in addition to the capture of the valence electron, excited or ionized. Capture of an inner electron can be followed by autoionization or related processes feeding transfer ionization channels instead of the capture channel.

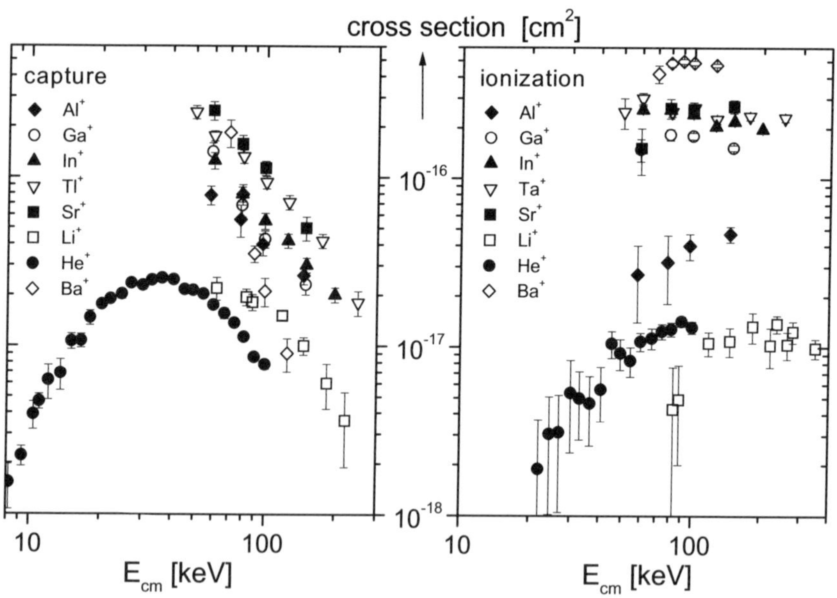

Fig. 12. Measured cross sections for electron capture (19) and ionization (20) in $H^+ + X^+$ collisions. $X^+ = He^+$: *Rinn* et al. [6, 7]; $X^+ = Li^+$: *Sewell* et al. [62]; $X^+ = Al^+, Ga^+, In^+, Ta^+$: *Watts* et al. [63]; $X^+ = Sr^+, Ba^+$: *Hopkins* et al. [64]

Vriens [65] suggested a classical scaling for ionization (20) of a singly charged ion X^+ by proton impact

$$\tilde{\sigma_{\mathrm{I}}} = \left(\frac{I}{Ry \cdot Z_{\mathrm{P}}}\right)^2 \cdot \frac{1}{x} \cdot \sigma_I \tag{21}$$

where I is the ionization potential, Ry the Rydberg energy, Z_{P} the projectile charge and x the number of electrons in the outermost shell. When plotted as a function of the collision velocity v_{rel}, measured in units of v_{e}, the scaled cross sections for ionization (20) and also for electron capture (19) in $\mathrm{H}^+ + X^+$ collisions form universal curves with deviations less than ± 50%. An exception is made by the He^+ capture and the Al^+ ionization data which exhibit slightly smaller scaled cross sections.

If a proton captures an electron from the ion X^+, the ion can emit additional electrons. Hughes et al. [66] showed that the total production of H^0 atoms in $\mathrm{H}^+ + \mathrm{Tl}^+$ collisions is an order of magnitude more efficient than the sole H^0 production by electron capture (19). The difference stems from transfer ionization where one electron is captured, and one or more electrons are ionized simultaneously.

4.3 Collisions Between Multiply Charged Ions

Ion–ion collision data required for various applications are sparse; for collisions between multiply charged ions they are practically absent. It is due to the recent development of powerful Electron Cyclotron Resonance (ECR) ion sources that *crossed-beam* experiments can be carried out with intense beams of multiply charged ions. Up to now only a few experiments, in which at least one ion carries a charge higher than one, have been performed.

Large cross sections for electron capture reactions can be expected for resonant ($Q = 0$) ion–ion collisions. *Brandau* et al. [67] measured cross sections for quasi-resonant electron capture in two mutually reverse reactions

$$\mathrm{C}^{2+} + \mathrm{B}^+ \longrightarrow \mathrm{C}^+ + \mathrm{B}^{2+} - Q, \tag{22}$$

$$\mathrm{B}^{2+} + \mathrm{C}^+ \longrightarrow \mathrm{B}^+ + \mathrm{C}^{2+} + Q, \tag{23}$$

where $Q = 0.77$ eV is small. The region of so-called nonadiabatic transitions is located at large internuclear distances R_{C}, where the energy difference between the initial and final states is comparable with the exchange interaction potential. In the simplest case of a two-level system, the exact solutions for model potentials as well as analytical approximations for more general potentials give cross sections close to $\sigma \approx \pi R_{\mathrm{C}}^2$ at the maximum. *Janev* and *Belic* [69] employed the *Rosen–Zener–Demkov* model [68] and some simplifying assumptions to calculate cross sections for (22) and predicted cross sections around 6×10^{-16} cm^2 for collision energies 0.5 keV $\leq E_{\mathrm{cm}} \leq$ 50keV which underestimated the measurement by Brandau et al. by a factor of four (see Fig. 13). The calculated oscillations were also not verified by the experiment. Subsequent calculations by *Uskov* et al. [67] employed analytical

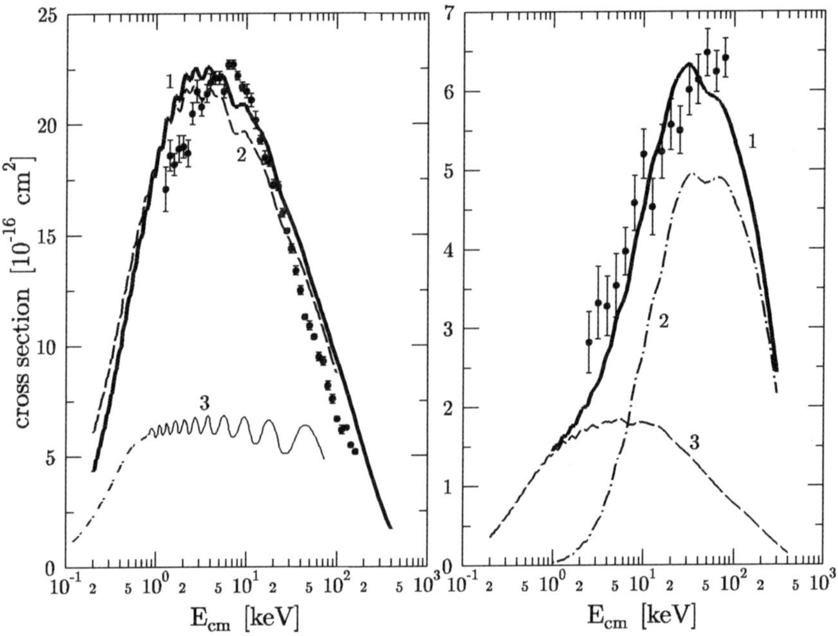

Fig. 13. Cross sections for electron capture (22) in $C^{2+} + B^+$ collisions (left) and for electron capture (23) in $B^{2+} + C^+$ collisions (right). Experimental data points by *Brandau* et al. [67] include 90% of statistical error. Curve 3, left: calculation by *Janev* and *Belic* [69]. The other curves are theoretical results by *Uskov* et al. [67]: Curve 1: total cross sections; curve 2, left: theoretical results with 20% of the incoming ions in the (metastable) $C^{2+}(2s\,2p\;^3P)$ state; curve 2, right: partial cross section for capture to the (metastable) $B^+(2s\,2p\;^3P)$ triplet state; curve 3, right: partial cross section for capture to the B^+ singlet states. From Ref. [67]

methods within the framework of nonadiabatic transitions and confirmed the experimental findings.

It is interesting to note that the cross sections σ_{C2B} for (22) and σ_{B2C} for (23) are not identical. If only single initial and final states were involved, the principal of detailed balance requires a cross section ratio given by the relative statistical weight of the molecular $^1\Sigma$ state in the ground configuration of the system $(CB)^{3+}$: $\sigma_{\mathrm{B2C}} = 1/12 \cdot \sigma_{\mathrm{C2B}}$. Obviously, curve 3 in Fig. 13 (right) is close to 1/12 of σ_{C2B}, which is given by curve 1 in Fig. 13 (left). The total cross section σ_{B2C}, however, is much larger than curve 3: metastable triplet states prevent a direct interrelation of the capture reactions (22) and (23).

Metastable ions in the triplet states $2s2p\;^3P$ will be populated up to a certain fraction within the incoming ion beams. Curve 2 in Fig. 13 (left) indicates the minor change of the obtained cross sections with 20% of the incoming C^{2+} ions being in the metastable $2s2p\;^3P$ state.

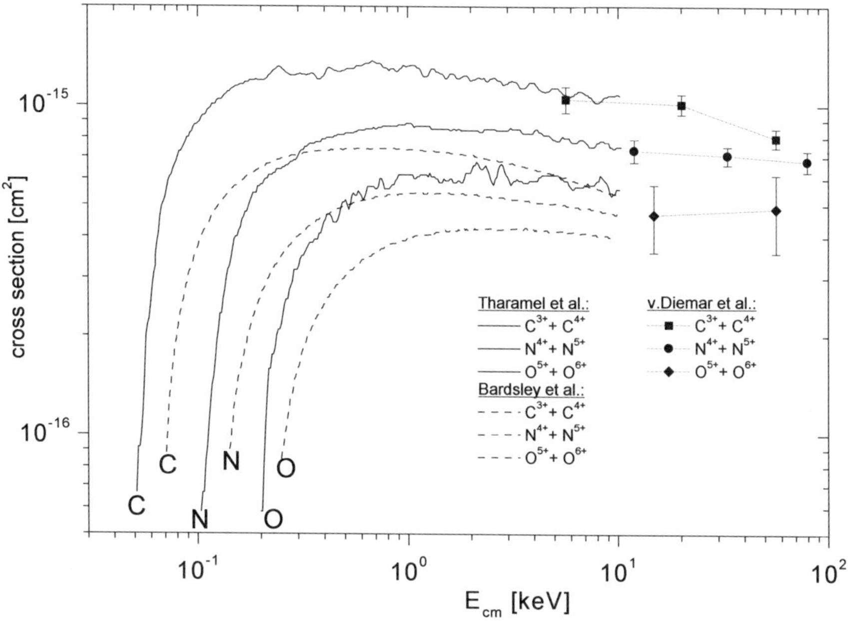

Fig. 14. Cross sections for resonant electron capture (24) in $X^{(Z-3)+} + X^{(Z-2)+}$ collisions for $X = \mathrm{C, N, O}$. Data points are preliminary results by *v.Diemar* et al. [70], connected by dotted lines to guide the eye; solid curves: theory by *Tharamel* et al. [71]; dashed curves: theory by *Bardsley* et al. [72].

Electron capture is fully resonant for collisions between multiply charged lithium- and helium-like ions

$$X^{(Z-3)+} + X^{(Z-2)+} \longrightarrow X^{(Z-2)+} + X^{(Z-3)+} \tag{24}$$

Bardsley et al. [72] and *Tharamel* et al. [71] have calculated in asymptotic theory cross sections for a sequence of isoelectronic collisions (24). The $1s^2$ electrons were treated as spectators in the electron capture process. Recently measured data points by *v.Diemar* et al. [70], plotted in Fig. 14, smoothly extend the curves calculated by Tharamel et al. to higher energies while the results by Bardsley et al. seem to underestimate the measured cross sections. In the Coulomb threshold region, where mutual repulsion between the collision partners prevents electron capture, the two calculations have not been checked by experiments so far.

Tharamel et al. [71] introduced a scaling law of the cross section σ_Z for the electron capture reaction (24)

$$\sigma_\mathrm{Z} = \frac{I_\mathrm{Li}}{I_\mathrm{Z}} \cdot \sigma_\mathrm{Li}, \tag{25}$$

where σ_Li is the cross section for electron capture in $\mathrm{Li}^0 + \mathrm{Li}^+$ collisions, and I_Li and I_Z are the ionization energies of the Li atom and the $X^{(Z-3)+}$ ion, respectively. For scaling purposes, σ_Z is plotted on the reduced energy scale u

$$u = \frac{m_X}{m_H} \cdot \frac{E_{cm}}{I_Z}, \tag{26}$$

where E_{cm} is the cm collision energy and m_X the ion mass, measured in units of the hydrogen mass m_H. For $u \geq 10$, cross sections for the different ion species form a common curve; below this limit Coulomb repulsion between the colliding multiply charged ions reduces the cross sections in comparison with $Li^0 + Li^+$ collisions. This effect is more pronounced for higher charge states.

Replacing the helium-like $X^{(Z-2)+}$ ion in reaction (24) by a bare He^{2+} ion, the electron capture is no longer resonant. Nonresonant capture in collisions between multiply charged ions

$$^3He^{2+} + ^{13}C^{3+} \longrightarrow {}^3He^+ + ^{13}C^{4+} \tag{27}$$

was measured by *Melchert* at al. [73] and calculated by *Sidky* and *Lin* [73]. The ^3He and ^{13}C isotopes were chosen to avoid contamination of the primary ion beams by molecular $^1H_2^+$ and $^4He^+$ ions, which have virtually the same charge-to-mass ratio as $^4He^{2+}$ and $^{12}C^{3+}$ ions, respectively. In their close-coupling calculation *Sidky* and *Lin* expand the active electronic wave function as a sum of atomic states on both charge centers. The two inner electrons of the lithium-like C^{3+} ion enter as a model screening potential for the outer electron. In Fig. 15 measured and calculated cross sections are compared. Up to a collision energy of 50 keV the capture goes primarily to the $He^+(1s)$ channel, since the $He^+(1s)$ level is closest to the initial C^{3+} binding energy. As soon as one reaches the matching velocity, where the incoming He^{2+} ion is traveling at about the average speed of the C^{3+} valence electron, the capture spreads to other states of He^+, and the 1*s* state represents only half of the total capture probability (at $E_{cm} = 150$ keV).

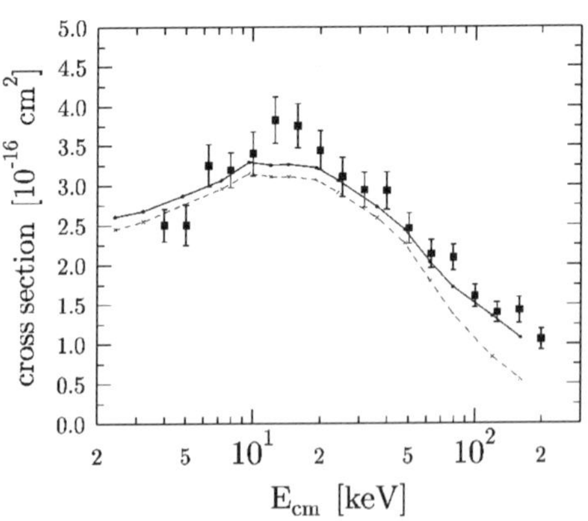

Fig. 15. Cross sections for electron capture (27) in $^3He^{2+}$ + $^{13}C^{3+}$ collisions. Experimental data points by *Melchert* et al. [73] include 90% of statistical error. Close coupling calculation by *Sidky* and *Lin* [73] for capture to all He^+ states (solid curve) and to the $He^+(1s)$ state only (dashed curve). From Ref. [73]

He^+ ions produced in the capture reaction (27) at $E_{cm} = 5.15$ keV were position-sensitively detected by Melchert et al. Taking into account the *crossed-beam* kinematics (see Sect. 2.2 and the discussion of Fig. 8), the Q-value of (27) can be estimated to be less than 36 eV. This experimental finding confirms that electron capture to the $1s$ ground state of the He^+ ion ($|Q| = 10$ eV) contributes mainly to the total cross section while capture to $n = 2$ ($|Q| = 51$ eV) or even higher states play a minor role.

5. Conclusion

Ion–ion collisions are one of the fundamental interactions between charged particles. As they occur within plasmas, they influence plasma modeling and plasma diagnostics. Interactions of ion beams with plasma targets and intra-beam scattering within ion beams are strongly related to ion–ion collisions. The unique ion–ion *crossed-beam* kinematics offers experimental advantages when observing single collision events.

ion–ion collisions in one- and two-electron systems are well described by modern theoretical approaches, at least as far as total cross sections are concerned. The measurement of differential ion–ion cross sections is still in its infancy. First angular differential measurements correspond with recent theories only. For multielectron systems, several experimental datasets are available but the theoretical descriptions of processes like electron capture or ionization are still rare. Several processes in collisions between various negative ions have been investigated experimentally, but are not described in this chapter. Pioneering theoretical work has been done for H^-+H^- collisions (see, e.g., the contribution by *Presnyakov* et al. in this volume).

Ion–ion collisions differ fundamentally from ion–atom collisions. Mutual Coulomb repulsion bends the particle trajectories, changes the potential energy of the system, and therefore influences the locations of potential energy curve crossings as well as the interaction energies. In certain kinematic situations, angular differential and state selective measurements can be performed simultaneously which is an advantage in comparison with ion–atom collisions. In general, cross sections for ion–ion collisions cannot be derived from results obtained in ion–atom collisions by the principle of detailed balance since the electronic levels populated in ion–ion collisions usually differ from those populated in ion–atom collisions.

In the future, modern ECR ion sources will allow the systematic investigation of collisions between multiply charged ions. Position-sensitive single-particle detection will enhance the information about the collision process itself: angular differential and state selective cross sections can be measured for appropriate collisions.

Acknowledgement. The author appreciates the fruitful collaboration with E. Salzborn and numerous colleagues and co-workers. Support by the Alexander-von-

Humboldt-Stiftung, Bonn, by the Deutsche Forschungsgemeinschaft (DFG), Bonn, and by the Bundesministerium für Erziehung, Wissenschaft, Forschung und Technologie (BMBF), Bonn, is gratefully acknowledged.

References

1. A. Dalgarno: In *Physics of Ion–Ion and Electron–Ion Collisions*, F. Brouillard, J.W. McGowan, eds. (Plenum Press, New York 1983) pp. 1–36
2. A.S. Schlachter, K.N. Leung, J.W. Stearns, R.E. Olson: In *Production and Neutralization of Negative Hydrogen Ions and Beams*, J.G. Alesi, ed., AIP Conference Proceedings No. 158 (Brookhaven 1986) p. 631
3. *HIBALL II: An Improved Conceptual Heavy-Ion-Driven Fusion Reactor Study*, Kernforschungszentrum Karlsruhe, Report KfK 3840 (July 1985)
4. D. Budicin, I. Hofmann, M. Conte, R. Schulze, F. Melchert, E. Salzborn: Il Nuovo Cimento Vol. **106 A (11)**, 1621 (1993)
5. F. Brouillard, W. Claeys: In *Physics of Ion–Ion and Electron–Ion Collisions*, J.W. McGowan, ed., Nato Advanced Study Institutes Series, Series B, Vol. 83 (Baddeck, Nova Scotia, Canada 1981) pp. 415–459
6. K. Rinn, F. Melchert, E. Salzborn: J. Phys. B **18**, 3783 (1985)
7. K. Rinn, F. Melchert, K. Rink, E. Salzborn: J. Phys. B **19**, 3717 (1986)
8. S. Krüdener, F. Melchert, K. v.Diemar, A. Pfeiffer, K. Huber, E. Salzborn, D.B. Uskov, L.P. Presnyakov: Phys. Rev. Lett **79**, 1002 (1997)
9. W.H. Richardson: J. Opt. Soc. Am. **62**, 55 (1972)
10. S. Krüdener: In *Proceedings of the XIXth International Conference on the Physics of Electronic and Atomic Collisions*, L.J. Dube, J.B.A. Mitchell, J.W. McConkey, C.E. Brion, eds., AIP Conf. Proc. 360 (Whistler, 1995) p. 677
11. S. Meuser, F. Melchert, S. Krüdener, A. Pfeiffer, K. v.Diemar, E. Salzborn: Rev. Sci. Instrum. **67**, 2752 (1996)
12. B. Peart, K. Rinn, K. Dolder: J. Phys. B **16**, 1461 (1983)
13. M.F. Watts, K.F. Dunn, H.B. Gilbody: J. Phys. B **19**, L355 (1986)
14. T.G. Winter: Phys. Rev. A **35**, 3799 (1987)
15. T.G. Winter: Phys. Rev. A **37**, 4656 (1988)
16. T.G. Winter, G.J. Hatton, N.F. Lane: Phys. Rev. A **22**, 930 (1980)
17. L.F. Errea, J.M. Gomez-Llorente, L. Mendez, A. Riera: J. Phys. B **20**, 6089 (1987)
18. T.G. Winter, S.G. Alston: Phys. Rev. A **45**, 1562 (1992)
19. W. Fritsch, C.D. Lin: J. Phys. B **15**, 1255 (1982)
20. W. Fritsch, C.D. Lin: In *Thirteenth International Conference on the Physics of Electronic and Atomic Collisions* (Berlin 1983), Abstract of Contributed Papers, p. 502
21. T.P. Grozdanov, E.A. Solov'ev: Phys. Rev. A **38**, 4333 (1988)
22. J.E. Miraglia: Phys. B **16**, 1029 (1983)
23. A.E. Martinez, R. Deco, D. Rivarola, P.D. Fainstein: Nucl. Instr. Meth. in Phys. Res. B **43**, 24 (1989)
24. D.R. Bates, G.W. Griffing: Proc. Phys. Soc. London **A 66**, 961 (1953)
25. C. Forster, R. Shingal, D.R. Flower, B.H. Bransden, A.S. Dickinson: Phys. B **21**, 3941 (1988)
26. K. Gramlich, N. Gruen, W. Scheid: J. Phys. B **22**, 2567 (1989)
27. M. Kimura: J. Phys. B **21**, L19 (1988)
28. W. Fritsch, C.D. Lin: Phys. Lett. A **123**, 128 (1987)

29. F. Melchert, K. Rink, K. Rinn, E. Salzborn, N. Gruen: J. Phys. B **20**, L223 (1987)
30. F. Melchert, K. Rink, K. Rinn, E. Salzborn: J. Phys. B **20**, L797 (1987)
31. B. Peart, K. Rinn, K. Dolder: J. Phys. B **16**, L361 (1983)
32. J.G. Murphy, K.F. Dunn, H.B. Gilbody: J. Phys. B **27**, 3687 (1994)
33. A. Pfeiffer, F. Melchert, E. Salzborn: unpublished (1994)
34. V.V. Afrosimov, A.A. Basalaev, G.A. Leiko, M.N. Panov: Sov. Phys. - JETP **47**, 837 (1978)
35. J. Hill, J. Geddes, H.B. Gilbody: J. Phys. B **12**, 3341 (1979)
36. G.W. McClure: Phys. Rev. **166**, 22 (1968)
37. M. Gealy, B. van Zyl: Phys. Rev. A **36**, 3100 (1987)
38. G. Poulaert, F. Brouillard, W. Claeys, J.W. McGowan, G. van Wassenhove: J. Phys. B **11**, L671 (1978)
39. S. Szücs, M. Karemera, M. Terao, F. Brouillard: J. Phys. B **17**, 1613 (1984)
40. B. Peart, M.A. Bennett, K. Dolder: J. Phys. B **18**, L439 (1985)
41. W. Schön, S. Krüdener, F. Melchert, K. Rinn, M. Wagner, E. Salzborn: J. Phys. B **20**, L759 (1987)
42. B. Peart, D.A. Hayton: J. Phys. B **25**, 5109 (1992)
43. W. Schön, S. Krüdener, F. Melchert, K. Rinn, M. Wagner, E. Salzborn, M. Karemera, S. Szücs, M. Terao, D. Fussen, R.K. Janev, X. Urbain, F. Brouillard: Phys. Rev. Lett. **59**, 1565 (1987)
44. F. Brouillard, W. Claeys, G. Poulaert, G. Rahmat, G. van Wassenhove: J. Phys. B **12**, 1253 (1979)
45. B. Peart, R.A. Forrest: J. Phys. B **12**, L23 (1979)
46. S. Krüdener, F. Melchert, E. Salzborn: unpublished (1991)
47. B. Peart, R. Grey, K. Dolder: J. Phys. B **9**, 3047 (1976)
48. D. Belkic: J. Phys. B **30**, 1731 (1997)
49. M.F. Watts, G.C. Angel, K.F. Dunn, H.B. Gilbody: J. Phys. B **17**, 1631 (1984)
50. B. Peart, R.A. Forrest, K.T. Dolder: J. Phys. B **14**, 3457 (1981)
51. B. Peart, R.A. Forrest, K.T. Dolder: J. Phys. B **14**, L603 (1981)
52. G.C. Angel, K.F. Dunn, P.A. Neill, H.B. Gilbody: J. Phys. B **13**, L391 (1980)
53. S. Krüdener: unpublished (1993)
54. B. Peart, R.A. Forrest, K.T. Dolder: J. Phys. B **14**, 1655 (1981)
55. K.F. Dunn, G.C. Angel, H.B. Gilbody: J. Phys. B **12**, L623 (1979)
56. P.A. Neill, G.C. Angel, K.F. Dunn, H.B. Gilbody: J. Phys. B **15**, 4219 (1982)
57. R.A. Forrest, B. Peart, K.T. Dolder: J. Phys. B **15**, L45 (1982)
58. F. Melchert, E. Salzborn, I. Hofmann, R.W. Mueller, V.P. Shevelko: Nucl. Instrum. Methods **A278**, 65 (1989)
59. A.M. Ermolaev, C.J. Noble, B.H. Bransden: J. Phys. B **15**, 457 (1982)
60. B. Peart, R.A. Forrest, K.T. Dolder: J. Phys. B **14**, L383 (1981)
61. S.J. Sramek, J.H. Macek, G.A. Gallup: Phys. Rev. A **22**, 1467 (1980)
62. E.C. Sewell, G.C. Angel, K.F. Dunn, H.B. Gilbody: J. Phys. B **13**, 2269 (1980)
63. M.F. Watts, C.J. Hopkins, G.C. Angel, K.F. Dunn, H.B. Gilbody: J. Phys. B **19**, 3739 (1986)
64. C.J. Hopkins, K.F. Dunn, H.B. Gilbody: J. Phys. B **24**, 2379 (1991)
65. L. Vriens: In *Case Studies in Atomic Collision Physics,* Vol. 1 (North Holland, Amsterdam 1969) p. 337
66. I.G. Hughes, K.F. Dunn, H.B. Gilbody: J. Phys. B **24**, L485 (1991)
67. C. Brandau, F. Melchert, S. Krüdener, S. Meuser, A. Pfeiffer, K. v.Diemar, T. Bartsch, E. Salzborn, D.B. Uskov, A.D. Ulantsev, L.P. Presnyakov: J. Phys. B **28**, L579 (1995)
68. Y.N. Demkov: Sov. Phys.- JETP **18**, 138 (1964)
69. R.K. Janev, D.C. Belic: J. Phys. B **15**, 3479 (1982)

70. K. v.Diemar, A. Pfeiffer, F. Melchert, E. Salzborn: unpublished (1998)
71. J. Tharamel, V.A. Kharchenko, A. Dalgarno: Phys. Rev. A **50**, 496 (1994)
72. J.N. Bardsley, P. Gangopadhyay, B.M. Penetrante: Phys. Rev. A **40**, 2742 (1989)
73. F. Melchert, S. Meuser, S. Krüdener, A. Pfeiffer, K. v.Diemar, E. Salzborn, E.Y. Sidky, C.D. Lin: J. Phys. B **30**, L697 (1997)

XVI Rearrangement Reactions in Ion–Ion Interactions

Leonid P. Presnyakov, E. Salzborn and H. Tawara

Summary. Recent experimental and theoretical results are discussed for ionic collisions with large cross sections. Single- and double-electron removal from H^- ions in collisions with other ions, and charge exchange between positively charged ions are considered at low and intermediate energies. For double ionization of He atoms by heavy ion impact, high energies of relative motion are also included. The theoretical methods are discussed from the viewpoint of general requirements of scattering theory.

1. Introduction

In hot hydrogenic plasmas with an impurity content of heavier elements, the emission spectrum and the distribution of the impurity ionization stages are largely determined by electron impact processes. There are, however, particular instances where transitions are caused preferentially by atom, proton and other ion impacts. In addition to modifying the plasma characteristics, such transitions have a special utility as diagnostics probes of the heavier component of the plasma. The importance of the proton impact processes had been pointed out by *Seaton* [1] and *Dalgarno* [2] long ago. The role of ion–atom collisions in fusion and astrophysical plasmas is also well understood, especially for emissivity and energy loss from the plasmas and for Neutral-Beam Injection (NBI) problems (see, for example, [3, 4]).

The ion–ion collisions may be important if their cross sections are large. Typical examples are the processes involving the negative ions in collisions with other ions (both positively and negatively charged). In collisions between two highly charged ions, the cross sections may be large for symmetric resonant and for quasi-resonant processes.

In the present chapter, we discuss single- and double-electron detachment in collisions of H^- ions with highly charged ions (HCI),

$$H^- + X^{q+} \quad \rightarrow \quad H^0 + \cdots, \tag{1}$$

$$H^- + X^{q+} \quad \rightarrow \quad H^+ + \cdots, \tag{2}$$

with other H^- ions,

$$H^- + H^- \quad \rightarrow \quad H^0 + H^- + e, \tag{3}$$

$$H^- + H^- \quad \rightarrow \quad H^0 + H^0 + 2e, \tag{4}$$

and quasi-resonant charge transfer,

$$X^{q+} + Y^{(q-1)+} \rightarrow X^{(q-1)+} + Y^{q+}, \quad (5)$$
$$Y^{q+} + X^{(q-1)+} \rightarrow Y^{(q-1)+} + X^{q+}, \quad (6)$$

at intermediate and low energies of the relative motion. Both experimental and theoretical results are considered. In order to illustrate one of the theoretical methods used, we discuss also single- and double-ionization of He atom under heavy-ion impact,

$$\mathrm{He} + X^{q+} \rightarrow \mathrm{He}^{+} + \cdots, \quad (7)$$
$$\mathrm{He} + X^{q+} \rightarrow \mathrm{He}^{2+} + \cdots. \quad (8)$$

The plan of this chapter is as follows. Section 2 contains a description of the recent experimental data jointly with theoretical calculations. The main attention is focused on the physical mechanisms of the processes. Section 3 is devoted to a discussion of the theoretical methods involved.

2. Experimental Data and Theoretical Calculations

At the present time, only a few sets of experimental data are published on ion–ion collisions, however; they demonstrate the most interesting features of the processes. The theory of ion–ion collisions is a natural continuation of the general theory of atomic collisions in which *fast* and *slow* collisions are analyzed with the help of one-center (atomic) and two-center (molecular) approaches, respectively. The theoretical analysis below is based on [5–16], and the original experimental data are taken from [14–19].

2.1 Single- and Double-Electron Detachment in Collisions of H^- Ions with Multiply Charged Ions

The experimental data are published for single-electron removal mainly. Medium-energy collisions with rare-gas ions ($q < 9$) have been investigated by *Salzborn*'s group [15], and slow collisions with He^+ by *Brouillard*'s group [17, 18] and by *Peart* and *Bennet* [19].

The data jointly with theoretical calculations for single-electron removal at intermediate energies [15] and for single- and double-electron detachment at slow collisions [5] are given in Fig. 1. Altogether, the resulting curves help to understand the physical mechanisms of the processes. At fast collisions, the one-center approach with non-stationary continuum-distorted waves [11, 15] shows that the leading mechanism is through- and over-barrier transmission of the electron in the time-dependent field of the projectile. The main process is the electron transition into the continuum; at slow collisions, however, it is charge transfer with capture of the electron into Rydberg states of the projectile. The two-center approach [5, 6] based on the exact analytic solution of the quasi-molecular problem (eigenfunctions and eigenstates) for the

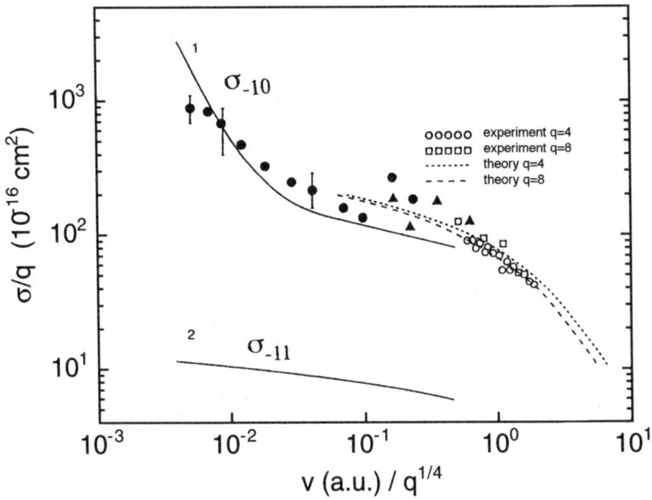

Fig. 1. Scaled cross sections of single- and double-electron removal in collisions of H^- with HCI versus scaled velocity of the relative motion. Curves (1) and (2) are calculations [5] for the single- and double-electron processes, respectively. Open circles and squares are experimental data [15], dashed curves are calculations [15]. Full circles [17, 18] and full triangles [19] are experimental data under He^{2+} impact

model Hamiltonian describes pseudo-crossings and other characteristics of the regions of nonadiabatic transitions (see Appendix).

Very large values of the single-electron cross sections at very small velocities have a simple kinematic explanation: in the Coulomb attraction field, the effective impact parameter is larger than the distance of closest approach. The double-electron capture can be considered as a product of two single-electron ones because the nonadiabatic regions for transitions of *inner* and *outer* electrons are well separated spatially. Thus, the problem is reduced to calculations of the transition amplitudes in H^0-HCI collisions.

The scaled velocity, $v/q^{0.25}$, in Fig. 2 is appropriate for charge-transfer in slow collisions.

For fast collisions, detachment into the continuum predominates, and the scaled velocity should be $v/q^{0.5}$ [11, 15]. The smallness of the ratios of double-to-single electron detachment (Fig. 2) shows that highly charged ions could be used for neutralization of H^- beams with high efficiency because the single-electron cross sections are very large.

2.2 Mutual Neutralization in Collisions of Two Negative Ions: Single- and Double-Electron Processes

In this case, the experimental data are available for both reactions (3) and (4) at center-of-mass energies between 2.5 and 100 keV [14]. The calculations have also been done for both processes at energies from 2.5 eV to 300 keV [5, 14]. At

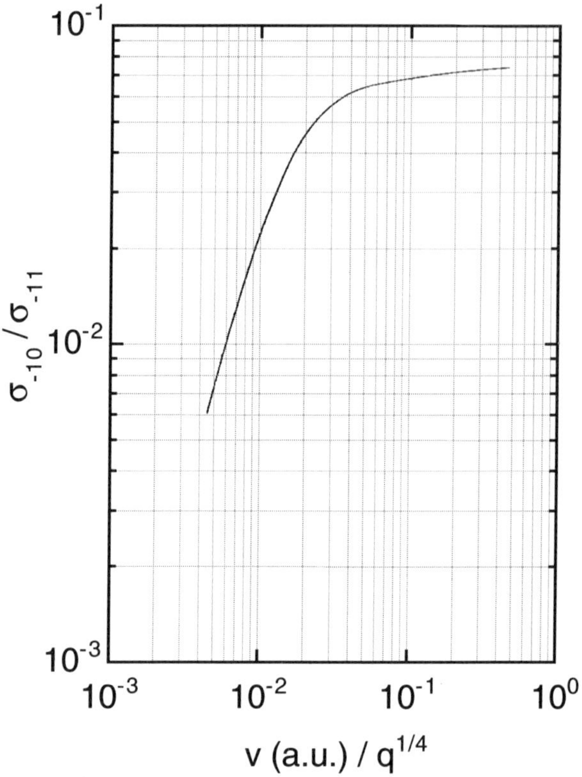

Fig. 2. Ratios of double-to-single electron removal according to curves (2) and (1) in Fig. 1

small velocities [5], the leading mechanism is the initial-state promotion into the continuum. When R is equal to or less than 36 a.u., there is promotion into the so-called *one-electron continuum*, after intersection with the ground state of the H^-H^- system. (It is worth noting that the negative hydrogen-molecule ion has no stable states, and was never observed.) At a distance of about 18 a.u., the H^- level intersects the ground-state level of the neutral hydrogen molecule, and goes into the *two-electron* continuum. The physical picture of single- and double-electron removal is exactly the same as described for medium velocities (see [14] and references therein).

In the present chapter (see Appendix), we employ just another way of calculating the autoionization widths on the basis of the two-center model Hamiltonian [5] instead of the modified perturbation theory [14]. The calculations for the *outer* electron detachment have been done using the Coulomb trajectory, and for the *inner* one, the straight-line path is appropriate. The results are given in Fig. 3 for cross sections and in Fig. 4 for the ratios. It is of interest that the ratio maximum follows from the calculations of the double-electron detachment [14] due to transitions to the double-electron continuum. When we compare only experimental results, the ratio curve is much smoother.

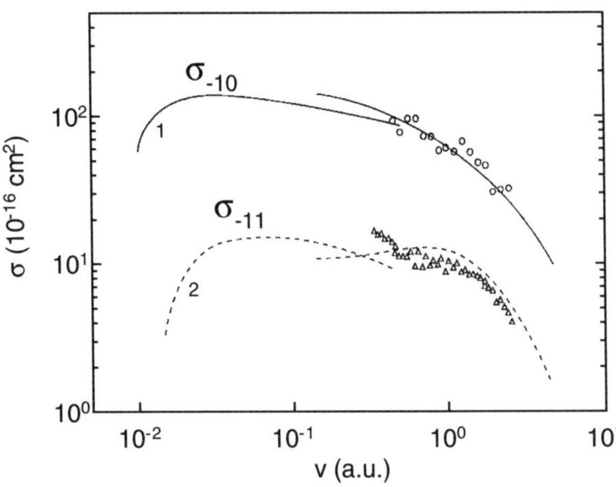

Fig. 3. Single- and double-electron detachment cross sections in collisions of two H^- ions versus velocity of the relative motion. Curves (1) and (2) are calculations [5]. The intermediate-energy experimental data and theoretical curves are taken from Ref. [14]

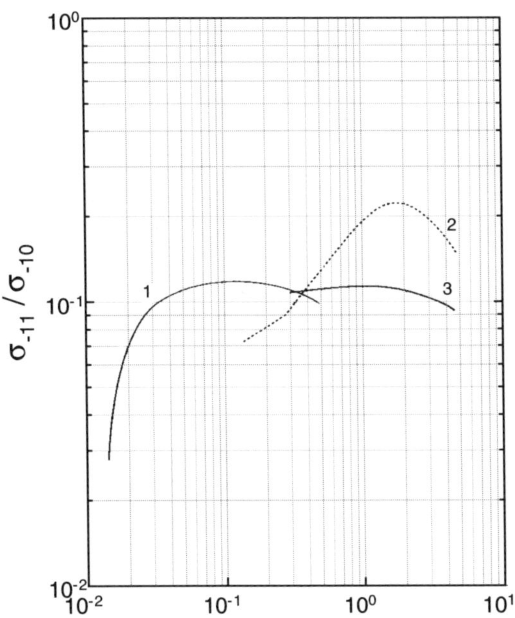

Fig. 4. Ratios of double-to-single electron detachment resulting from Fig. 3. Curve (1) represents results of the present calculations. Curves (2) and (3) are derived from calculations [14] and experimental data [14], respectively

2.3 Quasiresonant Charge Transfer Between Positively Charged Ions

Using a crossed-beam technique, *Salzborn*'s group [16] has measured absolute cross sections for the quasi-resonant charge transfer reaction,

$$C^{2+} + B^{+} \rightarrow C^{+} + B^{2+} - \Delta E, \tag{9}$$

by coincident detection of the product ions for cm energies between 1.3 and 160 keV. The inverse reaction,

$$B^{2+} + C^{+} \rightarrow B^{+} + C^{2+} + \Delta E, \tag{10}$$

was also investigated in the energy range between 2.5 and 80 keV. In this case, the energy difference (so-called *resonance defect*) is very small, $\Delta E = 0.77$ eV. The theoretical calculations [16] are in good agreement with the experimental data. The results for the direct reaction (9) and for the inverse one (10) are given in Fig. 13 of Chap. XV by *F. Melchert*. The calculations are based on the same principles of the theory of nonadiabatic collisions as was pointed out above: a) establishing the regions of nonadiabatic transitions, b) analysis of matrix elements and other parameters in those regions, and c) solving the close-coupling equations for the transition amplitudes. The problems a) and b) were solved analytically, and c) by means of numerical methods. The possible role of metastable ions in the beams was also investigated.

The main conclusions are as follows. The direct reaction (9) and the inverse (10) are not connected to each other by the statistical-weight ratio of the ground states of the colliding ions. This means that different channels are responsible for the direct and inverse reactions. In this case of many-electron ions, the reaction (9) involves mainly the singlet ground states of both ions, whereas the reaction (10) populates also the triplet (metastable) state of the B^{+} ion. For these processes, Coulomb repulsion is important at energies below 0.5 keV. Large values of the cross sections indicate the importance of the quasi-resonant processes for determining the characteristics of hot plasmas.

2.4 Ion–Atom Collisions: Single and Double Ionization of He by Heavy-Ion Impact

Concluding the brief overview of the experimental data available jointly with their theoretical interpretation, we discuss also the ratios of the cross sections of the processes (8) and (7). The corresponding cross sections and their ratios have been calculated recently [8] for projectiles with charge values from $q = 6$ to $q = 90$, at energies between 3 and 1300 MeV/u. For these calculations, a new modification of the non-stationary continuum-distorted wave functions [8] has been introduced. The results were obtained with the help of the second-order DWA with kinematic relativistic corrections. It is a typical example of fast collisions. The results show that the model of *independent electrons* (or *independent events*) can be used also for fast collisions. Figure 5 represents the calculated ratios jointly with the experimental data obtained by several groups. A list of references to the experimental work and earlier theoretical calculations is given in [8]. Recent calculations of absolute cross sections for double electron ionization [20] confirm the the applicability of non-stationary DWA.

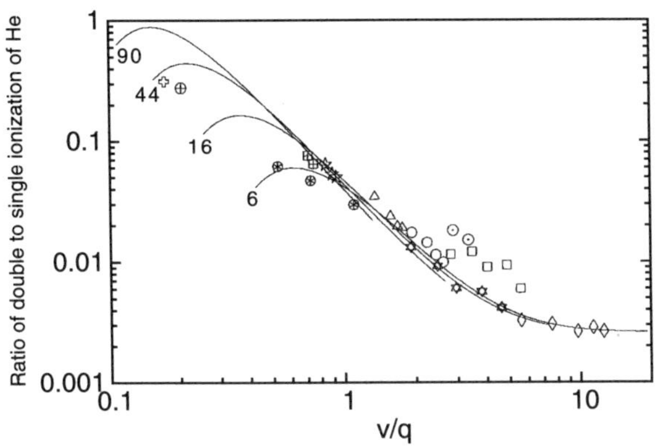

Fig. 5. Ratios of double-to-single ionization of He by several ions as a function of v/q (in atomic units). Full curves, theory (the q-values are indicated). Experimental data: open circles, C^{6+}, star within circle, U^{90+}. For other details and references, see [8]

2.5 Angular Differential Cross Sections for Charge Transfer Between a Bare and a Singly Charged He Ion

In many cases, the differential cross sections are extremely important for understanding physical features of ion–ion processes. Recently [21], the first measurements for charge transfer,

$$He^{2+} + He^{+}(1s) \rightarrow He^{+}(1s) + He^{2+}, \tag{11}$$

were reported jointly with the sequential quantum calculations. In principle, the physical picture of the process (11) is similar to that observed in ion–atom collisions with symmetric resonance. The oscillatory structure obtained is due to interference between scattering into *gerade* and *ungerade* molecular states, which arise due to the identity of the nuclear charges. However, there is an important difference resulting from Coulomb repulsion in the initial and final channels in ion–ion collisions. Both the gerade and ungerade potential energy curves have a repulsive behavior for the system $(He_2)^{3+}$. In the case of H^+ + H collisions, the potential curves have an attractive behavior at intermediate distances between the protons. There is the large potential well in the gerade state, and the comparatively small well in the ungerade state. The presence of attraction in the $(H_2)^+$ system leads to oscillatory rainbow scattering, which is absent in ion–ion collisions. As a result, it is hardly possible to establish a scaling law which could estimate the differential cross section in ion–ion collisions using the data available for ion–atom collisions. In both cases, the nonadiabatic rotational coupling gives noticeable quantitative corrections. The experimental data and calculations are given in Fig. 6.

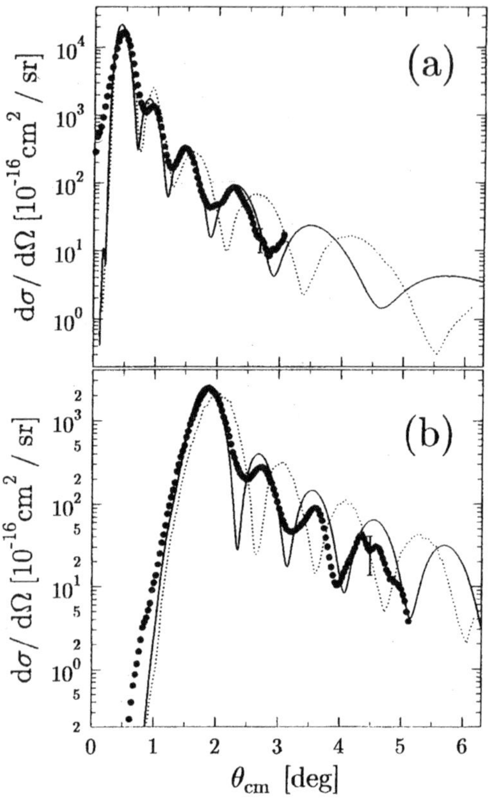

Fig. 6. Angular differential cross sections for the charge transfer process $He^{2+} + He^{+} \rightarrow He^{+} + He^{2+}$ at $E_{cm} = 2.5$ keV (a) and $E_{cm} = 0.5$ keV (b); solid circles – experiment [21], solid line – calculations [21], dotted line – semiclassical calculations [22]

3. Importance of Internal Theoretical Criteria

We discussed above typical experimental results on ion–ion collisions available up to now. The experimental data are limited, and the theory plays the essential role in the field. General requirements for the final results are the same for all branches of the theory of atomic collisions:

– well established range of applicability,
– estimated accuracy,
– description of existing experimental data, and (which is very important)
– prediction of future experimental results.

In all these cases, the analytic formulas and scaling laws provide a lot of valuable information on the physical pictures of the processes, even if those analytic expressions are approximate or derived as fits from numerical calculations.

Two quantum approaches are promising at present: the close coupling (CC), and nontraditional methods of continuum-distorted wave functions (CDW). In the CC approach, the following elements should be provided:

- symmetry properties in basis wave functions,
- unitarity of S-matrix (Hermitian character of the system of coupled equations),
- correct boundary conditions for rearrangement collisions (so called translation factors as the first approximation to the Jacobi coordinates),
- completeness of the basic set including the continuum states (in practice this means solving integro-differential equations instead of using pseudostates),
- analysis of the partial cross sections for estimation of the accuracy,
- parameterization of the equations and/or solutions in an analytic form for the development of scaling laws.

The CDW approach is subject to similar criteria, and should describe the asymptotic properties of the processes with respect to the natural physical parameters. Unification of both CC and CDW approaches promises interesting new results in the theory.

4. Conclusions

In the present chapter, the results for total cross sections are given in more detail in comparison with those for the differential cross sections which reflects the modern status of research in the field. Discussions above show that practical collaboration of several experimental and theoretical groups helps much in obtaining new results.

Here, we paid less attention to the very important problem of accuracy of the data discussed because of good agreement between experiment and theory. Such comparison represents a very important criterion, especially for the theory; however, it is one of the criteria. Other important criteria are listed in Sect. 3 of the present chapter.

Acknowledgement. The authors are grateful to many colleagues for effective collaboration during the research work and for discussions: Drs F. Melchert and S. Krüdener (Giessen University), Dr. I. Tolstikhina (NIFS), Drs. D. Uskov, V. Shevelko and A. Ulantsev (Lebedev Institute). L.P.P. acknowledges support by INTAS (project 4698) and RFBR (project 96-02-16090).

Appendix

The two-center Schrödinger equation for the electron moving in the Coulomb potential and in the short-range potential can be written in the form [10],

$$\left(\frac{1}{2}\nabla^2 + \frac{q}{r} + E\right)\Psi(\mathbf{r},\mathbf{R}) = \frac{2\pi}{\gamma}\left(\frac{\partial_{12}\Psi}{\partial r_{12}}\right)\delta(\mathbf{r}-\mathbf{R}) \tag{12}$$

subject to the boundary condition

$$\gamma = 2\pi\left(\frac{\partial_{12}\Psi}{\partial r_{12}}\right) \text{ at } r_{12}=0,\ r_{12}=\mid \mathbf{r}-\mathbf{R}\mid . \tag{13}$$

Here q is the charge of the incident ion, and the parameter γ is connected with the binding energy of the electron in the unperturbed H^- ion by

$$E = \frac{\gamma^2}{2},\ R \to \infty. \tag{14}$$

The solution of (12) is the Coulomb–Green's function found by *Hostler* and *Pratt* [12] in terms of Whittaker functions, M and W, with arguments and parameters depending on the electron energy. In our case, this energy is a function of the internuclear distance, R, and can be obtained from the solution of (13). This transcendental equation has the form,

$$\gamma = \frac{\mathrm{i}\Gamma(1-\mathrm{i}q/k)}{k}\left[\frac{2q}{R} + E(R))MW - M'W'\right], \tag{15}$$

$$k = k(R),\ E(R) = \frac{k^2(R)}{2}, \tag{16}$$

where the prime means differentiation with respect to the argument which is a function of the internuclear distance R and the energy. The M and W functions [12] satisfy the radial equation for the electron motion in the Coulomb field with the value of the orbital momentum equal to zero ($y = M$ or W):

$$\left[\frac{d^2}{dz^2} + 2\left(\frac{q}{z} + E\right)\right] y = 0. \tag{17}$$

It is easy to see that the coefficient in (16) has a linear zero when $q > 0$ and $E(R) < 0$, and also when $q < 0$ and $E(R) > 0$. The first case corresponds to the intersections of the negative-ion level with the Rydberg states of the positively charged ion. The second case corresponds to the promotion of the H^- ground-state level into the continuum in interactions with another negative ion. In both cases we can use the representation of the Whittaker functions in terms of the Airy functions for solving the transcendental equation (15). This approach was first proposed in Ref. [10] and later [13] was developed for the analysis of the nonadiabatic parameters at the level-crossing points in the discrete spectrum. When two negative ions collide, the physical picture is different; however, the mathematics is quite similar. The solution of (15) gives the real and imaginary parts of the energy. The latter describes the autoionization decay into the continuum. It is worth noting that the representations of the Whittaker functions in terms of the Airy functions have different forms in the discrete spectrum and the continuum, respectively, because the point $E(R) = 0$ is the branching point. However, the Airy-function techniques enable us to simplify this problem greatly.

References

1. M.J. Seaton: Proc. Phys. Soc. **68**, 457 (1955)
2. A. Dalgarno: Proc. Roy. Soc. A **262**, 131 (1961)
3. R.K. Janev, L.P. Presnyakov, V.P. Shevelko: *Physics of Highly Charged Ions* (Springer, Berlin, Heidelberg 1985)
4. H. Tawara: Report NIFS-DATA-25 (National Institute for Fusion Science, Nagoya 1995)
5. L.P. Presnyakov, H. Tawara: Phys. Scripta **T73**, 175 (1997)
6. L.P. Presnyakov, H. Tawara: Proc. ICPP-96 (Nagoya 1997), p. 706
7. L.P.Presnyakov: Phys. Scr. **T73**, 112 (1997)
8. L.P. Presnyakov, H. Tawara, I.Yu. Tolstikhina, D.B. Uskov: J. Phys. B **28**, 785 (1995)
9. L.P. Presnyakov, H. Tawara, D.B. Uskov: Nucl. Instrum. Methods B **98**, 332 (1995)
10. L.P. Presnyakov: Phys. Rev. A **2**, 1720 (1970)
11. L.P. Presnyakov, D.B. Uskov: Proc. P.N. Lebedev Institute, ed. by I.I. Sobelman (Nova Science Publishers, New York 1987), Vol. 179, p. 137
12. L. Hostler, R.H. Pratt: Phys. Rev. Lett. **10**, 469 (1963)
13. L.P. Presnyakov: in *Theory of Atomic Particle Collisions*, V.Galitsky, E.Nikitin and B.Smirnov, eds. Suppl. 6 (Nauka, Moscow 1981), p. 231 (in Russian)
14. F. Melchert, R. Schulze, S. Krüdener, S. Meuser, E. Salzborn, D.B. Uskov, A.D. Ulantsev, L.P. Presnyakov: J. Phys. B **28**, 3299 (1995)
15. F. Melchert, M. Benner, S. Krüdener, R. Schulze, K. Huber, E. Salzborn, D.B. Uskov, L.P. Presnyakov: Phys. Rev. Lett. **74**, 888 (1995)
16. C. Brandau, F. Melchert, S. Krüdener, S. Meuser, A. Pfeiffer, K. v Diemar, T. Bartsch, E. Salzborn, D.B. Uskov, A.D. Ulantsev, L.P. Presnyakov: J. Phys. B **28**, L579 (1995)
17. M. Terao, S. Szucs, M. Cherkani, F. Brouillard, J.R. Allan: Europhys. Lett. **1**, 123 (1986)
18. M. Terao, C. Harel, A. Salin, R.J. Allan: Z. Phys. D, **7**, 319 (1988)
19. B. Peart, M.A. Bennet: J. Phys. B **27**, 2551 (1995)
20. L.P. Presnyakov, D.B. Uskov: Sov. Phys. - JETP Lett. **66**, 22 (1997)
21. S. Krüdener, F. Melchert, K. v.Diemar, A. Pfeiffer, K. Huber, E. Salzborn, D.B. Uskov, L.P. Presnyakov: Phys. Rev. Lett. **79**, 1002 (1997)
22. C. Forster, R. Shingal, D.R. Flower, B.H. Bransden, A.S. Dickinson: J. Phys. B **21**, 3941 (1988)

XVII Hollow Atoms

Brigitte Ban-d'Etat and Jean Pierre Briand

Summary. The hollow atoms are (quasi) neutral atomic species having all their electrons in excited states while the innermost shells are empty. The technique to prepare hollow atoms is to ultimately strip given neutral atoms at low velocity and dress it, further, near or on a surface. By changing the experimental conditions one can prepare hollow atoms of very different configurations: below or at the surface capture occurs for Ar in the M and the N shells, and above it in high Rydberg states. By changing the surface target, conductor, semiconductor or insulator, we can study various phenomena. The way to study these ion–surface interactions can be Auger spectroscopy or x-ray spectroscopy. Finally, we will see that beyond their applications for the study of ion–surface interactions, the hollow atoms also constitute an interesting field of fundamental physics.

1. Introduction: Discovery of Hollow Atoms

The *hollow atom* is one of the latest and most exotic creations of atomic collision physics and represents a multiply excited or neutral atom with most or all of its electrons in external shells, while the inner shells remain empty. The existence of such exotic species, atoms with a fully inverted population, may, at first view, be imaginated using the straightforward extrapolation of the charge exchange process. In the charge exchange process, the most weakly bound electrons of a neutral atom are, for obvious energetic reasons, captured into highly excited states of the ion. Above a metal surface, i.e., an infinite reservoir of weakly bound electrons, one can then imagine that Z electrons could be captured into high Rydberg states of a bare ion of atomic number Z, as discussed by *Arifov* [1], but the existence of such processes suffers from some objections. Hollow atoms were discovered by *Briand* et al. in 1990 [2]. These authors, as discussed in more detail in Sect. 3.2, observed during the collision of Ar^{17+} ions on Ag targets, using high resolution x-ray spectroscopy, clear signatures of the formation of Ar atoms, in which all, or most, electrons were located in the M and N shells while the K and L shells remained empty; these they named *hollow atoms*. More highly excited hollow atoms (or ions) have now been identified as shown in Sect. 6.2. This principle was the result of an analysis of the emission mechanism during the ion–surface interaction.

Over the last few years, very intense beams of highly charged ions have been produced by an ion source, which can be of two types: an Electron Cyclotron Resonance Ion Source, ECRIS, or an Electron Beam Ion Source, EBIS. The transformation of a Highly Charged Ion (HCI) in a hollow atom and its relaxation by emission of electrons and/or x rays then became the subject of extensive theoretical and experimental studies [3].

In this chapter, we will discuss hollow-atom formation and its decay in the case of HCI–solid-target interaction. We will present the techniques of analysis: Auger spectroscopy and x-ray spectroscopy at low and very high resolution by using a crystal spectrometer, and finally we will discuss the very latest results obtained by varying the energy of the incident ion leading to the study of the interaction below, above or at a surface and by changing the target material, e.g., to insulator, conductor or semiconductor.

2. Formation of Hollow Atoms: Ion–Surface Interactions

The starting point of such an ion–surface interaction is the production of the Highly Charged Ion (HCI). One can note two techniques to produce these HCI from charged hydrogen up to fully stripped U^{92+} ions: the beam–foil technique and ion sources. The first one was developed in the framework of nuclear physics during the construction of particle accelerators. In 1975, bare Ar ions were produced at 5 MeV/u at CEV-Orsay. In 1984–1986, the increase of the acceleration energy of the ions led to the production of bare Kr and Xe ions at GANIL (Caen) and GSI (Darmstadt). In 1986 at BEVALAC (Berkeley), bare U ions were obtained. It was at the beginning of the 1980s that the new ion sources allowed the production of intense HCI beams at low energy: ECRIS and EBIS.

2.1 Production of Ions: ECR or EBIS Source

A high-intensity ion source based on electron cyclotron resonance in a magnetic mirror configuration, the first ECRIS, was built in Grenoble (France) in 1964 by *Geller* [4]. The structure of this first source was steadily improved. In addition to the MINIMAFIOS, built by the same group in 1985 [5], we can find now, all around the world, CAPRICE sources (compact and ferromagnetic ECR source).

In an ECRIS, highly charged ions are produced in a plasma step by step by ionization processes. In an ECRIS the plasma is heated by injection of a HF wave at the electron cyclotron resonance and magnetically confined inside. The structure is composed of two magnetic mirrors for the confinement and an additional radial magnetic field. By choosing the proper combination of magnetic field strength and the injected frequency, a closed ECR surface, along which the resonance condition is fulfilled, can be obtained inside the vacuum chamber. The ions produced are extracted electrostatically and then selected through a m/q selection magnet. Then the desired ion beam is guided to the experimental setup and interacts with any kind of target, solid or gas. In this way, quite intense bare Ar beams are now extracted from an ECRIS (from pA up to 25 nA at LBL(Berkeley)). Very recently, one can find ECRIS NANOGAN which are compact sources and can produce the beam for SPIRAL [6].

Whereas the first EBIS was built in Dubna (USSR) in 1969 by *Donets* [7], we can now find high-performance sources everywhere in the world. By using these sources we can produce, with the SuperEBIT in Livermore, bare uranium.

In an EBIS an electron beam emitted by a heated cathode is accelerated between the cathode and the anode. This electron beam, highly compressed electrostatically in the electron gun and magnetically in a superconducting solenoid, may ultimately strip, step by step, during a time τ, any kind of atom. On the one hand, the ions are radially trapped by the negative space charge of the electron beam and on the other, a potential distribution (bath shape) constitutes a longitudinal ion trap. The trapped ions are ejected by means of alternative voltages which confine and eject those ions from the source: an EBIS is a pulsed voltages system.

2.2 Neutralization and Decay of the HCI Approaching a Surface

Before the 1990s, the processes of describing the formation and decay of hollow atoms are quoted as a straightforward extrapolation of the Hagstrum model [8]. The neutralization holds when a slow ion approaches a metallic surface through a resonant charge-exchange process due to energy matching between an energy level of the ion close to the potential barrier and the electrons of the conduction band of the target or some Auger neutralization process. This extrapolation predicts, for HCI, resonant transfer from the electrons of the conduction band into very highly excited states of the ions, at large distances of the target. This prediction leads to the production of very excited hollow atoms whose properties and behavior, close to the surface, are questionable [9].

Outside the surface, the target electrons may either be captured by the ion, or ejected into the continuum [10, 11], as described by *Bardsley* et al. [12], or captured into the continuum of the ion.

(i) The Born–Oppenheimer approximation may be fulfilled for certain transfers with electrons from target inner shells and may not for weakly bound target electrons.

(ii) The ejection of electrons into the continuum is only limited, when the ion is outside the surface, by space charge considerations and the dynamics of the Auger cascades.

Several authors such as *Apell* [13] and *Burgdörfer* et al. [14] and *Bárány* and *Setterlind* [15] propose a description of the above surface charge exchange and relaxation based on the pioneering work of Hagstrum and a semi-classical extension to HCI, on the so-called *Classical Over-Barrier* model (COB model). The theory proposed by *Burgdörfer* (1993) [16] incorporates resonant multielectron capture of conduction electrons, resonant loss into unoccupied states of the conduction band, and intra-atomic Auger deexcitation. The effective barrier potential includes quantum corrections to the classical image potential. The Auger atomic rates are modified near the surface and a relaxation mechanism is added, namely dielectric loss to the conduction band. The very recent investigations in this field, taking account of the effect of the image charge of the ion approaching the surface (first introduced by *Winter* [17]) lead us to consider, in addition to the projectile velocity, the target material involved in the interaction. The amount of image charge acceleration gives information on the time of the first electron transfer from the target to the projectile and sets a lower limit to the projectile impact energy at the surface.

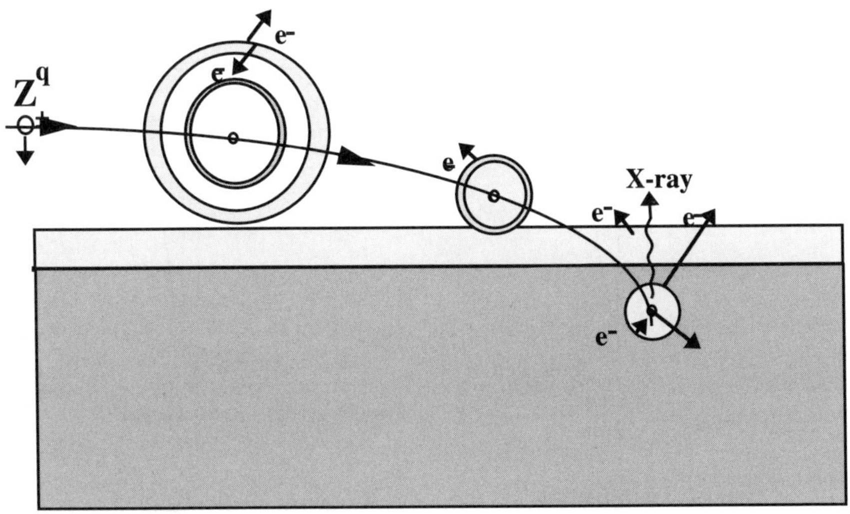

Fig. 1. HCI–surface interaction model

The approach of a slow HCI ($v_\mathrm{i} \ll v_\mathrm{F}$, v_F being the Fermi velocity of the electrons inside the metal target) towards a metal surface, see Fig. 1 (characterized in the jellium approximation) causes a collective response of the metal electrons, usually described at large distance by the classical image potential [18]. This potential accelerates the HCI towards the surface, and therefore leads to a lower limit to the impact velocity corresponding to an upper limit for the interaction time. In addition, the image interaction leads to a shift of the electron states of the ion and to a lowering of the potential barrier between the HCI and the surface (which is the sum of the ion potential, its image potential and the image potential of the captured electron). Approaching the surface, the ion's levels are shifted upwards in energy due to several electronic processes: image interaction, and screening of the charge's ion by already captured electrons (see Fig. 2). At that time, lower ion n-shells can be populated, by cascade of autoionization as by resonant neutralization. On the other hand, previously populated higher levels are emptied by resonant ionization into empty states of the conduction band, autoionization or promotion above the vacuum level. These processes continue all along the approach of the ion towards the surface and shift the population to lower n-states.

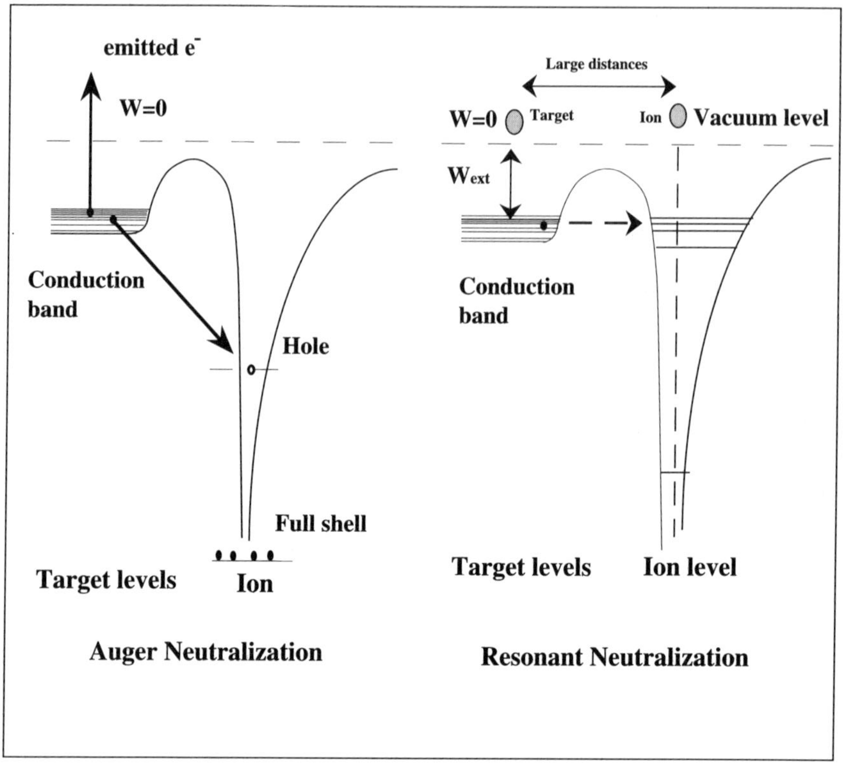

Fig. 2. Potential barrier between an HCI and a metal surface

These dynamical processes occur within a few femtoseconds only and all the atomic states can therefore not be considered as stationary states. The image attraction limits the interaction time above the surface. Then the relaxation of the hollow atom to its neutral ground state is not completed as it penetrates inside the target. The relaxation of this hollow atom in its final steps is governed by two competing processes.

At low projectile velocity, i.e., when a longer time to decay, prior to touching the surface, is allocated to the ion, Auger transitions fill the remaining L- or M-shell holes of the ion. At high velocity (below the surface), the filling of the L shell is ensured by Auger transitions (LMM) for the Ar ion (no side feeding), but it can also occur by collisions for lighter elements (N, O, ...).

3. Spectroscopic Analysis

Most experiments to date on ion surface interactions have been carried out in Auger spectroscopy. This is specially well suited for identification of light ions (C, N, O, ...) and surface atoms. It has extensively been used for the study of above-the-surface and at-the-surface interactions. Its sensitivity to the Doppler effect was first understood as an advantage by *De Zwart* et al. [19] to determine the direction of the ions at the time of the Auger emission. However, even at very high resolution, it is very difficult to well-identify the KL^x satellites, the fine structure of these lines merging into each other and may also be mixed with the electrons emitted below the surface whose width is very large. It is, by contrast with x-ray spectroscopy, the best technique for studying the emission of the surface target atoms [20].

3.1 Auger Spectroscopy

Auger spectroscopy can be used for any kind of ion or surface. This technique was used by *Hagstrum* in 1954 [8] to determine, for the first time, the essential parameters in the ions (1^+)–surface interactions. Most of the experiments with very highly charged ions to date have dealt with the study of the electrons ejected during the HCI–target interaction. The main concern of the first experiments was the counting of the number of ejected electrons per ion. This number has been studied as a function of the total potential energy of the ions and their velocities through potential emission or kinetic emission.

The spectroscopy of the electrons allows one to get an insight into the process under study. These electrons are mainly secondary electrons, following multiple capture of electrons, or ejected into the continuum [10]. These multiple capture processes lead to the formation of multiply excited states of the ions which mainly decay via Auger transitions. As capture occurs, with highly charged ions, at relatively large distances from the surface, the main problem was then to compare the lifetime of the Auger decay of the projectile with the time for the ion to reach the surface. Several authors then considered the Doppler shift of the observed lines or the intensity of the lines as a function of the normal velocity of the ion, i.e., by varying the ion kinetic energy or the angle of incidence. *Meyer* et al. [21], in their first interaction experiments between N^{6+} and a gold target, have observed the dependence of the line intensity over the time needed for the ion to reach the surface . The characteristic Auger spectrum of the target which comes from electron–hole transfers between the ion and the surface are also observed.

The Auger effect is the essential process for the deexcitation. The information about the first instants of the interaction can only be obtained by spectroscopy of the electrons but owing to their very low energy they cannot directly be identified. Nevertheless fast Auger electrons emitted inside the solid either from the projectile or the target atoms may also produce slow secondary electrons. A large number of authors [22–24] have investigated this field, Auger spectroscopy providing experimental evidence for the last steps of the relaxation cascade in a hollow atom. Previous Doppler-shift measurements of Auger lines as a function of observation angle lead to attribute the electrons to the approach of the ion towards the surface. These Auger electrons would be emitted after the ion penetrates the surface. In addition, the limitation due to the image charge acceleration has also to be considered. For HCI, the resulting acceleration leads to a sizable energy gain, ΔE, of the ions approaching the surface at low velocities. In a first approximation, the staircase model of sequential neutralization leads us to conclude that the energy gain due to the image acceleration, $\Delta E/W$, scales as $Z^{3/2}$ for $Z \rightarrow \infty$. A more accurate calculation [16], using the full simulation code for COB dynamics, finds, for very highly charged states, energy gains lower than predicted by the staircase model. There is a crossover at $Z = 15$ due to the fact that the hollow atom formed along the approach toward the surface can accommodate a large number of electrons due to incomplete Slater screening. As the initial charge rapidly changes as the result of multiple electron capture and loss processes, the resulting energy gain, ΔE, becomes a sensitive measure for where and how quickly a HCI is neutralized, i. e., where a hollow atom is formed above the surface. Recent studies in connection with these calculations have considered two types of interaction depending on the impact velocity. At very low impact velocities a substantial fraction of L-shell holes in the projectile below the surface are filled via Auger transitions (e.g., *Auger neutralization*). At high impact velocities vacancy transfer between the hollow projectile and target atomic levels via level crossing (Landau–Zener) and orbital promotion (Fano–Lichten) in close collisions provides a much faster filling of the L shell.

In the last section of this review we will review the interaction experiments done at low velocity, which is of great interest considering the modifications of the surface target (conductor, semiconductor and insulator) due to the interaction with slow highly charged ions or hyperthermal HCI.

3.2 X-Ray Spectroscopy at Low and High Resolution

The fluorescence yield is very small for light atoms and x-ray emission will be observable only from atoms with an atomic number greater than 16 (S). In 1985, *Donets* [7] first studied, at low resolution, the x rays emitted by Ar^{17+}ions impinging on various metallic surfaces. From this first result, Donets deduced that the Lyman-α line would be composed of 8 lines coming from the configuration, during the decay, of 1 K electron and a number of L electrons x lying between 1 and 8, KL^{x} lines. This was confirmed, in 1990 by *Briand* et al. [2], by using a crystal spectrometer at high resolution and a position sensitive detector. These authors have demonstrated that all the now well separated KL^{x} lines correspond to atoms having full M, N shells, thus proving the existence of hollow atoms.

Figure 3 presents the typical x-ray spectrum first observed at low resolution with a Si(Li) detector (FWHM=180 eV at 6 keV) when Ar^{17+} ions hit a metallic surface. This interaction experiment was done in Grenoble. The very intense (few nA for Ar^{17+}) ion beam was produced at low velocity by the ECR source Minimafios built by R. Geller. The large broadening of the lines observed in this spectrum clearly shows that the Ly_α and Ly_β rays come from a very complex array of unresolved lines.

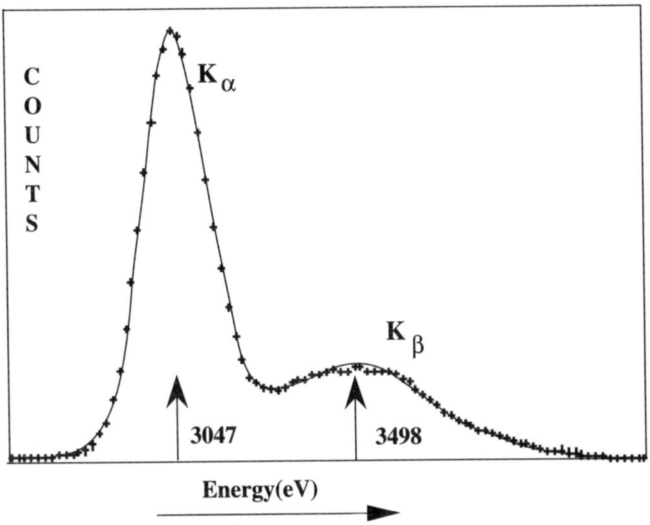

Fig. 3. K x-ray spectrum observed with Ar^{17+} at low resolution

In this case, Ar^{17+} impinging on a metal surface, the neutralization is due to the energy matching of the M shell of Ar (100–400 eV when taking account of screening) and bound electrons of the M-N shells of the target or the K shell of carbon due to surface contamination or Auger neutralization processes. While the M or N shells of the ions are quasi-instantaneously filled up when entering the solid, the L shell is fed more slowly via Auger transitions.

This interaction experiment was repeated a few months later at high resolution by using a mosaic graphite flat crystal spectrometer. The emitted x-ray beam was then focused to a point, symmetric to the normal of the crystal surface, on a *backgammon type* position sensitive detector. The observed spectrum is presented in Fig. 4.

This spectrum exhibits eight well-separated lines, of comparable intensities, whose energies correspond to the well known $2p \rightarrow 1s$ transitions, $1 \leq x \leq 8$ (KL^x). The energies of these lines, which are measured with an accuracy of 2 eV, are compared with *Bhalla*'s calculations [25], detailed Dirac–Fock calculations [26], and previously known values, in order to determine the number of M spectator electrons present at the time of the radiative decay. The energies of the four KL^x lines with $1 \leq x \leq 4$ correspond to ions whose M shell is incompletely filled. Detailed studies of the $KL\beta\gamma$ lines show that, in

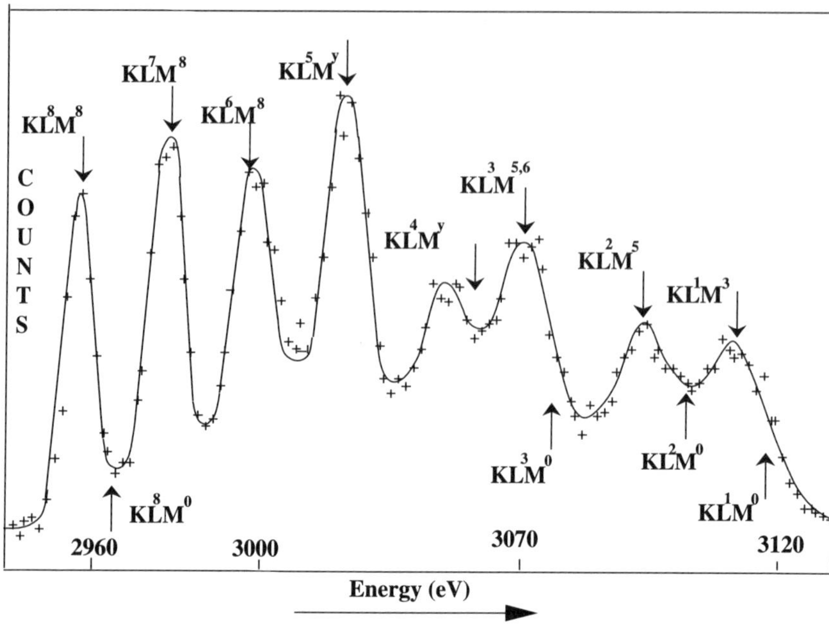

Fig. 4. K x-ray spectrum observed with Ar^{17+} at high resolution

that case, the outermost electrons were in the *N* shell. Their widths indicate some fluctuations in the number of *M* electrons. The energies of the four KL^x lines with $5 \leq x \leq 8$ exhibit similar shapes; their widths indicate that here the *M* shell is closed.

Similar experiments were performed by *Bonnet* et al. [27] in which evidence for x-ray emission above the surface was obtained. The x rays emitted above the surface appear to be particularly interesting because the x rays contain the information about the characteristics of ion–surface interactions, i.e., the population of very high *n* states and the subsequent deexcitation presumably via a large number of cascade transitions. The x rays emitted inside the solid, in contrast, reflect to a larger extent the characteristics of an ion–solid interaction, i.e., the population of lower *n* states and the deexcitation probably involving a smaller number of cascade transitions. These results led different authors such as *Schulz* et al. [28] and *d'Etat* et al. [29] to investigate the field of above-x-ray emission at grazing incidence and later on, at very low incident ion velocity. In the same way *Meyer* et al. [30] have observed Auger emission during the interaction of HCI with different target materials at grazing incidence.

4. Interaction Below, Above or At a Surface

Above a surface, the capture process feeds at large distances the Rydberg states of the ion referred to as *Rydberg hollow atoms.* Below or at the surface, when the collision holds at closer distances (with contact), the electrons are captured in lower excited states; these hollow atoms are named *surface hollow atoms.*

4.1 Below or at Surface Interactions

By looking at the x rays emitted directly, or in coincidence, by impinging bare or hydrogen-like ions of various atomic numbers on solid targets *d'Etat* (1993) [29] and *Briand* et al. [31] have studied the dynamics of the first capture and decay processes during the interaction. Below the surface, hollow atoms of high Z atomic numbers ($Z \geq 18$: argon, iron, krypton and xenon), prepared at the Livermore SuperEBIT, are formed by very fast capture processes of the target electrons into the M and N shells for Ar ions and N, O, P shells for the heavier ions. These highly excited electrons then fill the empty K and L shells mainly through a cascade of Auger transitions. The L shell is, therefore, quickly and completely stepwise filled via *LMM, LNN*, or *LNM* transitions, and at a given time of this cascade the K hole(s) is (are) filled and a characteristic K_α transition is observed.

The principle of the first experiment was to vary, for similar initial configurations of the ions, the K-shell lifetimes τ_K of the ions by changing either the atomic number of the projectiles or in a given ion the number of K holes (1 or 2 for a hydrogen-like or a bare ion). The results allowed us for the first time to measure, by using the atomic-clock property of the hollow atoms, the meantime for the filling of the L shell.

A second part of the experiment studies, directly and in coincidence, the actual configurations of the ions when the two K holes of a bare ion are stepwise filled. The Fe^{25+}, Fe^{26+}, Kr^{35+} and Kr^{36+} were produced by the EBIT source at LLNL, and the x-ray spectra observed with intrinsic germanium detectors at low resolution (180 eV). An interesting result was obtained by comparing the numbers of L spectator electrons for, e.g., Fe^{25+} and the first K transition (hypersatellite) for Fe^{26+}, which is found to be 3.8 for Fe^{26+} and 4.2 for Fe^{25+}. This can be explained by considering that the lifetime of the K hole is about two times shorter when two K holes are present instead of one. Since the L shell filling rate in this case is roughly the same for Fe^{26+} and Fe^{25+}, this means that the K_α line is emitted in a shorter time for Fe^{26+} than for and Fe^{25+} and then for fewer L electrons. A similar effect is also observed with krypton. The results summarized in Table 1, and a detailed study of the hypersatellite–satellite cascade then leads to a good estimation of the lifetimes.

The behavior of these hollow atoms below a surface proceeds via Auger transitions, each lasting at least 10^{-16} s, interrupted sometimes by capture

Table 1. Mean numbers of L electrons for Ar^{17+}, Fe^{25+} and Kr^{35+} ions, $K_{\alpha,\beta,\gamma}$ line relative intensities and changes in the mean numbers of L-spectator electrons during the hypersatellite-satellite cascade. N_b is the number of additional electrons in the satellite line compared to the hyper-satellite line

Ions	Ar^{17+}	Fe^{25+}	Kr^{35+}
$K_\alpha L^{<x>}$	$K_\alpha L^{<5>}$	$K_\alpha L^{<4.2>}$	$K_\alpha L^{<2.7>}$
$I(K_\beta)/I(K_\alpha)$	0.3	0.38	0.42
		$N_b.e^- M > N_b.e^- L$	$N_b.e^- M > N_b.e^- L$
$I(K_\gamma)/I(K_\alpha)$			$N_b.e^- N > N_b.e^- L$

Ions	Ar^{18+}	Fe^{26+}	Kr^{36+}
$K_\alpha$sat $L^{<x>}$	$K_\alpha L^{<5.4>}$	$K_\alpha L^{<5.3>}$	$K_\alpha L^{<3.2>}$
$K_\alpha$hsat$L^{<x>}$	$K_\alpha L^{<3.8>}$	$K_\alpha L^{<3.8>}$	$K_\alpha L^{<2>}$
N_b of add $e^- L$	2.6	2.6	2.2

processes whose mean times are of the order of magnitude of at least 10^{-17} s. Then during a very short time, there is a long series of very quick captures by spontaneous decays. The neutralization range found is, however, slightly larger than that estimated by *Hermann* et al. [32] (one monolayer) by extrapolating the energy losses of argon ions.

4.2 Above Surface Interactions

Many groups [1, 7, 27, 28, 30, 33] have been investigating in this field to determine whether or not it would be possible to observe some capture process from conduction electrons of the solid, far outside the surface, into highly excited states of the ion ($n \sim 20 - 30$ for the Ar^{17+}). Most of the time, such processes are quoted as the straightforward extrapolation of the Hagstrum model, which predicts, for highly charged ions, resonant transfer from the electrons of the conduction band into very highly excited states of the ions, at large distances from the target (about 20 Å if it is a metal or an insulator, as quoted by *Apell* [13] and *Snowdon* [34], and 17 Å for a pure metal *(Burgdörfer, Lerner* and *Meyer* [14])). This prediction would lead to the production of very excited hollow atoms whose properties and behavior, close to the surface, are questionable. The lifetime of these very excited ions to go to the ground state have to be shorter than the time needed by the ion to reach the surface; otherwise, these very weak species would explode while touching the target. According to the calculations done by *Desclaux* [35] and those of *Vaeck* and *Hansen* [36], this condition is very often difficult to fulfill, at least at normal incidence. The transit time has to be compared to the characteristic atomic time of these levels in order to be sure that the state which is formed is a real stationary state. Outside the surface, the target electrons may either be captured by the ion, ejected into the continuum or captured into the continuum of the ion. Such processes outside the surface have been studied at very low intensity by *Bottcher* and *Strayer* [37].

For highly charged ions Ar^{17+} this was studied by *d'Etat* et al. [29], while the ions interact with the metallic surface at grazing incidence. The observed x-ray spectrum was the sum of two spectra: one coming from ions which have captured many electrons in high n states and have enough time to decay by Auger cascades until a first electron reaches the L shell, prior to touching the surface; and another one, emitted by the ions whose outermost electrons are ionized when they touch the surface and which are reneutralized inside the target.

All the previous calculations lead to some general results:

1. the decay rate quickly increases when the energy of the Auger electron decreases;
2. the decay rate is then a maximum for the Auger transitions connecting the states of the closest n principal quantum numbers energetically allowed;
3. the smallest energetically allowed Δn Auger transition strongly depends on the number of electrons in the shells considered;
4. the decay rate is at a maximum for the $\Delta l = 0$ transitions;
5. the decay rate per electron slowly increases with the number of electrons in the considered shell.

Then, the actual number of electrons, at any time of the cascade, decreases from one step to the next in the cascade (triangular distribution). The capture processes then very likely occur outside the surface in a higher n state than inside the target, but with n lower the any value predicted by the extrapolation of the Hagstrum model, owing to the screening effects induced by the electrons moving toward the target.

Recently, *Briand* et al. [38–40] have studied HCI-metallic target interactions, at normal incidence and very low incident velocity. The 11 keV/q Ar^{17+} ions produced by the ECR sources of CEN at Grenoble, or GANIL at Caen were decelerated, in a simple gap, down to a tunable energy range of $0 < E < 14$ eV/q.

Figure 5 presents the x-ray spectrum observed; it is peaked on the KL^2 satellite which corresponds to the sum of the outside x-ray spectrum mentioned above, plus a signal of an *at-surface* interaction. When an ion touches a surface, at distances below 1 Å, the electrons are captured into the M and N shells of the ion instead of the Rydberg states. One or two electrons are then quickly captured into the M shell of the ion at the time of contact . When two more electrons are captured, a fast LMM Auger transition occurs, accelerating the still operative filling of the L shell which increases significantly the intensity of the KL^2 satellite and gives a signature of the contact of the ion. Between 0 and 14 eV/q, the relative intensity of the KL^x lines does not

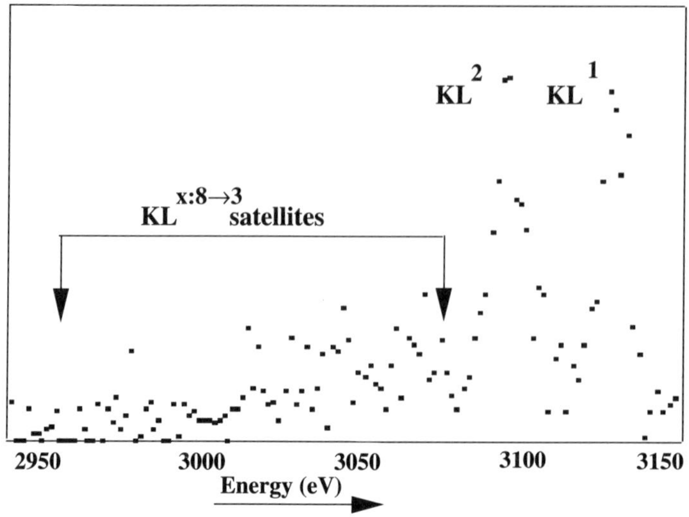

Fig. 5. K x-ray spectrum observed for 1 eV/q Ar^{17+} ions on Au surface

change. This invariance can be easily explained by the acceleration of the ion by its own image (for metallic targets). At very low initial kinetic energies the kinematics of the ion is then governed by its image and not by its initial kinetic energy ($E < 12$ eV/q). The ions have, however, been found in states much more ionized than previously expected, which means that the Auger reionization is always much faster than the capture refeeding processes.

5. Atomic-Clock Property of Hollow Atoms and Zeno's Paradox

These numerous and very fast Auger transitions from Rydberg states have not yet been directly observed because of the very low energy of the ejected electrons. At the end of the cascade the energy of the transitions is much larger, and one can observe either some Auger transitions or some x rays filling the K and L shells. During the stepwise filling of the L shell one K hole may also be filled and a characteristic x ray is emitted. In high resolution x-ray spectroscopy, the characteristic array of the eight satellite lines is observed [2]. The stepwise filling of the L shell plays the role of an hourglass and the K x ray emitted that of a flash. Therefore the relative intensity of these eight KL^{x} satellites depends on the relative rates with which the L and K shells are filled. If, for instance, the L shell filling rate is low compared with that of the K shell, as soon as the first electron reaches the L shell, a K_{α} line is emitted. One then observes a KL^{x} distribution strongly peaked at KL^{1}. On the contrary, if the L shell filling rate is comparable to or larger than that of the *K* shell, one observes a much more uniform distribution. The filling rates of the *L* shell have recently been calculated for both the *M* or *N* shells and the Rydberg states [35, 41], and from the relative intensity of these satellite lines one may deduce the states into which the electrons were originally captured. The clock property of these hollow atoms can be illustrated in two ways.

When two *K* holes are present in the ion there will be two successive *K* x rays emitted (hypersatellite and satellite). In this case one observes two well separated arrays of KL^x satellites. A detailed analysis of the two snapshots gives direct measurements of the filling of the *K* and *L* shells below a surface. This internal clock allows direct measurements of time of the order of one tenth of a femtosecond.

These very short times may also be converted into mean decay lengths by varying the hollow atoms' velocity. These hollow atoms have been studied over a large range of kinetic energies from 1 or 2 eV/q up to 4 MeV/q. At very high energy an Auger transition extends over a few tens of Å. This allows the preparation of specific hollow atoms in a carbon foil of 20 Å thickness and the study of their distant decay, outside, in vacuum. At very low velocities, a whole cascade of ten Auger transitions holds along about 1 Å. This was observed with fully decelerated Ar^{17+} ions above insulator surfaces. Thus, using the clock property, one can follow the behavior of a hollow atom along ranges of the order of 1 Å by varying its kinetic energy between 0 and 10 eV/q.

In a refeeding situation, near or below a surface, a hollow atom exhibits unusual features. One may consider that an atom capable of decaying slowly by a radiative transition may also, at the same time, experience faster Auger transitions or collisional refeeding in the very outermost shell. One could consider, separately, these two processes whereas the atom forms, as a whole, a quantum mechanical stationary state. The interaction between the outermost shell events and the innermost ones must yet be more or less efficient, depending on the angular momenta which couple the inner and outer electrons. As quantum mechanical stationary states, the hollow atoms must be considered as decaying via two channels of different branching ratios: Auger emission from large n states, or an x-ray emission between two small n states. This is a situation similar to *Zeno's paradox* [42]: a metastable quantum state which is very often weakly disturbed by an outside perturbation. After each external perturbation one restarts a new stationary state, and the lifetime of the mean level may be lengthened to infinity when the number of short perturbations increases indefinitely.

6. Interaction of HCI with Various Solid Targets

Taking account of the image-charge effect, the interaction of HCI with different target material (conductor, semiconductor, insulator) becomes the major field of investigation. In addition, since it has been possible to decrease the velocity of the projectile by working at grazing incidence as by decelerating the HCI beam, the study of what happens about the formation and the decay of the hollow atoms above various target materials became of great interest.

6.1 Below Interaction with Conductor and Insulator Surfaces

Figure 6 presents the K x-ray spectra observed with the same Ar^{17+} ions at 10 keV/q on gold and silicon targets, at normal incidence (most of the rays are emitted below the surface). One clearly sees the dramatic changes in the relative intensity of the eight KL^x lines which show that different kinds of hollow atoms are also formed below metal and insulator surfaces. The general trend of the evolution of the relative intensity of the eight KL^x lines, observed on many different targets, is that the spectra for metals are peaked on the satellites corresponding to a large number of L spectator electrons. In the framework of the atomic clock property of the hollow atoms this means that the L shell of the ions is more quickly populated in metals than in insulators. The L shell being filled by the M and N electrons through the fast *LMM (LMN ...)* Auger transitions, these findings mean, in turn, that the M and N shells of the ions are filled faster in metals than in insulators (by conduction electrons). Some additional weaker and monotonic changes in the relative intensities of the KL^x satellites have also been found in various insulators or semiconductors.

At low resolution, the characteristic x-ray spectra of argon ions of different charge states (from 8+ up to 17+) on different targets (C, SiH or SiO_2) was studied [43]. Below a carbon surface, the capture of weakly bound L electrons through a Landau–Zener mechanism is not possible, the distance of interaction being larger than the interatomic distances in the lattice. Near a highly charged ion the bound character of an L electron of C is also questionable and does not make possible any electron promotion process. Some other mechanisms must then be found to explain the observed transfer. The most likely process would be the Auger neutralization process to describe the close interaction (contact) of singly charged ions on surfaces. The velocity of the ions being comparable, or smaller than that of the L or valence (V) electrons, the ion may be quasi-bound in the solid. These target electrons may then participate in Auger *MVV* transitions filling the M shell. Inside a solid the argon ions having no discrete state above $n = 5$ (dissolution into the continuum due to the external screening), the highest ion levels to be filled in this collision are consequently the M and N shells. The Auger transitions being more probable for the ejected electrons of the lowest energy, this may also explain the high selectivity of the population into the M and N shells of argon.

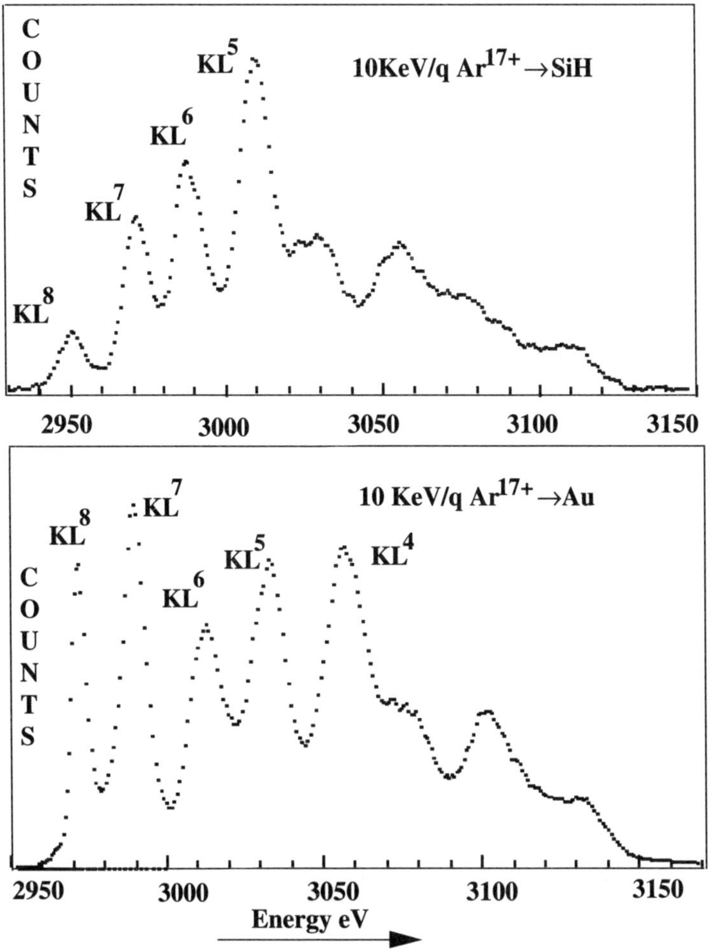

Fig. 6. KL^x x-ray spectra observed of Ar^{17+} ions impinging on SiH and Au surfaces

6.2 Above Interaction with Conductor and Insulator Surfaces

For slow HCI interacting with insulator surfaces one expects the above described scenario of Fig. 2 to change considerably. The interaction of the slow HCI with an insulator target is difficult to study because of the static electric charges that may appear on the surfaces. The absence of a sufficient number of conduction electrons in these materials makes the formation of such outside hollow atoms or ions by highly charged ions, flying over insulator surfaces, also questionable.

For these studies, LiF has been chosen by a large number of authors [44–46] because at elevated temperatures it exhibits an ionic conductivity which is sufficient to avoid changing of the surface by impact of low intensity ion beams. Whereas in metals the conduction band is divided at the Fermi level into electronic bands of occupied and unoccupied states, in LiF, the 2*p*-electrons in F form a completely occupied valence band with binding energies of about 12 eV (bulk value; surface value can differ somewhat). This valence band exhibits a flat dispersion curve which results in a lattice. A broad energy gap of about 14 eV extends well above the vacuum level to the first empty allowed states. Not only the electronic structure but also the dielectric response of LiF clearly differs from a perfect conductor. The classical over-barrier condition has to be recalculated by taking into account a dynamic image potential, which involves the permittivity $\epsilon\ (\omega)$ of LiF between about $\epsilon(0) = 8.6$ (static value) and $\epsilon(\infty) = 1.9$ (optical value). By using this scenario, *Limburg* et al. [44] concluded that there is no formation of outside hollow atoms when N^{6+} (low charge state) ions interact with the LiF surface.

For heavier ions, *Briand* et al. [39] claimed that there is creation of outside hollow ions during the interaction of Ar^{17+} ions on SiO_2 (insulator) surfaces. Two different experiments were performed:

1. At low resolution, the interaction of 20 keV/q Ar^{17+} ions at grazing incidence on SiO_2 target was studied and the signature of the formation of hollow atoms already found above metal surfaces were observed.
2. Ar^{17+} were fully decelerated and their interaction at high resolution and normal incidence were studied on metal and semiconductor. They observed, namely, the signature of the touchdown of the ions on metals which does not appear for Si, and they tentatively explain these results by considering the backscattering of ions above insulator surfaces due to a transient and local buildup of positive charges. Figure 7 shows the differences in relative intensity of the KL^1 and KL^2 lines during the interaction on SiH and Au. The x-ray spectrum observed with the SiH target displays the very characteristic array of x-ray satellites expected and observed for a pure outside decay.

These results may be more likely explained by considering the insulator character of SiH and its effect on the image acceleration first observed by Winter. With an insulator or semiconductor, the removal of valence electrons generates localized positively charged centers, the lifetimes of which are long enough to overcome, at normal incidence, the image acceleration. Then the ion may be decelerated and backscattered. One would only observe the characteristic pure outside decay of the hollow atoms. With metals, the ions gain some energy and hit the surface. So they experience some surface interaction leading to direct capture of a few electrons into the *M* shell of the ion like inside the bulk. More recently [40] these authors gave a proof of this backscattering of the very slow ions above insulators and semiconductors. They compared the behavior of the x-ray spectra of the same Ar^{17+} ion above a metal as a function of the initial kinetic energy of the ions and demonstrated that the image acceleration vanishes (is overcome) above insulators and semiconductors (the backscattered ions have also been directly

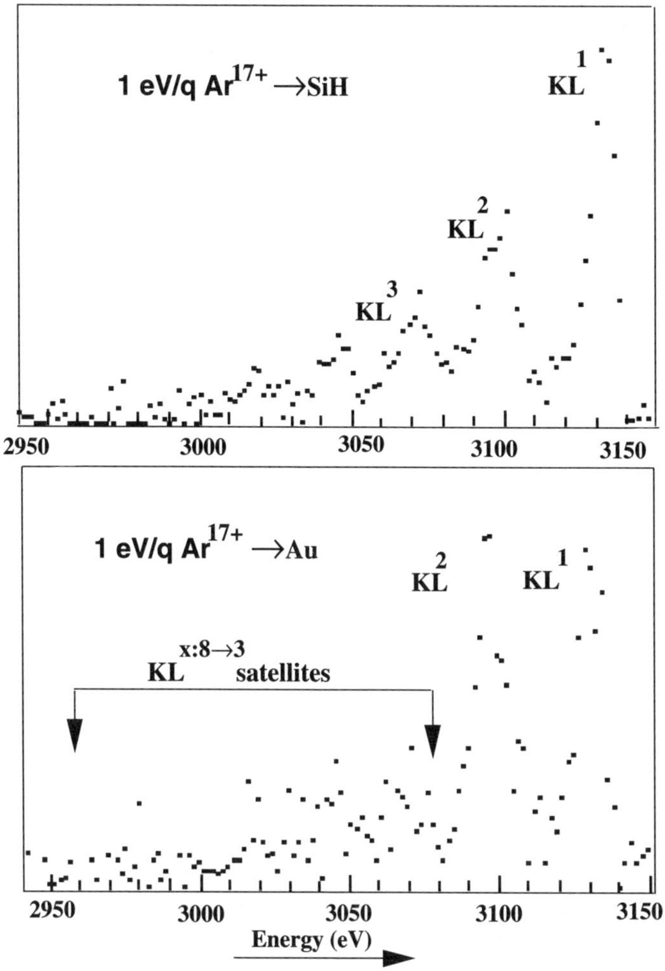

Fig. 7. KL^x ray spectra emitted by fully decelerated Ar^{17+} ions (1–2 eV/q) on SiH and Au surfaces

observed by the same authors [47]). As can be seen, the formation of hollow atoms or hollow ions above an insulator is still under study [48].

7. Conclusion and Outlook

In this chapter we have discussed some aspects of the interaction of highly charged ions leading to the formation of so-called hollow atoms. The very latest experiments are essentially focused on the comparison of the interaction with targets of different material, i.e., conductor, semiconductor and insulator. This last point is studied by looking at the Auger electrons as well as the x rays emitted during the interaction. In both analyses the classical over-barrier model is used to describe the decay of the hollow atoms. We have seen that, for very highly charged ions, from 17+ up to 54+, x-ray spectroscopy leads to an accurate study of the decay and allows us to establish a very accurate time scale. New technologies have lead most authors to investigate, now, the target effects, some looking at electronic sputtering and others looking at the nano-signature on insulators and semiconductors left by the ions during their impact on the surface.

Acknowledgement. Brigitte Ban-d'Etat would like to thank Paul Indelicato for his permanent support and his constant help in understanding and using the Latex system.

References

1. U.A. Arifov, E.S. Mukhamadiev, E.S. Parilis, A.S. Pasyuk: Sov. Phys. - JTP **18**, 240 (1973)
2. J.P. Briand, L. de Billy, P. Charles, S. Essaaba, P. Briand, R. Geller, J.P. Desclaux, S. Bliman, C. Ristori: Phys. Rev. Lett. **65**, 159 (1990)
3. A. Arnau, F. Aumayr, P.M. Echenique, M. Grether, W. Heiland, J. Limburg, R. Morgenstern, P. Roncin, S. Schippers, R. Schuch, N. Stolterfoth, P. Varga, T.J.M. Zouros, HP. Winter: Surf. Sci. Rep. **27**, 113 (1997)
4. R. Geller: Annu. Rev. Nucl. Part. Sci. **40**, 15 (1990)
5. B. Jacquot, P. Briand, F. Bourg, R. Geller: Nucl. Instrum. Methods. A **269**, 1 (1988)
6. L. Manoury, A.C.C. Villari, J.C. Angelique, F. Clapier, G. Dhilly, G. Di Bartolo, M. Ducouturieux, T. Ethvighot, P. Jardin, N. Lecesne, R. Leroy, M. Lewitowicz, C.F. Liang, R. Lightenthaler, C. Marry, J. Obert, N.A. Orr, J.Y. Pacquet, N. Pauwels, J.C. Putaux, E. Robert, M.G. Saint-Laurent, P. Sortais, O. Tengblad: Rapport du Ganil (1997)
7. E.D. Donets: Nucl. Instrum. Methods. B **9**, 522 (1985)
8. H.D. Hagstrum: Phys. Rev. A **96**, 325 (1954)
9. J.P. Briand: Review talk, XV Conf on X-Ray, AIP Conf. Proc. **215**, T.A. Carlson, M.O. Krause, T. Manson, eds. (AIP, New York 1990), p. 513 and Inner-Shell Processes (X 90), Knoxville (1990)
10. M. Delaunay, M. Fehringer, R. Geller, P. Varga, HP. Winter: Europhys. Lett. **4**, 377 (1987)
11. H. Winter: Z. Phys. D **21**, S129 (1991)
12. J.N. Bardsley, B.M. Penetrante: Comm. At. Mol. Phys. **27**, 43 (1991)
13. P. Apell:Nucl. Instrum. Methods. B **23**, 242 (1987)
14. J. Burgdörfer, P. Lerner, F. Meyer: Phys. Rev. A **44**, 5674 (1991)
15. A.Bárány, C.J. Setterlind: Nucl. Instrum. Methods. B **98**, 184 (1995)
16. J. Burgdörfer: in *Fundamental Processes and Applications of Atoms and Ions*, C.D. Lin, ed. (World Scientific, Singapore 1993)
17. H. Winter: Europhys. Lett. **18**, 207 (1997)
18. F. Aumayr: XIX ICPEAC, Whistler, Canada (1995) AIP Conf. Proc. Vol. **360**, L.J. Dube, J.B.A. Mitchell, J.W. Conkey, C.E. Brion, eds. (AIP, New York 1995), p. 631

19. S.T. De Zwart, T. Fried, U. Jellen, A.L. Boers, A.G. Drentje: J. Phys. B **18**, L623 (1985)
20. G. Giardino: *Accelerator-Based Atomic Physics Techniques and Applications*, Chapter 11 (AIP, Chapel Hill, North Carolina Press 1997)
21. F.W. Meyer, C.C. Havener, P.A. Zeijlmans van Emminchoven: Phys. Rev. A **48**, 4476 (1993)
22. J. Das, M. Morgenstern: Comments At. Mol. Phys. **29**, 205 (1993)
23. R. Köhrbrück, M. Grether, A. Spieler, N. Stolterfoht, R. Page, A. Saal, J. Bleck-Neuhaus: Phys. Rev. A **50**, 1429 (1994)
24. F.W. Meyer, S.H. Overbury, C.C. Havener, P.A. Zeijlmans van Emminchoven, D.M. Zehner: Phys. Rev. Lett. **67**, 723 (1991)
25. C.P. Bhalla: Phys. Rev. A **8**, 2877 (1993)
26. J.P. Briand, L. de Billy, P. Charles, S. Essaaba, P. Briand, R. Geller, J.P. Desclaux, S. Bliman, C. Ristori: Phys. Rev. A. **43**, 565 (1991)
27. J.J. Bonnet, A. Fleury, M. Bonnefoy, M. Chessevent, T. Lamy, A. Brenac, A. Simionovici, H.J. Andrä, S. Andriamonje: Z. Phys. D **21**, s 343 (1991)
28. M. Schulz, C.L. Cocke, S. Hagmann, M. Stöckli, H. Schmidt-Bocking: Phys. Rev. A **44**, 1653 (1991)
29. B. d'Etat, J.P. Briand, G. Ban, L. de Billy, J.P. Desclaux, P. Briand: Phys. Rev. A **48**, 1098 (1993)
30. F.W. Meyer, L. Folkerts, H.O. Folkerts, S. Schippers: Nucl. Instrum. Methods. B **98**, 441 (1995)
31. J.P. Briand, B. d'Etat-Ban, D. Schneider, M.A. Briere, V. Decaux, J.W. McDonald, S. Bardin: Phys. Rev. A **53** 2194 (1996)
32. R. Hermann, C.L. Cocke, J. Ullrich, S. Hagmann, M. Stöckli, H. Schmidt-Bocking: Phys. Rev. A **50**, 1435 (1994)
33. L. Folkerts, R. Morgenstern: Europhys. Lett. **13**, 377 (1990)
34. K.J. Snowdon: Nucl. Instrum. Methods. B **34**, 309 (1988)
35. J.P. Desclaux: Proceedings HCI 1994, F. Aumayr, G. Betz and HP. Winter, eds. **18** (1995); Nucl. Instrum. Methods. B **98**, 18 (1995)
36. N. Vaeck, J.E. Hansen: Z. Phys. D **21**, S221 (1991)
37. C. Bottcher, M.R. Strayer: *The Physics of Electronic and Atomic Colliions*, A. Dalgarno, R. S. Freund, P. M. Koch, M. S. Lubell, T. B. Lucarto, eds. (AIP, New York 1989)
38. J.P. Briand, G. Giardino, G. Borsoni, M. Froment, M. Eddrief, C. Sebenne, S. Bardin, D. Schneider, J. Jin, H. Khemliche, Z. Xie, M. Prior: Phys. Rev. A **54** 4136 (1996)
39. J.P. Briand, S. Thuriez, G. Giardino, G. Borsoni, M. Froment, M. Eddrief, C. Sebenne: Phys. Rev. Lett. **77**, 1452 (1996)
40. J.P. Briand, S. Thuriez, G. Giardino, G. Borsoni, V. Le Roux, M. Froment, M. Eddrief, C. de Villeneuve, B. d'Etat-Ban, C. Sebenne: Phys. Rev. A. **55**, R2523 (1997); J.P. Briand, L. de Billy, J. Jin, H. Khemliche, M. Prior, Z. Xie, M. Nectoux, D. Schneider: Phys. Rev. A **53** R2925 (1996)
41. N. Vaeck, J. Hansen: J. Phys. B **28**, 3523 (1995)
42. W.M. Itano, D.J. Heingen, J.J. Zollinger, D.J. Wineland: Phys. Rev. A. **41**, 2295 (1990)
43. J.P. Briand: Proc. 14th Int'l. Conf. Appl. Acc. Res. Ind. (Denton. Texas 1996) J.L. Duggan, J.L. Morgan, eds. (Texas 1996), p. 137
44. J. Limburg, S. Schippers, R. Hoekstra, R. Morgenstern, H. Kurz, F. Aumayr, HP. Winter: Phys. Rev. Lett. **75**, 217 (1995)
45. J. Limburg, S. Schippers, R. Hoekstra, R. Morgenstern, H. Kurz, M. Vana, F. Aumayr, HP. Winter: Nucl. Instrum. Methods. B **115**, 237 (1996)

46. F.W. Meyer, Q. Yan, P. Zeijlmans van Emminchoven, I.G. Hughes, G. Spierings: Nucl. Instrum. Methods. B **125**, 138 (1997)
47. J.P. Briand: Invited talk at the ISIAC meeting, Budapest (1997) (unpublished)
48. F. Aumayr, J. P. Briand: Comments Phys. Rev. Lett. **79**, 2590 (1997)

XVIII Large Charged Clusters in Plasmas

Boris M. Smirnov

Summary. Charging processes of clusters in plasmas are considered. The cluster equilibrium charge state increases with cluster size and may be quite high. Clusters are positively charged in an arc plasma and carry a negative charge in a glow discharge plasma. The stability conditions for multicharged clusters are analyzed.

1. Introduction

In contrast to multicharged atomic ions which are characterized by high ionization potentials, the binding energies of electrons in charged clusters are comparable with the ionization potentials of atoms or the work functions of materials. Usually, clusters in a plasma are charged. In the case of gas discharges with a low electron density and a small gaseous temperature as takes place in a glow discharge, clusters are negatively charged because the electron mobility is higher than that of ions. A high negative charge of particles in a dusty plasma can lead to specific phenomena. Charged particles can be captured by various electric traps in a gas discharge plasma, and then particles of negative charge can form crystal structures in such a trap [1–6]. The negative charge of solid dielectric particles in this crystal can reach up to 10^3–10^5 e [7, 8], where e is the electron charge. Due to a large charge of dielectric particles, the plasma consisting of charged particles and electrons is nonideal in spite of the large distances between particles.

A cluster charge in a discharge plasma can be both positive and negative depending on processes which establish this charge in a plasma. In order to increase the cluster charge, a unipolar plasma of a corona discharge is used. The cluster charge increases with cluster size and is high for large clusters. Interaction of charges in clusters of high charge leads to cluster instability, so that a cluster of a given size has a limited charge. In this chapter, we consider the processes which establish the charge of clusters or small particles in plasmas, and analyze instabilities of large charged clusters.

2. Charging of Small Dielectric Particles in Ionized Gases

A cluster is a system of bound atoms, and large clusters differ from small bulk particles by the existence of so-called *magic numbers* [9, 10] at which the cluster parameters (the ionization potential, the electron affinity, the binding energy) have extrema. The magic numbers correspond to completed cluster structures, and because the cluster structure is not essential for the phenomena under consideration, below we will not distinguish clusters from small particles. We will consider charged clusters formed in a gas discharge plasma as a result of attachment of electrons and ions to clusters, and ionization of clusters. These processes together with recombination processes involving clusters, establish a certain cluster equilibrium charge state in a plasma. The character of these processes depends on the properties of the clusters under consideration which determines theoretical models for the description of cluster charging.

Below, we consider two cases of cluster charging depending on mechanisms of recombination processes. In the case of small metallic clusters presented in a dense plasma, recombination of electrons by a positive charged cluster results from three body processes, while in the case of a large dielectric cluster or particle and a small electron density, the recombination process corresponds to a pair process of attachment of an electron or ion to a particle.

In the case of a dielectric particle, one can consider the process of electron and ion attachment as a result of their interaction with active centers located on the particle surface. In other words, electrons and ions of the plasma may collide directly and be captured by surface molecules of the particle which are active centers. Then, a charge of these active centers is concentrated on them and does not spread over the whole particle surface, as takes place for metallic particles. We consider the regime of charging when the total particle charge is small compared to the total number of active centers on the particle surface. Then one can consider the creation of the particle charge as a result of the moving of positive and negative charges towards the particle, and these currents can be analyzed independently.

In order to determine currents of charged atomic particles of a certain type towards the particle, we account for the motion of ions determined by its diffusion in a gas and drift under action of an electric field of a charged particle. Then, the ion density equals zero on the particle surface and tends to the equilibrium value at large distances from the particle. Taking the particle charge to be Ze, where Z is the particle charge expressed in electron charges, we get for the current of positive ions to the particle at a distance R

$$J = 4\pi R^2(-D_+ \mathrm{d}N/\mathrm{d}R + K_+ E N)e. \tag{1}$$

The first and second terms correspond to diffusion and drift motion, respectively, N is the number density of ions, D_+, K_+ are the diffusion coefficient and the mobility of positive ions, and $E = Ze/R^2$ is the electric field strength created by the particle. Using the Einstein relationship between the diffusion coefficient and mobility of ions $D_+ = eK_+/T$, where T is the gaseous temperature, we obtain for the positive ion current on the particle surface

$$J = -4\pi R^2 D_+ e \left(\frac{\mathrm{d}N}{\mathrm{d}R} - N\frac{Ze^2}{TR^2} \right). \tag{2}$$

Taking into account that ions do not recombine in space because of the small density, we come to the conclusion that the ion current is independent of R. Then one can consider the above relation for the ion current as the equation for the ion number density. Solving this equation with the boundary condition $N(r) = 0$, where r is the particle radius, we obtain

$$\begin{aligned} N(R) &= \frac{J}{4\pi D_+ e}\int\limits_r^R \frac{\mathrm{d}R'}{(R')^2}\exp\left(\frac{Ze^2}{TR'} - \frac{Ze^2}{TR}\right) \\ &= \frac{JT}{4\pi D_+ Ze^2}\left[\exp\left(\frac{Ze^2}{Tr} - \frac{Ze^2}{TR}\right) - 1\right]. \end{aligned} \quad (3)$$

Using the condition that the ion number density tends to the equilibrium value N_+ in a plasma at large R far from the particle, we obtain the following expression for the ion current

$$J_+ = \frac{4\pi D_+ N_+ Ze^3}{T[\exp(Ze^2/Tr) - 1]}. \quad (4)$$

This is so-called *Fuks equation* which describes the current of positive ions. However, in the same manner one can find the current of negative ions or electrons in the form

$$J_- = \frac{4\pi D_- N_- Ze^3}{T\,[1 - \exp(-Ze^2/Tr)]}. \quad (5)$$

This formula follows from (4) as a result of the $Z \to -Z$ change and the change of the parameters of positive ions to those of negative ions.

In the limit of small particle charge, these equations give the *Smolukhowski formula* for the diffusion flux of particles on the surface of an absorbed sphere

$$J/e = 4\pi DNr \quad (6)$$

with parameters for neutral atomic particles. In the limit $Ze^2/(rT) \gg 1$, (5) transforms to the *Langevin formula* for the recombination rate constant of the charged particle and an ion of the opposite sign if the ion mass is small compared to the particle mass:

$$J_- = 4\pi Ze^3 D_-/T = 4\pi Ze^2 K_-. \quad (7)$$

Currents of positive and negative ions, given by (4) and (5), to the particle surface are identical because the plasma state is stationary. The equilibrium particle charge state is then given by

$$Z = \frac{rT}{e^2}\ln\frac{D_+}{D_-}. \quad (8)$$

Thus, the particle has a negative charge if $D_+ < D_-$, i.e. the mobility of positive ions is lower than that of negative ions or electrons. Note that these formulas are valid under the following condition

$$r \gg \lambda, \quad r \gg \frac{e^2}{T} \tag{9}$$

where λ is the mean free path of the ions.

Let us consider the case when a plasma consists of electrons and positive ions, i.e. the negative plasma charge with respect to electrons. It takes place in a glow gas discharge where the electron distribution function on velocities differs from a Maxwellian one. Then the Fuks formula for the current of electrons to a small particle without using the Einstein relation between the mobility and diffusion coefficient of attaching charged particles has the form

$$J_e = \frac{4\pi K_e N_e Z e^2}{1 - \exp(-Ze^2/T_e r)}, \tag{10}$$

where T_e is the effective electron temperature, and $T_e = eD_e/K_e$, and K_e, D_e are the mobility and diffusion coefficient of electrons in a gas. We take $-Z$ for the charge of a small particle and use the fact that a typical electron energy eD_e/K_e exceeds significantly a typical ion energy. According to the Langevin formula (6), the ion current to the particle surface equals

$$J_+ = 4\pi Z e^2 K_+ N_+, \tag{11}$$

where N_+ is the density of positive ions far from the particle. Equalizing the fluxes of charged particles and taking into account the plasma quasi-neutrality $N_e = N_+$, we get for the charge of a small particle

$$Z = r \frac{D_{rme}}{eK_e} \ln \frac{K_e}{K_+}. \tag{12}$$

According to this formula, we have a condition $Ze^2/(rT) \gg 1$, where T can correspond to the temperature of both electrons and ions.

Note that here we have three parameters of the length dimension, r, λ and e^2/T, and the particle charge depends on the relation between these parameters. A high particle charge corresponds to the case

$$r \gg e^2/T \tag{13}$$

which we consider now. Above, the case of a dense gas $r \ll \lambda$ was studied, and now another limiting case $r \gg \lambda$ will be analyzed. If the particle has a charge Ze, and the mean free path strongly exceeds the parameter e^2/T, i.e., $\lambda \gg e^2/T$, collisions with atoms are not essential for collisions of ions with the particle. Because the distance of closest approach r_o is connected with the impact parameter of collision ρ by the relation

$$1 - \rho^2/r_o^2 = Ze^2/r_o\varepsilon, \tag{14}$$

where ε is the collision energy in the center-of-mass system of axes, we have the cross section of ion collision with the particle surface:

$$\sigma = \pi r^2 \left[1 - Ze^2/(r\varepsilon)\right]. \tag{15}$$

If $Z > 0$, i.e., the charges of the particle and ions have the same sign, it is necessary to account for $\varepsilon \leq Ze^2/r$, the probability of contact between an ion and particle equals zero, because in this case the potential energy of repulsion of the particle and colliding ion exceeds their kinetic energy near the particle surface. For the rate constant of contact between a colliding ion and a particle after averaging over a Maxwellian energy distribution, one has

$$k = \langle v\sigma \rangle = k_o \exp\left[- |Z| e^2/(rT)\right], \tag{16}$$

where r is the particle radius, $k_o = \pi r^2 \sqrt{8T/(\pi m)}$, and m is the ion mass. In the case of attraction between the particle and the ion, the mean rate constant of their collision is

$$k = k_o[1 + |Z| e^2/(rT)]. \tag{17}$$

Introducing the reduced parameter $x = |Z| e^2/(rT)$ and the probability ξ that an ion attaches to the particle as a result of their contact, we join the above formulas for the limit $\lambda \ll r$

$$J = \begin{cases} \xi k_o N_i(1 + x), & Ze^2 \leq 0, \\ \exp(-x), & Ze^2 \geq 0. \end{cases} \tag{18}$$

Equation (18) is valid for an arbitrary ion sign. In particular, let us consider the equilibrium particle charge in a plasma consisting of electrons and ions. Taking $\xi = 1$ in (18), we obtain the relation for the $-Ze$ particle charge

$$Z = \frac{rT}{e^2} \ln\left[\sqrt{\frac{T_e}{T}\frac{m_i}{m_e}}\left(1 + \frac{Ze^2}{rT}\right)\right], \tag{19}$$

where m_i, m_e are the ion and electron masses, and T, T_e are the ion and electron temperatures, respectively. One can apply this expression in the case when the electron distribution function differs from the Maxwellian one, but the electron current to the particle is determined by fast electrons, i.e. the particle potential greatly exceeds the typical thermal energy of electrons. Then instead of the electron temperature T_e in (19) we use the parameter $\tau = -f(\varepsilon_o)/f'(\varepsilon_o)$, where $f(\varepsilon)$ is the electron distribution function on energies ε, and (19) transforms to

$$x = \ln\left[(1 + x)\sqrt{\frac{\tau}{T}\frac{m_i}{m_e}}\right], \tag{20}$$

where $x = Ze^2/(rT)$. Let us consider a typical argon plasma of a glow discharge for which the parameters of formula (20) are $T = 500$ K, $\tau = 3$ eV. Then for particles of radius $r = 1\ \mu$m, we obtain $Z = 300$. Note that in a unipolar plasma higher particle charges can be reached. That is why for the generation of highly charged particles a uniform plasma of a corona discharge is used.

3. Charged Metallic Clusters in Plasmas

Let us consider the other character of equilibrium of charged clusters in a plasma. In the case of a dielectric particle, charges are located on active centers of the cluster surface that provides a high cluster charge. In the case of metallic clusters, electrons belong to the whole cluster and propagate over the whole cluster volume. This leads to the other character of the charge distribution, and the ionization equilibrium between charged clusters results from collisions with electrons

$$A_n^{+Z} + e \leftrightarrow A_n^{+Z+1} + 2e, \tag{21}$$

where n is the number of atoms in a cluster and Z is its charge. Because the work function of metals is lower than the ionization potential of corresponding atoms, the ionization potentials of clusters and small metallic particles lie between these values. Hence, ionization of metallic clusters occurs at relatively low electron temperatures T_e of the plasma which determines the cluster charge. For such an equilibrium, the probability $P_\mathrm{Z}(n)$ for a cluster consisting of n atoms and having a charge Z is determined by the Saha formula [11]:

$$\frac{P_Z(n)N_\mathrm{e}}{P_{Z+1}(n)} = 2\left(\frac{m_\mathrm{e}T_\mathrm{e}}{2\pi\hbar^2}\right)^{3/2} \exp\left[-\frac{I_Z(n)}{T_\mathrm{e}}\right], \tag{22}$$

where m_e is the electron mass, N_e is the electron number density, and $I_\mathrm{Z}(n)$ is the ionization potential of the cluster.

For large clusters with $n \gg 1$, the difference $I_{|rmZ+1}(n) - I_\mathrm{Z}(n)$ corresponds to the energy required for removal of an electron from the cluster surface to infinity. Hence we have

$$I_{Z+1}(n) - I_Z(n) = (Z + 1/2)e^2/r. \tag{23}$$

We account for an uncertainty in this difference to be of the order of e^2/r and make it symmetric. From this it follows that

$$I_Z(n) = \frac{Z^2e^2}{2r} + I_0(n), \tag{24}$$

where $I_0(n)$ is the ionization potential of a neutral particle consisting of n atoms.

Let us define the mean cluster charge Z according to the symmetric relation $P_{Z-1/2}(n) = P_{Z+1/2}(n)$. Then from (22) we have

$$Z = \sqrt{\frac{2T_\mathrm{e}r}{e^2}\left\{\ln\left[\frac{2}{N_\mathrm{e}}\left(\frac{m_\mathrm{e}T_\mathrm{e}}{2\pi\hbar^2}\right)^{3/2}\right] - \frac{I_0(n)}{T_\mathrm{e}}\right\}} + \frac{1}{2}. \tag{25}$$

Taking the cluster ionization potential in (25) as $I_0(\infty) = W,\ I_0(1) = I$, where I is the atom ionization potential, and writing the cluster ionization potential for a large number of atoms n in the form $I_0(n) = W + \text{const}/n^{1/3}$, we have const $= I - W$, and

$$Z = \sqrt{\frac{2T_e r}{e^2} \ln \left[\frac{2}{N_e} \left(\frac{m_e T_e}{2\pi\hbar^2} \right)^{3/2} \right] - \frac{W}{T_e} - \frac{I - W}{T_e n^{1/3}}} + \frac{1}{2}. \tag{26}$$

It follows from (25) and (26) that the basic dependence of the cluster charge on its size for large clusters is given by $Z \sim n^{1/6}$, and the proportionality coefficient depends on the cluster temperature. In particular, according to (26) the charges of tungsten clusters of size $n = 10^3$ for the electron number density $N_e = 10^{13}\ \text{cm}^{-3}$ and the electron temperatures $T_e = 5000$ K and 8000 K are correspondingly $Z = 3.2$ and 4.9. For rhenium clusters at these parameters, the charges are equal to 3.0 and 4.8, respectively. It is seen that the positive charge of metallic ions in an arc discharge is small compared to the negative charge of dielectric clusters in a glow discharge.

From (22) it follows that metallic clusters can be both positively and negatively charged depending on the electron temperature. Let us denote by N_+ the density of positively singly charged clusters, and by N_- the number density of clusters of a single negative charge. Evidently, the sign of the mean cluster charge coincides with the sign of the value $\ln(N_+/N_-)$. In other words, if $N_+/N_- > 1$, the mean cluster charge is positive and for $N_+/N_- < 1$ it is negative. Let us use the Saha relation [11] for the ratio of the number densities of positively charged N_+ and neutral N_o clusters

$$\frac{N_+ N_e}{N_o} = 2 \left(\frac{m_e T_e}{2\pi\hbar^2} \right)^{3/2} \exp \left[-\frac{I_0(n)}{T_e} \right], \tag{27}$$

where N_e is the electron number density, and $I_0(n)$ is the ionization potential of a neutral cluster consisting of n atoms. In the same way, one can express the ratio of the number densities of neutral and negatively charged clusters through the cluster electron affinity.

Let us express the cluster ionization potential $I_0(n)$ through the atom ionization potential $I_0(1) = I$ and the work function of the corresponding surface W on the basis of the above formula $I_0(n) = W + (I - W)/n^{1/3}$. Similarly, we bound the cluster electron affinity with the atomic affinity EA. Then we have

$$\frac{N_+}{N_-} = \zeta^2 \exp \left(-\frac{\Delta}{T_e n^{1/3}} \right), \tag{28}$$

where

$$\zeta = \frac{2}{N_e} \left(\frac{m_e T_e}{2\pi\hbar^2} \right)^{3/2} \exp \left(-\frac{W}{T_e} \right), \tag{29}$$

and $\Delta = I + EA - 2W$.

Let us consider as an example the copper case, when $I = 7.73$ eV, $EA = 1.23$ eV and $W = 4.4$ eV. Then $\Delta = 0.16$ eV, and the ratio $\Delta/(T_e n^{1/3})$

is small for large clusters and the considered electron temperatures. Hence, the sign of the mean cluster charge is determined by the sign of the $\ln \zeta(T_e)$ function which is monotonic on the electron temperature T_e. Therefore, at high electron temperatures clusters are positively charged, and at low electron temperatures they have mean negative charge. The transition electron temperature T_* is determined by the equation $\zeta(T_*) = 1$. In particular, in the copper case and for large clusters, it is equal to $T_* = 2380$ K for $N_e = 1 \times 10^{11}$ cm^{-3}, $T_* = 2640$ K for $N_e = 1 \times 10^{12}$ cm^{-3}, and $T_* = 2970$ K for $N_e = 1 \times 10^{13}$ cm^{-3}.

Evidently, the cluster negative charge increases as the electron temperature decreases. But the maximum negative cluster charge is restricted. Indeed, the electron affinity of a negative cluster is lower than that of a neutral cluster because of the interaction of the removal electron with the Coulomb field of the charged cluster. Formula (16) can be rewritten for a negatively charged cluster as

$$EA_Z(n) = EA(n) - \frac{Z^2 e^2}{2r}, \qquad (30)$$

where $EA(n)$ is the electron affinity of a neutral cluster of n atoms, and $EA_Z(n)$ is this value for such a cluster of charge Z. Evidently, the maximum negative cluster charge follows from the equation $EA_Z(n) = 0$, which gives for the maximum negative charge of a metallic cluster

$$Z_{\max} = Z_o n^{1/6}, \qquad Z_o = \sqrt{\frac{2EA \cdot r_W}{e^2}}, \qquad (31)$$

where r_W is the Wigner–Seits radius. In particular, for a large copper cluster when the cluster electron affinity coincides with the metal work function, this formula gives $Z_o = 0.95$. Note that the maximum charge of a dielectric particle or cluster varies with increase of its size as $n^{1/3}$, whereas for metallic clusters this dependence according to (25), (26) and (30) has the form of $Z \sim n^{1/6}$.

4. Peculiarities of Dusty Plasmas with Highly Charged Particles

The character of charging of dielectric particles differs from that of metallic ones. In the case of a metallic particle surface, valence electrons are spread over the whole particle surface. Hence, the interaction of electrons is essential, and the ionization potential of neutral metallic clusters, as well as the electron affinity of these clusters, tends to the surface work function in the limit of large cluster sizes. In the dielectric particle case, there are active centers on its surface which constitute traps for electrons. The capture or charge exchange of atomic or molecular negative ions and these centers is possible. Other active centers of the particle surface can capture positive atomic or molecular ions. These centers can also transfer an electron to them, so that the particle obtains a positive charge. In Sect. 2 we considered the process of charging of small dielectric particles in the regime when it is limited by fluxes of negative and positive ions to its surface. Now we analyze this process from the other standpoint.

The basic difference between ionization equilibrium for metallic and dielectric particles lies in the character of processes which establish this equilibrium. In the case of metallic particles the ionization equilibrium is due to the process (21), which is the three-body recombination process. It takes place for clusters and particles of submicron sizes. In the case of dielectric particles, the ionization equilibrium results from the processes

$$e + A_n^{-Z} \quad \rightarrow \quad \left(A_n^{-(Z+1)}\right)^{**}, \tag{32}$$

$$\left(A_n^{-(Z+1)}\right)^{**} + A \quad \rightarrow \quad A_n^{-(Z+1)} + A, \tag{33}$$

$$B^+ + A_n^{-(Z+1)} \quad \rightarrow \quad B + A_n^{-Z} \tag{34}$$

so that an autodetachment state $\left(A_n^{-(Z+1)}\right)^{**}$ is quenched by collisions with surrounding atoms. Because the rate constant of pair attachment of an electron to a dielectric particle exceeds significantly the ionization rate constant of the particle by electron impact, these particles are charged negatively.

In contrast to metallic particles, the binding energies of active centers are independent of the particle size because the action of each center is concentrated in a small space region. It is clear that the number of such centers is proportional to the area of the particle surface, and for particles of micron sizes this value is large compared to that occupied by charges. Hence, above and below we consider the regime of charging of a small dielectric particle far from saturation of active centers. Then positive and negative charges can exist simultaneously on the particle surface. They transport over the surface and can recombine there.

Usually the binding energy of electrons in negative active centers are in the EA = 2–4 eV range, and the ionization potential for positive active centers is about $I \approx 10$ eV. Hence, attachment of electrons is more profitable for electrons of a glow discharge, and a small dielectric particle has a negative charge in a glow gas discharge.

Due to the charge Z, the particle obeys the electric potential $\varphi = Ze/r$, where r is the particle radius. If $e\varphi < EA$, the electron state is stable, while in the case $e\varphi > EA$ a barrier transition is possible which leads to decay of the electron state. The system exists under these conditions, if new electrons attach to the particle. An isolated charged particle emits electrons until it reaches the limiting charge

$$Z_* = r \cdot EA/e^2. \tag{35}$$

In particular, for a dielectric particle of $r = 1\ \mu$m radius and $EA = 3$ eV, this charge is $Z_* = 2 \times 10^3$ and the particle electric potential equals 3 V.

Evidently, this regime, when charging of a small dielectric particle is limited by fluxes of electrons and positive ions to the particle surface, takes place if electrons or ions colliding with the particle surface are found in the field of action of active centers. Then they attach the surface or recombine on a center of the opposite charge sign. In this case the number of active centers must be large compared to the particle charge, i.e. the particle size should be large enough. It is valid for micron particles. In particular, the above numerical example corresponds to distances between neighboring charged centers of about 0.3 μm, that is larger by one–two orders of magnitude than the typical distance between neighboring active centers.

One more condition is required in the case when the particle negative charge exceeds the value Z_*. Then the binding states of captured electrons become autodetachment states and can decay as a result of the barrier electron transition. In this case a high rate of capture of electrons is required. The electron flux to the particle surface must be so large that the lifetime of captured electrons with respect to their barrier transitions must be larger compared to a typical time of electron capture per electron charge of the particle.

Let us estimate the dependence on the parameters of the lifetime of a negatively charged center on the surface of a dielectric particle. Its decay has a barrier character, and the probability per unit time for escaping of an electron through the potential barrier has the following exponential dependence [12]

$$\frac{1}{\tau} \sim \exp(-2S), \quad S = \int_r^{R_c} \mathrm{d}R \sqrt{\frac{2m_\mathrm{e}}{\hbar^2}\,[EA - V(r) + V(R)]}. \qquad (36)$$

Here EA is the electron binding energy, $V(R) = Ze^2/R$ is the interaction potential of an electron with the Coulomb field of the particle at a distance R from the particle center, and R_c is the turning point, i.e.

$$R_c = \frac{r}{1 - EA/\varepsilon_\mathrm{o}}, \qquad (37)$$

where $\varepsilon_\mathrm{o} = Ze^2/r$. Thus we have

$$S = \frac{\pi}{2} r \cdot \sqrt{\frac{2m_\mathrm{e}}{\hbar^2} \cdot \frac{\varepsilon_\mathrm{o}}{1 - EA/\varepsilon_\mathrm{o}}}. \qquad (38)$$

Assuming ε_o to be of the order of a typical atomic value, we obtain $S \sim r/a_\mathrm{o}$, where a_o is the Bohr radius. Being guided by small particles of micron size, we obtain a very high lifetime of surface negative ions with respect to their barrier decay time. Hence, the last condition is satisfied for micron particles located in a plasma, and the particle charge exceeds that of metallic clusters and the value Z_* in (35).

5. Stability of Large Clusters

Usually charges are found on the cluster surface. Their interaction creates tensions in the cluster which can lead to cluster decay. Below we consider instabilities of clusters as a result of interaction of its charges.

If a cluster is found in the liquid state and is like a liquid drop, its decay is determined by the Rayleigh instabilities of a charged liquid drop. In this case the surface tension energy $4\pi r^2\gamma$ averts from the drop decay due to the Coulomb interaction $Z^2e^2/(2r)$ of charges. Here r is the drop radius, Z is the drop charge expressed in electron charges e, and α is the surface tension. We assume drop charges to be distributed uniformly over the drop surface.

The Rayleigh instability can be applied to a drop dividing into two identical drops due to large deformations or to decay of the drop owing to small drop vibrations. Below, we restrict our consideration to drop decay due to small deformations whose threshold has the form

$$Z^2e^2 = 16\pi\gamma r^3. \tag{39}$$

This criterion corresponds to the case when the charge is distributed uniformly over the drop surface and assumes the Poisson coefficient to be zero, so that a deformation in one direction does not create deformations in other directions. Note that stable clusters with respect to this criterion can be unstable for large deformations, i.e. these clusters are found in metastable states.

Let us consider clusters with a short-range interaction of atoms or molecules when the cluster binding energy is determined mostly by interaction of nearest neighbors. Then the cluster has the structure of close packing, and an internal atom has 12 nearest neighbors for the solid cluster state. Then we have for the cluster radius $r = r_W n^{1/3}$, where $r_W = [3m/(4\pi\rho)]^{1/3}$ is the Wigner–Seits radius, m is the atom mass, ρ is the bulk density, n is the number of drop atoms, and the surface energy is represented in the form $4\pi r^2\gamma = An^{2/3}$ with the value A being the specific surface energy of the cluster. These parameters for clusters of rare gases obtained on the basis of data [15] are given in Table 1. Using these parameters, the criterion (35) can be represented in the form

$$n_c(Z) = \frac{Z^2e^2}{4Ar_W}. \tag{40}$$

Here $n_c(Z)$ is the critical size of the cluster starting from which the cluster can have charge Z and is stable with respect to small deformations. The values of n_c/Z^2 are also given in Table 1. Note that there are different versions of the Rayleigh instability [14–18] depending on the instability character, the distribution of charges in the drop, the Poisson coefficient of the drop and its polarizability. Let us represent formula (36) in the form

$$n_c(Z) = C\frac{Z^2e^2}{4Ar_W}. \tag{41}$$

Table 1. Parameters of rare gas clusters

Cluster	r_W,Å	A, meV	n_c/Z^2
Ne	1.74	15.3	135
Ar	2.08	53	33
Kr	2.15	73	23
Xe	2.40	97	15

Let us assume that this formula describes experimental data. Then, comparing this formula with the measured reliable values of n_c for krypton and xenon [17–21] with $Z = 2 - 4$, we get $C = 0.8 \pm 0.1$. Let us consider once again the case of the stability of charged clusters when the cluster size is very small. Below we will study a charged cluster of M_3^{++} ions which, evidently, have the structure of a linear chain $M^+ - M - M^+$. We analyze this system on the basis of a simple model [16] taking account of the exchange interaction of $M^+ - M$ and the Coulomb interaction of ions. Then the interaction potential of this system equals

$$U(r_1, r_2) = V(r_1) + V(r_2) + \frac{e^2}{r_1 + r_2}. \tag{42}$$

Here r_1, r_2 are the distances between the middle atom and ions, and $V(r)$ is the ion–atom interaction potential. This interaction potential for this system admits two types of chain vibrations. Symmetric vibrations correspond to $r_1 = r_2$, while antisymmetric vibrations have $r_1 + r_2 =$ constant. It is seen that an antisymmetric vibration conserves the distance between ions, i.e., the Coulomb interaction potential of ions is conserved for these vibrations. Hence, the ion stability is determined by symmetric vibrations. Evidently, the condition of cluster stability has the form

$$\frac{\mathrm{d}^2 U(R,R)}{\mathrm{d}R^2}(R = R_\mathrm{e}) \geq 0, \tag{43}$$

where the equilibrium distance R_e given by

$$\frac{\mathrm{d}U(R,R)}{\mathrm{d}R}(R = R_\mathrm{e}) = 0. \tag{44}$$

Let us represent $V(r)$ in (38) in the standard form

$$V(r) = \varepsilon \left[\left(\frac{R_\mathrm{e}}{r} \right)^{n/2} - \frac{n}{4} \left(\frac{R_\mathrm{e}}{r} \right)^{n} \right], \tag{45}$$

where ε is the binding energy of the bond $M^+ - M$. Then the criterion (40) can be written as

$$\frac{\varepsilon R_\mathrm{e}}{e^2} \geq C, \qquad C = \frac{(n-1)^{(n-1)/(n-4)}}{4n(n-4)3^{3/(n-4)}}. \tag{46}$$

In particular, we have $C = 0.10$ for $n = 8$, $C = 0.065$ for $n = 10$ and $C = 0.047$ for $n = 12$. Taking the equilibrium ion–atom distance as $R_e =$ 2–3 Å, we have from the condition (34) $\varepsilon > 0.2$–0.6 eV. Because the typical binding energy for a metallic diatomic ion is ~ 1 eV, we conclude that a metallic ion M_3^{++} can exist. Such ions have been detected for some metallic ions, for example, and do not exist for Si [22], Ni [23], Ge [24], As [25], Sn [26], Sb [27], W [23], Au [23, 28, 29], and Bi [27]. These ions have not been observed for Ga [30], Ag [31, 32], In [33], Hg [34], and Pb [33, 35].

References

1. J.H. Chu, I. Lin: Phys. Rev. Lett. **72**, 4009 (1994)
2. H. Thomas, G.E. Morfill, V. Demmel, J. Goree, B. Feuerbacher, D. Mölmann: Phys. Rev. Lett. **73**, 652 (1994)
3. Y. Hayashi, K. Tachibana: Japan J. Appl. Phys. **33**, L804 (1994)
4. A. Melzer, T. Trottenberg, A. Piel: Phys. Lett. A **191**, 301 (1994)
5. T. Trottenberg, A. Melzer, A. Piel: Plasma Sourc. Sci. Technol. **4**, 450 (1995)
6. G.E. Morfill, H. Thomas: J. Vac. Sci. Technol. A **14**, 490 (1996)
7. V.E. Fortov, A.P. Nefedov, O.F. Petrov, A.A. Samarin, A.V. Chernyshev, A.M. Lipaev: Sov. Phys. - JETP Letters, **63**, 187 (1996)
8. V.E. Fortov, A.P. Nefedov, V.M. Torchinskii, V.I. Molotkov, A.G. Khrapak, O.F. Petrov, K.F. Volykhin: Sov. Phys. - JETP Letters, **64**, 92 (1996)
9. O. Echt, K. Sattler, E. Recknagel: Phys. Rev. Lett. **47**, 1121 (1981)
10. W.D. Knight, K. Clemengerger, W.A. de Heer, W.A. Saunders, M.Y. Chou, M.L. Cohen : Phys. Rev. Lett. **52**, 2141 (1984)
11. L.D. Landau, E.M. Lifshitz: *Statistical Physics* (Pergamon, Oxford 1980)
12. L.D. Landau, E.M. Lifshitz: *Quantum Mechanics* (Pergamon, Oxford 1965)
13. L.D. Landau, E.M. Lifshitz: *Electrodynamics of Continuous Media* (Pergamon, Oxford 1984)
14. D.R. Lide (ed.): *Handbook of Chemistry and Physics*, 76th edn. (CRC Press, Boca Raton 1995–1996)
15. K. Emeleus, A. Bresin: Intern. J. Electron. **54**, 195 (1983)
16. A.V. Eletskii, B.M. Smirnov: Sov. Phys.- Uspekhi **32**, 763 (1989)
17. A. Ding, J. Hesslich: Chem. Phys. Lett. **94**, 54 (1983)
18. O. Echt, D. Kreisle, E. Recknagel, J.J. Saenz, R. Casero, J.M. Soler: Phys. Rev. A **38**, 3236 (1988)
19. P. Schreier, T.D. Märk: J. Chem. Phys. **87**, 1456 (1987)
20. P. Schreier, A. Stamatovic, T.D. Märk: J. Chem. Phys. **89** , 2956 (1988)
21. M. Lezius, P. Schreier, A. Stamatovic, T.D. Märk: J. Chem. Phys. **91**, 3240 (1989)
22. T.T. Tsong: J. Vac. Sci. Technol. **B3**, 1425 (1985)
23. T. Jentsh, W. Drachsel, J.H. Block: Chem. Phys. Lett. **93**, 144 (1982)
24. P. Joyes, P. Sudraud: Surf. Sci. **156**, 451 (1985)
25. T. Sakurai, T. Hashizume, A. Jimbo, T. Sakata, C. Colliex, R. Ohana, P. Sudraud, J. van de Walle: J. de Phys. **45**, C9, 453 (1984)
26. A. Dixon, C. Colliex, R. Ihana, P. Sudrand, J. van de Wille: Phys. Rev. Lett. **46**, 865 (1981)
27. G.L. Kellog: Surf. Sci. **120**, 319 (1982)
28. P. Sudraud, C. Colliex, J.van der Wille: J. de Phys. **40**, L207 (1979)

29. A.R. Waugh: J. Phys. D **13**, L203 (1980)
30. R.J. Colbertson, G.H. Robertson, Y. Kuk: J. Vac. Sci. Technol. **17**, 203 (1980)
31. W. Schulze, B.Winter, J.Urban, J.Goldenfeld : Z. Phys. D **4**, S.379 (1987)
32. C. Bréchignac, M. Broyer, Ph. Cahuzac, G. Delacretaz, P. Labastie, L. Wöste: Chem. Phys. Lett. **118**, 174 (1985)
33. A. Hoareau: Chem. Phys. Lett. **143**, 602 (1988)
34. K. Sattler, J. Mühlbach, O. Echt, P. Pfau, E. Recknagel: Phys. Rev. Lett. **47**, 160 (1981)
35. A. Hoareau, A. Melinon, P. Cabaud: Phys. Rev. D **18**, 1731 (1985)

Subject Index

Druck: Strauss Offsetdruck, Mörlenbach
Verarbeitung: Schäffer, Grünstadt